Porous Media

Fluid Transport and Pore Structure

Porous Media
Fluid Transport and Pore Structure
SECOND EDITION

F. A. L. DULLIEN

Department of Chemical Engineering
University of Waterloo
Waterloo, Ontario, Canada

ACADEMIC PRESS, INC.
Harcourt Brace Jovanovich, Publishers
San Diego   New York   Boston
London   Sydney   Tokyo   Toronto

Copyright © 1992, 1979 by ACADEMIC PRESS, INC.
All Rights Reserved.
No part of this publication may be reproduced or transmitted in any form or
by any means, electronic or mechanical, including photocopy, recording, or
any information storage and retrieval system, without permission in writing
from the publisher.

Academic Press, Inc.
San Diego, California 92101

United Kingdom Edition published by
Academic Press Limited
24–28 Oval Road, London NW1 7DX

Library of Congress Cataloging-in-Publication Data

Dullien, F. A. L.
    Porous media : fluid transport and pore structure / F.A.L.
Dullien. -- 2nd ed.
        p.   cm.
    Includes bibliographical references and index.
    ISBN 0-12-223651-3
    1. Transport theory.   2. Porous materials.   I. Title.
QC175.2D84   1991
530.4'75--dc20
                                                        91-18975
                                                           CIP

PRINTED IN THE UNITED STATES OF AMERICA
91 92 93 94   9 8 7 6 5 4 3 2 1

To my wife Ann
for her moral and professional support

Contents

Preface

Since the publication of the first edition of this book a great deal of research has been done in the author's field of interest. As a result, major parts of the book have been completely revised and many additions have been made. As this text is very strongly research oriented, it is fair to say that the second edition is a new progress report on this field. Owing to limitation of space, time, and the author's own interests, this review of the field of transport in porous media is not exhaustive. The author apologizes to those colleagues whose work, regrettably, receives less attention than their importance warrants.

The unique property of a porous medium, the one that distinguishes it from other solid bodies on the one hand and from simple conduits on the other, is its complicated pore structure. The vast majority of porous media contain an interconnected three-dimensional network of capillary channels of nonuniform sizes and shapes, commonly referred to as *pores*.

Fluid flow, diffusion, and electrical conduction in porous media take place within extremely complicated microscopic boundaries that in the past made a rigorous solution of the equations of change in the capillary network practically impossible. However, this situation has changed recently because high-powered digital computers now permit the solution of these equations in small samples of the pore network. The past state of affairs is one of the reasons why some of the brilliant and successful practitioners in the field of "flow through porous media" have tried, as much as possible, to stick with the continuum approach in which no attention is paid to pores or pore structure. Another reason is that the continuum approach is often adequate for the phenomenological description of macroscopic transport processes in porous media. The continuum approach, however, fails to provide a clue to help explain any of a multitude of observations that depend on the properties of the microscopic channels and the behavior of the fluids on the microscopic scale. The desire of scientists and engineers to be able to understand and then explain their observations has always been a powerful driving force for progress. Therefore there have been numerous attempts over the past sixty years or so to explain the flow phenomena in terms of the microscopic structure as accurately as possible. The results have seldom been entirely satisfactory, but with every step a further penetration into an immensely complicated territory has been achieved. There is a great deal of information available in the technical literature on the role played by pore structure in determining transport phenomena in pore spaces.

This book has been written with the primary purpose of presenting in an organized manner the most pertinent information available on the role of pore structure and then putting it to use in the interpretation of experimental data and the results of model calculations.

Pore structure is inseparable from the convective, diffusive, and interfacial effects that take place in the pores; these effects are all interrelated so that there is little point in trying to evaluate their relative merits with the aim of deciding which of them is most important.

Existing books on "flow through porous media" have been written with an emphasis on the fluid mechanical aspects. Interfacial effects, such as interfacial tension and wettability, have been under intensive and productive investigation for quite some time. In this book the author has made an attempt to show that there are benefits to be gained by trying to think about the phenomena in porous media in terms of interactions among the three main factors, i.e., transport phenomena, interfacial effects, and pore structure. The book contains many examples of applications of this concept, and it is the hope of the author that many readers will find this approach useful as well as an inspiration and motivation to do more fundamental research on the role played by pore structure.

The author wishes to acknowledge an operating grant from the National Research Council of Canada. The perfect typing of the manuscript of the second edition and a great deal of dedicated secretarial help by Susie Bell and the outstanding professional artwork of Rinze Koopmans are gratefully acknowledged. Many discussions with the author's colleagues, especially Professor I. Chatzis and L. Catalan, have aided the process of formulating the author's ideas. The moral as well as professional support throughout all the author's work and life by his wife Ann, to whom this book is dedicated, has been invaluable.

List of Symbols

Latin Letters

a	cross-sectional area; speed of sound; edge length of cube
a	$= (1/8)R^2$ [Eq. (6.3.59)]
a_{ijmn}	medium's dispersivity tensor [Eq. (6.3.101)]
a_L	$= D_L/v_{DF} = \sigma_x^2/2x$ = geometrical dispersivity
$a_{L,\,eff}$	effective dispersivity
a_T	average transverse dispersivity
a_w	activity of water
$2a$	wall-to-wall distance
A	cross-sectional area, constant in Eq. (3.2.18)
$(A_v)_i$	unfilled cross-sectional area in pore corner
A_A	area fraction
A_∞	longitudinal macrodispersivity [Eq. (6.3.5)]
δA	incremental surface area
b	constant; constant in Eq. (1.1.7); number of branches
$2b$	wall-to-wall distance
B	constant in Eq. (3.2.19); hydraulic conductivity; channel conductance
Bd	bulk density (Table 1.5)
c	compressibility; constant parameter; molar concentration; concentration
$\bar{c}$	average tracer concentration
c_m^*	molar tracer concentration
c_0	initial tracer concentration
C	connectivity; dimensionless coefficient in Eqs. (6.3.15) and (6.3.16)
C	$= (R_p - R)R_p$ [Eq. (2.5.17)]
C'	$\equiv T$ = shift factor

Ca	capillary number
CA_{imb}	capillary number defined by Eq. (5.3.83)
CA_{mob}	capillary number defined by Eq. (5.3.88)
C_D	drag coefficient
$C(L)$	covariance
d	distance
D	pore or capillary diameter; dimension; darcy
Da	Darcy number
DI	difficulty index [Eq. (5.3.71)]
D_b	breakthrough diameter
D_{ct}	dispersion coefficient in a capillary tube
D_e	pore entry diameter
D_f	fiber diameter
D_F	fractal dimension
D_H	hydraulic diameter
D_ℓ	diameter of large capillary
D_L	longitudinal dispersion coefficient
D_p	sphere or particle diameter
rD_p	interparticle distance
$\bar{D}_p$	effective average particle or fiber diameter
$\bar{D}_{p2}$	surface average sphere diameter
$\bar{D}_{pr}$	average sphere diameter defined by Eq. (3.7.10)
D_s	diameter of small capillary
D_T	container diameter; transverse dispersion coefficient
D_v	volume average diameter
D_v^+	dimensionless volume average diameter
D_2	diameter of 3-D object
$\bar{\bar{D}}$	hydraulic dispersion tensor
$\mathscr{D}$	tracer or mutual diffusion coefficient
$\mathscr{D}_{AB}$	mutual diffusion coefficient
$\mathscr{D}_{eff}$	$= \mathscr{D}/X =$ effective diffusion coefficient
$\mathscr{D}_{KA}$	Knudsen diffusion coefficient of A
$\mathscr{D}_m$	moisture or hydraulic diffusivity
$(\mathscr{D}_{eff})_{ss}$	$= \mathscr{D}\phi/X =$ effective diffusion coefficient measured in steady-state experiment
E	modulus of elasticity
E_D	inherent efficiency of conversion of work to the creation of surface
$E(-x)$	$= -\int_x^\infty \dfrac{e^{-u}\,du}{u}$ Eq. (1.4.1)
$E[\ldots]$	expected value
f	separation factor (Eq. 4.2.10)
f_i	frequency of nodes of type i
f_p	friction factor
f_r	relative frequency of pores of type r in the network
$f_T(P'')$	adsorption isotherm
f_ℓ	friction factor defined by Eq. (3.3.10)
f_w	fractional flow function of water
f_ϕ	non-Newtonian friction factor
$f(a, b, c)$	joint density function of distribution of lengths a, b, and c
$f(A, \ell, \alpha)$	number of tubes per unit volume in the intervals $A \rightarrow A + dA$, $\ell \rightarrow \ell + d\ell$ and $\alpha \rightarrow \alpha + d\alpha$
$f(D)$	density function of distribution of pore diameters
$f(R)$	pore size distribution defined by Eq. (3.4.10)

$f(2a)$, $f(2b)$, $f(2c)$	cumulative size distributions in the x-, y-, and z-directions, respectively
$f(\phi)$	porosity function [Eq. (3.2.6)]
F	flow contribution to molar flux
F	$= R_o/R_w =$ formation factor
F_c	capillary force
F_e	effective formation factor at partial saturation
F_s	fractional flow of solvent
F_v	viscous force
F_w	fractional flow of water
$F(a, b, c)$	cumulative joint distribution of lengths a, b, and c
$F(D_2)$	density function of distribution of diameters of 3-D objects
$F(D_e)$	density function in model of formation factor
$F(t)$	surface force potential function [Eq. (2.5.26)]
$F(\alpha^2 R^2, \alpha R)$	function in Eq. (3.3.62)
$F(\tau_\phi)$	function defined by Eq. (3.5.2)
$F(\tau_w)$	function defined by Eq. (3.5.1)
g	gravitational acceleration constant
g_1	conductance of capillary segment of length ℓ_1
g_3	$= g_1/\sqrt{3} =$ conductance of capillary segment of length $\ell_3 = \ell_1\sqrt{3}$
$g(r^*)$	density function of distribution of r^*
$g(\delta; \theta; \phi; \Omega, D_2)$	density function of distribution of section lengths δ of an object of diameter D_2 in an orientation (θ, ϕ, Ω)
G	Gibbs energy; molar flow rate; genus
G_m	mass flow rate per unit area
$G(p_b)/G^o$	relative conductivity of network
h	variable height [Eq. (1.8)]; net thickness [Eqs. (1.4.1), (1.4.2)]; elevation; half width of channel
h_c	$\equiv \psi =$ capillary pressure head
H	intermittency (Hurst) exponent
H	$= R_{max}/R_{min} =$ heterogeneity factor
I	$= F_e/F =$ resistivity index
$\mathbf{j}_i^v$	mass flux of i with respect to $\mathbf{v}^v$
J_m^*	molar flux of tracer
$J(S_w)$	Leverett J-function [Eq. (2.3.6)]
k	Darcy permeability coefficient [Eq. (3.1.1)]; mass transfer coefficient
k'	Kozeny constant [Eq. (3.3.7)]
$\overline{\overline{k}}$	permeability tensor
$\tilde{k}$	"pure water permeability constant" [Eq. (4.2.14)]
$\underline{k}$	mass transfer coefficient (Eq. (4.2.23))
k_{CK}	permeability predicted by Carman–Kozeny model
k_H	hydraulic conductivity [Eq. (3.1.4)]
k_i	effective or phase permeability
k_{ij}	component of permeability tensor
k_M	$= \mathscr{D}K_H =$ permeability coefficient [Eq. (4.2.8)]
k'_n, k''_n	directional permeabilities
k_{ri}	relative permeability
k_0	shape factor [Eq. (3.3.7)]
k_1	permeability predicted by 1-D capillaric model
$k_{1,2,3}$	constants in Eqs. (3.2.38), (3.2.39), and (3.2.40)
k_2/μ	2-D network conductivity [Eq. (3.3.46)]
k_3	permeability predicted by 3-D pseudo capillaric network model
$k_{11}, k_{22}; k'_{11}, k'_{22}$	principal phase permeability coefficients
$k_{12} = k_{21};$ $k'_{12} = k'_{21}$	interaction phase permeability coefficients

K	quantity defined by Eq. (4.2.19); K-factor [Eq. (6.3.11)]
K_H	Henry's law constant
ℓ	chord length; length of a step [Eq. (6.3.57)]; length of capillary; characteristic pore scale; oil blob length
ℓ	$= v_p/n$ [Eq. (6.3.48)]
ℓ_ℓ	aggregate length of large capillaries in the model
ℓ_s	aggregate length of fine capillaries in the model
ℓ_1	lattice constant of 1-D capillaric model
ℓ_3	lattice constant of 3-D capillaric model
$\delta\ell$	elemental length
L	length of sample; intercept length, length, length scale of permeability correlation
L_A	length of lines in a plane per unit test area
L_e	average effective path length of flow
L_L	length fraction
L_m	mixing length, i.e., the distance in the macroscopic flow direction over which $\bar{c}/c_0$ changes from 0.9 to 0.1 or from 0.8 to 0.2
L'_m	dimensionless mixing length defined by Eq. (6.2.14a)
L_v	length of lines per unit test volume
$L(R)$	total length of pores with radii between R and $R + dR$ [Eq. (2.5.11)]
m	mass; cementation factor; molality
m	$= k_{2r}\mu_1/k_{12}\mu_2$ = mobility ratio
mD	millidarcy
$(\Sigma m)_r$	number of pores connected to a pore of type r at both ends
M	Mach number; molecular weight
$M(\phi)$	effective viscosity
M	$= (3/2)(v_p\ell\mu/D_{ct})$ [Eq. (6.3.83)]
n	number of capillaries in parallel; mole number; number of nodes in a network; number of lines per unit area; number of points; number of dust particles; exponent in Archie's law; number of steps taken in unit time [Eq. (6.3.48)]
$\mathbf{n}$	unit normal vector of surface
$\mathbf{n}_i$	mass flux of i with respect to solid matrix
n_x	flux of tracer
$n(L)$	density function of distribution of intercept length
$n(\delta)$	density function of distribution of circle diameters
N	number of steps; number of separate networks; total number of bonds; number of occurrences of an event; Avogadro number
N_A	number of features per unit test area
N_A, N_B	molar flux of A and B, respectively
N_{As}	molar flux of A per unit cross section of sample
N_d	deflection number [Eq. (3.2.44)]
N_e	effective pore number [(Eq. (3.2.37)]
N_p	number of pores per unit section
N'_{Re}	non-Newtonian Reynolds number
N_s	number of spheres in unit volume of bed
N_T	total molar flux
N_w	molar flux of water
$N(D)$	density function of distribution of sphere diameters or pore diameters
$N(D_p)$	density function of distribution of particle diameters
p	probability of a particle traveling a distance x_p in the characteristic time τ [Eq. (6.3.52)]; fraction of void space that is accessible through pores of diameters less than a given value (p. 50)
$\tilde{p}$	random quantity in Eq. (6.3.59)

p_b	fraction of open bonds
p_{cr}	critical percolation probability
p_s	fraction of open sites
$p(u)\,du$	probability that a molecule has velocity between u and $u + du$
$p(u, t/u_0/t_0)\,du$	probability that a molecule has velocity u at time t if it had velocity u_0 at time t_0
$p(x, t)\,dx$	probability of tracer to be between x and $x + dx$ at time t
P	hydrostatic pressure; number of points; denoting a point on a surface
P'	hydrostatic stress intensity; absolute value of macroscopic pressure gradient; pressure on the convex side of the interface
P''	pressure on the concave side of the interface
P_A	number of points per unit test area
P_c	capillary pressure
P_c^*	$= P_c/P_{cb}$
P'_{cb}	$= P_{cb}/4\sigma \cos\theta$ reduced breakthrough or bubbling capillary pressure
Pd	particle density (Table 1.5)
P_{I1}	capillary pressure defined by Eq. (5.3.69)
P_{I2}	capillary pressure defined by Eq. (5.3.70)
P_L	number of intersections per unit length of test lines with features in the plane of polish
P_m	arithmetic mean pressure
P_0	vapor pressure of bulk liquid
P_p	capillary pressure in piston type displacement
P_s	snap-off capillary pressure
P_v	number of points of intersection between surfaces and lines per unit test volume
Pe	$= v_{DF}\overline{D}_p/\mathscr{D} =$ Peclet number in porous media
Pe'	$= v_{DF}\overline{D}_p/D_L =$ dynamic Peclet number in porous media
Pe_{ct}	$= D\overline{u}/\mathscr{D} =$ Peclet number in capillary tube
Pe_R	$= \overline{R}v_{DF}/\mathscr{D}$ (p. 516)
Pe'_{ct}	$= \overline{u}D_p/D =$ dynamic Peclet number in capillary tube
Pe'_{ft}	Peclet number in Eq. (6.3.56)
Pe'_L	$= v_p L/D_L =$ a dynamic Peclet number
Pe'_t	Peclet number for transverse dispersion in Eq. (6.3.56)
Pe'_T	$= v_p \ell/D_T =$ dynamic Peclet number for transverse dispersion
P	$= P + \rho g z$
P^*	$= P/\rho v^2 =$ dimensionless pressure
q	$= \ell/t =$ velocity of tracer particle in a step; probability that a randomly chosen pore will not exceed a given size (p. 50); production rate
q_{av}	$= dQ/dN$ [Eq. (5.2.10)]
q_i	particle size distribution parameter [Eq. (3.2.1)]
q_k	number of fraction of penetrated pores with no exit (dendritic pores)
Q	volumetric flow rate
δQ	incremental volume flow
r	radial coordinate; radius; principal radius of curvature; ratio of true-to-apparent area of solid surface
$\mathbf{r}$	position vector
r_1, r_2	radii of rotation
$r(c, t)$	rate of production of solute per unit volume of solution [Eq. (6.3.31)]
R	radius of curvature
R	$= D/2$
$\mathbf{R}$	force
R_{eq}	equivalent pore radius defined by Eq. (5.3.86)
R_o	resistance of saturated sample

R_p	cumulative oil recovery
R_w	resistance of electrolytic solution of the same geometry as the sample
R_1	constriction radius
R_1^*	dimensionless constriction radius
$\mathscr{R}$	universal gas constant
Re	Reynolds number
Re_c	Reynolds number defined by Eq. (4.1.7)
Re_f	fiber Reynolds number
Re_k	Reynolds number defined by Eq. (6.3.20)
Re_p	particle or superficial Reynolds number
Re_v	Reynolds number based on D_v
Re_{50}	Reynolds number defined by Eq. (6.3.19)
s	distance along axis of pore
S	specific surface area per gram adsorbent; saturation; surface area of solid spreading coefficient
S'	constriction factor
S_{eff}	saturation based on drainable porosity
S_k	saturation predicted by model in step k
S_k^*	saturation predicted by model in step k if all pores were independent domains
S_0	specific surface area per unit solids volume
S_p	aggregate surface area of particles
S_t	surface area defined by Eq. (2.5.25)
S_v	specific surface area per unit bulk volume
$S_v(\theta)$	unfilled fraction of pore cross section
S_w	water saturation
$S(D_k)$	cumulative volume fraction of pores of size $D \geq D_k$
S^2	function expressed by Eq. (6.3.68)
Sc	$= \nu/\mathscr{D}$, Schmidt number
t	time; multilayer thickness; geometrical factor in Eqs. (3.3.46) and (3.3.47); duration of a step [Eq. (6.3.58)]
t_1, t_0	the time in which the mean square displacement of molecules by molecular diffusion is equal to the radius R and the length ℓ of a tube, respectively [Eq. (6.3.62)]
t_D	$= \eta t/r^2$ (Fig. 1.61)
T	absolute temperature
T	$= L/\bar{u}$ [Eq. (6.2.10)]
T	$= t/N =$ time interval [Eq. (6.3.40)]
T	$\equiv (L_e/L)^2 =$ hydrodynamic tortuosity factor
T_{ij}	component of tortuosity tensor
T_{ij}^*	component of tensor related to the medium's tortuosity
T_n	time of displacement of a tracer particle [Eq. (6.3.58)]
u^*	ratio of average velocity of ganglion to the interstitial velocity of water
$\mathbf{u}^*$	$= \mathbf{u}/v =$ dimensionless velocity in a capillary
$\bar{u}$	average velocity in individual pore
u_i	$= u(d\xi_i/d\sigma)$ [Eq. (6.3.91)]
$\bar{u}_m$	mean molecular speed
u_r^*	dimensionless r velocity component in a periodically constricted tube
u_z^*	dimensionless z velocity component in a periodically constricted tube
$\langle \bar{u}_z \rangle$	$\equiv v_p =$ average pore velocity in porous medium
$U(t)$	longitudinal component of the velocity of a marked particle relative to the mean velocity [Eq. (6.3.76)]

$v, \mathbf{v}$	Darcy or superficial velocity
v^*	ganglion volume expressed in units of the number of pores occupied by the ganglion
$\mathbf{v}^*$	mass average particle velocity
v_{DF}	$\equiv v/\phi$ = average pore velocity defined by the Dupuit–Forchheimer assumption
$\mathbf{v}_i$	Darcy or superficial velocity of phase i
v_n	velocity component taken in the direction of ∂P
v_p	average pore velocity in porous media
$\mathbf{v}^v$	volume average velocity of mixture with respect to the solid matrix
$v(D)$	random part of velocity of a marked particle due to dispersion
V	volume; volume of pores with entry diameters $< D_e$
V'	molar volume of liquid
V_a	adsorbed volume of adsorbate in mL gas STP per grain of adsorbent
V_B	bulk volume of porous medium
$V^*_{\ell k}$	filled volume of a pore in class ℓ in penetration step k ($\ell = 1, 2, 3, \ldots, k$)
V_m	volume of gas in mL STP that should be able to cover the whole surface with a monolayer
$\overline{V}_m$	partial molar volume
V_o	STP number
V_p	aggregate volume of particles; pore volume
V_T	total pore volume of porous medium
V_v	volume fraction
$V(D)$	volume of pore of diameter D
$V'(D)$	volume of cylindrical capillary
$V(R)$	cumulative distribution of pore volume with R
$V(t)$	transverse component of velocity of a marked particle [Eq. (6.3.83)]
x	position coordinate
x	$\equiv D_\ell/D_s$ [Eq. (3.3.9)]
x	$\equiv P''/P_0$ [Eqs. (2.5.6) to (2.5.8)]
x'	$= x - ut$ or $x - v_{DF,x}t$
x_i	position coordinate
X	distance of plane from center of sphere [Eq. (1.2.12)]; adsorbed liquid volume [Eq. (2.5.23)]; "electrical" tortuosity factor
X_b	volume fraction of network consisting of bonds
X_n	coordinate of displacement of marked particle
y	position coordinate
y	$\equiv v_\ell/v_s$ [Eq. (3.3.9)]
y	$\equiv \ell_\ell/\ell_s$ [Eq. (3.3.10)]
y'	$= y - v_{DF,y}t$
y_A	mole fraction of A
y_k	fractional number of open bonds $j \le k$ that are penetrated
Y_k	fractional number of bonds that are penetrated if bonds $j \le k$ are open
Y_n	coordinate of displacement of marked particle
z	coordination number; average number of branches (bonds) meeting at a node [Eq. (2.5.30)]; position coordinate; distance measured vertically upward
z'	$= z - v_{DF,z}t$
z_i	number of bonds meeting at a node of type i [Eq. (2.5.30)]
Z	coordination number defined by Eq. (2.5.34a)
Z_r	quantity defined by Eq. (2.5.34b)
$Z(x_i)$	value of property at x_i

Greek Letters

α	bivariate density function [Eq. (1.2.16)]
α	$= 1/k$ [Eq. (3.2.17)]
$\alpha_{\mathrm{p}}(D)$	volume density function of size distribution of capillaries in parallel [Eq. (3.3.20)]
$\alpha_{\mathrm{s}}(D)$	volume density function of size distribution of capillaries in series [Eq. (3.3.21)]
$\alpha(D_{\mathrm{e}})\,dD_{\mathrm{e}}$	fraction of pore volume characterized by entry diameters between D_{e} and $D_{\mathrm{e}} + dD_{\mathrm{e}}$
αR	porosity function [Eq. (3.3.62)]
β	bivariate density function Eq. (1.2.12), Eq. (1.2.23); dimensionless larger spacing Eq. (1.2.33); inertia parameter Eq. (3.2.17); exponent in Eq. (2.5.41); angle of directional cosine; constant in Eq. (7.3.50); second-order memory fluid parameter Eq. (3.5.4)
β	$= \gamma^2 - 1$ in Eq. (6.2.22)
$\beta(D, D_{\mathrm{e}})\,dD\,dD_{\mathrm{e}}$	fraction of pore volume characterized by diameters between D and $D + dD$ and pore entry diameters between D_{e} and $D_{\mathrm{e}} + dD_{\mathrm{e}}$
$\Gamma(\tau)$	shear rate; function defined by Eq. (6.3.53)
γ	dimensionless coefficient in Eqs. (6.3.15) and (6.3.16); angle of directional cosine
γ	$= b/a$ in Fig. 6.9c
γ	$= c_{\mathrm{p}}/c_{\mathrm{v}}$
$\gamma(D_{\mathrm{e}})$	density function of pore entry diameter
$\gamma(L)$	variogram
Δ	difference
δ	diameter; diameter of 2-D features; length of "yardstick" in fractals
$\tilde{\delta}$	$= \bar{R}/\ell$ [Eq. (6.3.101)]
η	hydraulic diffusivity [Eq. (1.4.2)]; individual fiber collection efficiency
η_{ϕ}	Darcy viscosity [Eq. (3.5.7)]
θ	contact angle; polar angle
θ	$= \phi S_{\mathrm{w}} = $ volumetric moisture content
κ	$= R_{\mathrm{i}}/R_{\mathrm{o}}$ of an annulus
κ	$= \mu_{\mathrm{o}}/\mu_{\mathrm{w}} = $ viscosity of original fluid/viscosity of injection fluid $= $ viscosity ratio
κ	$= \mu_2/\mu_1$ (2 $=$ displacing fluid) (Fig. 5.82)
Λ	characteristic length
λ	ratio of interface velocities in the two branches of a pore doublet; mean free path; pore size distribution index [Eq. (5.2.21)]; dimensionless coefficient in Eqs. (6.3.15) and (6.3.16)
λ_i	kriging factor
μ	dynamic viscosity; chemical potential
μ	$\equiv \cos\theta$ in Eq. (6.3.80)
μ_{ik}	number of penetrated tubes of size i in Eq. (2.5.31); mean value
$\bar{\mu}_{\mathrm{o}}$	"oil phase" viscosity in Eq. (6.3.114)
ν	kinematic viscosity
$\nu_{\ell}, \nu_{\mathrm{s}}$	number of large and fine capillaries, respectively
ν_{p}	$= t/T = $ number of pore volumes of effluent collected at time t
ξ	constant in Eq. (6.3.49)
ξ	$\equiv \ell \left/ \left(\dfrac{R_{\mathrm{e}}^2}{r_{\mathrm{dr}}} - \dfrac{R^2}{r_{\mathrm{imb}}} \right) \right.$ Eq. (5.3.87)
ξ_i	local coordinate in the fixed coordinate system x_i [(Eq. (6.3.91)]
π	osmotic pressure
π	$= \sigma/\mu\upsilon$ (p. 424)

ρ	mass density; radial distance
ρ_w	resistivity of water (electrolyte)
Σ_i	total nimber of ions given by one mole of electrolyte (Eq. (4.2.13b))
σ	surface tension or interfacial tension; retention; distance measured along streamline
σ_k^2	variance of permeability
σ_x	standard deviation of tracer concentration in x-direction
σ^0	area of an adsorption site
σ^2	variance
τ	$= t\mathscr{D}/R^2$ [Eq. (6.2.14)]
τ	$= V/Q$ (mean) residence time
τ_w	wall shear stress in capillary tube
τ_ϕ	wall shear stress in bead pack
ϕ	porosity; azimuthal angle; piezometric head; angle; volume fraction; osmotic coefficient; half cone angle in Eq. (2.3.1)
ϕ^*	volume fraction
ϕ_c	capillary head
ϕ_{eff}	drainable porosity [Eq. (5.2.12)]
ϕ_i	angle defined in Fig. 2.60; volume fraction concentration of i
ψ	any point function [Eq. (3.3.57)]; viscosity level parameter [Eq. (3.5.11)]
Ω	angle of rotation
ω	angular velocity

Subscripts

a	denotes advancing; air; adsorbed layer; adsorption
b	denotes breakthrough or bubbling pressure; denotes bond; bulb
BL	denotes Buckley–Leverett
c	denotes capillary quantity; critical value; calculated quantity
ct	denotes capillary tube
d	denotes desorption
dr	denotes drainage
e	denotes entry
eff	denotes effective value
exp	denotes measured quantity
f	denotes pore fluid; front
g	denotes gas
H	denotes hydraulic diameter or conductivity
HP	denotes Hagen–Poiseuille
i	denotes "irreducible" saturation
i, j, k, ℓ	dummy index denoting pore size category
imb	denotes imbibition
ℓ	denotes liquid; large quantity
liq	denotes liquid
m	denotes mean value; monolayer
ma	denotes rock matrix
mt	denotes multilayer
n	denotes narrow
nw	denotes nonwetting phase
o	denotes entrance or inlet to tube
p	denotes pore

r	denotes receding; reference phase
s	denotes solid; site; small quantity
v	denotes vapor
w	denotes water or wetting phase; well; wide
1	denotes initial state
1, 2, 3	denotes dimension in space
2	denotes final state
λ	denotes wavelength

Special Symbols

∇	"del" or "nabla" operator
∇^*	$D\nabla$
∇^{*2}	$D^2\nabla^2$
δ	denotes a small quantity

Porous Media
Fluid Transport and Pore Structure

Introduction

In this brief introduction, the scope and importance of transport phenomena in porous media are emphasized as they relate to the many materials familiar to most of us. A guide to the organization of the book and its logical order of development is then presented.

Porous materials are encountered literally everywhere in everyday life, in technology, and in nature. With the exception of metals, some dense rocks, and some plastics, virtually all solid and semisolid materials are "porous" to varying degrees.

A material must pass both of the following two tests to be of interest for this book.

(1) It must contain relatively small spaces, so-called pores or voids, free of solids, imbedded in the solid or semisolid matrix. The pores usually contain some fluid, such as air, water, oil, etc., or a mixture of different fluids.

(2) It must be permeable to a variety of fluids; that is, fluids should be able to penetrate through one face of a septum made of the material and emerge on the other side. In this case one refers to a "permeable porous material."

All solids and semisolids contain interstitial spaces of the size of ordinary molecules that can be penetrated by some molecules by a molecular diffusion mechanism. The distinction between a "porous solid" and just any solid

1

remains clear-cut, however, if permeation by viscous flow mechanism is stipulated as a condition for the material to qualify as a porous medium. In other words, a "true" porous material should have a specific permeability, the value of which is uniquely determined by the pore geometry and is independent of the properties of the penetrating fluid.

There are many examples of porous materials in everyday life and the environment. Textiles and leathers are highly porous; they owe their thermal insulating properties, as well as the property that they "breath," to their pore structure. Paper towels and tissue paper are also highly porous; these owe their absorbency partly to their porous structure and partly to the property that they are strongly wetted by water.

Building materials such as bricks, concrete, limestone, sandstone, and lumber all are porous. They are better thermal insulators because of their porous nature. Home insulating materials uniquely owe their insulating properties for the most part to their very high porosity, which enables them to trap large amounts of air. Wood can be impregnated because of its porosity.

Soil is capable of performing its function of sustaining plant life only because it can hold water in its pore spaces, and plants absorb this water by the action of their capillaries. Water flows into wells from the pores in the ground, and water is purified by filtering it through porous beds of sand. *All* true filters owe their filtering action to pores that pass liquids or gases, but retain solid particles. Snow on the ground is highly porous; as a consequence, it can hold a large volume of water as well as provide an excellent insulating layer. Structural steel in reinforced concrete structures (e.g., in bridges) corrodes only because road salt can get to it through the pores and microcracks in the concrete. Water freezing in the pores of rocks, concrete, and building stones causes these to crack and crumble.

A human breathes partly through the pores in the skin. Lungs and bones both have unique and elaborate pore structures, and even hair is porous. Bread, cakes, and pastry are highly porous, as are dried meats and vegetables.

There are many examples where porous media play important roles in technology and, conversely, many different types of technology that depend on or make use of porous media. The most important areas of technology that, to a great extent, depend on the properties of porous media are (1) hydrology, which relates to water movement in earth and sand structures such as dams, flow to wells from water-bearing formations, intrusion of sea water in coastal areas, filter beds for purification of drinking water and sewage, etc.; and (2) petroleum engineering, which is mainly concerned with petroleum and natural gas production, exploration, well drilling, and logging, etc. Despite the great similarity of the physical systems and processes in these two fields of technology, each of them has a distinct technical literature and terminology.

In chemical engineering, heterogeneous catalysis is an important technology where pore diffusion of gases as well as impregnation of porous catalysts with catalyst precursor and distribution of molten catalyst are processes that depend on pore structure. Chromatography and, in particular, gel permeation chromatography are important processes, the latter of which depends on both diffusion and flow through porous media. Other separation processes use porous polymer, biological, and inorganic membranes. Filtering of gases and liquids and drying of bulk goods are equally important technologies based on flow or diffusion through porous media. The impregnation of plastics with plasticizers also depends on pore structure.

In medicine and biochemical engineering, biological membranes and filters, the flow of blood and other body fluids, and electroosmosis are a few examples where the role played by porous media is critical.

In electrochemical engineering, porous electrodes and permeable and semipermeable diaphragms for electrolytic cells play important technological roles in the struggle to obtain improved current efficiencies.

Sintering of granular materials is a very large tonnage technology where pore structure is of importance, as is the manufacture of ceramic products, paper, leather, textiles, rubber, etc.; the erection of concrete structures, such as bridges, high-rise buildings, parking garages, etc.; the surfacing of freeways; and so forth.

Porosity also plays an important role in metallic, plastic, and enamel coatings, where its presence is definitely undesirable.

Coverage of the subject matter is limited to nondeformable porous media and to phenomena that take place in the pore space. Mechanical, electrical, and other physical properties of the solid matrix of the medium are not covered.

In Chapters 1–3, the three foundations of the physical phenomena in the pore space of porous media are laid down. Chapter 1 introduces important concepts of pore structure used throughout the book. Chapter 2 is devoted to capillarity and pore structure, and Chapter 3 to single-phase transport phenomena—fluid flow, molecular diffusion, and electrical conduction.

In Chapter 4, a few selected operations utilizing flow and/or diffusion through a porous medium are reviewed briefly. In Chapter 5, simultaneous flow of immiscible fluids and immiscible displacement are discussed with a special emphasis on the influence of pore structure, wettability, and capillary versus viscous forces. Finally, in Chapter 6, hydrodynamic dispersion is reviewed, emphasizing the role played by the pore structure in determining dispersion. Diffusion, covered in Chapter 3, is the special case of hydrodynamic dispersion in the absence of convection.

1 | *Pore Structure*

Pores are invisible to the naked eye in the majority of porous media. The porous nature of a material is usually established by performing any one of a number of experiments on a sample and observing its behavior, because porous materials behave differently from nonporous ones in a number of respects.

Appropriate experiments lead to the determination of various macroscopic parameters, which are often uniquely determined by the pore structure of the sample and play a similar role in the characterization of porous media as, for example, density, specific conductance, dielectric constant, yield strength, and heat of vaporization play in other branches of science and technology.

In atomic and molecular physics and physical chemistry, phenomenal progress has been made in the interpretation of macroscopic material properties in terms of the microscopic (atomic or molecular) structure of matter. Insight into the microscopic pore structure has been similarly helpful in understanding and interpreting the properties of porous media.

Not very long ago there was great resistance to atomic and molecular theories, but that all belongs to the past now. At present there still exists some skepticism regarding the interpretation and prediction of the macroscopic properties of porous media with the help of microscopic pore structure models. The practical advantages to be gained from being able to explain various, seemingly disconnected, macroscopic phenomena by the same micro-scopic pore structure model are too obvious, however, to be overlooked. As the microscopic models keep improving and become increasingly reliable and useful, the skepticism will gradually fade away.

In this chapter first the macroscopic pore structure parameters are pre-sented and discussed, along with the experimental methods used for their determination. The rest of this chapter is devoted to a review of the state of the art of microscopic pore structure research. At the end of the chapter, a few important selected natural and manufactured porous materials are reviewed.

1.1. MACROSCOPIC PORE STRUCTURE PARAMETERS

All macroscopic properties of porous media are influenced, to a greater or lesser degree, by the pore structure. *Pore structure parameters are those properties that are completely determined by the pore structure of the medium and do not depend on any other property.* Macroscopic pore structure parame-ters represent average behavior of a sample containing many pores. The most important macroscopic pore structure parameters are the "porosity," the "permeability," the "specific surface area," the "formation resistivity factor," and the reduced "breakthrough (displacement) capillary pressure" or "bub-bling pressure."

1.1.1. Porosity

"Porosity" (also called "voidage") ϕ is the fraction of the bulk volume of the porous sample that is occupied by pore or void space. Sometimes, this word is used inaccurately with a different meaning; for example, in the expression "graded porosity" of a filter consisting of layers of different grain sizes, "porosity" is not a measure of void fraction but rather of void size. Depending on the type of the porous medium, the porosity may vary from near zero to almost unity. For example, metals and certain types of volcanic rocks have very low porosities, whereas fibrous filters and thermal insulators are highly porous substances.

Types of void space: It is important to distinguish between two kinds of pore or void space, one that forms a continuous phase within the porous medium, called "interconnected" or "effective" pore space, and the other that consists of "isolated" or "noninterconnected" pores or voids dispersed over the medium. Noninterconnected void or pore space cannot contribute to transport of matter across the porous medium, only the interconnected or effective pore space can.

"Dead-end" or "blind" pores are interconnected only from one side. Even though these can often be penetrated, they usually contribute only negligibly to transport.

Experimental methods: The various experimental methods used to determine porosities have been adequately discussed by Collins (1961) and Scheidegger (1974), who subdivided these in the following categories:

(a) Direct method. This method consists of measuring bulk volume of a porous sample and then somehow destroying all the voids and measuring the volume of only the solids.

(b) Optical methods. The porosity of a sample is equal to the "areal porosity," provided that the pore structure is "random." The areal porosity is determined on polished sections of the sample. It is often necessary to impregnate the pores with some material such as wax, plastic, or Wood's metal in order to make the pores more visible and/or to distinguish between interconnected and noninterconnected pores. When impregnating the sample from the outside, evidently only the interconnected pores will be penetrated (cf. Underwood, 1970).

Whenever very small pores are present along with larger ones, it is difficult to make sure that all the small pores have been accounted for by the measurement. This is one of the reasons why porosities determined by optical methods may differ significantly from the results obtained by other methods (Dullien and Mehta, 1971/1972). Various optical instruments are available for rapid determination of areal porosities.

(c) Imbibition method. Immersing the porous sample in a preferentially wetting fluid under vacuum for a sufficiently long time will cause the wetting fluid to imbibe into all the pore space. The sample is weighed before and after imbibition. These two weights, coupled with the density of the fluid, permit calculation of the pore volume. When the sample is completely saturated with the wetting fluid, a volumetric displacement measurement in the same wetting fluid gives directly the value of the bulk volume of the sample. From the pore volume and the bulk volume the porosity can be directly calculated.

Imbibition, if done with sufficient care, will yield the best values of the effective porosity.

(d) Mercury injection method. The bulk volume of the sample is determined by immersion of the sample in mercury. As most materials are not wetted by mercury, the liquid will not penetrate into the pores.

After evacuating the sample, the hydrostatic pressure of mercury in the chamber containing both the sample and the mercury is increased to a high value. As a result, the mercury will enter the pore space and, provided that the pressure is high enough, it will penetrate even into very small pores.

Nevertheless, the penetration is never quite complete because it takes infinite pressure to perfectly fill all the edges and corners of the pores. It is conceivable that very high pressures may cause changes in the pore structure of the sample.

(e)   Gas expansion method. This method also measures the effective porosity. The bulk volume of the sample is determined separately. The sample is enclosed in a container of known volume, under known gas pressure, and is connected with an evacuated container of known volume. When the valve between the two vessels is opened, the gas expands into the evacuated container and the gas pressure decreases. The effective pore volume V_p of the sample can be calculated by using the ideal gas law:

$$V_p = V_B - V_a - V_b[P_2/(P_2 - P_1)], \tag{1.1.1}$$

where V_B is the bulk volume of the sample, V_a is the volume of the vessel containing the sample, V_b is the volume of the evacuated vessel, P_1 is the initial pressure, and P_2 is the final pressure.

A gas expansion porosimeter has been described by, among others, Beeson (1950).

(f)   Density methods. Density methods depend on determining the bulk density of the sample and the density of the solids in the sample. Since the mass of a porous medium resides entirely in the solids matrix, we have the following:

$$m = \rho_s V_s = \rho_B V_B, \tag{1.1.2}$$

where m is the mass of the sample, ρ_s the density of the solids in the sample, and ρ_B the bulk density of the sample.

By the definition of porosity ϕ, there follows that

$$\phi = 1 - (V_s/V_B) = 1 - (\rho_B/\rho_s). \tag{1.1.3}$$

The density methods yield total porosity.

1.1.2.   Permeability

"Permeability" is the term used for the conductivity of the porous medium with respect to permeation by a Newtonian fluid. "Permeability," used in this general sense, is of limited usefulness only because its value in the same porous sample may vary with the properties of the permeating fluid and the mechanism of permeation. It is both more useful and more scientific to separate out the parameter that measures the contribution of the porous medium to the conductivity and is independent of both fluid properties and flow mechanisms. This quantity is the "specific permeability" k, which in this monograph is called "permeability," for short, unless otherwise stated. Its value is uniquely determined by the pore structure.

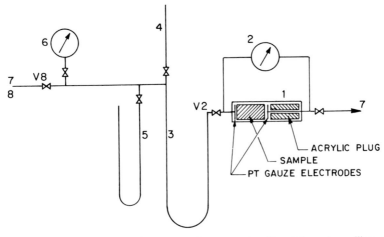

Figure 1.3. Schematic diagram of apparatus for measuring "breakthrough capillary pressures": 1, Hassler core holder; 2, galvanometer; 3, mercury reservoir; 4, burette; 5, mercury manometer; 6, pressure gauge; 7, vacuum pump; 8, nitrogen cylinder (El-Sayed, 1978).

The breakthrough capillary pressure P_{cb} of penetration of a nonwetting phase, as defined in Chapter 2, is an important macroscopic parameter of the porous medium. It corresponds to the first appearance of a nonwetting phase (e.g., mercury in an evacuated sample or air in a sample saturated with a wetting fluid) on the outlet face of a plug sample (see Fig. 1.3). The breakthrough capillary pressures obtained for a variety of different sandstone samples, using mercury, are listed in Table 1.1. The detection of breakthrough by electrical means is extremely sensitive and the experimental procedure is straightforward.

The physical meaning of the breakthrough capillary pressure, as explained in detail in Chapter 2, corresponds to the incipient formation of a continuum of the nonwetting phase through a pore network of arbitrarily large size.

The reduced breakthrough capillary pressure P'_{cb},

$$P'_{cb} = P_{cb}/4\sigma \cos \theta, \qquad (1.1.9)$$

determined with the help of suitable fluids such as mercury or air-wetting fluid, is a characteristic of the pore structure of the sample. It determines, together with σ and θ, the minimum value of P_c at which the penetrating fluid becomes hydraulically (or electrically) conductive in a macroscopic sample. It is often convenient to compare the pore structures of different materials with the help of this quantity. "Bubbling" pressures have been used (e.g., Macmullin and Muccini, 1956; Katz and Thompson, 1986; Dullien, 1991) to predict permeabilities of various materials.

TABLE 1.1. Experimentally Determined Breakthrough Pressures[a]

Sample	$D_{\text{inf.pt.}}$ of Hg porosimetry curve (μm)	Pressure of inf.pt. of Hg porosimetry curve (psia)	Breakthrough pressure, P_{cb} (psia)	Breakthrough diameter, D_b (μm)
Boise	47	4.5	2.0	104.0
Bartlesville	27	4.8	6.7	31.5
Berea 108	26	8.1	7.4	28.6
Berea BE1	22.5	9.4	8.3	25.5
Noxie 47	24	8.8	7.5	28.2
St. Meinrad	14	15.1	11.6	18.3
Torpedo	10	21.2	17.1	12.4
Big Clifty	9.0	23.5	11.7	18.2
Noxie 129	8.5	24.9	18.0	11.8
Cottage Grove	11	19.3	16.4	12.9
Clear Creek	8	26.5	14.0	15.1
Bandera	5	42.4	39.0	5.4
Whetstone	2	106.0	72.0	2.94
Belt Series	0.44	482.0	150.0	1.41

[a] From El-Sayed (1978).

1.1.4. Specific Surface Area

The "specific surface" of a porous material is defined as the interstitial surface area of the voids and pores either per unit mass (S) or per unit bulk volume (S_v) of the porous material. The specific surface based on the solids' volume is denoted by S_0.

Specific surface plays an important role in a variety of different applications of porous media. It is a measure of the adsorption capacity of various industrial adsorbents; it plays an important role in determining the effectiveness of catalysts, ion exchange columns, filters; and it is also related to the fluid conductivity or permeability of porous media.

The major methods of determining specific areas have been reviewed by Collins (1961) and Scheidegger (1974):

(a) Adsorption. These methods are discussed in Chapter 2. The surface area is usually obtained based on unit mass of the sample.

(b) Quantitative stereology ("optical" methods). These methods use photomicrographs of polished sections of the sample with sufficient contrast to clearly distinguish the pores from the solid matrix.

(c) Fluid flow. Measurements of permeability and porosity have been related to the specific surface of the sample, particularly when this consists of a bed of randomly packed reasonably monosized rotund particles (see Chapter 3).

1.1.5. Formation Resistivity Factor

The "formation resistivity factor" F is defined as the ratio of the electrical resistance R_0 of the porous sample saturated with an ionic solution to the bulk resistance R_w of the same ionic solution occupying the same space as the porous sample, thus:

$$F = R_0/R_w \qquad (1.1.10)$$

It is evident that F is, by definition, always greater than unity. The formation resistivity factor measures the influence of pore structure on the resistance of the sample. Description of the conductivity cell and the experimental procedure used in the determination of formation resistivity factors has been given, for example, by Amyx *et al.* (1960) and El-Sayed (1978). More discussion of the formation resistivity factor can be found in Chapter 3.

1.1.6. Macroscopic System, Macroscopic (Statistical) Homogeneity and Heterogeneity

It has been assumed throughout the discussion of the various macroscopic pore structure parameters that the reader has an intuitive understanding of the meaning of "macroscopic parameter." Indeed, as long as the porous medium is statistically homogeneous, the meaning of "macroscopic system" is relatively straightforward. Unfortunately, porous media are often macroscopically heterogeneous rather than homogeneous. This fact necessitates a tightening up of the definition of "macroscopic system."

A macroscopic porous medium may be defined by a smooth permeability (or porosity) variation as a function of the sample volume, as illustrated in Fig. 1.4. One can think of starting with a very large sample. After measuring its permeability, the sample is divided in a random manner into two halves; one half is discarded, and the permeability of the remaining half is determined. Then half of this is discarded and the permeability of the remaining half is determined, and so forth. In order to show the variation on all scales it is convenient to use semilog representation. As the samples become very small, the averages are taken only over a small number of pores and, hence, they will fluctuate with the size of sample. Such samples are submacroscopic.

Network modeling of behavior of porous media has indicated that a cubic network of $15 \times 15 \times 15$ nodes, containing randomly distributed node sizes, the nodes representing pores, can be considered macroscopic for all practical purposes and even networks of much smaller sizes exhibit only relatively

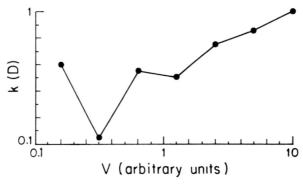

Figure 1.4.   Illustration of the definition of "Macroscopic Porous Medium."

small fluctuations of properties. Such network models, above a certain size, represent homogeneous macroscopic media, because they contain a statistically representative sample of randomly distributed different nodes. A homogeneous medium is necessarily macroscopic and, conversely, submacroscopic media are always heterogeneous. In other words, their properties fluctuate in dependence of sample size, because they do not contain a statistically representative distribution of pores. Real macroscopic media, however, are

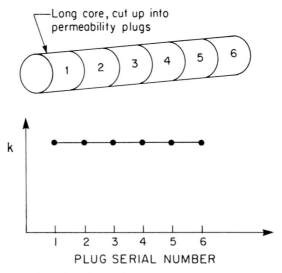

Figure 1.5.   Illustration of the definition of "Homogeneous Porous Medium."

often heterogeneous owing to the spatial variability of pore morphology. Operationally, a homogeneous medium is defined, as illustrated in Fig. 1.5, by the requirement that all macroscopic samples of a scale that is comparable with the scale of the medium under observation have the same permeability, porosity, etc., which is a necessary condition of homogeneity. Theoretically speaking, it is possible that macroscopic samples have, over a certain range of scales, different properties and, therefore, over that range of scales the medium is heterogeneous; but on a larger scale the macroscopic samples have the same properties and, hence, on that larger scale the medium is homogeneous.

Another important point to note is that certain properties, such as the permeability, reflect heterogeneous morphology to a much greater extent than some other properties, such as the porosity. Homogeneity versus heterogeneity in terms of pore structure is discussed in Section 1.2.4.

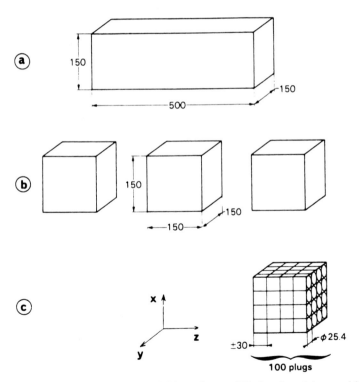

Figure 1.6. Schematic diagram of subdivision scheme of blocks of sandstone and limestone (Henriette *et al.*, 1989).

1.1.7. Anisotropy of Pore Structure

"Anisotropy," as customarily applied to porous media, means that some properties of the medium do not have the same value when measured in different directions. In an anisotropic porous medium, the permeability, the formation resistivity factor, and the breakthrough capillary pressure depend on direction.

Anisotropy of porous media is usually caused by layering. That is to say, a less permeable layer lies parallel to a more permeable one. It is seldom an intrinsic property as is the case, for example, with properties of crystals.

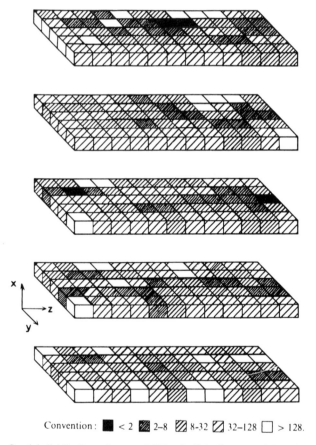

Convention: ■ < 2 ▨ 2–8 ▨ 8-32 ▨ 32–128 ☐ > 128.

Figure 1.7. Spatial distribution of permeabilities (md) in limestone block (Henriette *et al.*, 1989).

Anisotropy of porous media has received particular attention with reference to anisotropy of the permeability. A more detailed discussion of this question and related problems is given in Chapter 3.

1.1.8.   Measured Permeability Distributions

There are only a few reports in the literature on laboratory measurements of the spatial distribution of permeabilities and porosities in rock samples. An interesting example is provided by Henriette *et al.* (1989) who investigated a sandstone and a limestone block, each $150 \times 150 \times 500$ mm large. The blocks were subdivided, as shown in Fig. 1.6. The permeabilities (in the

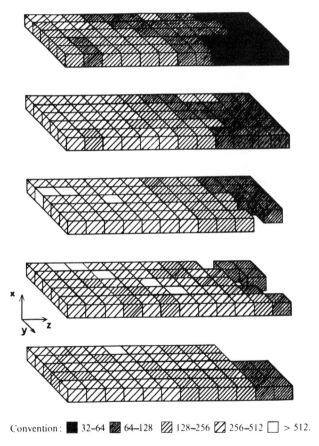

Convention: ■ 32–64 ▨ 64–128 ▨ 128–256 ▨ 256–512 □ > 512.

Figure 1.8.   Spatial distribution of permeabilities (md) in sandstone block (Henriette *et al.*, 1989).

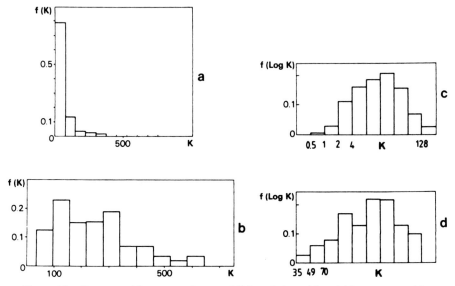

Figure 1.9.   Frequency histograms of permeabilities of plugs. (a) and (c): limestone; (b) and (d): sandstone. (Class intervals in (a) and (b) are linear; in (c) and (d) they follow geometric series with quotients of 2 and $\sqrt{2}$, respectively.) (Henriette *et al.*, 1989.)

z-direction) and the porosities were measured first of the entire block, then of each one-third block, and finally of each 1/300th of the original block (each cube, shown in Fig. 1.6, was sliced into four layers in the z-direction and each layer was cut into $5 \times 5 = 25$ plugs). The permeability data are all tabulated in the reference, the spatial distributions are shown graphically in Figs. 1.7 and 1.8, and the frequency distributions are shown in Fig. 1.9. The enormous variation in permeabilities should be noted. Autocorrelation coefficients were also calculated: In the sandstone blocks the permeability was correlated over two to three small plugs along the coordinate axes, but in the limestone blocks there was no correlation past one small plug. The permeabilities correlated very poorly with the porosities in both blocks. As permeabilities were measured only in the z-direction, no inference on the anisotropy of the blocks studied is warranted. The distribution of z-permeabilities in the x, y-planes gives no indication of what x- and y-permeabilities might have been like.

1.2.   MICROSCOPIC PORE STRUCTURE PARAMETERS

In this section the microscopic pore structure picture is reviewed. This is an extremely difficult subject because of the great irregularity of pore geometry. Owing to the irregular variations in the shape of capillaries, pore "diameter" or pore "size" is an intuitive simplification of reality. Notwithstanding the formidable difficulties faced by pore structure researchers, a great deal of good work has been done in this field, and the results, although imperfect and semiquantitative, nevertheless have helped to explain and correlate various phenomena in porous media.

This monograph contains many examples of the insight gained into the various phenomena in porous media through an improved understanding of pore structure.

1.2.1. Capillaries, Voids, and Pores—Visual Studies of Pore Structure

Direct proof of the existence of empty spaces (called voids, pores, or capillaries, etc.) in solid media is obtained by examining, under an optical or electron microscope, prepared sample sections or surfaces of the medium. For some time, scanning electron microscopes have permitted a study of three-dimensional views of surface pores. Figure 1.10 shows photomicrographs of various different materials.

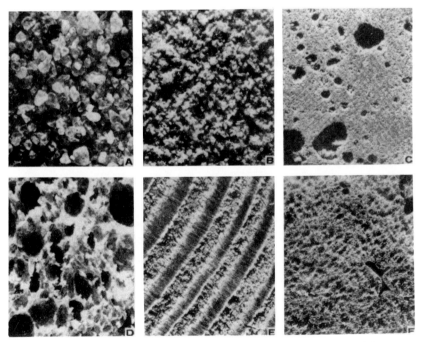

Figure 1.10. Examples of natural porous materials ($\times 10$): (A) beach sand; (B) sandstone; (C) limestone; (D) rye bread; (E) wood; (F) human lung (Collins, 1961).

It is often useful to distinguish between "macro" and "micro" pores, although the dividing line between the two classes is not always sharp.

1.2.2.   Topology of Pore Structure

The parameters characterizing the topology of networks of capillaries are the dimensionality of the network, the connectivity, and the microscopic pore topology.

1.2.2.1.   *Connectivity and Genus*

The topological parameters, characterizing the interconnectedness of "shapes" or "structures," are the "connectivity" and the "genus." They have been reviewed in the literature, for example, by Fischmeister (1974).

"Connectivity" is a topological parameter (Cairns, 1961) that measures the degree to which a structure is multiply connected. It is defined as the number of nonredundant closed-loop paths by which all regions inside the shape can be inspected (see Fig. 1.11). Redundant loops are those that may be transformed into another by deformation, that may be shrunk into a point without passing out of the closed surface, or that do not give access to any new part of the shape. A basic theorem of topology states that the "connectivity" C of a closed surface or shape is equal to its "genus" G. "Genus" is the largest number of cuts that may be made through parts of the shape without totally disconnecting any part from the rest.

The doughnut in Fig. 1.11a has genus and connectivity one because its whole exterior can be explored through one closed-loop path and no more than one cut can be made without the part being disconnected. Twisting or stretching the doughnut would not change the topology. In Fig. 1.11b the

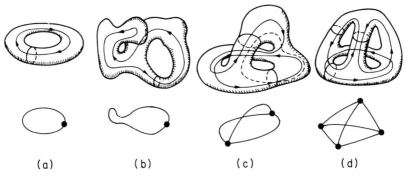

Figure 1.11.   Illustration of the concepts of "connectivity" and "genus" (a) $b = 1$, $n = 1$, $G = C = 1$; (b) $b = 1$, $n = 1$, $G = C = 1$; (c) $b = 3$, $n = 2$, $G = C = 2$; (d) $b = 6$, $n = 4$, $G = C = 3$ (Fischmeister, 1974).

appendix attached to the doughnut does not change the topology. The shape in Fig. 1.11c has genus and connectivity two because there are two nonredundant loops; the third one shown as dotted line is redundant. Figure 1.11d shows an example of genus and connectivity three. The lines in Fig. 1.11 are called "deformation retracts" of the shapes. It is apparent from these examples that the connectivity C and the coordination number z are two different parameters.

A general theorem of topology states

$$G = C = b - n + N, \tag{1.2.1}$$

where b is the number of branches, n is the number of nodes, and N is the number of separate networks. It is noted that open-ended branches not connected to a node should not be counted in b since they may be shrunk away without changing the genus.

The simple shapes shown in Fig. 1.11 are vastly simpler than the structure of a typical porous medium. For real porous media the method of finding the connectivity depends on whether the structure of the medium is regular or irregular. In the case of regular arrays of voids, it can be readily computed by inspecting the symmetry of the system.

In the case of irregular pore structure, the only known way of determining the connectivity is by reconstructing the deformation retract or "branch-and-node chart" of pore structure from a series of parallel sections obtained by "serial sectioning." The consecutive parallel sections must be spaced close enough to allow each capillary branch to be followed from one section to the next, as shown in Fig. 1.12a and b. As an unknown number of the retract lines that pass through the boundaries of the volume studied is connected to nodes, there is a considerable uncertainty in the result, unless a large enough volume is studied so that the number of lines leaving the volume represents only a small fraction of the number of branches inside the volume.

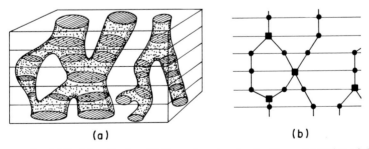

(a)   (b)

Figure 1.12. (a) Serial sectioning; (b) "branch-and-node chart" representation of the structure (Fischmeister, 1974).

It has been shown (e.g., Macdonald *et al.*, 1986) that, whereas the connectivity C is a unique property of the sample, there is, in general, room for different branch-and-node chart representations, resulting in different values of b and n.

In a homogeneous, macroscopic, porous medium, the numerical value of the connectivity is proportional to the size of the sample, whereas the coordination number is independent of the size. Fischmeister (1974) has recommended that the absolute value of the connectivity be referred to a unit volume of the sample—that is, to use the "specific connectivity," which is independent of the size of the sample. This would, however, be not representative, as the specific connectivities of two geometrically similar samples (i.e., samples that differ only in degree of magnification) are in inverse ratio to each other as their degrees of magnification, whereas their coordination numbers are identical. Two different samples may also have the same specific connectivities but may be characterized by quite different topologies. For example, one sample may contain islands of a network with a small unit cell, whereas the other may contain a continuum of a network with the same coordination number, but larger unit cell. Provided that the pore topology of different samples is represented by *standard* branch-and-node charts, the connectivity per node may, however, be used to compare the topologies of the samples.

Another measure of the interconnectedness of pore structure is the coordination number z, the average number of branches (or bonds) meeting at a node. For the determination of z, however, distances and sizes make a difference and also introduce an element of arbitrariness. Therefore there cannot exist a unique relationship between z and C.

It is noted that whereas the connectivity of an open chain consisting of distributed pore sizes is equal to zero, the coordination number z is equal to two. To make z a better measure of the degree of "branchedness" of a network, a "modified coordination number" could be defined for which the minimum number of branches (bonds) meeting at a node counted is three.

1.2.2.2. *Determination of Connectivity of Macropores in Small Sandstone Samples*

Serial sectioning of samples has been performed for some time by metallographers for the purpose of determining the genus of a phase (Fischmeister, 1974; De Hoff and Rhines, 1968; De Hoff *et al.*, 1972; De Hoff, 1983). More recently this technique has been also applied to porous media (Pathak *et al.*, 1982; Lin and Cohen, 1982; Lin and Perry, 1982; Kaufmann *et al.*, 1983; Macdonald *et al.*, 1986). The data used by Macdonald *et al.* (1986) has formed the basis of a number of pore structure studies. The data were generated by first injecting, under high pressure, molten Wood's metal into a previously evacuated sandstone sample. After solidification of the metal the

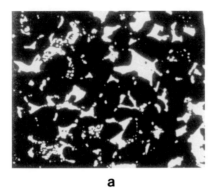

a

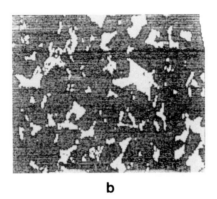

b

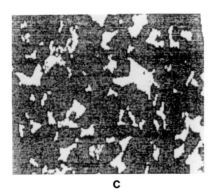

c

Figure 1.13. Representative serial section: (a) photomicrograph; (b) original digitized picture; (c) digitized picture after filtering to remove nonconnected features (after Kwiecien *et al.*, 1990).

FEATURE 65 'POINTS TO' FEATURE 132

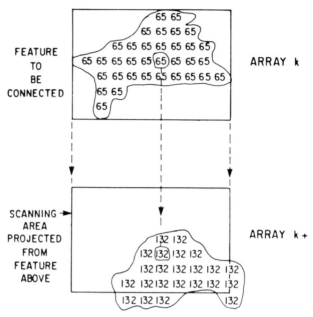

Figure 1.14. Connection of features on adjacent serial sections by the overlap criterion (after Macdonald *et al.*, 1986).

rock matrix was dissolved with hydrofluoric acid and replaced with clear epoxy resin. A piece of this sample of dimensions $1310 \times 1040 \times 762$ μm was encased in epoxy resin, and its surface was successively ground and polished to remove layers of about 10 μm thickness. Seventy-eight layers were removed, and each layer was photographed. A sample photomicrograph is shown in Fig. 1.13. The white features represent the pore space of macropores because the metal is reflective; and the black regions represent the rock matrix because the epoxy is clear.

The micrographs were entered into the computer, where they were digitized; then connection between features in neighboring photomicrographs was established based on the assumption that two features are connected if and only if they overlap. Connection of two pixels between two features present in two neighboring micrographs k and $k + 1$ is illustrated in Fig. 1.14. For the purpose of topological studies, where distances and directions play no role, this kind of a treatment of the data results in a branch-node chart of the sample from which the genus may be readily determined. The genus is defined as the number of nonintersecting cuts that can be made upon a surface without separating it into disconnected parts. It has been shown by Barrett and Yust (1970) that the genus of the enclosing surface is numerically equal to the connectivity of the branch-node network derived from that surface. The connectivity is a measure of the number of independent paths between two points in the pore space and, hence, of the degree of interconnectedness of the pores. The surface nodes introduce complications because it is not known how they are connected on the outside of the sample. The surface nodes in the plane of polish are visible, but the lateral surface nodes are not and they must be obtained by means of the overlapping criterion.

Much of the study deals with the problem of edge effects in the construction of the branch-node chart, which are very important owing to the small size of the sample. A larger sample, however, would result in much increased computer time and decreased resolution of the features. The maximum possible value of the genus G_{max} is obtained by connecting all the surface nodes to one external node. The least possible value of the genus, G_{min}, is calculated by not connecting any of the surface nodes to an external node. In the Berea sandstone sample $G_{max} = 593$ and $G_{min} = 420$. Dividing these values by the sample volume of $10.4 \times 10^8 \ \mu m^3$ the genus per unit volume can be calculated. This yields a genus of about 5×10^{-7} per μm^3, or a genus of 2 per $4 \times 10^6 \ \mu m^3$, which corresponds to a sphere of a diameter of about $200 \ \mu m^3$, which is the size of an average grain in the Berea sandstone sample. The pertinent data on the Berea sandstone sample (Berea 2c) and on another smaller, preliminary sample (Berea 1xx) are listed in Table 1.2. The genus was determined versus the volume of the sample section-by-section, with the interesting result that past a certain minimum volume the genus is a linear function of the volume, as shown in Fig. 1.15. The slope of the line gives the best estimate of the genus per unit volume. It is logical that linearity could not exist if the pore topology of the sample had varied in the direction of grinding and polishing—that is, normal to the planes of sectioning. Although pore geometry (e.g., pore sizes) may be different for the same

TABLE 1.2.   Berea Sandstone Samples[a]

Property	Berea 2c	Berea 1xx
Cross section	$1310 \ \mu m \times 1040 \ \mu m$ $= 1.36 \times 10^6 \ \mu m^2$	$1350 \ \mu m \times 950 \ \mu m$ $= 1.28 \times 10^6 \ \mu m^2$
Total depth	$762 \ \mu m$	$514 \ \mu m$
No. of serial sections	78	50
Average spacing, $\Delta \bar{z}$	$9.9 \ \mu m$	$10.5 \ \mu m$
Volume	$10.4 \times 10^8 \ \mu m^3$	$6.6 \times 10^{-8} \ \mu m^3$
No. of features in sample	3564	2583
Average grain size	$\approx 200 \ \mu m$	$\approx 200 \ \mu m$

[a]After Macdonald *et al.* 1986.

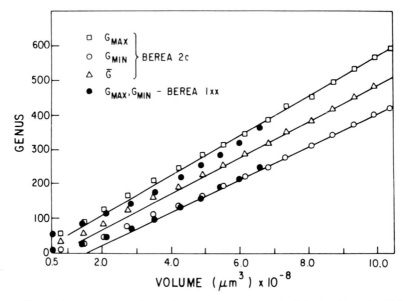

Figure 1.15.   Genus vs. volume of Berea sandstone samples (Macdonald *et al.*, 1986).

topology, nevertheless in one and the same sample direct proportionality between genus and volume is a strong indication of uniform topology of the sample.

The work of Macdonald *et al.* (1986) is an improvement over that of Pathak *et al.* (1982), who performed a manual trace and count of branches and nodes. No applications of the available techniques of mathematical morphology could be found (e.g., Serra, 1982).

1.2.3.   The Problem of Defining Pore Size

"Pore size" has been customarily used in conjunction with experimentally determined "pore size distribution," $\alpha(D)\,dD$, meaning the distribution density of pore volume by some length scale assigned to it. Every method of "pore size distribution" determination (Dullien and Batra, 1970) defines "pore size" in terms of a pore model best suited to the quantity measured. The general procedure used for "pore size distribution" determination consists of measuring some physical quantity in dependence on another physical parameter that is under the control of the experimenter and is varied in the experiment. For example, in mercury porosimetry, the volume

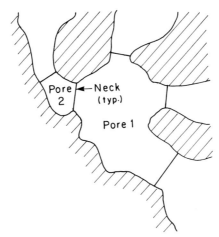

Figure 1.16. Two-dimensional representation of three-dimensional pores and pore throats.

of mercury penetrating the sample is measured as a function of the pressure imposed on the mercury. The "pore size" is calculated from this pressure by Laplace's equation of capillarity and, using the "bundle of capillary tubes" model of pore structure, the volume of mercury is assigned to this "pore size."

Before attempting the definition of "pore size," the object "pore" must be defined. While a completely general definition of a pore is probably not possible, a "pore" is defined as a portion of pore space bounded by solid surfaces and by planes erected where the hydraulic radius of the pore space exhibits minima, analogously as a room is defined by its walls and the doors opening to it. If the local minima, the pore throats (the doors), can be located and imaginary partitions erected at these positions (the doors closed) then the pores are defined, as illustrated schematically in Fig. 1.16, and their size can be determined by any arbitrary definition of "size" of an irregularly shaped object. Such definitions have been used for a long time in the field of particle size measurement.(The concept of "void particles" was introduced by Rumpf (see Debbas and Rumpf, 1966).) In the final analysis, the problem of pore size measurement has been reduced to the well-established problem of particle size measurement. The same analogy applies also to the determination of pore shape.

The reason for using the hydraulic radius r_H as a measure of pore throats is that r_H is a useful measure of "size" also in the case of irregularly shaped cross sections, as shown by the comparisons listed in Table 1.3. Here $2/r_m$ is related to the capillary pressure P_c; that is, the pressure difference across the

TABLE 1.3.   List of Comparative Values to Show Equivalence of the Reciprocal Hydraulic
Radius $(1/r_H)$ and Twice the Reciprocal Mean Radius of Curvature
$2/r_m = [(1/r_1) + (1/r_2)]$ in a Capillary[a,b]

Cross section	$(1/r_1) + (1/r_2)$	$1/r_H$
Circle	$2/r$	$2/r$
Parallel plates	$1/b$	$1/b$
$\quad a:b = 2:1$	$1.50/b$	$1.54/b$
Ellipse $a:b = 5:1$	$1.20/b$	$1.34/b$
$\quad a:b = 10:1$	$1.10/b$	$1.30/b$
Rectangle	$1/a + 1/b$	$1/a + 1/b$
Equilateral triangle	$2/r_i$	$2/r_i$
Square	$2/r_i$	$2/r_i$

[a]After Carman, 1941.
[b]r_i is the radius of the inscribed circle.

fluid/fluid interface of mean radius of curvature r_m, in mechanical equilibrium, by Laplace's equation of capillarity

$$P_c = \frac{2\sigma}{r_m} \qquad (1.2.2)$$

where σ is the interfacial tension. For the case of nonzero contact angle θ, r_m must be replaced by R, according to the relation

$$R = r_m \cos \theta. \qquad (1.2.3)$$

The definition of hydraulic radius r_H of a capillary of uniform cross section is

$$r_H = \frac{\text{volume of capillary}}{\text{surface area of capillary}}. \qquad (1.2.4)$$

For the case of a variable cross section the above definition can be generalized for any normal cross section of the capillary as follows:

$$r_H = \frac{\text{area of cross section}}{\text{length of perimeter of cross section}}. \qquad (1.2.5)$$

The applicability of Eq. (1.2.5), however, is limited to the special case of capillaries of a rotational (axial) symmetry that have normal cross section.

For the general case of irregular capillaries, the minimum value of the ratio given by the right-hand side of Eq. (1.2.5) must be found by varying the orientation of the sectioning plane about the same fixed point inside the

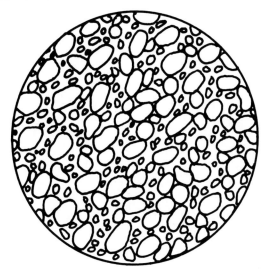

Figure 1.17. Homogeneous distribution of grains illustrated as monolayer. (Reprinted with permission from Morrow, 1971.)

capillary. The minimum value of this ratio is, by definition, the hydraulic radius r_H of the irregular capillary at the fixed point. The value of r_H of a section is assigned to the center of gravity of the section. Both definitions (i.e., r_m and r_H) are best suited to the case of pore throats that control both capillary penetration by a nonwetting fluid into the porous medium and the flow rate of fluids through the porous medium.

Locating pore throats is difficult and, therefore, initially it was assumed by the author and coworkers that the probability of a sectioning plane to pass through a pore throat is negligibly small. Much later, after pore throats were finally located, it was found that this assumption introduced serious errors.

1.2.4. Homogeneity and Heterogeneity in Terms of Microscopic Properties

A sufficient condition of homogeneity of the medium is that macroscopic samples taken from the medium display statistically indistinguishable pore morphologies, as illustrated in Fig. 1.17. That is, the samples contain statistically indistinguishable samples of pore bodies and pore throats that are distributed in space uniformly in either a disordered (like molecules in a gas) or an ordered (like molecules in a crystal) fashion. Evidently, this definition of homogeneity permits correlations between the positions and/or types (shapes, sizes) of pores over distances involving a few pores, because samples

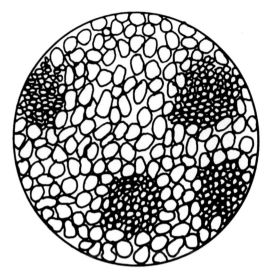

Figure 1.18. Heterogeneous distribution of grains illustrated as monolayer. (Reprinted with permission from Morrow, 1971.)

containing only a few pores are submacroscopic (and, therefore, heterogeneous).

Real porous media never meet the above-stated sufficient condition of homogeneity perfectly. They often meet, however, to a certain degree, the necessary condition of homogeneity, given in Section 1.1.6.

A necessary condition of macroscopic heterogeneity of the medium is that macroscopic samples taken from the medium display statistically different pore morphologies, which may be due to differences between the samples of pore bodies and/or pore throats or to correlations between the spatial positions of pores, or to both of these causes—one possible realization of which is illustrated in Fig. 1.18.

One can, in principle, assign an average length scale to the heterogeneities, meaning a representative distance over which there is a significant statistical variation of pore morphology of the medium. A sufficient and necessary condition of macroscopic heterogeneity of the medium is that the length scale of heterogeneities of pore morphology be comparable with the scale of the macroscopic medium under observation.

As a corollary it may be concluded that a sufficient and necessary condition of homogeneity of the macroscopic medium under observation is that the length scale of heterogeneities of pore morphology be much less than the scale of observation.

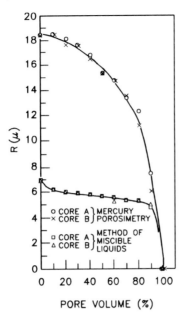

Figure 1.19. Cumulative pore size distribution curves of Bentheim sandstone. Permeability: 975×10^{-11} cm^2. Porosity: 0.289. (After Klinkenberg, 1957.)

1.2.5. Pore Size Distribution Determination

As pointed out above, the most representative approach to pore size distribution determination involves measurement of the size of all the pores in a macroscopic sample. All other methods are indirect and the results obtained by them depend on *a priori* assumptions made on the pore structure. The same model (e.g., the bundle of capillary tubes model) may lead to widely different pore size distributions, depending on the type of experiment used. This is illustrated in Fig. 1.19 by the example presented by Klinkenberg (1957), who interpreted the miscible displacement breakthrough curve observed in a sandstone in terms of miscible displacement in a bundle of capillary tubes of distributed diameters. The results indicate pore sizes that are smaller than the smallest (macro) pore in the sandstone.

1.2.5.1. *Miscellaneous Methods of Pore Size Distribution Determination*

The widely used methods of mercury porosimetry and sorption isotherms are reviewed and discussed in detail in Chapter 2. An interesting idea of using diffusion measurements has been presented by Brown and Travis

(1983). The method is promising enough to warrant experimental testing on samples of known pore size distribution.

Miscellaneous methods of pore structure investigation include radiography. Ritter and Erich (1948) wrote:

> The intensity of scattered radiation depends on a difference in scattering power between the two phases, and the resultant size determination is a measure of the distance one can travel in the scattering boundary without passing a phase volume in either direction. In the case of a dispersed system of solid and air, when the particles are widely separated, the scattering due to their distances apart will be suppressed and one will obtain a particle size distribution. When the particles are squeezed together so closely that the interstices are small compared with particle size, the small-angle scattering will be indicative of pore size.... The situation when pore and particle size are about equal is the worst case.

Notwithstanding these problems, Ritter and Erich obtained surprisingly good agreement with results of mercury porosimetry for a Fuller's earth sample.

Clark and Liu (1957) described a method for quantitative determination of porosity by x-ray absorption. Edwards an Simpson (1955) gave a method of neutron-derived porosity determination for thin bed boreholes of oil wells. Imelik and coworkers (1965) reviewed three existing theories of central x-ray scattering. These theories make it possible to obtain the specific surface of porous media. As a result of an improvement made in these theories, the authors could also obtain a mean pore diameter.

Clarke (1964) used gamma radiation to determine the surface area of graphite. The amount of gas adsorbed was determined by measuring the gamma radiation intensity. Bonnemay and coworkers (1964) discussed possible applications of radiography to the study of structure and operation of porous electrodes.

Plavnik and Dubinin (1966) applied small-angle, x-ray scattering to the investigation of the pore structure of activated charcoal. Dimensions of the pores and their relative volumes were determined. Micropores of about 6 Å radius were found.

Yarnton and Simpson (1961) used an interesting method in determining the size distribution of micropores in powder particles. They used liquids consisting of molecules of different sizes, which gave rise to different densities of the powder particles. From a knowledge of the molecular sizes the pore size distribution of the micropores could be derived.

Schlogl and Schurig (1961) used an unusual method to determine the pore sizes of ion exchange resins: measurement of the diffusion coefficient in the resin at various temperatures. The smaller the pore, the smaller the value of the diffusion coefficient because the water "freezes" in the smaller pores. The size distribution of the pores was calculated using spheres as models for the pores.

Rosenberg and Shombert (1961) studied the reaction of oxygen with photoexcited adsorbed dyes. They found that the penetration of oxygen through the pores was rate determining in some cases.

Pierce *et al.* (1971) found that a transverse magnetic field retards the flow rate of mercury in bead packs and natural rock samples. The magnitude of the effect increased with increasing bead size and field intensity.

Frequency response methods have been applied to pore structure investigation by Turner (1958, 1959), who gave a method to study the flow structure (distribution of residence times) of a medium or its pore structure in terms of a model chosen (see also Turner, 1972). The part of Turner's work dealing with the treatment of channels with dead spaces (pockets) has attracted considerable attention (*Turner structure;* see Chapter 6). Turner's method makes possible measurement of the distribution (by volume) of pockets of different depths and, hence, the relative amount of "dead" liquid contained in the pore (or void) space.

It is noted that in packed beds and many other porous media (e.g., sandstones) there is practically no "dead space," or "dead-end pore volume" in the strict sense of the word. In many relatively wide spaces, however, the flow may be much slower than in the "necks" or "windows" and also in the central cores of the wider spaces. The possibility of measuring dead-end pore volume by frequency response was pointed out also by Fatt (1959).

Felch and Shuck (1971) used the effect of flow on the net rate of solute transport across a porous diaphragm in a "diaphragm diffusion cell" [see, e.g., Dullien *et al.* (1974)] to deduce a pore size distribution on the basis of the "bundle of capillary tubes" model. The results depend on the form assumed for the distribution function.

1.2.5.2. *Macropore Size Distribution Determination by Quantitative Stereological Approach*

The term "macropore" was coined to distinguish the pores constituting the main skeleton of a pore network from much smaller pores present, called "micropores." The first attempts by the author and coworkers to obtain realistic pore size distributions used photomicrographs taken of plain surfaces of polished samples impregnated with a low-melting alloy, commonly called Wood's metal. A representative photograph is shown in Fig. 1.20. The white regions, representing the macropores, were regarded as random sections through separate "pore particles." The problem was how to reconstruct the size distribution of the 3-D "pore particles" from the sections.

Quantitative Stereology. Before discussing the stereological techniques of measuring pore size distributions, it is appropriate to review the fundamen-

Figure 1.20. Photomicrograph of section of sandstone saturated with Wood's metal.

tals of this branch of science. Extensive literature references can be found in the very readable texts by Underwood (1970) and Underwood *et al.* (1976).

Stereology deals with a body of methods for the exploration of three-dimensional space, when only two-dimensional sections through the bodies of their projections on a surface are available. Quantitative stereology has evolved from quantitative and stereometric microscopy and draws heavily on some fundamentals of geometrical probability [e.g., Kendall and Moran (1963). See also Weibel (1979), (1980) and Serra (1982)].

Fundamentals of quantitative stereology: There are various approaches to stereological problems. The statistical–geometrical approach depends on measuring and classifying a large number of two-dimensional shapes or images. It is applicable when the objects are randomly distributed in space. In such cases, a single section or projection, if large enough to contain a statistically significant number of features, may suffice to obtain valid results.

If objects of similar shape are arranged regularly in space, many section planes or visual angles may be necessary. Whether the section plane must be placed randomly throughout the material or just displaced systematically parallel to itself depends on the type of structural orientation involved.

Under certain conditions, serial sectioning is necessary. In the case of complicated, interconnected, nonconvex shapes this is the only known method to obtain all the information that is necessary to characterize a body. This method is used also to find the functional relationships among various parameters characterizing an irregularly shaped object (such as a potato).

Finally, there are cases in which a small number of similar objects of complicated shape are available, but serial sectioning is impractical. In these cases many random sections may be prepared in order to build up a picture of the average three-dimensional shape. Random projections might also be used here instead of sections. The resulting average figure, reconstructed under the conditions just described, is called a "stereogram."

Quantitative stereology deals with the quantitative characteristics of points, lines, surfaces, and volumes. Exact expressions are known that relate measurements on two-dimensional sections to the three-dimensional structure. The basic equations apply regardless of whether the features occur randomly or in some ordered arrangement; in other words, averages can be obtained from either random or oriented structures. Additional information about oriented structures may be obtained by special methods. It may be that the extent of departure from the average value is the important consideration, or perhaps the directional nature of some measurement is desired.

"Point counting" refers to counting the number of grid points P_α falling over the phase α, which is then divided by the total number of grid points P_T to give the ratio denoted by P_P.

Another type of point count measures the number of intersections generated per unit length of test lines P_L with traces of surfaces on the plane of polish.

TABLE 1.4.   Relationship of Measured ($\bigcirc$) to Calculated ($\square$) Quantities

Microstructural feature	Dimensions of symbols (arbitrarily expressed in terms of millimeters)			
	mm^0	mm^{-1}	mm^{-2}	mm^{-3}
Points	$\textcircled{$P_P$}$	$\textcircled{$P_L$}$ $\longrightarrow$	$\textcircled{$P_A$}$ $\longrightarrow$	$\boxed{P_V}$
	$\downarrow$	$\downarrow$	$\downarrow$	$\nearrow$
Lines	$\textcircled{$L_L$}$	$\textcircled{$L_A$}$	$\boxed{L_V}$	
	$\downarrow$	$\downarrow$	$\nearrow$	
Surfaces	$\textcircled{$A_A$}$	$\boxed{S_V}$		
	$\downarrow$			
Volumes	$\boxed{V_V}$			

A measurement similar to P_L is N_L, defined as the number of intercep-
tions (rather than true points) of objects or features per unit length of the
test lines. If the features happen to be lines, then, of course, N_L is identical
to P_L.

Another type of point count determines the number of points P_A per unit
area of the plane of polish that meet a certain requirement.

An important count is N_A, the number of objects per unit test area.

"Areal analysis" refers to the measurement of the lineal fraction L_L of the
chords of intercept lengths with the two-dimensional features.

Underwood (1970) has summarized the relationships of measured quantities in the two-dimensional section to calculated quantities in three-dimensional space (see Table 1.4). All symbols in the table are interrelated by
equations except those in the first and last three columns:

$$V_V = A_A = L_L = P_P, \tag{1.2.6}$$

$$S_V = (4/\pi) L_A = 2P_L, \tag{1.2.7}$$

$$L_V = 2P_A, \tag{1.2.8}$$

$$P_V = \tfrac{1}{2} L_V S_V = 2 P_A P_L. \tag{1.2.9}$$

Here V_V is the volume fraction, S_V is the surface-to-volume ratio, L_V is the
length of lines in space per unit test volume, L_A is the length of lines in a
plane per unit test area, and P_V is the number of points of intersection
between surfaces and lines per unit test volume.

In addition, there is an important relationship for the mean intercept length $\bar{L}_3$, defined for a single particle as

$$\bar{L}_3 = (1/N) \sum_{i=1}^{N} (L_3)_i, \qquad (1.2.10)$$

where the subscript 3 refers to the dimensionality of the particle and $(L_3)_i$ are the intercepted lengths from N random penetrations by a straight test line. Note that the definition allows for the possibility of a particle being intercepted more than once by a single penetration of the test line. Thus the value of $\bar{L}_3$ is applicable to concave bodies as well as to convex ones. For one particle we have the relationship

$$\bar{L}_3 = 4(V/S), \qquad (1.2.11)$$

where V is the volume and S is the surface of the particle. The same expression is valid for the average volume-to-surface-area ratio for a number of particles. From the point of view of the mean intercept length, the entire connected pore space can be regarded as a single particle.

Note that $\bar{L}_3$ is equal to the hydraulic diameter D_H of a conduit of arbitrary but uniform cross section.

Particle and grain size distributions: Quantitative stereology deals also with the determination of distribution of particle diameters. The methodology is very well developed in the case of spherical particles, and recently progress has been made in the area of more general shapes as well.

The main difficulty encountered in applying quantitative stereology to pore size distribution measurements is in the lack of sphericity of most "pore particles." This fact has given some incentive to the development of methods applicable to more general shapes.

The various stereological methods in use for the determination of size distribution of spherical particles may be divided into two groups: (1) "section analysis," methods that use the size distribution of the circles obtained by sectioning the distribution spheres with a plane; and (2) "chord analysis," methods that are based on the distribution of chords or intercept lengths obtained by intercepting the distribution of spheres with straight lines.

"Section analyses" are usually subdivided into two categories, depending on whether the size of a circle in the section is characterized by its diameter or its area. Accordingly, we have "diameter analyses" and "area analyses."

Technically, "chord analyses" are also performed in sections, and it can be shown (Dullien *et al.*, 1969/1970) that intercepting the distribution of spheres with straight lines results in the same chord length distribution as the one obtained by first sectioning the spheres and then intercepting the circles in the sections with straight lines.

Section methods are based on the fact that a sphere is cut by a random plane with equal probability at any distance from its center. Let us suppose that uniformly distributed spheres of a given probability density distribution of diameters are intersected with equidistant parallel planes. The number of spheres with diameters between D and $D + dD$ that are intercepted by a plane at a distance between X and $X + dX$ from their centers is equal to the number of sphere centers that are at distances between X and $X + dX$ from the plane (on either side of it). This number is equal to $2\,dX\,AN(D)\,dD$, where A is the area of the plane and $N(D)\,dD$ is the number of spheres of diameters between D and $D + dD$ per unit volume of space. Per unit area of the plane, the number is $2N(D)\,dD\,dX$.

As we cannot conveniently measure X, it must be replaced with a measurable quantity that can be either the diameter δ or the area A of a circle.

The number of circles per unit area of test plane with diameters between δ and $\delta + d\delta$ is

$$\beta\,dD\,d\delta = 2N(D)\,dD\,|(\partial X/\partial \delta)_{\mathrm D}|\,d\delta. \tag{1.2.12}$$

However, for a sphere we have the following relationships:

$$X^2 = \tfrac{1}{4}(D^2 - \delta^2), \qquad |(\partial X/\partial \delta)_{\mathrm D}| = \tfrac{1}{2}\delta\big/(D^2 - \delta^2)^{1/2}, \tag{1.2.13}$$

giving

$$\beta\,dD\,d\delta = N(D)\,dD\,\delta\,d\delta\big/(D^2 - \delta^2)^{1/2}. \tag{1.2.14}$$

In order to find the total number of circles $n(\delta)\,d\delta$ with diameters between δ and $\delta + d\delta$, one must integrate Eq. (1.2.14) over all values of D such that $D \geqslant \delta$:

$$n(\delta)\,d\delta = \int_{D=\delta}^{D_{\max}} \frac{N(D)\,dD\,\delta\,d\delta}{(D^2 - \delta^2)^{1/2}}. \tag{1.2.15}$$

The size distribution of the spheres $N(D)$ is found by solving Eq. (1.2.15), using the known distribution $n(\delta)\,d\delta$. This equation has been solved numerically by Scheil (1935). Other solutions of this problem, which are generally considered superior to the Scheil solution, have been given by Schwartz (1934), Saltykov (1958), and Wicksell (1925, 1926), with "area analysis" being used by Saltykov (1967).

The basis for the "chord analysis" methods is as follows: A sphere can be regarded as a target (like one used in target shooting) when it is intercepted with straight lines. Each sphere offers an area equal to $2\pi X\,dX$ as a target at distances between X and $X + dX$ from its center. Supposing that there are $N(D)\,dD$ spheres per unit volume of space with diameters between D and $D + dD$, then the total cross-sectional area these spheres offer at distances between X and $X + dX$ from their centers is $2\pi XN(D)\,dD\,dX$. It is shown subsequently that this is also equal to the number of intersections per unit

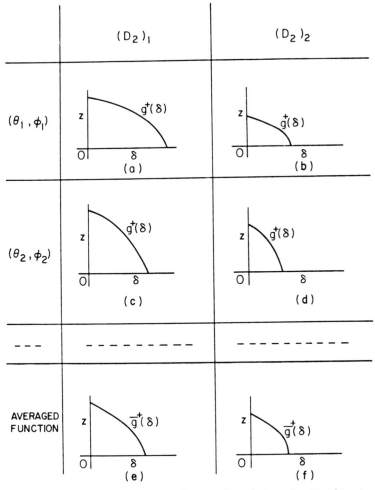

Figure 1.23. Functions $z = g(\delta; \theta, \phi, D_2)$ and $\bar{g}(\delta; D_2)$ for spheroids. (Reprinted with permission from Chang and Dullien, 1976.)

differentiation operation,

$$\overline{\left(\frac{\partial g}{\partial \delta}\right)^-} = \left(\frac{\partial \bar{g}}{\partial \delta}\right)^-, \qquad \overline{\left(\frac{\partial g}{\partial \delta}\right)^+} = \left(\frac{\partial \bar{g}}{\partial \delta}\right)^+. \qquad (1.2.27)$$

The averaging procedure is shown diagrammatically, for the decreasing part of the function, in Fig. 1.23. Let us suppose that we fix the value of D_2, say, at $(D_2)_1$. Then, depending on the orientation (θ, ϕ, Ω), we have a different function for $g^+(\delta)$. This is shown in Fig. 1.23a and c. If we now change D_2 to $(D_2)_2$, then we have different $g^+(\delta)$ again, depending on the orientation (θ, ϕ, Ω), as shown in Fig. 1.23b and d. The averaging process must be performed over (θ, ϕ, Ω) for each fixed D_2. In Fig. 1.23 this is to be done columnwise, giving the functions $\bar{g}^+(\delta)$ of (e) and (f). Because of the assumption that the objects have similar shape and only have different sizes, $g^+(\delta)$ in (b) is similar to $g^+(\delta)$ in (a) and $g^+(\delta)$ in (d) is similar to $g^+(\delta)$ in (c), etc. Consequently $\bar{g}^+(\delta)$ in (f) is similar to $\bar{g}^+(\delta)$ in (e).

A similar, but nonrigorous treatment was applied by Dullien and Dhawan (1974) to give the size distribution determination of pores in sandstone samples where the "pore particles" are not convex objects and do not have similar shapes. In addition, serial sectioning of individual "pore particles" was considered impractical. The 'pore particles" were treated as if they had been convex objects of similar shape, and it was assumed that a function obtained from the two-dimensional features in the plane of polish of the sample can be substituted for $\bar{g}(\delta)$.

Before introducing the g function or "shape function" method, Dullien and Dhawan used the sphere model for the void particles. Both the section diameter and the section chord analyses were used with varying degrees of success, depending on the type of porous material analyzed.

Dullien and Mehta (1971/1972) determined the particle size distribution of common salt by standard sieve test, prepared a packed bed of the salt, saturated the void space with Wood's metal, and dissolved the salt with water. Quantitative stereological analysis was performed on photomicrographs taken of polished sections of the bed. The voids in these used to be salt particles, which are shown in white in Fig. 1.24. The results of the sieve test and the photographic analysis are shown in Fig. 1.25. They are in reasonably good agreement, considering the great difference between the two types of tests. Sieve tests tend to measure a diameter that is less than the mean caliper diameter; and, on the other hand, the photomicrographs contain features that correspond to aggregates of salt particles. The sphere model, however, was adequate to represent this porous medium, as the comparison between Mehta's original results (1969) and the subsequent calculations by Dhawan (1972), who used the g-function approach, has shown.

The sphere model approach has worked better in the case of the section diameter method than with the section chord method. The main problem with the sphere model was encountered in that part of the pore size distribution function that corresponds to relatively small diameters. The reason for this lies in the irregular shapes of the features (see Fig. 1.20). Performing lineal analysis on the photomicrograph results in an excessively large number of short chords or intersections with these features and, as a result, the calculations will yield an excessively large number of smaller pores. Using the section diameter method, Batra (1973) determined the pore

Figure 1.24. Photomicrograph of section of salt pack saturated with Wood's metal (Mehta, 1969).

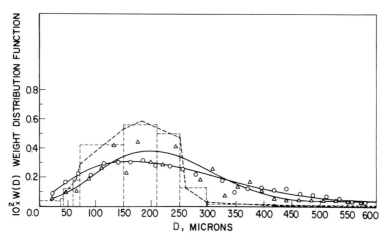

Figure 1.25. Comparison of weight distributions obtained by the following different methods of analysis: packed bed (○); monolayer of particles suspended in oil (△); sieve (---) (Mehta, 1969).

size distribution of a large number of sandstones in terms of the spherical pore model.

As can be seen in Fig. 1.26a, use of the "shape function" method corrects the size distribution quite drastically in case of a packed bed of beads where the void particles are shaped somewhat like stars (see Fig. 1.26b). Mercury porosimetry tends to measure the volume of the "pendular rings" or "toroidal" spaces in packed beds of beads quite accurately, and the excessively large volume assigned to these spaces by the sphere model is erroneous. An example of a shape function determined by Dhawan (1972) is shown in Fig. 1.27. The difference between the shape function of a sphere and the shape function of highly irregular voids is apparent from this diagram.

In the case of natural porous media the pore size range is often very wide. For example, in the case of sandstones the pores inside the cementing materials (e.g., clays) are very small. These micropores usually remain undetected in the photomicrographs.

Crisp and Williams (1971) used optical methods to measure pore size distributions in sandpacks and natural deposits. They, however, "bypassed" the use of quantitative stereological methods by using "thick sections" of the samples which, when viewed under the microscope, permitted them to see pores in three dimensions. Thus, they saw the dimensions of the pores. Two measurements were made on each pore, one in the direction of the greatest dimension and the other at right angles to it. The average of the two was

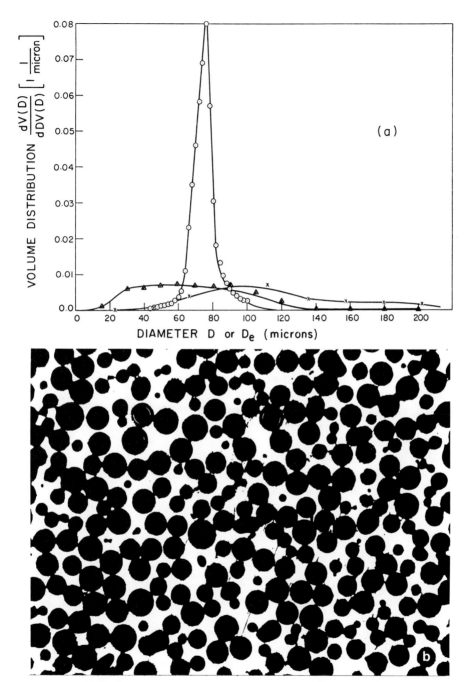

Figure 1.26. (a) Comparison of pore size distribution in a random pack of 250-μm glass beads obtained by the following different methods: mercury porosimetry ($\odot$); shape function model ($\times$); sphere model ($\blacktriangle$). (b) Photomicrograph of a section of a random pack of 250-μm glass beads. (Reprinted with permission from Dullien and Dhawan, 1974.)

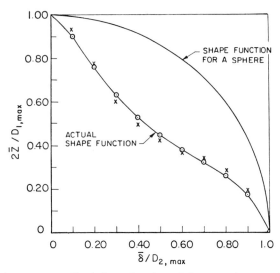

Figure 1.27. Average normalized shape function of the pores in a Bartlesville sandstone showing experimental (○) and fitted valued (×). (Reprinted with permission from Dullien and Dhawan, 1974.)

used as pore diameter. In the almost monometric sands, the mean pore diameter was 30–40% of the mean particle diameter in agreement with the Carman–Kozeny equation. In polymorphic shell gravels, the mean pore size was only 15–20% of the mean particle size, and it diminished with increased content of fine particles.

Crisp and Williams (1971) estimated the fraction p of void space that is accessible through pores of pore diameter less than or equal to a given value. The binomial formula was used to calculate the "accessibility" $p = 1 - q^z$, where q is the probability that a randomly chosen pore will not exceed a given size (i.e., the fractional number of pores smaller than a given size), read from the cumulative pore size distribution curve, and z is the coordination number. Numerical calculations were performed for values of z ranging from 1 to 8.

In this enlightened and interesting work, the effects of interplay between pores of different diameters were neglected (see Chatzis and Dullien, 1977).

Combined Use of Mercury Porosimetry and Quantitative Stereology: Bivariate Pore Size Distribution.   Dullien and Dhawan (1975) used combined "mercury porosimetry" and "photomicrography" in the following series of experiments. A sandstone core was sectioned into a number of cylindrical plugs. Using an equipment resembling a mercury porosimeter submerged in a hot temperature bath, they injected Wood's metal into the first sample under increasing

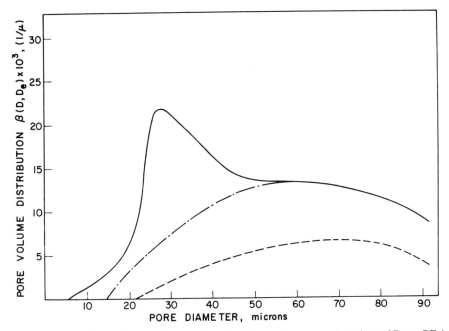

Figure 1.28. Photomicrographic pore size distributions of penetrated portions of Berea BE-1 sandstone at various injection pressures of Wood's metal (Dhawan, 1972).

pressures until breakthrough was registered. The capillary pressure was noted, then the sample was sealed off by means of valves and it was removed from the bath. After the Wood's metal in the sample solidified and the sample cooled down to ambient temperatures, it was treated with hydrofluoric acid at one end, washed with water very thoroughly, dried, and impregnated with epoxy resin. It was necessary to proceed in this manner to obtain a surface of reasonably uniform hardness for a good polish. Photomicrographs were prepared of the polished surface, and the size distribution of the pores impregnated in the sample was determined. The section diameters were measured with the help of an automatic image analyzer, and the pore size distribution was calculated, using the shape function determined from the shapes in the photomicrographs.

The same procedure was followed in the case of the second and third plug except, using a higher capillary pressure, more Wood's metal was injected into the second plug than into the first. The third plug was practically saturated with Wood's metal. The series of pore size distribution curves obtained in this way is shown in Fig. 1.28.

Inspection of these curves shows that, in complete agreement with the results of the theoretical analysis of penetration into a network containing

Bivariate pore size distribution of Berea sandstone.

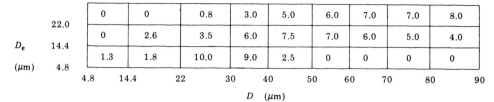

D_e (μm)									
22.0	0	0	0.8	3.0	5.0	6.0	7.0	7.0	8.0
14.4	0	2.6	3.5	6.0	7.5	7.0	6.0	5.0	4.0
4.8	1.3	1.8	10.0	9.0	2.5	0	0	0	0

D (μm): 4.8 14.4 22 30 40 50 60 70 80 90

Figure 1.29. Example of bivariate pore size distribution function (Dullien, 1981).

randomly distributed capillaries of different diameters, at breakthrough some, but not all, of the largest pores have been penetrated. All the largest pores are completely penetrated in the second plug. The volume of the smallest pores penetrated for the first time in each plug is extremely small. Most of the penetration in the third plug took place into pores that were blocked by the smallest pores. The bivariate pore size distribution $\beta(\overline{D}_e, \overline{D}) \Delta D_e \Delta D$, giving the fractional volume of pores in the diameter range $D \rightarrow D + \Delta D$, the entry to which is controlled by pore throats in the diameter range $D_e \rightarrow D_e + \Delta D_e$, derived from these data is presented as a two-dimensional histogram in Fig. 1.29.

Dendritic Pore Fractions. When a nonwetting fluid (e.g., mercury or Wood's metal) is forced to penetrate a porous sample, at the point of breakthrough only a fraction of the injected fluid contributes to conduction (either electrical or hydraulic) through the sample, because the rest of the fluid is in the form of tree-like (dendritic) micro-fingers from which there is no exit. As the applied capillary pressure is increased further an increasing number of these micro-fingers merge with the conducting channels of the fluid. Evidently, the pore throats controlling flow through these micro-fingers are the "pore exit throats" that have a smaller diameter than the entry pores through which the fingers were originally penetrated.

The "dendritic pore fractions," that is, the fraction of pore volume occupied by dendritic fluid, was determined experimentally as a function of applied capillary pressure and saturation by El-Sayed and Dullien (1977). The sandstone sample was first evacuated, then mercury was admitted to one face of the sample, and the pressure was gradually raised until breakthrough was achieved. Subsequently, all the mercury that could be miscibly displaced was flushed out from the sample, using radioactive mercury. The amount displaced was found in three widely different sandstone samples to be about 50% of the breakthrough saturation. The radioactive mercury introduced into the pore space was immediately displaced by ordinary mercury, using a

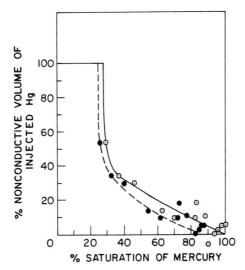

Figure 1.30. Dendritic pore volume fractions as a function of saturation (Berea sandstone, BE-1). Values shown are based on saturations at 117 psia (○) and 15,000 psia (●) (El-Sayed, 1978).

similar miscible displacement process. It was found afterward that no radioactive mercury was left in the sample, indicating that there was no measurable penetration of the radioactive mercury into the dendritic mercury.

The pressure on the mercury was raised in stages, and the same sequence of steps was repeated at each level of penetration until the entire pore space was filled with mercury. The dendritic pore fractions, determined by this technique, are shown in dependence on the saturation of the sample in Fig. 1.30.

1.2.5.3. *Pore Size Distribution Determination from Reconstructed 3-D Pore Structure*

Introduction. The main shortcomings of the stereological method may be summed up as follows:

(1) No information on pore throat sizes.

(2) Counting two or more pores as one pore.

(3) All "pore-particles" are assumed to be geometrically similar.

(4) The pore shape function derived from 2-D photomicrographs is assumed to represent the shape function of 3-D "pore-particles."

(5) A great deal of the detailed geometry of the pore space remains unknown.

Therefore, it was decided to serially section small samples of Berea sandstone and reconstruct the 3-D pore structure after digitizing the 2-D images taken of the serial sections, as already described in Section 1.2.2.2.

In a series of articles Lin and coworkers (1982, 1982, 1983) presented a deterministic approach to modeling the three-dimensional pore and grain geometry and pore network topology, based on computer reconstruction of serial sections. Lin and Perry (1982) used a pore (or grain) surface triangulation technique as a shape descriptor, which gives the following parameters: surface area, Gaussian curvature, genus, and aspect ratio of the pore. The aspect ratio was obtained by using a spheroidal model. In their article, however, they pointed out that their method is not suitable for modeling the pore network. The method used by Lin and Cohen (1982) is similar to the one described by De Hoff *et al.* (1972) and Pathak *et al.* (1982). In another study, Lin (1983) carried out three-dimensional measurements in the pore space in the direction of the three orthogonal axes and then used these parameters for pore models, consisting of ellipsoids, or elliptical cylinders or double elliptical cones.

Random Point Generating Approach. The same set of 78 photomicrographs, representing serial sections through a Berea sandstone sample, that were processed in Macdonald *et al.* (1986), were used for locating, at random, points in the digitized three-dimensional pore space and measuring, in three orthogonal directions, the lengths of straight lines passing through each point (Yanuka *et al.*, 1986). The set of the three orthogonal lengths measured was stored in the form of a joint distribution function $f(a, b, c)\, da\, db\, dc$, with $2a$, $2b$, and $2c$ being the wall-to-wall distances measured in the three orthogonal directions, as illustrated in Fig. 1.31. In addition to the Berea sandstone sample, 80 serial sections were also prepared of a $4.3 \times 3.5 \times 1.4$ mm random glass bead pack at 15–20 μm increments and then photographed and digitized. The glass beads were in the 177-to-350 μm size range. Finally, three regular packings or uniform size spheres, (a) simple cubic, (b) orthorhombic, and (c) rhombohedral, were also tested. In this case there were no physical samples because the media could be described as continuous functions mathematically. They were chosen to test the method, owing to their known pore structures.

Using the number of random points generated, the sample porosities ϕ were calculated as follows

$$\phi = \frac{\sum\limits_{i=1}^{n} f(x, y, z)}{n} \tag{1.2.28}$$

where n is the total number of points and $f(x, y, x) = (0, 1)$, where 1 represents pore space and 0 represents solid space. The results of the porosity determinations are given in Table 1.5.

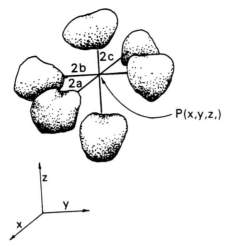

Figure 1.31. Determination of pore size in three orthogonal directions at a randomly chosen point P(x, y, z) (after Yanuka *et al.*, 1986).

TABLE 1.5.   Porosities of the Different Porous Media[a]

Number of counts and repetitions[b]	Simple cubic packing (%)	Ortho-rhombic Packing (%)	Rhombo-hedral Packing (%)	Pack of glass beads (%)	Berea sandstone (%)
			Type of medium		
				From a total count of the digitized data	
	47.64 (exact)	39.54 (exact)	25.95 (exact)	38.4	23.6
				Experimental values obtained in bulk samples	
				38–39[c]	22–23.2[d]
100 × 10	46.30 ± 4.33	39.40 ± 4.09	23.30 ± 3.20		
1000 × 10	47.94 ± 2.44	38.91 ± 1.91	25.10 ± 1.60	38.89 ± 1.72	23.80 ± 1.40
10,000 × 10	47.55 ± 0.52	39.45 ± 0.51	25.80 ± 0.46	38.56 ± 0.35	23.69 ± 0.44
50,000 × 10	47.75 ± 0.23	39.60 ± 0.21	25.95 ± 0.16	38.69 ± 0.21	23.86 ± 0.15

[a]After Yanuka *et al.*, 1986.

[b]Experiment repeated 10 times (e.g., an experiment of 100 points repeated 10 times).

[c]Calculated from the measured bulk density of the pack of glass beads and the density of the particles Bd = 1.50 to 1.53 and Pd = 2.45 g/cm³, respectively.

[d]Calculated by taking the ratio of the measured volume of water filling the pore space under vacuum to the total volume of the sample.

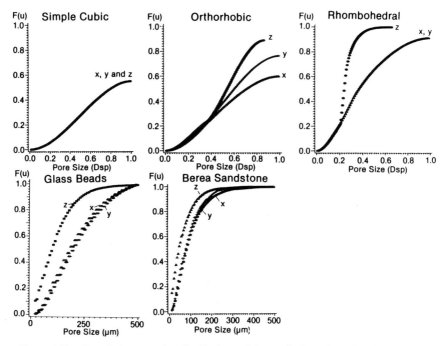

Figure 1.32. Cumulative pore size distributions of the media investigated in the x, y, and z directions ($u = 2a, 2b, 2c$) (after Yanuka *et al.*, 1986).

The cumulative pore size distributions found in the samples in the x, y, and z coordinate directions, $f(2a)$, $f(2b)$ and $f(2c)$, are plotted in Fig. 1.32. The z-direction is perpendicular to the plane of polish of the samples. It is apparent from the figure that the pore sizes range beyond 200 μm in contrast with the maximum pore size of about 70 μm assumed in the network simulation studies by Chatzis and coworkers (1985, 1987, 1988, 1988), which yielded good agreement with experimental drainage capillary pressure and relative permeability curves. The large wall-to-wall lengths measured are probably due to the presence of relatively large pore throats through which the line could pass, resulting in the combined size of several pores. Aniso-metric pore geometry may also contribute to this effect.

The joint distribution function was used also to obtain the minimum and the maximum harmonic mean pore radius $R_{\min}$ and $R_{\max}$ by forming the three possible combinations of pairs of the lengths (a, b), (a, c), and (b, c)—that is, $[\frac{1}{2}(1/a + 1/b)]^{-1}$, $[\frac{1}{2}(1/a + 1/c)]^{-1}$, and $[\frac{1}{2}(1/b + 1/c)]^{-1}$. Choosing the minimum and the maximum values of these gave frequency distribution densities $f(R_{\min})$ and $f(R_{\max})$, respectively. These were trans-formed to volume-based size distributions $V(R_{\min})$ and $V(R_{\max})$ by assuming

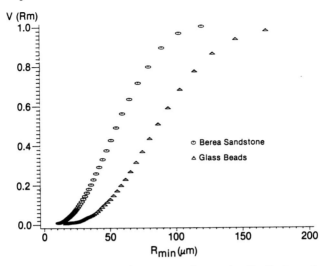

Figure 1.33. Cumulative normalized (volume-based) pore size distributions of a bead pack and a Berea sandstone sample (after Yanuka *et al.*, 1986).

pores of ellipsoidal shape. The pore size distributions of the Berea sandstone and the glass bead pack $V(R_{min})$ have been reproduced in Fig. 1.33. It is evident from this figure that for the sandstone the values of R_{min} extended beyond 100 μm, consistently with the distributions shown in Fig. 1.32.

The cumulative joint distribution function $F(a, b, c)$ was used to generate a model of the porous medium composed of ellipsoids distributed randomly in space. Random points were generated in a cube-shaped space that were used as centers of ellipsoids. Values of $F(a, b, c)$ between 0 and 1 were generated by a uniformly distributed random number generator, and values of a, b, and c were obtained by taking the inverse of the function $F(a, b, c)$. The ellipsoids thus generated often intersected with each other, and in each case the volume of intersection was excluded. The random process of generating ellipsoids continued until the total volume of ellipsoids generated (excluding the volumes of intersection) yielded the known porosity of the sample. The intersection between two ellipsoids was used to calculate the throat size between the two pores by calculating the radius of a sphere of the same volume as the volume of intersection between the two ellipsoids. The sphere radius r was assumed to represent the throat radius. The throat radius frequency distribution densities $f(R)$ of the different media are shown in Fig. 1.34. For the Berea sandstone the peak of distribution is at about 20 μm radius and the maximum radius is about 60 μm. These values are again much greater than the throat diameters, ranging from about 5 μm to about 42 μm, used by Chatzis and coworkers (1985, 1987, 1988, 1988), which

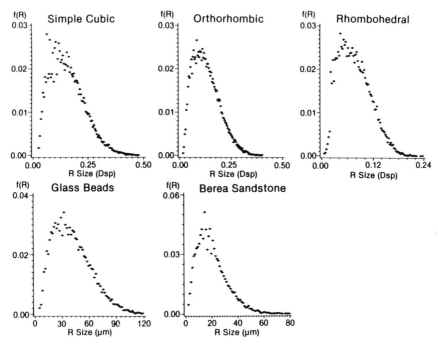

Figure 1.34. Pore throat radius distributions calculated from volumes of intersection between randomly chosen ellipsoids (after Yanuka *et al.*, 1986).

resulted in realistic predictions of the drainage capillary pressure and the relative permeability curves.

The average coordination number z of the Berea sandstone sample[1] was found to be 2.8, which is very close to the value of 2.9 calculated from the relation (Ridgway and Tarbuk, 1967):

$$1 - \phi = 1.072 - 0.1193z + 0.004312z^2. \qquad (1.2.29)$$

Equation (1.2.29) has also predicted the average coordination number of the random glass bead pack (4.6 vs. 4.3) and the (exact) coordination numbers of the three regular sphere packings (simple cubic: 5.8; orthorhombic: 4.7 vs. 4.6; and rhombohedral: 3.1 vs 3.3). This relation, therefore, appears to be quite reliable for both regular and random structures.

The validity of the modeling approach used by Yanuka *et al.* (1986) was checked also by comparing radii of the circles inscribed in the narrow passages of the three different regular sphere packings as calculated by Kruyer (1958) with the average throat radii found by Yanuka *et al.* (1986),

---

[1]In Equation (1.2.29) z is the coordination number of the pore space. In the original relation of Ridgway and Tarbuk [Eq. (1.2.31)], however, z was the coordination number of the packing of spheres.

expressed in units of sphere diameter (simple cubic: 0.207 vs. 0.156; or-thorhombic: 0.142 vs. 0.130; and rhombohedral: 0.077 vs. 0.077). Evidently, this agreement is quite good and it seems to indicate that the method used may have some validity as long as the straight lines ("yardsticks") in the pore space cannot pass through pore throats. This was indeed the case for the three regular sphere packs, where the throats were situated on the faces of the unit cell of the packing.

One important lesson learned from the study reported by Yanuka *et al.* (1986) is that the correct pore body sizes cannot be found in the computer reconstruction of pore structure unless first the throats are located and then partitions are erected at the throats that separate adjacent pore bodies. This procedure is analogous to closing the doors in a building which were originally wide open: As a result every room will be a separate, isolated entity, whereas with the doors open one could walk freely from room to

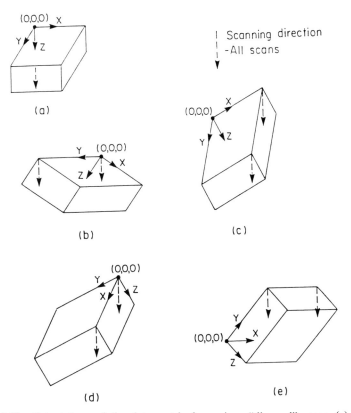

Figure 1.35. Orientations of the data matrix for various "diagonal" scans; (a) original orientation, (b) scan with planes parallel to the *y*-axis (XMIN scan), (c) scan with planes parallel to the *y*-axis (XMAX scan), (d) scan with planes parallel to the *x*-axis (YMIN scan), (e) scan with planes parallel to the *x*-axis (YMAX scan) (after Kwiecien *et al.*, 1990).

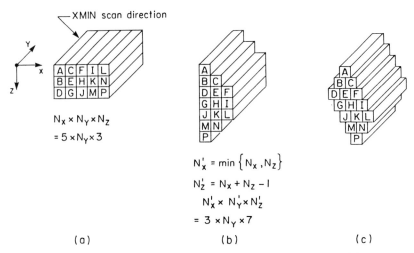

Figure 1.36. Example of data matrix transformation for "diagonal" scans; (a) original data matrix, (b) new data matrix for XMIN scan: columns filled with zeroes are added (not shown) to fill out rectangular array, (c) overlap relationship of pixels for the data matrix in (b) (after Kwiecien *et al.*, 1990).

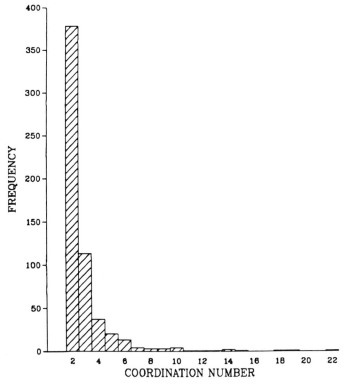

Figure 1.37. Frequency distribution of coordination numbers (after Kwiecien *et al.*, 1990).

room. Similarly, with partitions erected at all the pore throats the yardstick used to measure pore body sizes cannot inadvertently measure the combined size of more than one pore body.

Method of Locating Pore Throats in Computer Reconstruction. The digitized serial sections (photomicrographs) were used as follows (Kwiecien, 1987; Kwiecien *et al.*, 1990). Each pixel was assumed to be the top surface of a volume element (voxel) with a cross section equal to the pixel area and a depth equal to the spacing between the two consecutive serial sections. When both the pixel and the one immediately below it are pore space pixels, the two two-dimensional pore space features containing these pixels were assumed to be connected. This is the same "overlap" criterion used by Macdonald *et al.* (1986).

The approach followed was to first locate the pore throats and then, by symbolically closing them, define the pore bodies. A pore throat is defined as a local minimum in the "size" of pore space which thus separates two pore bodies from one another. As discussed in Section 1.2.3, the most practical definition of pore radius is twice the minimum value of the ratio: area of cross section passing through a fixed point in the pore space to the perimeter of this section [see Eq. (1.2.5)].

The ideal way of locating pore throats, described in Section 1.2.3 was replaced with the practical way of scanning the computer reconstruction of pore structure with a few sets of parallel planes of distinct, different orientations. The first and obvious plane is the plane of polish, or serial sectioning, of the sample. This plane is perpendicular to the z-axis. Next, the scanning planes perpendicular to the x-axis and the y-axis were used. In addition to these relatively simple cases, four more scans were made: two parallel to the y-axis and another two parallel to the x-axis, as illustrated in Fig. 1.35. Had both the sample and the voxels been exactly cube shaped, then all these scans would have been parallel to diagonal planes passing through two opposite edges of the cube. As the scanning was carried out in terms of pixels (or voxels) the diagonal scans are best understood by the example shown in Fig. 1.36, where XMIN denotes one of the two diagonal scans parallel to the y axis. N_x, N_y, N_z and N'_x, N'_y, N'_z denote the number of pixels in the three coordinate directions in the original arrangement and in the diagonal arrangement, respectively. Figure 1.36 shows the new overlap criterion for the diagonal scanning. (It should be noted that the pixel shape was not square and was nonuniform because of the unequal spacings between consecutive serial sections. The pixel size in the x-direction was 5.20 μm, and in the z-direction it varied from 6.5 μm to 17.8 μm.) For each scan there is a set of potential pore throats. These sets are compared in order to identify the true throats, using the principles outlined earlier. At the time of writing, the work of improving this technique is still in progress, because a number of throats

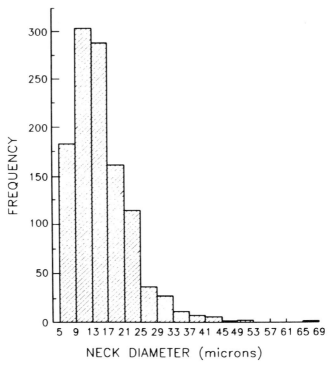

Figure 1.38. Frequency distribution of neck (pore throat) diameters (after Kwiecien, 1988).

appear to have been missed by the scanning and some other throats exhibit anomalous behavior.

After identifying a pore throat, a set of solid matrix voxels was introduced in its place, thus separating the two adjacent pore bodies. The coordination number—that is, the number of throats belonging to each pore—was determined. The volume of each pore body was directly obtained by adding up the volumes of the voxels contained in it. In addition, the dimensions of the smallest rectangular parallelepiped completely containing each pore body were determined. Some of the results are presented in Figs. 1.37–1.39.

Figure 1.37 shows the frequency distribution of the coordination numbers. The average coordination number is 2.9, which is about the same as the value obtained in the ellipsoidal model by Yanuka *et al.* (1986) and its exactly the same value as calculated by Eq. (1.2.29). The shape of the distribution is also very similar to the one obtained by Yanuka *et al.* (1986).

The frequency distribution of pore throat diameters is presented in Fig. 1.38. The average neck diameter was found to be 15.5 μm. The throat size

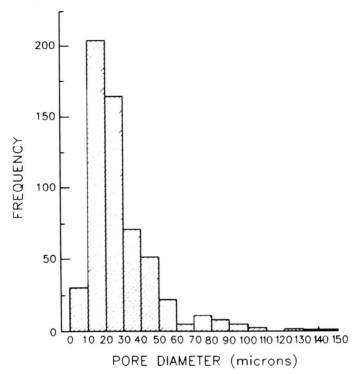

Figure 1.39. Frequency distribution of pore body diameters. Pore bodies modeled as cubes (after Kwiecien, 1988).

distribution found is very close to the corresponding distribution used with good results in the network simulations by Chatzis *et al.*, (1982, 1985, 1987, 1988, 1988) and Diaz *et al.* (1987).

The frequency distribution of pore bodies, modeled as cubes, is shown in Fig. 1.39. The number average pore body diameter was found to be about 29 μm. Modeling the pore bodies as spheres has yielded an average pore body diameter of about 36 μm. These distributions are very close to the distributions used successfully in network modeling transport properties of Berea (and other) sandstones (see Fig. 2.60).

1.2.6. Three-Dimensional Pore Structure Reconstructing from Two-Dimensional Photomicrographs Using Stochastic Theory

Pore structure reconstruction from serial sections is a time and labor-intensive process. Therefore, there is a great incentive to accomplish the same end by only using the polished surface of a sample impregnated with a suitable material for the purpose of creating a good contrast between the pore space and the solid matrix. Mathematical procedure to this end has been developed by Quiblier (1984), based on the work of Joshi (1974). Unfortunately, the work that has been done so far used "thin sections" of samples rather than truly two-dimensional sections; this has resulted in certain difficulties in precise feature identification, related to the depth effect of thin sections.

The procedure depends on the measurement of the porosity ϕ and an autocorrelation function $R_z(\mathbf{u})$, $(\mathbf{u} = \mathbf{x}_2 - \mathbf{x}_1)$ of the pore space (the same function also for the solid matrix). Details of the mathematics cannot be reproduced here for reasons of space. The procedure is currently under active study by Adler, Jacquin, and Quiblier (1990) and also in the author's laboratories. If reconstructed pore structure has the same statistical properties of pore structure as the original sample, it can be used to predict the permeability, the formation factor, capillary pressure, and relative permeability curves by methods available for this purpose and discussed in Chapters 2 and 3.

1.2.7. Fractals and Pore Morphology

The concept of fractals has gained widespread popularity over the past decade, and it has been used in relation to flow through porous media in a number of instances. Therefore, it is necessary to make reference to these applications in this monograph.

According to Mandelbrot (1986), the "father" of fractals, "a fractal is a shape made of parts similar to the whole in some way." Mandelbrot (1987) has also pointed out that a neat and complete characterization of fractals is still lacking.

The basic idea of fractals originated apparently from the problem of trying to determine the length of a coastline. It appeared that the shorter the yardstick used, the longer became the measured length of a coastline (see Fig. 1.40). In fact, plotting the logarithm of the measured length L as a function of the logarithm of the length of the yardstick used, δ, a straight line of negative slope of 0.52 was obtained as shown in Fig. 1.41, suggesting that the measured length would increase indefinitely as the length of the yardstick was reduced beyond any limit. Doing the same exercise in the case of any smooth curve, the slope of the log-log plot is zero and one can write

$$\log L(\delta) = \log a + 0 \times \log \delta$$

$$= \log a + \log \delta^0$$

$$= \log a + \log \delta^{1-D},$$

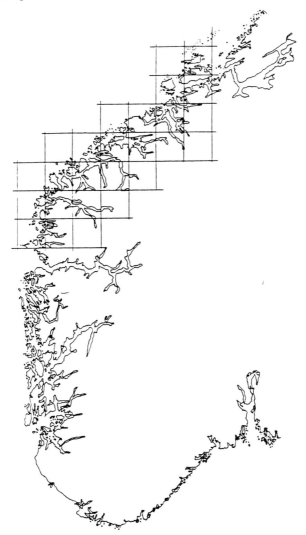

Figure 1.40. Part of the coastline of Norway. The square grid has a spacing of $\delta \simeq 50$ km (Feder, 1988).

with

$$D = 1, \text{ because the dimension of a smooth curve is 1.}$$

Analogous interpretation of the results obtained with the length of the coastline yields

$$\log L(\delta) = \log a - 0.52 \log \delta$$
$$= \log a + \log \delta^{1-1.52}$$
$$= \log a + \log \delta^{1-D},$$

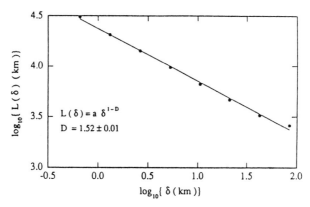

Figure 1.41.   The measured length of the coastline shown in Fig. 1.40, as a function of δ (Feder, 1988).

with

$$D_{\mathrm{F}} = 1.52,$$

or

$$L(\delta) = a\,\delta^{1-D}.$$

D_{F} is the "fractal dimension" of the curve. It should be noted that the observed increasing measured coastline length with decreasing yardstick length does not imply the observed linearity of the log-log plot. Only because of statistical self-similarity of the coastline is this linear behavior observed.

In the real physical world, there are limits—cutoff points of the fractal—because δ below a certain value (lower cutoff) and above a certain value (upper cutoff) has no physical meaning. Therefore, it is physically meaningless to say that length of the coastline is infinite, in the limit.

Mathematically speaking, however, there are no such limitations and the numerous ingenious fractal curves invented by mathematicians do have infinite lengths in the limit $\delta \to 0$.

A few of these curves are shown as examples in Figs. 1.42 and 1.43. The construction of the triadic Koch curve is simple, the particular version of the "triangle sweep," however, is complicated. What is particularly interesting about this fractal is that its fractal dimension is $D_{\mathrm{F}} = 2$. This curve is "plane-filling."

For the "orderly" fractal curves such as those shown in Figs. 1.42 and 1.43 there are rules for the calculation of the fractal dimension D_{F}. Owing to space limitations of this monograph, it is not possible to go into this subject in any greater detail. The interested reader is referred to the extensive

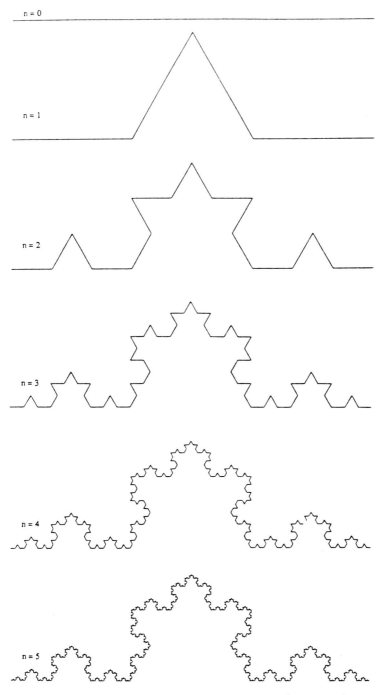

Figure 1.42.   Construction of the triadic Koch curve (Feder, 1988).

Figure 1.43.   Triangle sweep with $D_F = 1.944$. For the first few generations the previous generations are also shown as dashed lines (Feder, 1988).

literature of fractals and particularly to the very readable book by Feder (1988), in addition to the classic works by Mandelbrot (1977, 1982).

The present chapter deals with pore structure and therefore the pertinent question is what relationship, if any, is there between pore morphology and fractals.

The most obvious application of fractals to pore morphology appears to be to the pore surface, which in the case of natural porous media is usually rough. A plane section normal to the pore surface will show irregular outlines of the surface that may have fractal properties analogous to the case of coastlines. Determination of the fractal dimension D_F, along with the lower cutoff value of δ, would permit determination of the "true" pore surface area.

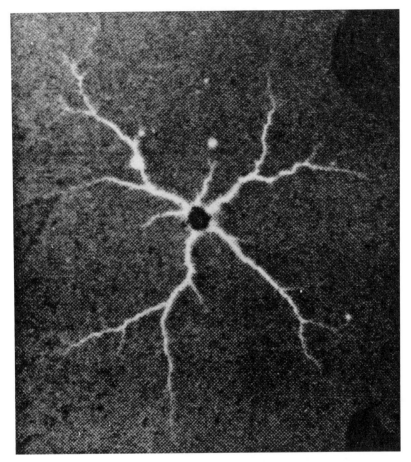

Figure 1.44. Water injected into a 1 mm thick plaster of Paris disk (Daccord and Lenormand, 1987).

It is interesting to note at this point that the specific surface S_v (i.e., the surface-to-volume ratio for spheres) is

$$S_v = \frac{6}{D} \qquad (1.2.30)$$

where D is the sphere diameter. Hence $\log S_v$ plots as a straight line versus $\log D$ with slope of -1, implying a "fractal dimension" $D_F = 2$. This, however, does not necessarily mean that specific surface is a fractal.

Cracks and fractures in a solid matrix appear likely candidates to be fractals (see Fig. 1.44) in a certain sense.

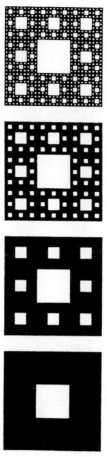

Figure 1.45.   Construction of the Sierpinski carpet ($D_F = \ln 8/\ln 3 = 1.89\ldots$) (Feder, 1988).

Certain pore morphology (e.g, vuggy carbonates) may be pre-fractal, along the same lines as the Sierpinski carpet shown in Fig. 1.45, where the carpet would represent a plane section through the idealized vuggy pore structure. It is important to note that the Sierpinski carpet becomes a fractal (i.e., linear log (perimeter) vs. log δ plot) only as $\delta \to 0$, but the pre-fractal behavior approximates the fractal behavior well after a finite number of generations.

Fractals may also have applications in characterizing large heterogeneous bodies of porous media, such as aquifers and petroleum reservoirs. In this case, the structure is not a visible geometry but it represents a density function of the spatial distribution of a property such as, for example, porosity or permeability. The underlying idea is that a small region of the

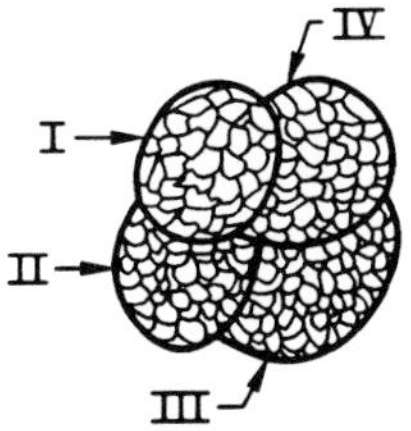

REGION	<k>	STANDARD DEVIATION
I	1500 mD	700 mD
II	500 mD	300 mD
III	700 mD	400 mD
IV	1000 mD	600 mD

Figure 1.46.   Diagrammatic representation of possible spatial distribution of permeabilities.

porous body may be characterized by a certain permeability or a certain porosity that is significantly different from the surrounding regions. We can imagine that distinct regions are defined by using a set of objective statistical rules and then their boundaries are drawn. This results in a representation of the large body of the porous medium in terms of many small building blocks. If it is found, on analyzing the spatial distribution of these building blocks of different porosities and/or permeabilities, that this is uniformly random, then the scale of heterogeneity of the medium is very small and the large porous body as a whole will display homogeneous behavior. On the other hand, if spatial correlations can be found between the building blocks of different porosities and permeabilities, then there may be justification for the creation of larger regions, each consisting of building blocks of similar porosities and/or permeabilities. After dividing the medium into such larger regions by using a set of objective statistical rules, boundaries can be drawn around these regions. This process can be continued by creating even larger regions, provided that the existence of spatial correlations justifies the creation of such regions. A "cartoon" of two such classes of regions of different sizes is shown in Fig. 1.46.

A sophisticated approach is to test if the spatial distribution density of a property of the medium is a "multifractal", so called because, in this case, the fractal property is not a length or an area but a continuous function of the spatial coordinates. The criterion of being a fractal is unchanged: It is the property of self-similarity—that is, parts of the spatial distribution density "map" are similar to the whole in some way.

Such "maps" are difficult and time-consuming to determine experimentally, not to mention the work involved in testing the maps for multifractal properties. Lenormand *et al.* (1990) took the simpler approach, consisting of generating two-dimensional multifractal maps by computer, also including anisotropy, which they interpreted as porosity distributions and which they compared with pictures taken of layered Berea sandstone cores during gas injection, as "seen" by a CT-Scanner. Not altogether surprisingly, similar streaky patterns were obtained as those present in the two-dimensional anisotropic multifractal maps. Unfortunately, *a priori* predictions are not possible and the merit of this work lies in showing *qualitatively* the interesting fluid/fluid displacement patterns that may arise as a result of spatial correlations of permeability.

1.2.8. Packing of Particles and the Nature of Interstices

In consolidated porous media, *a priori* knowledge of the pore structure is usually impossible. (An exception to this rule is formed by certain filters where the pores have been prepared by piercing the filter material by bombardment with nuclear particles. In this case it seems to be a valid assumption that the pores are nearly cylindrical capillaries of uniform cross section.) By contrast, packed beds of particles are often amenable to theoretical analysis. A great deal of work has been done in this area. One of the most general features of packed beds of particles is that several neighboring particles surround a void space. Neighboring voids are separated from each other by necks or "windows."

1.2.8.1. *Regular and Irregular Packs*

Structural properties of packed beds have been reviewed by Haughey and Beveridge (1969). The next discussion follows this review rather closely.

Packing of particles is determined by several factors which include the shape of the particles, the distribution of particle sizes, whether the packing is "ordered" (regular) or "disordered" (irregular). In the case of an ordered packing of uniform spheres, one can specify the rule whereby the particles are arranged in regular arrays. In the case of disordered packs of uniform spheres, however, the bulk porosity of the pack and the average number of contacts per sphere determine the structure in a very complicated way.

The term "random" packing is used generally instead of "disordered" packing. However, packings are not formed by placing particles randomly in place, and therefore, the term "disordered" or "irregular" packing may be preferred. Unlike the unique position of each sphere in a regular packing, the location of any sphere in a "random" bed can be only expressed by a probability distribution.

Most notable of the regular packs are the cubic, orthorhombic, tetragonal–sphenoidal, and rhombohedral packings. Morrow and Graves (1969) plotted the attainable porosities for various unit cells as a function of the average acute face angle.

It is evident that for any ordered packing of spheres or, for that matter, any other shapes, the resulting pore structure will be anisotropic. This means that intersecting the pack with uniformly distributed parallel planes will result in different distribution of "free area fractions" (area porosities) in the sections, depending on the orientation of the planes. A useful practical criterion of a disordered pack is the independence of the "free area fraction" at a "point" in the section of the orientation of the sectioning plane, where the term "point" has been used in the macroscopic sense of the word.

Of practical importance are stable packings only, that is to say, packings where one sphere touches at least four others in such a manner that at least four of the contact points are not contained in the same hemisphere. Open-structured regular packings are relatively loose arrangements with at least some of the spheres free to move with respect to each other, as in beds formed by crushing solids, sedimentation, and flocculation. The loosest forms discussed have a mean bulk porosity possibly greater than 0.877 (Hilbert and Cohn-Vossen, 1932; Heesch and Laves, 1933).

1.2.8.2. *Porosities and Coordination Numbers*

The "thinnest" common regular packing is the cubic with a "coordination number" (points of contact per sphere) being equal to 6 and the mean bulk porosity to 0.4764. The "densest" regular packing, the rhombohedral, has a coordination number equal to 12 and a mean bulk porosity of 0.2595.

Just as each regular packing has a characteristic mean bulk porosity, it has been found that in the random packing of identical spheres a characteristic range of mean porosity values is associated with particular methods of formation. Four modes have been distinguished by Haughey and Beveridge (1969):

(1) *Very loose random packing.* When the fluid velocity is slowly reduced in a fluidized bed, the spheres settle to a porosity of about 0.44. Similar values have been obtained by sedimentation of spheres or by inversion of the bed container.

(2) *Loose random packing.* Porosities of 0.40 and 0.41 are obtained by packing the spheres so that they roll individually into place over similarly placed spheres, by individual random hand-packing, or by dropping the spheres into the container as a loose mass. Monte Carlo models usually produce values representative of these packings.

(3) *Poured random packing.* When spheres are poured into a container, mean porosities from 0.375 to 0.391 are found.

(4) *Close random packing.* The minimum porosities from 0.359 to 0.375 are obtained when the bed is vibrated or shaken down vigorously. The number of points of contact between a given sphere and adjacent spheres and the angular distribution of these points are also of interest, in addition to the porosities. For regular packings the number of such points, the "coordination number," is indicative of the packing employed. Cubic, orthorhombic, tetragonal–sphenoidal, and rhombohedral packings have, respectively, 6, 8, 10, and 12 contact points. No such correspondence occurs in random packing, where a range of values of the coordination number is found. Values for the closest and the loosest random packings have been obtained that distinguish between close contacts of spheres and near contacts. The dependence

of mean porosity ϕ on the coordination number z could be fitted as follows (Ridgway and Tarbuck, 1967):

$$\phi = 1.072 - 0.1193z + 0.004312z^2. \qquad (1.2.31)$$

Debbas and Rumpf (1966) derived theoretical relationships for the statistical characterization of random beds of spheres and irregular particles. A method was proposed for the determination of the coordination number. Experiments were conducted to test the theoretical relationships.

There are several different definitions of "porosity," in addition to the mean bulk porosity ϕ, which is the ratio of the pore (or void) volume to the bulk volume of the bed. The local porosity ϕ' is the fractional free volume in an element of bed volume. As the width of an element decreases, the local porosity tends to the local free area fraction (local area porosity) ϕ'_A. The local mean porosity $\bar{\phi}$ is the fractional free volume in a local region of volume V, consisting of the local elements between a reference point or plane and the outer limits of the region of interest.

Local variations in porosity are present in all beds. In the cubic system, for example, the local mean porosity $\bar{\phi}$ relative to a reference plane between sphere layers shows variations between unity and 0.41, while the free area fraction ϕ'_A correspondingly will vary from unity to 0.215 even though the bulk mean porosity ϕ remains constant at 0.476. Similar results are found for other packings. In a random packing such local variations also occur, but in terms of probability distributions. In a random packing, however, ϕ'_A, if taken over a large enough area, equals the bulk mean porosity. Microscopic heterogeneities in packed beds have been discussed by Morrow (1971).

1.2.8.3. Layer Spacings, Radial Distributions, Specific Surface Areas

Every regular packing can be thought of as consisting of parallel layers of identical structure. For any orientation, two successive layers of spheres will lie a fixed distance apart, the distance between sphere centers being termed the "layer spacing" βD_p, where D_p is the sphere diameter. For the common packings, β will lie in the range $0.707 \leqslant \beta \leqslant 1.00$, the limits being at rhombohedral and cubic, respectively.

Although a layer has little practical significance in random packing of identical spheres, the concept is useful. Average values of β can be calculated by the following formula, based on a rhombohedral unit cell in which the spheres are moved equally apart to give a mean porosity ϕ corresponding to the random packing:

$$\beta = \sqrt{2/3}\left\{\pi\Big/\left[3\sqrt{2}\,(1 - \phi)\right]\right\}^{1/3}. \qquad (1.2.32)$$

The distance between the center of a base sphere and the centers of surrounding spheres is called "interparticle distance" rD_p, which can be

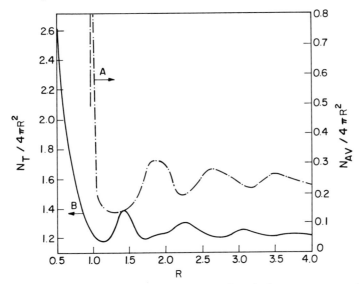

Figure 1.47.   The radial distribution of the average number of sphere centers per unit area in a spherical shell (curve A) (Scott, 1962) and the total number of spheres cut per unit area by a spherical envelope (curve B) (based on data by Scott, 1962) (Haughey and Beveridge, 1969).

found theoretically for regular packings. Thus, in cubic packing, six spheres touch the base spheres, so that $r = 1$, while there are 12 spheres around the base sphere at distances $r = \sqrt{2}$ and eight at $r = \sqrt{3}$, with others at greater distances. In the case of random packings, the radial distributions of interparticle distances can be found from a set of coordinate data. The distribution of average values shows a damped waveform, while at any given value of r there will be a distribution about this mean (see Fig. 1.47).

The number of spheres N_s in unit volume of bed is given quite generally as

$$N_s = 6(1 - \phi)\big/\pi D_p^3, \qquad (1.2.33)$$

so that the specific surface are S_v of the bed, that is to say the ratio of the internal area to the bulk volume, is given quite generally as

$$S_v = 6(1 - \phi)\big/D_p. \qquad (1.2.34)$$

Mason (1971) described random sphere packing in terms of tetrahedral subunits. The geometrical properties generated by computer agreed well with known properties of random sphere packings.

1.2.8.4. Void Shape and Void Size

The problem of void shape and void size in regular packings of identical spheres has been discussed by numerous investigators. Graton and Fraser (1935) distinguished two different types of voids, for example, in the rhombohedral packing—the larger one having a square shape ("concave cube") and the smaller a triangular shape ("concave tetrahedron").

The interconnected channels in a pack have been characterized by the size of a sphere that can pass through the windows in the channel. For rhombohedral packing this maximum sphere size is $[(1/\sqrt{3}) - 0.5]D_p \simeq 0.0773D_p$.

In accounting for the tortuous void path, a "tortuosity factor" has been defined as the average length of flow path relative to the length along the average direction of flow. The problems encountered in the definition of "tortuosity factor" are discussed in Chapter 3.

Various measures for void size in packed beds have been recommended, such as the size of the largest interstitial sphere possible within the void that can be calculated for regular packing. An equivalent spherical void diameter D_3 may also be defined as the size of a spherical void of volume equal to the local mean pore volume associated with a sphere in the bed, so that

$$\frac{1}{6}\pi D_3^3 = \left[\bar{\phi}/(1 - \bar{\phi})\right]\frac{1}{6}\pi D_p^3. \qquad (1.2.35)$$

The simplest approximation is the use of an equivalent ("hydraulic") circular tube diameter D_H as four times the ratio of the pore volume to the wetted surface area associated with a given sphere, so that

$$D_H = \left(4/\pi D_p^2\right)\frac{1}{6}\pi D_p^3\left[\bar{\phi}/(1 - \bar{\phi})\right] = \frac{2}{3}\left[\bar{\phi}/(1 - \bar{\phi})\right]D_p. \qquad (1.2.36)$$

This expression has also been applied to the bed as a whole by replacing $\bar{\phi}$ by ϕ, the mean bulk porosity.

1.2.8.5. Influence of Particle Shape and Size Distribution

Most practical packings involve random assemblies but will also contain a size distribution of particles that are likely to be nonspherical. Such a situation can usually only be handled empirically.

Random packings of identical spheres show mean bulk porosities in the range from 0.36 to 0.44, and similar values are also being found for packings of a narrow size distribution and not too irregular shape. Most common loose or compact granular materials have values lying between 0.3 and 0.5. With nonuniform angular particles, however, almost any degree of porosity is obtained, the deviation from the quoted range being a function of the nonsphericity. With the smaller particles, wider size distributions are found and, with any tendency to fibrous or platelike shapes, it is probable that porosity values will lie within the extremes of 0.1 or 0.9, with platelike materials falling in a narrower range. The actual size has no influence on the structure of regular packings, but with natural materials and random beds, this is not always so. Thus, sand has been found to possess porosities from 0.39 to 0.41 for a coarse grade through 0.51 to 0.54 for fine (Muskat, 1946).

Particle shape and size distributions are the two factors most likely to affect packing structure and its properties. Dense and loose random packing porosities for various particle shapes have been measured, the porosity increasing in the following order: cylinder, spheres, granules, Raschig rings, and Berl saddles (Oman and Watson, 1944).

It is rare to find uniformly sized irregular particles, except for manufactured particles such as Raschig rings or Berl saddles. As a result, the smaller particles tend to fill voids, reducing porosity. The converse effect of bridging, leading to higher porosities, is aided by an angularity in shape.

Most practical applications even with closely graded materials will involve a particle size variation of between two and one hundredfold, the range being smaller at the larger sizes.

It was shown by Horsfield (1934) that the filling of the voids of a rhombohedral packing by five successive specific sizes would give a minimum porosity of 0.1481. Further addition of finer particles can reduce this to 0.0384. Similar studies have been made in other packings. Many workers have made experimental studies of the porosity with mixtures of two or more particle sizes, mainly aimed at finding conditions for minimum porosity of packing. When smaller particles are mixed into a bed of larger particles, opposing effects are found that jointly determine the composition giving the minimum porosity. The smaller particles, on the one hand, tend to increase the porosity by forcing the larger particles apart but, on the other hand, can decrease the porosity by falling in the voids between the larger particles, the latter predominating for a size ratio greater than about 3:1. The porosity obtained will depend not only on the particle shape and size ratio but also on the amount of each size fraction present. Since smaller particles have a lower free-fall velocity and mass-to-surface ratio, when they fall into a bed they will produce less particle roll, which would otherwise distribute the particles into more stable positions. The degree of bridging yielding larger voids is thus greater in small particle beds. Such high-arched beds are inherently unstable, and the particles can reorientate themselves settling into a dense configuration upon refluidization, vibration, or tapping. Changes in the porosity of 5–20% are possible. These studies are of interest in the preparation of aggregates (e.g., concrete).

Sohn and Moreland (1968) found that porosity was a function only of the dimensionless standard deviation of size distribution and was independent of particle size. For finer mixtures of densely distributed systems they found that the porosity depended on mixture composition, mean size ratio, and porosities of individual components with minima at compositions of 55–75% of the larger component.

It is noted that in many practical cases segregation of the particles according to size will occur, a factor that can have a significant effect on the packing structure.

1.2.8.6. *Wall Effects*

When the packing is not sufficiently large in extent, any external surfaces such as bed support or walls have a significant effect on the packing properties.

A base plane, while not affecting regular packings, does introduce some local order into an irregular packing, because the first particle introduced must lie in the same plane layer. Even if the first layer proved perfectly regular (based on a triangular structure), the second layer can be built from several focal points and an irregular form would result if these growths do not mesh. The base layer, however, usually contains imperfections with a mixture of local clusters of square and triangular units. Randomness will increase with increasing distance from the base layer with a resultant disappearance of distinct layers. Wadsworth (1960, 1961) found that while the bottom layer displayed almost perfect rhombohedral packing, this rapidly disappeared in rising up through higher layers. By the eighth layer no spheres with 12 points in contact remained and the distribution assumed a Gaussian form, typical of the rest of the bed.

It is obvious that adjacent to an external surface there will be a region of relatively high porosity due to the discrepancy between the radii of curvature of the wall and the particles. Graton and Fraser (1935) and Rendell (1963) present findings that the effect of a container wall, especially when curved, is propagated a considerable distance into the interior of a packing of uniformly sized spheres. A cyclic variation of the local porosity with distance from a cylindrical wall has been measured, extending some three to four particle diameters into the packing.

It is apparent that errors in the determination of the bulk mean porosity will be found if care is not taken to avoid these outside influences. Distinguishing between the overall porosity ϕ_0 measured in a container of diameter D_T and the bulk mean porosity ϕ, we have $\phi = \phi_0$ when the wall effects become insignificant as $D_T/D_p \to \infty$. The effect of the wall is to increase the overall porosity due to the increase in local porosity in the region near the wall. The effect of the D_T/D_p ratio has been studied in many different systems. It is generally concluded that the wall effect is locally negligible if $D_T/D_p > 10$.

1.3. PORE STRUCTURE OF SOME IMPORTANT MATERIALS

In this section, a few selected natural and industrial porous media are discussed. A detailed coverage of this field is not possible because of limitations of space.

1.3.1. Natural Porous Media

1.3.1.1. *Geological Materials*

This subject has been reviewed by Davis (1969), whose presentation is followed here. For geological materials one distinguishes among dense rocks, volcanic rocks, indurated sedimentary rocks, and nonindurated sediments.

Dense rocks: A large number of dense rocks have porosities of less than 2% and permeabilities of less than 10^{-2} darcy. Granite, dolerite, quartzite, slate, and gabbro are but a few of the common dense rocks that are abundant in the earth's crust. Typical values of porosity and permeability for dense rocks have been tabulated by Davis (1969).

Volcanic rocks: Rocks that have originated from the solidification of a magna at or near the surface of the earth are classified as volcanic rocks. Typical values of permeability and porosity have been given by Davis (1969).

When molten rocks cool in dikes and sills a few hundred to a few thousand feet below the surface, the resulting rocks are dense and almost nonpermeable. The same materials reaching the surface as lava flows can form some of the most permeable aquifers known.

Weathering will tend to increase porosity of volcanic rocks but may drastically reduce the permeability. Thick tropical soils may commonly exceed 60% porosity, but permeabilities of the water-saturated soils are mostly below 1 darcy.

Permeability of basalt and related rocks is commonly anisotropic when considered in large volumes, although the anisotropy may not be too evident in small laboratory samples.

Indurated sedimentary rocks: Indurated sedimentary rocks become hard because of the action of cementation.

More is known about the hydraulic characteristics of permeable sedimentary rocks than of any other major group of rocks. This is due to the economic interest in these rocks by the petroleum industry.

Most sedimentary rocks can be classified broadly as shales, sandstones, and carbonate rocks. The total amount of salt, gypsum, conglomerate, and other minor sedimentary types is probably less than 5% of the overall volume of sedimentary rocks.

Shales as used in the broad sense constitute roughly 50% of all sedimentary rocks. Laboratory work as well as aquifer tests suggest that permeabilities of unfractured shale are always less than 10^{-3} darcy and most commonly less than 10^{-5} darcy. Porosities of shales are much easier to determine than permeabilities. Shales under the surface will have porosities from 10 to 25%. Older shales that have been compacted by burial and have also undergone some cementation may have porosities from 2 to 10%.

Sandstones make up roughly 25% of the sedimentary rocks of the world. Although compositions vary widely, quartz is the predominant mineral in most sandstones. In general, sandstones are considerably more permeable than shales. Some sandstones nevertheless are so firmly cemented by quartz or other mineral matter that they have very low permeabilities.

Porosities of sandstone are almost entirely a function of the kind of cementing material between the sand grains and the extent to which the larger grains interdigitate. Nonindurated sands will have porosities of between 30 and 45%. For indurated sands, porosities from 10 to 25% are more common. From the standpoint of geological processes, two main reasons account for loss of porosity. The first is cementation by calcite, silica, and other minerals introduced into the rock by circulating ground water. The second is compaction that squeezes fine-grained matrix material into open pores and interdigitates larger grains by grain-boundary solution at points of highest stress. Porosity of sandstone tends to diminish systematically with depth, so that sandstone below about 10 km has less than 5% porosity. This fact has profound effects on the economics of petroleum exploration because not only is it unreasonable to expect significant permeability at depths of more than 10 km, but the total amount of fluids in storage also decreases rather rapidly with depth.

The permeability of sandstone is a direct function of the size of the interconnected pores. For sandstones with only a moderate amount of cementing material, permeability can be correlated directly with grain size (Johnson and Greenkorn, 1963). On the other hand, if a formation has about the same grain size but varying amount of cement, then a direct relation between porosity and permeability has been claimed (Levorsen, 1954). In general, however, just a simple measure of either grain size or porosity will not prove to be a reliable guide to the permeability of sandstones.

The presence of horizontal stratification in most sandstones means that the permeability of very large samples, say more than 100 m^3, should be considered to be uniformly anisotropic. The gross effect of this permeability stratification is that the effective vertical permeability of large masses of sandstone is very low even though the horizontal permeability may be quite high.

As cementation and compaction take places, more of the permeability will be due to fractures within the rock that cut through the original layers of low permeability with the overall effect of reducing the permeability parallel with bedding and increasing the vertical permeability by fracturing.

Several studies have shown that the average permeability of even small samples will tend to be much greater parallel with the bedding than perpendicular to the bedding. In a study by Greenkorn *et al.* (1964), 81 out of a total of 142 small samples of sandstone had significant anisotropy. Only 63 of the samples had recognizable stratification. Of the 23 samples in which vertical permeability was the greatest, 17 either had no visible stratification or the bedding made an angle of more than 45° with the horizontal plane.

The effective or average permeability in heterogeneous sandstone has been taken as a value between the arithmetic mean and the harmonic mean of the various permeability measurements made in a given stratum (Cardwell and

Parsons, 1945) or as the geometric mean (Greenkorn *et al.*, 1964; Warren and Price, 1961; Bennion and Griffiths, 1966).

Carbonate rocks consist mostly of calcite and dolomite with clay and quartz as common secondary minerals. The terms "limestone" and "dolomite" can be used for rocks primarily composed of the minerals calcite and dolomite, respectively.

Geologically young carbonate rocks mostly have porosities ranging from about 20 to more than 50%. Permeabilities of young reef and fragmental limestone are of the order of 1000 darcys. Permeabilities of unfractured portions of older carbonate rocks are generally less than 1 darcy and commonly less than 0.01 darcy. High permeabilities are developed in older carbonate rocks only through fracturing and a solution of the carbonate minerals.

Bennion and Griffiths (1966) have compiled permeability data from several thousand samples of conglomerate sandstone and limestone units and have found log-normal distributions that were skewed slightly to the right, so that it appears to be a fairly general rule that the statistical distributions of permeabilities within given geologic units are log-normal.

Nonindurated sediments: Countless measurements of permeability and porosity have been performed on samples of nonindurated sediments of diverse geologic origins. According to Davis (1969), laboratory tests have the advantage of giving more reliable information of porosity and on the extent of heterogeneity and anisotropy than can be gained from field tests. Unfortunately, laboratory studies have two serious disadvantages. First, the number of samples needed to define average conditions in nature is usually excessive. Second, obtaining samples that are undisturbed and hence representative of natural conditions is a most difficult task.

Masch and Denny (1966), in a review of work on artificially mixed samples of sand-sized material, have concluded the following:

(1) Permeability increases with increase in median diameter.

(2) For a given median diameter, the permeability decreases with an increase in the standard deviation of particles sizes.

(3) Permeability is most sensitive to changes in standard deviation in samples with larger median grain sizes.

(4) Other factors being equal, unimodal distributions have higher permeabilities than bimodal distributions.

(5) Permeabilities can be estimated by a correlation, provided a size analysis of the sediment sample is available.

1.3.1.2. *Living Organisms*

Living organisms are not usually referred to as porous media, although cellular structure is a special kind of porous medium filled with a variety of different fluid substances.

Certain parts of living organisms such as the skin, bones, and certain glandlike structures have been recognized, however, as having a definite pore or void structure. Blood vessels form conduits, the geometry of which has been the subject of considerable investigation recently.

The study of the structure of porous organs, part of the science called biomorphology, has been performed almost entirely by optical means, using increasingly in recent years the methods of quantitative stereology. A good brief review of the state of the art in this field has been given by Weibel (1976).

Pores and tubes form an important part of plant tissues. These play a decisive role in the ascent of sap and the movement of soluble carbohydrates in the stems of plants. A good review of this field has been given by Preston (1958). Sneck et al. (1956) and Stayton and Hart (1965) have reported on methods of measuring pore size distribution in woods.

1.3.2. Industrial Porous Materials

These materials comprise a large variety of different substances. They can be subdivided into various, more or less arbitrary, categories.

1.3.2.1. Paper, Textile, and Leather

The control of pore structure in paper making has been reviewed by Rance (1958).

The porosity of paper is controlled by beating of the fibrous stock. The unbeaten fibers have an open structure with a porosity as high as 80%; blotting paper and air filters are examples. On the other hand, highly beaten stock is dense, transparent, and hard, with porosity approaching 0; tracing paper is an example. Between these extremes lies a wide range of types of paper made for a large variety of applications. Throughout this range control of the properties of the product is, to a large extent, exercised through control of the porous structure.

In paper technology the term "porosity" is used rather loosely. The so-called porosity of paper is determined by measuring either the rate of penetration of some liquid into or the air of permeability of the paper. From air permeability measurements the equivalent pore radius has been calculated using a modified form of the Hagen–Poiseuille equation. Pore sizes have also been calculated based on capillary rise of aqueous liquids in a paper sheet. Pore radii obtained by these methods range from 0.2 to about 4 μm in different sorts of papers.

Corte (1957) described several methods for determining pore size distribution in paper. One of these is due to Wakeham and Spicer (1949) who used mercury porosimetry. Another method consists of stepwise permeation of nitrogen through the pores of the paper. The sample is clamped between metal rings and covered with dioxane, which fills the tube above the sample. Nitrogen is pumped to exert pressure on the bottom of the sheet. Each rise in pressure opens up more and more of the smaller pores. From the resulting plot of flow rate against pressure it is possible to calculate the pore size distribution. It is noted that the distribution curves obtained by the mercury and the dioxane methods are not identical, partly because one is based on volume penetration and the other on rate of flow. Indeed, the dioxane method yields a so-called relative permeability curve of the nonwetting phase.

Corte has also reported some results of direct pore size distribution by optical means. Contact photographs of very thin sheets of paper were taken and the holes were measured. The distribution was roughly log-normal.

In addition to these methods there have been some interesting studies of the penetration of paper by aerosols. The nature of the pore size distribution is vitally important in this field and the equivalent pore radius is virtually useless as an index of filtering efficiency. The same equivalent pore radius can be found for paper with a few large pores as for a paper with many smaller pores. The former paper would be useless for filtering smokes, whereas the latter would be highly efficient. High-efficiency smoke filters must be reasonably permeable to gaseous flow, otherwise filtration is impossible; yet they must filter out particles as low as 0.5 μm diameter.

Empirical relationships between the pore structure and many other physical properties of paper have been widely studied and recorded.

To the author's knowledge, pore structure control and research has played a far lesser role in the textile and leather industries than in paper technology. Pore size distributions in various fabrics have been determined by Honold and Skau (1954) and Honold *et al.* (1956). The effect of pore size distribution on the windproofing and water resistivity of cotton fabrics has been studied by Wakeham and Spicer (1949). Porosities of fibers and yarns have been studied by Quynn (1963) and Wagner (1967).

Macropores in leather have been determined with mercury porosimetry by Kanagy (1963). Stromberg (1955) studied the pore size distribution in collagen and leather by mercury porosimetry. The porosities in leather range from about 0.56 to 0.59.

1.3.2.2. *Concrete, Cement, and Plaster*

A vast amount of literature is available on the subject of pore structure of concrete and related materials. The pore structure of concrete has received a great deal of attention in recent years. According to Podvalny and Prozenko (1974), the structure of concrete consists of clinker grains that are separated from each other by a hydrated mass of calcium hydrosilicates, in the central portion of which there run thin veins of capillaries which form an intercon-

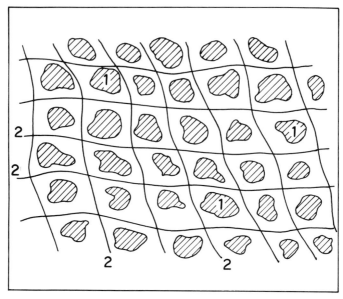

Figure 1.48. Scheme of concrete structure: (1) impermeable component; (2) permeable component (Podvalny and Prozenko, 1974).

nected network, as shown schematically in Fig. 1.48. The permeability of good concrete is under 0.1 mdarcy and the porosities range from 6 to 10%.

According to the mercury porosimetry pore size distribution curve shown in Fig. 1.49, the pore radii in a typical hydrated cement range up to about 10 μm, but the bulk of the pore volume consists of pores of radii less than 1 μm and reaches into the region under 100 Å.

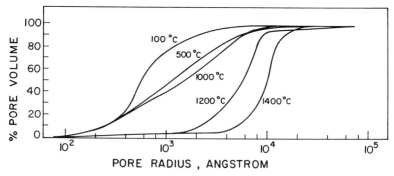

Figure 1.49. Variation of pore distribution of Secar 250 hydrated cement as a function of firing temperature (Jung, 1973).

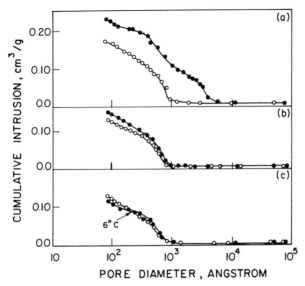

Figure 1.50. Cumulative pore size distributions at 0.4 w:c cement pastes hydrated at 6°C (●) and at 40° C (○); (a) 4 days; (b) 28 days; (c) 360 days. (Reprinted with permission from Diamond, 1973.)

Collepardi (1973) and Young (1974) have investigated the morphology and pore size distribution in tricalcium silicate pastes. The pore walls consist of dense rays of needles that intermesh when calcium chloride has been added to the paste. Several other studies on the influence of various additives on the structure of concrete and cement pastes can be found in the same reference (Ratinov *et al.*, 1974; Chiarioni and Reverberi, 1974; Delvaux *et al.*, 1974). Butt *et al.* (1973) and Chekhovsky *et al.* (1973) have studied the influence of hardening time and conditions, composition of cement, water–cement ratio, chemical additives, and the type of compacting on the pore structure of hardened cement paste and concrete. Diamond (1973) studied the degree of hydration, density, porosity, and mercury intrusion porosimetry pore size distribution as functions of hydration temperature. The temperatures used were 6 and 40°C. Initially, the pastes hardened at low temperature had much coarser pore size distributions, but in a month the differences were small. After a year the distributions were identical, except in the range below several hundred angstroms where the paste hydrated at the low temperature showed a negligible content of pores. Typical pore size distributions are shown in Fig. 1.50.

According to June (1973), the formation of pore systems by hydration in concrete follows an exact law and it is possible to calculate the porosity of hydrated cement and concrete by a formula. Jung calculated the porosity of hydrated cements and determined experimentally the mercury intrusion porosimetry pore size distributions as a function of the firing temperature, ranging from 20 to 1400°C. In general, the calculated porosity increased and the pore size distribution changed, as shown in Fig. 1.49.

Feldman (1973) presented a review of research on certain structural characteristics of hydrated Portland cement, associated with the instability of and the presence of several types of water in the pore structure. Application of several techniques, including a novel "helium flow technique," has shown that the spaces in hydrated Portland cement approximate the interlayer spaces as in some clays and are not small-necked, fixed-dimension pores. As water is removed they partially collapse. They reopen on reexposure to water, and the water occupying these interlayer positions possesses the property of being able to stiffen the material. During exposure to higher humidities, interlayer volume is increased, leading to dimensional instability and such phenomena as creep.

Romer (1973) has worked out a microscopic method to differentiate all pore and void spaces in concrete from the solids by impregnating the sample with a liquid polymer containing a fluorescent tracer. The samples are viewed in ultraviolet light and even pores under 1–2 μm become visible. The various pores are divided into the following categories: water capillaries, microcracks and cracks, surrounded aggregates with high mortar capillarity, peripheral cracks associated with aggregates, and air-void water pores-entrapped void. Stroeven (1974) applied a fluorescent spraying technique to make microcracks visible in sections and at the surface of concrete samples. Quantitative stereological methods were applied to characterize the cracks. A comprehensive literature survey on concrete pore structure research has been done by Modry *et al.* (1972).

1.3.2.3. *Sintered Materials*

The structural characteristic that is uniquely associated with sintered materials is residual porosity. Extremely few sintered materials can be produced without any porosity remaining after sintering (Hirschhorn, 1969). Mercury porosimetry has been used extensively for porosity investigations in sintered products (see Mondry *et al.*, 1972).

Sintered materials usually contain a relatively large amount of interconnected pores and therefore they have permeability. The mechanical properties of sintered materials depend very strongly on the porosity. The permeability of sintered materials influences the applicability of the material for certain uses.

The process of sintering consists of a gradual elimination of the voids that are initially present between the particles. Sintering usually is divided into the following six stages: (1) initial bonding among particles, (2) neck growth, (3) pore channel closure, (4) pore rounding, (5) densification or pore shrink-

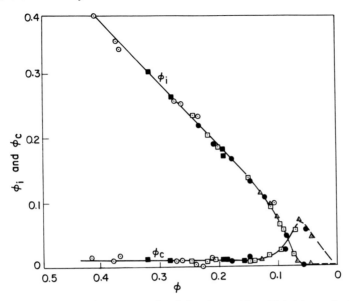

Figure 1.51. Variation of interconnected and closed porosities with total porosity for copper compacts pressed at various pressures and sintered at 1832°F. The values shown, measured in psis, are 11,200 (⊙); 22,400 (■); 33,600 (▫); 44,800 (●); 67,200 (△) (Arthur, 1954–55).

age, and (6) pore coarsening. The actual distribution of the total porosity into the interconnected and isolated types may be determined experimentally by several methods and some such data for copper compacts are given in Fig. 1.51. It is noted that with porosities greater than about 10% most of the porosity is in an interconnected form, and with porosities less than about 5–10% most of the porosity is of the closed or isolated type. The reason for this behavior can be found in statistical laws that have been used quantitatively in percolation probability theory. Qualitative relationships have been proposed to predict the change of porosity with sintering (see, for example, Ivensen, 1973). A great deal of modern microscopic and electron microscopic methodology regarding the pore structure of sintered materials during sintering can be found in the work of Hirschhorn and Roll (1970).

Gusman (1973) has claimed that in reaction-sintering densification is completely determined by the pore structure change. Correlation between initial and final porosity and volume changes has been given for reaction-sintering of ceramic materials and exemplified by the Si–C–O–N system. Srbek *et al.* (1973) investigated the effects of the shaping process (e.g., extrusion dry pressing without initial compression, dry pressing with initial compression, wet pressing, and slip casting) on the pore structure of electrical porcelain.

1.3.2.4.   Coal and Coke

The porosity of coal is low, usually only a few percent. Zwietering and van Krevelen (1954) established the fact that in coal there are two separate pore systems, one extending from about 60 Å to 4 μm in pore radius, as determined by mercury porosimetry, and the other so small that nitrogen and methane molecules are not capable of penetrating into them unless they possess a certain activation energy. Helium atoms, however, have free access to the fine pores even at low temperatures. The so-called macropore system is due to cracks and, therefore, in this case the diameter of the pore necks is about the same as that of the pores themselves. The volume due to the crack varies over wide limits because the amount of cracks is more or less accidental. The micropore volume appears to be practically constant, however.

Dickinson and Shore (1968) and Baker and Morris (1971) found that the interim pore structure of graphitized carbon can be permanently damaged by mercury at high pressure. The effect of the damaging process was to increase the volume of pores accessible to mercury of size below 2 μm. As a result, with some graphites, porosity derived from mercury penetration at the maximum pressure was found to be greater than that computed from helium density measurements. Bohra (1974), however, found that up to a pressure of 5000 psi the mercury intrusion volume was greater than the pore volume determined by helium displacement.

Matthews (1974) used gas diffusion measurements to establish that the pore structure of graphite contains a bimodal pore network, consisting of a macrovoid structure pore size greater than 0.1 mm, superimposed on a micropore structure (100–0.1 μm). For very thin specimens the macrovoid structure may not form an interconnecting network.

Hewitt (1967), by determining the permeability of graphite samples to mercury at various saturations, demonstrated that the principal pore structure consists of a network of larger cracks. The internal surface of graphite is mostly associated with micropores. The mercury permeability–saturation curves were used by Hewitt also to calculate "differential tortuosities." The value of this quantity ranged from about 1 to about 100.

1.3.2.5.   Porous Crystals

The types of pore systems in porous crystals have been reviewed by Barrer (1958), whose presentation is followed below. Porous crystal adsorbents differ from microporous carbons or silica gels in that the latter normally exhibit a spectrum of pore sizes, whereas in crystals the pore systems are often quite rigid and all of one type, having diameters of molecular dimensions. Accordingly, porous crystal adsorbents can behave toward appropriate mixtures of molecules as perfect molecular sieves.

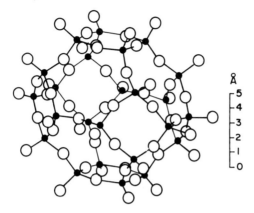

Figure 1.52. The arrangement of (Al, Si)O$_4$ tetrahedra that gives the cubooctahedral cavity found in some felspathoids and zeolites (Barrer, 1958).

Barrer distinguished the following four types of intracrystalline pore space:

(1) *Layer-type porous crystals* where the pore spaces in the crystals are parallel-sided slits. The slits have widths comparable to molecular dimensions. A range of porous layer–lattice crystals were developed from bentonite clays by a simple ion exchange process.

(2) *Porous crystals with isolated cavities.* An example is β-quinol in which the cavities may be filled in part by molecules such as argon or by other species of comparable dimensions. Water is notable in forming two lattices, each containing large cavities of two kinds. These lattices form only if there is partial occupancy of the cavities by guest molecules that may be of great diversity. In such structures the host lattice is stabilized by the guest molecules.

(3) *Porous crystals with interconnected cavities.* These are the group of porous crystals, the "zeolites," within which suitable molecules may diffuse readily. These may often be regarded as having continuous frameworks built by stacking together cages of various types. One such cage is the cubooctahedron, part of which is shown in Fig. 1.52. This has 14 faces, six of which are made from rings of four (Al, Si)O$_4$ tetrahedra and eight of which are made from rings of six such tetrahedra. In these structures often windows lead from one super cage to each of the other super cages.

(4) *Channel patterns in some porous crystals.* The stacking of polyhedral cavities having wide windows leading from one cavity to others produces a regular channel network of some diversity. A pattern of channels intersecting in three directions at or nearly at right angles is found also in some of the fibrous zeolites.

1.3.2.6. *Active Carbons and Silica Gels*

Active carbons are highly porous materials (Dacey, 1967). Their porosity can be as high as 80–85%. Some of the pores, however, are so small that the pore volume available to different molecules will depend on their size. Therefore, the concept of porosity is somewhat arbitrary but is best taken to be that determined by helium displacement.

The pores in most active charcoals usually fall into three groups: the macropores, 5000–10,000 Å in diameter; the transitional pore, 100–200 Å; and the micropores, 8–20 Å. Hence a typical pore size distribution of an active carbon is trimodal, as can be seen in Fig. 1.53.

The macropores may be observed with an ordinary microscope and their volume can be determined directly with a mercury porosimeter. The transitional pores may be seen with an electron microscope and their volume can be determined with a mercury porosimeter at very high pressures. Transitional pore volumes are usually obtained from sorption isotherms. The micropore volume is very difficult to determine accurately. The most accurate method consists of use of the adsorption isotherms. At the point on the isotherm at which capillary condensation occurs, the micropores are already filled while the macro- and transitional pores have a monolayer coverage only.

Macropores are generally thought to arise from fissures and interstitial spaces present in the original char, reflecting the structure of the starting material such as wood or shell. The transitional pores may arise from the carbonization process and are enlarged into pores during activation. Micropores arise from the activation process.

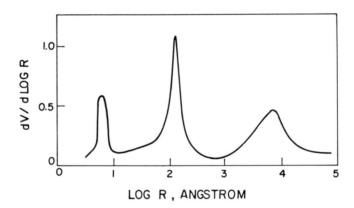

Figure 1.53. Differential pore volume distribution curve for active carbon (Dubinin, 1955).

It is expected that any pore varies in diameter and cross-sectional shape along its length. The hysteresis observed for most active carbons is explained in part by this "ink-bottle" effect.

In contrast to activated charcoals made from natural products, Saran charcoal has a remarkably uniform pore structure consisting almost entirely of micropores less than 15 Å in their narrowest dimension.

The surface area of activated charcoals cannot be determined from adsorption isotherms directly with any certainty. It is believed that before a complete monolayer is formed some of the micropores will be filled. However, with the knowledge of the micropore volume, an estimate of the overall surface area may be made.

Wynne-Jones (1958) showed that adsorption on carbon of a number of gases with molecules of considerably different sizes leads to a definite determination of the surface area, which for some carbons was found to be so large that a high proportion of all the carbon atoms must be at the surface. This is analogous to the behavior of zeolitic molecular sieves.

Carbon blacks are less porous than activated charcoal, but they have a high surface area because of their small particle size. The pores in some carbon blacks are micropores from 10 to 15 Å in diameter.

Foster and Thorp (1958) investigated the factors governing the pore size distribution in silica gels by determining sorption isotherms for water and ethyl alcohol on gels prepared by a number of different methods. It has been found in agreement with other workers that the higher the pH at the time of gel formation, the larger the pore radius of the resulting gel. The range of pore radii found was from about 10 to about 30 Å.

1.4. FIELD LEVEL CHARACTERIZATION OF POROUS MEDIA

1.4.1. Introduction

Reservoir description is of great practical importance for the numerical simulation of both oil and gas reservoirs and aquifers. The purpose of reservoir simulation is to predict the performance of the reservoir as accurately as possible. There are two basic approaches: the one using finite difference formulation and the other using finite element formulation. In the finite difference formulation the reservoir is discretized into a finite number of blocks; on each block a material balance is written into which flow equations, equations of state, and capillary pressures are introduced. Details of reservoir simulation techniques may be found in specialized monographs, for example, Schlumberger (1972) and Peaceman (1972).

For the input to a simulation program fluid data and rock data are required, of which only the latter relates to reservoir characterization. The required rock data are as follows (Schlumberger, 1972):

(1) Formation elevations
(2) Formation thickness
(3) Porosity
(4) Permeability
(5) Capillary pressure
(6) Relative permeability
(7) Formation fluid saturations
(8) Compressibility

Every cell of the discretized model must be identified by a given value of each of these parameters. The reservoir description techniques used are the same for both oil and gas reservoirs and aquifers and they mostly consist of coring, logging, and pressure transient tests. All of these techniques can be applied only after wells have been drilled. Each reservoir is heterogeneous in the sense that the permeabilities of core plugs obtained even from the same core always vary and the permeabilities measured in the same well sometimes vary by orders of magnitude. The number of wells that may be drilled is limited, for economic reasons, and hence it is obvious that construction of a detailed, accurate permeability or porosity map of a reservoir poses a very serious problem.

1.4.2. Formation Elevation and Formation Thickness

The size of the reservoir is usually characterized by a formation elevation map that determines both the areal extent and the thickness of the reservoir. The data used in determining the size of a reservoir include knowledge of the geological formations, sometimes geophysical (e.g., seismic) data, in addition to core data, well logs, and results of pressure tests. Formation elevations and thicknesses are obtained from drilling records, including rates of penetration and the nature of cuttings, cores, and electrical resistivity logs. A typical strip log is shown in Fig. 1.54. Cores are cut by a hollow core barrel that goes down around the rock core as drilling proceeds (see Fig. 1.55). In well logging various sensors are lowered into the well bore that permit measurements on those parts of the formation adjacent to the well bore. A schematic diagram of the basic spontaneous or self-potential (S.P.) logging circuit is shown in Fig. 1.56.

The S.P. curve is a recording of the difference between the potential of a movable electrode in the bore hole and a reference potential of an electrode on the surface. The S.P. is seen in the left-hand track of the log shown in Fig. 1.57. The S.P. curve differentiates between impervious shales and permeable sands. The deflections on the S.P. curve correspond to electric currents flowing in the mud in the bore hole, which are caused by electromotive forces

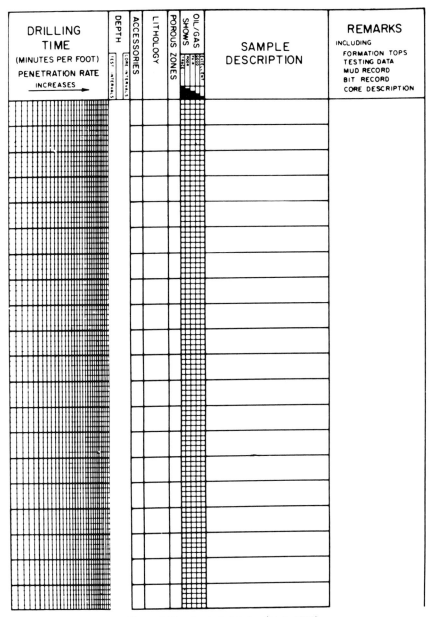

Figure 1.54. Typical strip log (Link, 1982).

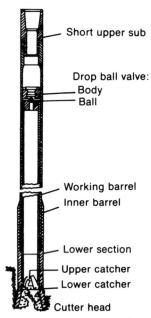

Figure 1.55.   Rotary core barrel (Link, 1982).

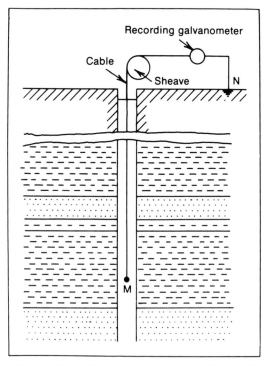

Figure 1.56.   Self-potential logging circuit (Link, 1982).

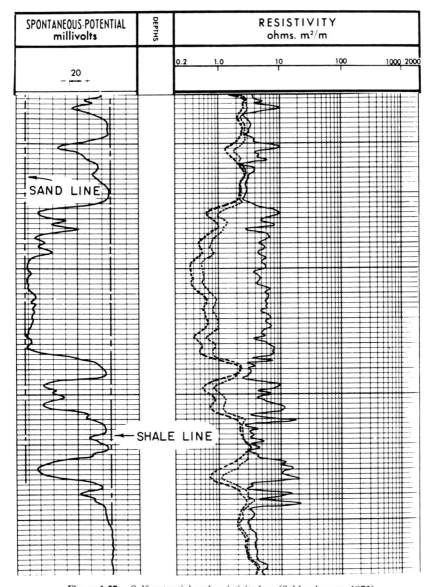

Figure 1.57. Self-potential and resistivity logs (Schlumberger, 1972).

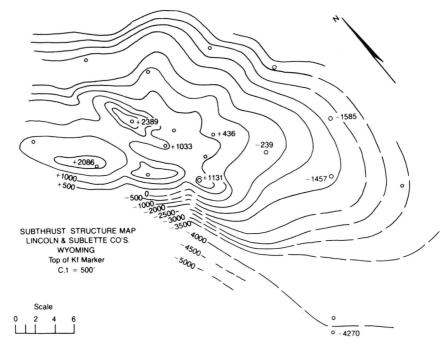

Figure 1.58.   Subthrust structure map, overthrust belt, Wyoming (Link, 1982).

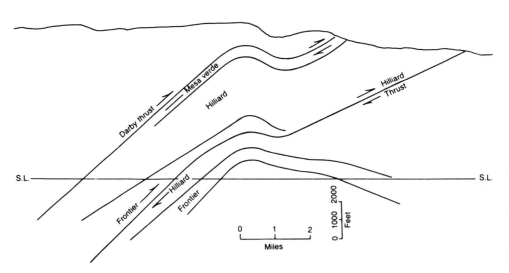

Figure 1.59.   Overthrust belt, Wyoming, cross section (Link, 1982).

of electrochemical and electrokinetic origins (Schlumberger, 1972; Wyllie, 1963).

In conventional resistivity logs currents are passed through the formation between electrodes and the resulting voltages are measured between other electrodes (see right-hand track of Fig. 1.57).

The well data are used to prepare isopach contour maps of formation elevations and thicknesses. An example of an isopach map is shown in Fig. 1.58 (note that a subthrust trap is a fault trap that occurs in overthrust belts where the deformed reservoir forms a trap below a thrust fault, as shown schematically in Fig. 1.59).

1.4.3. Determination of Reservoir Boundaries by Pressure Drawdown Tests

An important method for the determination of reservoir boundaries is the pressure drawdown test. Ideally this test consists of producing a well at a constant rate, with the pressure being initially uniform throughout the reservoir. In most practical cases the shut-in pressure prior to the test is obtained by extrapolation.

It is logical that a plane boundary at a distance d from the well would have the same effect on the pressure as the presence of a second well, producing at the same rate, at a distance $2d$ from the first well in a reservoir without boundaries (infinite acting reservoir). For the latter case, however, an analytical solution can be found for the bottom-hole pressure P_w in which the effect of the boundary is represented by the effect of the second (image) well (Slider, 1983):

$$P_w = P_i - \frac{0.141q\mu}{kh}\left(\frac{1}{2}\right)(\ln t_{Dreal} + 0.809)$$

$$- P_{skin} - \frac{0.141q\mu}{kh}\left(\frac{1}{2}\right)\left[E_i\left(\frac{-1}{4t_{Dimage}}\right)\right], \qquad (1.4.1)$$

where

P_i = initial pressure, psia

q = production rate, res b/day

μ = viscosity, cp

k = permeability, darcy

h = net thickness, ft

t_{Dreal} = dimensionless time for real well

t_{Dimage} = dimensionless time for image well

$$E(-x) = -\int_x^\infty \frac{e^{-u}\,du}{u}$$

It is evident from Eq. (1.4.1) that without the effect of the second well (or boundary) P_w is a straight line function of $\ln t_{\text{Dreal}}$. The initial portion of the experimental plot P_w versus $\ln t_{\text{Dreal}}$ is also a straight line, because initially the presence of the second well (or boundary) is not felt. Hence the difference between the measured value P_w, at a later time, and the extrapolated value P'_w, to the same time is equal to (Slider, 1983):

$$P'_w - P_w = \frac{0.141 q\mu}{kh}\left(\frac{1}{2}\right)\left[-E_i\left(\frac{-1}{4t_{\text{Dimage}}}\right)\right], \qquad (1.4.2)$$

where

$$t_{\text{Dimage}} = \frac{6.33kt}{\phi\mu c(2d)^2} = \frac{\eta t}{(2d)^2}, \qquad (1.4.3)$$

where

d = distance to boundary, ft

ϕ = porosity

c = compressibility, psi^{-1}

t = time, days

η = hydraulic diffusivity, ft^2/day

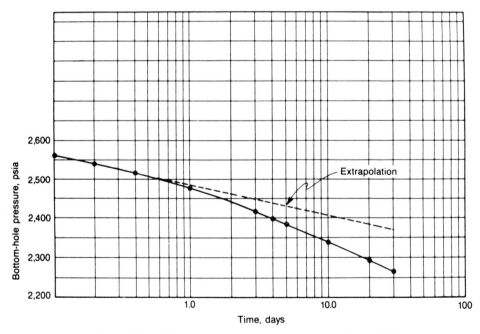

Figure 1.60.   Effect of a plane barrier on a drawdown (Slider, 1983).

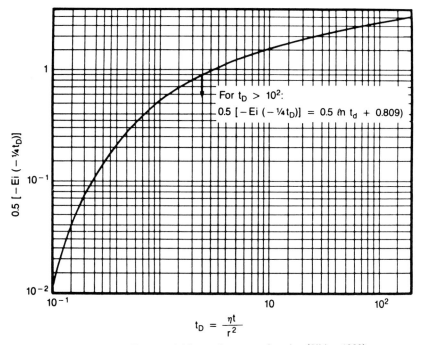

The chart shows the exponential-integral pressure function with axis label $0.5 [-\mathrm{Ei}(-\tfrac{1}{4}t_D)]$ on the vertical axis (values 10^{-2}, 10^{-1}, 1) and horizontal axis (10^{-1}, 10, 10^2). Annotation:

For $t_D > 10^2$:

$0.5 [-\mathrm{Ei}(-\tfrac{1}{4}t_D)] = 0.5 (\ln t_d + 0.809)$

$$t_D = \frac{\eta t}{r^2}$$

Figure 1.61. Exponential-integral pressure function (Slider, 1983).

A sample plot of bottom-hole pressure versus time is shown in Fig. 1.60. For given values of q, μ, k, h, ϕ, and c, d may be calculated from this plot by means of Eqs. (1.4.2) and (1.4.3) and the plot of E_i versus t_D shown in Fig. 1.61.

This procedure can be used also in the case of more than one barrier, as long as the different barriers are at sufficiently different distances from the test well, so that the distance to each barrier can be determined, one after the other, without the interference of the other barriers situated at greater distances from the well. More details of these calculations and the discussion of their limitations can be found in Slider (1983). It is noted that the drawdown test does not indicate in which direction a boundary lies from the well. This direction is usually found from geological evidence of the presence of a fault somewhere without knowing its exact location.

1.4.4. Porosity

Porosities can be measured in the core analysis lab on core samples obtained from wells and by obtaining various logs, including a sonic log, a formation density log, or a neutron log (Schlumberger, 1972). It is important to note that logs only measure the properties of the formation close to the bore hole.

The following empirical relation has been proposed by Wyllie (1963) for the porosity:

$$\phi = \frac{\Delta t - \Delta t_{ma}}{\Delta t_f - \Delta t_{ma}}, \tag{1.4.4}$$

where Δt is the time for an acoustic wave to travel through one foot of formation along a path parallel to the bore hole, measured by the sonic tool; and Δt_f and Δt_{ma} are the corresponding transit times in the pore fluid and the rock matrix, respectively. This relation is limited in its applications to clean, compacted formations of intergranular porosity, containing only liquids. Secondary porosity, such as vugs, presence of shale, fractures, or gas, introduce errors.

In the density log a radioactive source at the bore hole wall emits medium-energy gamma rays into the formation, resulting in Compton scattering, the extent of which is directly related to the number of electrons per unit volume—the electron density of the formation. This in turn is related to the true bulk density, determined by the density of the rock matrix, the porosity, and the density of the fluids filling the pores.

The porosity is calculated from the formula:

$$\phi = \frac{\rho_{ma} - \rho_b}{\rho_{ma} - \rho_{liq}}, \tag{1.4.5}$$

where ρ_b is the bulk density of the clean, liquid-filled formation, obtained from the density log; ρ_{ma} is the matrix density, and ρ_{liq} is the density of the liquid. The presence of shale or gas in the formation introduces an error.

In neutron logging, radioactive sources mounted in a sonde are used to emit continuously neutrons that collide with nuclei of the atoms present in the formation. Neutrons colliding with hydrogen nuclei are slowed down more than those colliding with other nuclei. Those neutrons that have slowed down to "thermal" velocities may be captured by nuclei of other atoms, thus making it possible to count them with a detector. The counting rate at a fixed source-detector spacing may be used to measure the hydrogen concentration of the formation. If the pore space is liquid filled, the response is basically a measure of porosity. Presence of shale and gas affect the readings also in neutron logging.

The porosity of clear water-bearing formations may also be estimated from resistivity logs and Archie's formula:

$$F \equiv \rho_o/\rho_w = a/\phi^m, \tag{1.4.6a}$$

where

F = formation resistivity factor, ratio

ρ_o = resistivity of a nonshaly formation sample 100% saturated with brine, ohm m

ρ_w = resistivity of brine, ohm m

m = "cementation factor"

a = empirical constant

The so-called "Humble" formula (Winsauer *et al.*, 1952) has been widely used:

$$F = 0.62/\phi^{2.15}. \qquad (1.4.6b)$$

For more accurate work the values of a and m must be determined for the formation in question.

The readings of the sonic, density, and neutron logs depend, in addition to the porosity, also on the lithology—that is, if the formation consists of sandstone, limestone, dolomite, etc., on the shaliness and the presence of gas and light hydrocarbons. As the sonic, density, and neutron logs are affected to a different degree by the matrix composition and the presence of gas or light oils, a combination of several logs, usually in the form of cross plots, will give better porosity results (Schlumberger, 1972).

1.4.5. Permeability

Absolute permeabilities (i.e., permeabilities measured relative to a fluid with which the formation is 100% saturated) can in general be measured only on core samples in the lab. As even permeabilities measured on core samples taken from the same well often vary widely, and samples are available only on a minute fraction of the formation, constituting the reservoir, the serious problem faced in reservoir description is what permeabilities to assign to the vast majority of the reservoir. Simultaneously the question arises that, supposing there is a significant variation of permeability on a scale of a few centimeters, would it then be necessary for adequate characterization of the reservoir, to specify the permeability on a 3-D grid with 1-centimeter spacings? There does not appear to be data available on the basis of which even a qualified answer to this question could be given. The practice used in reservoir description follows the practice of production, which is controlled by wells. The custom has been to use average effective permeabilities, on the scale of well spacings, which are seldom less than 100 feet and are often much greater than that. This procedure is conditioned by the practice of obtaining permeabilities in the field by pressure transient testing of reservoirs by wells.

Single well testing uses mostly pressure build-up tests, involving shutting in a well and then monitoring the well pressure as a function of time, or constant-rate drawdown tests to determine permeabilities (actually permeability times formation thickness is determined directly). These tests indicate the value of an average effective permeability only in the vicinity of a well at the saturations existing there.

In pressure build-up analysis the well pressure is assumed to have stayed constant during the shut-in time had the well not been shut in.

Shutting-in the well has the same effect as injecting into the reservoir at the same rate as the production rate at the time of shut-in (= negative production). Consequently the pressure increase ΔP_q caused by a negative rate $-q$ acting for a time Δt is the same absolute value as the pressure drop for a constant rate drawdown (Slider, 1983), that is,

$$\Delta P_{\mathrm{q}} = \frac{0.141 q \mu}{kh} \left[\left(\frac{1}{2} \right) (\ln t_{\mathrm{D}} + 0.809) \right] + \Delta P_{\mathrm{skin}}, \qquad (1.4.7)$$

where the reduced time is

$$\Delta t_{\mathrm{D}} = \frac{6.33 k \, \Delta t}{\phi \mu c r_{\mathrm{w}}^2}, \qquad (1.4.8)$$

$$r_{\mathrm{w}} = \text{well radius, ft.}$$

Equation (1.4.7) applies only after Δt_{D} becomes greater than 100, and it can be applied only until the effect of shut-in reaches the outer drainage boundary (Slider, 1983).

Accordingly, the well pressure P_{w} during the time of shut-in is

$$P_{\mathrm{w}} = P_{\mathrm{wf}} + \frac{0.141 q \mu}{kh} \left[\left(\frac{1}{2} \right) (\ln \Delta t_{\mathrm{D}} + 0.809) \right] + \Delta P_{\mathrm{skin}}, \qquad (1.4.9)$$

as shown in Fig. 1.62.

$$P_{\mathrm{wf}} = \text{bottom hole pressure, flowing, psig.}$$

By combination of Eqs. (1.4.8) and (1.4.9) and changing to decadic logarithms, Eq. (1.4.9) may be written as follows (Slider, 1983):

$$P_{\mathrm{w}} = \text{constant} + \frac{0.1625 q \mu}{kh} \log \Delta t. \qquad (1.4.10)$$

Hence, a plot of well pressure versus the logarithm of shut-in time yields a

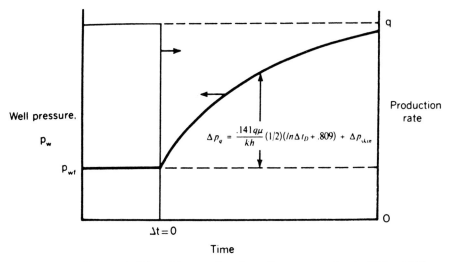

Figure 1.62. Pressure buildup with an unchanging well pressure at shut-in (Slider, 1983).

straight line with the slope

$$m = \frac{0.1625q\mu}{kh}, \qquad (1.4.11)$$

as illustrated in Fig. 1.63. From Eq. (1.4.11) the permeability k may be calculated. The plot illustrated in Fig. 1.63 is known as the Miller, Dyes, and Hutchinson (M.D.H.) pressure build-up plot (1950).

It is noted that the same slope m is obtained for the semilog plot of the well pressure versus producing time for a constant rate drawdown test.

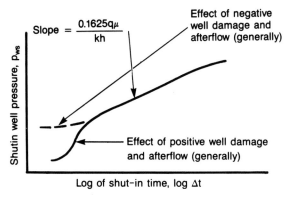

Figure 1.63. Pressure buildup for a well with an unchanging flowing pressure at shut-in (Slider, 1983).

Another widely used method of analyzing the pressure build-up data is the Horner plot, an excellent critical discussion of which is given in (Slider, 1983).

As opposed to single well tests, in multiple well tests an average value of permeability times thickness between wells is obtained. Using an "active" well, in which a disturbance (signal) is introduced, and an observation well, where the effect of disturbance is registered, amounts to an "interference" test. Usually the flow rate in the active well is changed for an extended period of time, which creates a pressure interference in the observation well.

Another method called "pulse" testing consists of a series of alternating flow and shut-in periods in the active well, causing a corresponding sequence of pressure pulses in the observation well. In this case analysis of the data is usually easier than in the single interference test (Slider, 1983).

According to reservoir engineering monographs (e.g., Crichlow, 1977; Greenkorn, 1983) single well tests are still the standard procedure to measure reservoir permeabilities.

1.4.6. Interpolation of Reservoir Rock Data

Formation elevation formation, thickness, porosity, and permeability data are all obtained at or in the vicinity of well bores (i.e., at discrete locations in the reservoir). In practice contouring computer programs are used to interpolate between these points and thus arrive at contour maps such as the isopach map (Fig. 1.58) already mentioned. The isopach map is prepared by using all the available geological information and the contour lines parallel the known geologic trends. This technique is used almost invariably when preparing formation elevation and formation thickness contour maps.

Isoporosity and isopermeability maps (see Fig. 1.64), on the other hand, are usually prepared by "mechanical" contouring, which means that the data is fitted, extrapolated, or interpolated between the known points by using some mathematical model. Most of the isoporosity and isopermeability contour maps are two-dimensional and do not attempt to take into account porosity and permeability variations in the vertical direction. It is however well known that many or most formations are stratified; that is, they contain layers of different permeabilities. It is not usually known how these layers are connected between the wells, where their presence has been established. It is generally admitted that it is an oversimplification to assume that stratified reservoirs consist of parallel layers of different permeability, like a layer cake. Another question is the degree of cross flow between the layers. If the layers are separated by shale streaks there is no cross flow. Often it has been found that the vertical permeability of a formation is several orders of magnitude less than the horizontal permeability. In such a case it follows that the layers can be treated as practically independent systems. If this were the case then the two-dimensional isopermeability maps should use permeabilities aver-

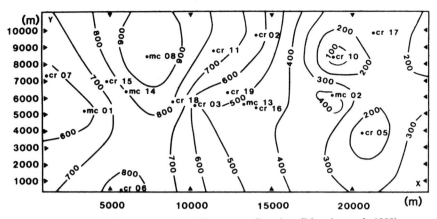

Figure 1.64. Reservoir permeability map at Germigny (Meunier *et al.*, 1989).

aged in the vertical direction, in some suitable manner. A good discussion of this problem is given in Craig (1971).

A very recent review (Raghavan, 1989) concludes by stating that "although layering is the most common form of reservoir heterogeneity, it is interesting to find that there is a paucity of information on layered reservoir performance." The same review also points out that "unfortunately the statistical concerns of characterizing variation in permeability with depth are often comingled with the geological aspects of the problem."

It is known (Slider, 1983) that if a very homogeneous appearing sandstone sample (e.g., a 6 inch long, 3 1/2 inch diameter core) is cut into as many 1

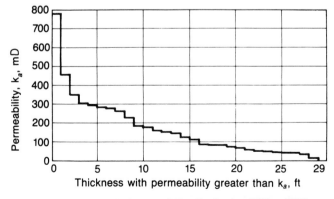

Figure 1.65. Typical permeability distribution (Slider, 1983).

inch long and 3/4 inch diameter permeability plugs as possible, the measured permeabilities will often vary by as much as 300%. In cores taken from a well bore the permeability variation is usually much greater than that. It is customary to plot the cumulative distribution of permeabilities as a function of formation thickness, without regard to the position of the sample in the well bore, as shown in Fig. 1.65 as an example. (For example, 10 feet of the formation has permeability greater than 190 mD.) The permeability distributions tend to be log normal.

1.4.7. Geostatistics

Over the past few decades a new and superior method of interpolating between the discrete values of rock properties obtained at well bores and showing irregular variations has been developed (Matheron, 1969; Matheron, 1971; Delfiner and Matheron, 1979, Renard, 1984; Davis and Culhane, 1984; Matheron, 1980; daCosta, 1985; Journel and Huijbregts, 1978). Details of this technique called "kriging" may be found in the references cited.

It is assumed that the property of the reservoir considered (e.g., permeability) that varies from point to point in the particular manner existing in the reservoir is a particular realization of a random function. In other words, it is assumed that there are an infinite number of other possible realizations of the same random function.

The simpler and more specialized case of kriging is "stationary kriging," when the mean and the higher moments of the property considered are stationary in space; that is, they are the same when calculated at any point in the medium. In other words, "stationarity" means that there is no discernible trend of, for example, the mean value of the property over the reservoir. The value of the property of course varies from point to point within the reservoir—only the mean is the same everywhere.

It is also assumed that the unique realization of the variation in space of the property available is characterized by the same probability density function (pdf) as the pdf of the random function, representing the property for all realizations. This property is called "ergodicity."

If the porosity and the permeability of a reservoir is a stationary and ergodic phenomenon then the reservoir may be called "homogeneous" in the stochastic sense of the word, even though these properties of the reservoir vary from point to point. This definition of homogeneity of a reservoir is probably the only one that any real reservoir is ever going to meet, to a reasonable approximation, because space invariant porosity and permeability do not exist, even approximately, in any known reservoir.

In the knowledge of the values of the property at a number of discrete points of known coordinates in the reservoir, kriging makes it possible to estimate the value of the property at any other point with a quantified confidence limit.

If $Z(x_i)$ are known values of the property at given points x_i in the neighborhood of point x, then the estimate of the true and unknown value of $Z(x)$, denoted by $Z^*(x)$, is formed by

$$Z^*(x) = \sum_i \lambda_i Z(x_i), \qquad (1.4.12)$$

where λ_i = kriging factors.

λ_i are found by two conditions, which are

(1) that the estimate is unbiased, i.e., the expected value $E[Z^*(x) - Z(x)] = 0$; and

(2) that the variance $\mathrm{var}[Z^*(x) - Z(x)]$ is minimum.

For stationary kriging $E[Z(x)] = m$, the mean value of the property of the reservoir is independent of x. For condition (1) we have

$$E\left[\sum_i \lambda_i Z(x_i) - Z(x)\right] = 0, \qquad (1.4.12a)$$

whence

$$\sum_i \lambda_i = 1. \qquad (1.4.13)$$

For condition (2) we want

$$E\left[\{Z^*(x) - Z(x)\}^2\right] = E\left[\left\{\sum_i \lambda_i Z(x_i) - Z(x)\right\}^2\right]$$

to be minimum.

The above expression may be written in terms of the so-called variogram:

$$\gamma(x_i - x_j) = \tfrac{1}{2}E\left[\{Z(x_i) - Z(x_j)\}^2\right] \qquad (1.4.14)$$

as follows

$$E\left[\{Z^*(x) - Z(x)\}^2\right] = -\sum_i \sum_j \lambda_i \lambda_j \gamma(x_i - x_j) + 2\sum_i \lambda_i \gamma(x_i - x)$$

$$(1.4.15)$$

Equation (1.4.15) is a quadratic function of the unknowns λ_i, which is minimized by the method of Lagrange multipliers, resulting in a linear system of $n + 1$ equations in terms of the variogram $\gamma_{ij} = \gamma(x_i - x_j)$, which can be solved for the $n + 1$ unknowns λ_i ($i = 1 \ldots n$) and μ the Lagrange multiplier.

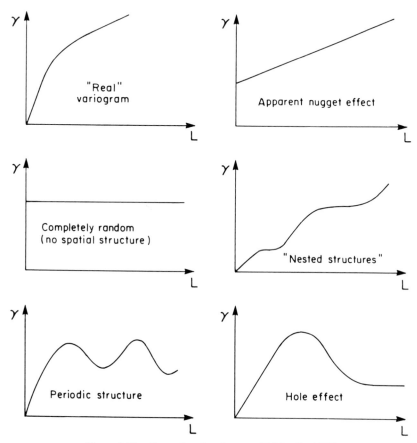

Figure 1.66.   Examples of variagrams (deMarsily, 1982).

It is apparent that the measured values of the property must be processed to obtain the variogram of the system. The variogram is calculated as follows. First a certain number of distance classes between the points of measurement are defined, the number of entries in each class is counted, and the arithmetical average distance corresponding to each class is computed. Finally the arithmetical average value of $\{Z(x_i) - Z(x_j)\}^2$ is calculated in each class. These values are plotted versus the average distance corresponding to each class.

A large number of different types of variograms are known that the diagnostic to the expert of the type of spatial structure of the property. For example when the property is distributed uniformly randomly in space such that the values of the property at different points are independent of each other, the variogram is a horizontal line. (See Fig. 1.66.)

When $Z(x_i)$ is a stationary random function, the variogram is related to the customary covariance C as follows:

$$\gamma(h) = C(0) - C(h)$$

where $h = x_i - x_j$ and $C(0) = \sigma^2 = E[Z(x)^2 - m^2]$, the variance.

The variance of the values estimated by kriging can be calculated in the usual manner of calculating variances.

When stationarity cannot be assumed, first the deterministic trend $m(x)$ is calculated by least squares and then this trend is suitably taken into consideration in making the estimates by what is called universal kriging. For details the reader is referred to the literature cited.

Kriging has not been widely used, until now, partly because of insufficient measured data being available in most cases. Some very recent publications (Meunier *et al.*, 1989; Carr *et al.*, 1989) show an increased interest in the application of this technique to reservoirs.

A very recent approach to the characterization of heterogeneous pore structure on the field scale is fractal geostatistics (Hewett, 1986; Emanuel *et al.*, 1989; Hewett and Behrens, 1990a; Hewett and Behrens, 1990b; Hewett and Behrens, 1991). The basic assumption of this approach is that geological properties, including porosity and permeability, exhibit fractal distributions which are characterized by a power-law-variogram model of the form

$$\gamma(h) = \gamma_0 h^{2H} \tag{1.4.16}$$

where $h = x_i - x_j$, $\gamma_0 = $ a characteristic variance scale at a reference-unit lag distance, and $H = $ the fractal codimension equal to the difference between the Euclidean dimension in which the distribution is described and the fractal dimension of the distribution. H values derived from measurements of topographic features of the earth's surface typically fall in the range $0.7 < H < 0.9$. $H = 0.5$ corresponds to complete lack of correlation. The information used in constructing the model is obtained from well logs. Permeability and porosity maps are constructed by stochastic (fractal) interpolation between the wells in the horizontal direction. Because reservoir description by this technique is not unique the technique is called "conditional simulation" of reservoir heterogeneity. Various realizations of the reservoir are screened by matching the predicted tracer breakthrough with available field data and the realization is selected that matches the data best. The technique looks promising for simulation of reservoir performance.

1.4.8. Other Rock Data

Additional rock data that are normally required of a reservoir are capillary pressure curves, relative permeability curves, rock compressibility data, and fluid saturations. These are usually either measured on core samples in the laboratory or determined by logging.

REFERENCES

Adler, P. M., Jacquin, C. G., and Quiblier, J. A. (1990). Paper accepted by *Physico Chemical Hydrodynamics.*
Amyx, J. W., Bass, D. M., Jr., and Whiting, R. L. (1960). "Petroleum Reservoir Engineering." McGraw-Hill, New York.
Arthur, G. (1954–55). *J. Inst. Metals* **83**, 329.
Bach, G. (1967). *In* "Stereology" (H. Elias, ed.). (Proc. Int. Congr. Stereol., 2nd), p. 174. Springer-Verlag, Berlin and New York.
Baker, D. J., and Morris, J. B. (1971). *Carbon* **9**, 687.
Barrer, R. M. (1958). *In* "The Structure and Properties of Porous Materials" (D. H. Everett and I. S. Stone, eds.), 6–34, Butterworths, London.
Barrer, R. M. (1967). *In* "The Solid–Gas Interface" (A. E. Flood, ed.), Vol. **II**, 557–609. Dekker, New York.
Barrer, R. M., and Barrie, J. A. (1952). *Proc. R. Soc. London Ser.* A **213**, 250.
Barrer, R. M., and Grove, D. M. (1951). *Trans. Faraday Soc.* **47**, 826.
Barrett, L. K., and Yust, C. S. (1970). *Metallography* **3**, 1.
Batra, V. K. (1973). Ph.D. Dissertation, Univ. of Waterloo, Canada.
Bear, J. (1972). "Dynamics of Fluids in Porous Media." American Elsevier, New York.
Beeson, C. M. (1950). *Trans. AIME* **189**, 313.
Bennion, D. W., and Griffiths, J. C. (1966). *Trans. AIME* **237**, 9.
Bohra, J. N. (1974). *In* "Pore Structure and Properties of Materials" (S. Modry and M. Svata, eds.). (*Proc. Int. Symp. RILEM/UPAC*, Sept. 18–21, 1973), Final Report, Part II, pp. C401–405. Academia, Prague.
Bonnemay, M., Bronoel, G., Levert, E., and Peslerbe, G. (1964). *J. Electrochem. Soc.* **111**, 265.
Bourbie, T., and Walls, J. (1982). *SPEJ* (October), 719.
Brace, W. F., Walsh, J. B., and Frangos, W. T. (1968). *J. Geophys. Res.* **73**, 2225.
Broekhoff, J. C. P., and de Boer, J. H. (1968b). *J. Catal.* **10**, 377.
Brooks, R. H., and Corey, A. T. (1964). "Hydraulic Properties of Porous Media." Hydrology Paper No. **3**, Colorado State University, Fort Collins, Colorado.
Brown, L. F., and Travis, B. J. (1983). *Chem. Eng. Sci.* **38**, No. 6, 843.
Butt, Yu. M., Kolbasov, V. M., Berlin, L. E., and Melnitsky, G. A. (1974). *In* "Pore Structure and Properties of Materials" (S. Modry and M. Svata, eds.). (*Proc. Int. Symp. RILEM/UPAC*, Prague, Sept. 18–21, 1973), Final Report, Part II, pp. C402–C413. Academia, Prague.
Cahn, J. A., and Fullman, R. L. (1956). *Trans. AIME* **206**, 610.
Cairns, S. S. (1961). "Introductory Topology." Ronald Press, New York.
Cardwell, W. T., Jr., and Parsons, R. L. (1945). *Trans. AIME* **169**, 34.
Carman, P. C. (1941). *Soil Sci.* **52**, 1.
Carr, L. A., Benteau, R. I., Corrigan, M. P., and Van Doorne (1989). *SPE Formation Evaluation* **4**, No. 3, 335.
Chang, K. S., and Dullien, F. A. L. (1976). *J. Microsc.* **108**, 61.
Chatzis, I., and Dullien, F. A. L. (1977). *J. Can. Pet. Technol.* **16**, 97.
Chatzis, I., and Dullien, F. A. L. (1981). *Powder Technol.* **29**, 117.
Chatzis, I., and Dullien, F. A. L. (1982). Revue de l'Institut Français du Pétrole, March/April, 183.
Chatzis, I., and Dullien, F. A. L. (1985). *Int. Chem. Eng.* **25**, 47.
Chatzis, I., and Kantzas, A. (1988). *Chem. Eng. Comm.* **69**, 191.
Chatzis, I., and Kantzas, A. (1988). *Chem. Eng. Comm.* **69**, 169.

Chekhovsky, Ju. V., Berlin, L. E., and Brusser, M. I. (1973). *In* "Pore Structure and Properties of Materials" (S. Modry and M. Svata, eds.). (*Proc. Int. Symp. RILEM/UPAC*, Prague, September 18–21, 1973), Preliminary Report, Part I, pp. B51–B71. Academia, Prague.

Chen, T., and Stagg, P. W. (1984). *Soc. Pet. Eng. J.* (December), 639.

Chiaroni, M., and Reverberi, A. (1974). *In* "Pore Structure and Properties of Materials" (S. Modry and M. Svata, eds.). (*Proc. Int. Symp. RILEM/UPAC*, Prague, September 18–21, 1973), Final Report, Part I, pp. B187–B195. Academia, Prague.

Clark, G. L., and Liu, C. H. (1957). *Anal. Chem.* **29**, 1539.

Clarke, J. T. (1964). *J. Phys. Chem.* **68**, 884.

Collepardi, M. (1973). *In* "Pore Structure and Properties of Materials" (S. Modry and M. Svata, eds.). (*Proc. Int. Symp. RILEM/UPAC*, Prague, September 18–21, 1973), Preliminary Report, Part I, pp. B25–B40. Academia, Prague.

Collins, R. E. (1961). "Flow of Fluids through Porous Materials." Von Nostrand-Reinhold, Princeton, New Jersey.

Corte, H. (1957). *In* "Fundamentals of Papermaking Fibres" (*Trans. Symp.*, Cambridge, 1957) (F. Bolam, ed.), p. 301. Technical Section of the British Paper and Board Makers Association, Kenley.

Craig, F. F. (1971). "The Reservoir Engineering Aspects of Waterflooding." Soc. of Petr. Eng. of AIME, Dallas, Texas.

Crichlow, H. B. (1977). "Modern Reservoir Engineering—A Simulation Approach." Prentice-Hall Inc.

Crisp, D. J., and Williams, R. (1971). *Mar. Biol.* **10**, 214.

Daccord, G., and Lenormand, R. (1987). *Nature* **325**, 41.

Dacey, J. R. (1967). *In* "The Solid–Gas Interface" (E. A. Flood, ed.), pp. 995–1024. Dekker, New York.

Da Costa, A. J. (1985). Paper SPE 14275 presented at the 1985 SPE Annual Technical Conference and Exhibition, Sept. 22–25, Las Vegas.

Davis, S. N. (1969). *In* "Flow through Porous Media" (R. J. M. de Wiest, ed.), pp. 54–90. Academic Press, New York.

Davis, M. W., and Colhane, P. G. (1984). *Proc. NATO/ASI Geostatistics for Natural Resources Characterization* (G. Very, ed.). Reidel, Dordrecht, The Netherlands, 539.

Debbas, S., and Rumpf, H. (1966). *Chem. Eng. Sci.* **21**, 583.

De Hoff, R. T. (1962). *Trans. AIME* **206**, 610.

De Hoff, R. T. (1983). *J. Microscopy* **131**, 259.

De Hoff, R. T., Aigeltinger, E. H., and Craig, K. R. (1972). *J. Microscopy* **95**, 65.

De Hoff, R. T., and Rhines, F. N. (1968). "Quantitative Microscopy." McGraw-Hill, New York.

Delfiner, P., and Matheron, G. (1979). "Les fonctions aléatoires intrinsèques d'ordre k." Ecole nationale Superieure des Mines de Paris, Fontainebleu/CIG.

de Marsily, G. (1982). "Mechanics of Fluid in Porous Media." *NATO/ASI Proceedings*, July 1982, University of Delaware.

Delvaux, P., Baes, E., and della Taille, M. (1974). *In* "Pore Structure and Properties of Materials" (*Proc. Int. Symp. RILEM/UPAC*, Prague, Sept. 18–21, 1973) (S. Modry and M. Svata, eds.), Final Report, Part I, pp. B175–B185. Academia, Prague.

Dhawan, G. K. (1972). Ph.D. Dissertation, Univ. of Waterloo, Canada.

Diamond, S. (1973). *In* "Pore Structure and Properties of Materials" (*Proc. Int. Symp. RILEM/UPAC*, Prague, September 18–21), (S. Modry and M. Svata, eds.), Preliminary Report, Part I, pp. B73–B87.

Diaz, C. E., Chatzis, I., and Dullien, F. A. L. (1987). *Transport in Porous Media.* **2**, 215.

Dickenson, J. M., and Shore, J. W. (1968). *Carbon* **6**, 937.

Dubinin, M. M. (1955). *Quat. Rev.* **9**, 101.

Dullien, F. A. L. (1981). *Powder Technol.* **29**, 100.

Dullien, F. A. L. (1991). Unpublished note.

Dullien, F. A. L., Rhodes, E., and Schroeter, S. (1969/1970). *Powder Technol.* **3**, 24.

Dullien, F. A. L., and Batra, V. K. (1970). *In* "Flow through Porous Media," pp. 2–30, *American Chemical Society*, Washington, D.C.

Dullien, F. A. L., and Mehta, P. N. (1971/1972). *Powder Technol.* **5**, 179.

Dullien, F. A. L., and Dhawan, G. K. (1974). *J. Interface Colloid Sci.* **47**, 337.

Dullien, F. A. L., and Dhawan, G. K. (1975). *J. Interface Colloid Sci.* **52**, 129.

Eijpe, R., and Weber, K. J. (1971). *Amer. Assoc. Pet. Geol. Bull.* **55**, No. 2, 307.

El-Sayed, M. S. (1978). Ph.D. Dissertation, Univ. of Waterloo, Canada.

El-Sayed, M. S., Dullien, F. A. L., and Batra, V. K. (1977). *J. Colloid and Interface Sci.* **60**, 497.

Edwards, J. M., and Simpson, A. L. (1955). *Pet. Trans. AIME* **204**, 132.

Emanuel, A. S., Alameda, G. K., Behrens, R. A., and Hewett, T. A. (1989). SPE Reservoir Engineering, Aug. 1989, 311.

Fatt, I. (1959). *Pet. Trans. AIME* **216**, 449.

Feder, J. (1988). "Fractals." Plenum, New York.

Felch, D. E., and Shuck, F. O. (1971). *Ind. Eng. Chem. Fundamentals* **10**, 299.

Feldman, R. F. (1973). *In* "Pore Structure and Properties of Materials" (*Proc. Int. Symp. RILEM/UPAC*, Prague, Sept. 18–21), (S. Modry and M. Svata, eds.), Preliminary Report, Part I, pp. C101–C115. Academia, Prague.

Fischmeister, H. F. (1974). *In* "Pore Structure and Properties of Materials" (*Proc. Int. Symp. RILEM/UPAC*, Prague, September 18–21, 1973), Part II, pp. C435–C476. Academia, Prague.

Foster, A. G., and Thorp, J. M. (1958). *In* "The Structure and Properties of Porous Materials" (D. H. Everett and I. S. Stone, eds.), pp. 227–260. Butterworths, London.

Freeman, D. L., and Bush, D. C. (1983). *SPEJ* (December), 928.

Graton, L. C., and Fraser, H. J. (1935). *J. Geol.* **43**, 785.

Greenkorn, R. A. (1983). "Flow Phenomena in Porous Media." Marcel Dekker, Inc.

Greenkorn, R. A., Johnson, C. R., and Shallenberger, L. K. (1964). *Trans. Soc. Pet. Eng.* **231**, 124.

Gusman, J. Ja. (1973). *In* "Pore Structure and Properties of Materials" (*Proc. Int. Symp. RILEM/UPAC*, Prague, September 18–21), (S. Modry and M. Svata, eds.), Preliminary Report, Part I, pp. B119–B135. Academia, Prague.

Haskett, S. E., Narahara, G. M., and Holditch, S. A. (1986). "A Method for the Simultaneous Determination of Permeability and Porosity in Low Permeability Cores." *PSE* 15379.

Haughey, D. P., and Beveridge, G. S. G. (1969). *Can. J. Chem. Eng.* **47**, 130.

Heesch, H., and Laves, F. (1933). *Z. Kristallogr.* **85**, 443.

Henriette, A., Jacquin, C. G., and Adler, P. M. (1989). *Physicochemical Hydrodynamics* **11**, No. 1, 63.

Hewett, T. A. (1986). "Tactal Distributions of Reservoir Heterogeneity and Their Influence on Fluid Transport." 61st Ann. Tech. Conf. of SPE, New Orleans, LA. Oct. 5–8, 1986.

Hewett, T. A., and Behrens, R. A. (1990). "Considerations Affecting the Scaling of Displacements in Heterogeneous Permeability Distributions." SPE 20739. 65th Ann. Tech. Conf. of SPE, New Orleans, LA, Sept. 23–26, 1990.

Hewett, T. A., and Behrens, R. A. (1990). SPE Formation Evaluation, Sept. 1990, 217.

Hewett, T. A., and Behrens, R. A. (1991) in "Reservoir Characterization II." (L. W. Lake, H. B. Carroll, Jr., and T. C. Wesson, eds.). Academic Press, 1991, 402.

Hewett, G. F. (1967). *In* "Porous Carbon Solids" (R. L. Bond, ed.), p. 203. Academic Press, New York.

Hilbert, D., and Cohn-Vossen, S. (1932). "Anschauliche Geometrie," Chapter 2. Springer, New York.

Hillard, J. E. (1967). *In* "Stereology" (*Int. Congr. Stereol.*, 2nd) (H. Elias, ed.), p. 211. Springer-Verlag, Berlin and New York.

Hilliard, J. E. (1968). *Trans. AIME* **242**, 1373.

Hirschhorn, J. S. (1969). "Introduction to Powder Metallurgy." American Powder Metallurgy Institute, New York.

Hirschhorn, J. S., and Roll, K. H. (eds.) (1970). "Advanced Experimental Techniques in Powder Metallurgy," Vol. 5, Perspectives in Powder Metallurgy. Plenum Press, New York.

Honald, E., and Skau, E. L. (1954). *Science* **120**, 805.

Honald, E., Boucher, R. E., and Skau, E. L. (1956). *Textile Res. J.* **26**, 263.

Horsfield, H. T. (1934). *J. Soc. Chem. Ind.* **53**, 107T.

Imelik, B., Weigel, D., and Renouprez, A. (1965). *J. Chem. Phys.* **62**, 125.

Ivensen, V. A. (1973). "Densification of Metal Powders During Sintering" (Translation by E. Renner). Consultants Bureau, New York.

Johnson, C. R., and Greenkorn, R. A. (1963). *Geol. Soc. Am. Proc.* **73**, 180.

Joshi, M. (1974). Ph.D. Thesis, Univ. of Kansas.

Journel, A., and Huijbregts, C. (1978). Mining Geostatistics, Academic Press.

Jung, M. (1973). *In* "Pore Structure and Properties of Materials" (*Proc. Int. Symp. RILEM/UPAC*, Prague, September 18–21), (S. Modry and M. Svata, eds.), Preliminary Report, Part I, pp. B89–B107. Academia, Prague.

Kanagy, J. R. (1963). *J. Am. Leather Chem. Assoc.* **58**, 524.

Katz, A. J., and Thompson, A. H. (1986). *Phys. Rev.* **B34**, 8179.

Kaufmann, P. M., Dullien, F. A. L., Macdonald, I. F., and Simpson, C. S. (1983). *Acta. Stereol.* **2**, (Suppl. 1), 145.

Kendall, M. G., and Moran, P. A. P. (1963). "Geometrical Probability," Hafner, New York.

Klinkenberg, L. J. (1941). *A.P.I. Drill Prod. Pract.*, 200.

Klinkenberg, L. J. (1957). *Petroleum Transactions*, *AIME* **210**, 286.

Kruyer, S. (1958). *Trans. Faraday Soc.* **54**, 1758.

Kwiecien, M. J. (1987). M.A.Sc. Thesis, University of Waterloo, Waterloo, Ontario.

Kwiecien, M. J., Macdonald, I. F., and Dullien, F. A. L. (1990). *J. Microsc.* **159**, 343.

Lenormand, R., Kalaydjian, F., Bieber, M. T., and Lombard, J. M. (1990). "Use of a Multifractal Approach for Multiphase Flow in Heterogeneous Porous Media: Comparison with CT-Scanning Experiments." *SPE* 20475. Presented at 65th SPE Conference, Sept. 23–26, New Orleans.

Levorsen, A. I. (1954). "Geology of Petroleum," Freeman, San Francisco, California.

Lin, C. (1983). *J. Math. Geology* **15**, 3.

Lin, C., and Cohen, M. H. (1982). *J. Appl. Phys.* **53**, 4152.

Lin, C., and Perry, M. J. (1982). IEEE Workshop on Computer Vision, 38.

Link, P. K. (1982). "Basic Petroleum Geology." OGCI Publications, Tulsa, Oklahoma.

Lord, G. W., and Willis, T. F. (1951). ASTM Bull. **177**, 56.

Macdonald, I. F., Kaufmann, P., and Dullien, F. A. L. (1986). *J. Microsc.* **144**, 297.

Macmullin, R. B., and Muccini, G. A. (1956). *AIChE J.* **2**, 393.

Mandelbrot, B. B. (1977). "Fractals: Form, Chance and Dimension." W. H. Freeman, San Francisco.

Mandelbrot, B. B. (1982). "The Fractal Geometry of Nature." W. H. Freeman, New York.

Mandelbrot, B. B. (1986). "Fractals in Physics," (Pietronero and Tosatti, eds.), North Holland, Amsterdam.

Mandelbrot, B. B. (1987). *In* "Encyclopedia of Physical Science and Technology" **5**, 579. Academic Press.

Masch, F. D., and Denny, K. J. (1966). *Water Resources Res.* **2**, 665.

Mason, G. (1971). *J. Coll. Interface Sci.* **35**, 279.

Matheron. G. (1969). "Le Kriegeage universal." Les Cahiers du Centre de Morphologie Mathematique, Fasc. 1, CG Fontainebleau.

Matheron, G. (1971). "The theory of regionalised variables." Les Cahiers du Centre de Morphologie Mathematique, Fasc-5, CG Fontainebleau.

Matheron, G. (1980). Splines and Kringing—Their Formation Equivalence. *In* "Down to Earth Statistics" (D. F. Merriam, ed.), Syracuse University, Geology Contributions, 77.

Matthews, J. F. (1974). *In* "Pore Structure and Properties of Materials" (*Proc. Int. Symp. RILEM/UPAC*, Prague, September 18–21, 1973), (S. Modry and M. Svata, eds.), Final Report, Part I, pp. A111–C119. Academia, Prague.

Mehta, P. N. (1969). *M.A. Sc. Dissertation*, Univ. of Waterloo, Canada.

Meunier, G., Coulomb, C., and Laille, J. P. (1989). *SPE Formation Evaluation* **4**, No. 3, 327.

Miller, G. C., Dye, S. A. B., and Hutchinson, C. A. (1950). *Pet. Trans. AIME* **189**, 91.

Modry, S., Svata, M., and Jindra, J. (1972). "Bibliography on Mercury Porosimetry." House of Technology, Prague.

Morrow, N. R. (1971). *Bull. A.A.P.G.* **55**, 514.

Morrow, N. R., and Graves, J. R. (1969). *Soil Sci.* **107**, 102.

Muskat, M. (1946). "Flow of Homogeneous Fluids Through Porous Systems," J. W. Edwards, Ann Arbor, Mich.

Newitt, D. M., and Conway-Jones, J. M. (1958). *Trans. Inst. Chem. Eng.* **36**, 422.

Odeh, A. S., and McMillen, J. M. (1972). *SPEJ* (October), 403.

Oman, A. O., and Watson, K. M. (1944). *Natl. Petrol. News* **36**, R795.

Pathak, P., Davis, T., and Scriven, L. (1982). "Dependence of Residual Nonwetting Liquid on Pore Topology." 57th Annual SPE, New Orleans.

Paulus, M. (1962). *Coros. Ind.* **37**, 448.

Paulus, M. (1963). *Corros. Ind.* **38**, 449.

Peaceman, D. W. (1972). "Fundamentals of Numerical Reservoir Simulation." Elsevier Scientific Publishing Co.

Pierce, C. I., Headley, L. C., and Sawyer, W. K. (1971). *Soc. Pet. Eng. J.* **11**, 223.

Plavnik, G. M., and Dubinin, M. M. (1966). *Izv. Akad. Nauk SSSR, Ser Khim*, 628.

Podvalny, A. M., and Prozenko, M. P. (1974). *In* "Pore Structure and Properties of Materials." (*Proc. Int. Symp. RILEM/UPAC*, Prague, September 18–21, 1973) (S. Modry and M. Svata, eds.), Final Report, Part I, p. A13–A31. Academia, Prague.

Preston, R. D. (1958). *In* "The Structure and Properties of Porous Materials" (D. H. Everett and F. S. Stone, eds.), Vol. X of Colston Papers, pp. 360–382. Butterworths, London.

Quiblier, J. A. (1984). *J. Colloid Interface Sci.* **98**, 84.

Quynn, R. G. (1963). *Textile Res. J.* **33**, 21.

Raghavan, R. (1989). *SPE Formation Evaluation* **4**, No. 2, 219.

Rance, H. F. (1958). *In* "The Structure and Properties of Porous Materials" (D. H. Everett and F. S. Stone, eds.), Vol. X of Colston Papers, pp. 302–321. Butterworths, London.

Ratinonov, V., Rozenberg, T., Kozlenko, T., Kucherjaeva, G., and Tokar, V. (1974). *In* "Pore Structure and Properties of Materials" (S. Modry and M. Svata, eds.). (*Proc. Int. Symp. RILEM/UPAC*, Prague, September 18–21, 1973), Final Report, Part I, p. B205–B215. Academia, Prague.

Renard, D. (1984). *Proc. NATO/ASI, Geostatistics for Natural Resources Characterization* (G. Very, ed.). Reidel, Dordrecht, The Netherlands, 679.

Rendell, M. J. (1963). *J. Ramsey Soc. Chem. Eng.* **10**, 31.

Ridgway, K., and Tarbuck, K. J. (1967). *Brit. Chem. Eng.* **12**, 384.

Ritter, H. L., and Erich, L. C. (1948). *Anal. Chem.* **20**, 665.

Romer, B. (1973). *In* "Pore Structure and Properties of Materials" (*Proc. Int. Symp. RILEM/UPAC*, Prague, September 18–21) (S. Modry and M. Svata, eds.). Preliminary Report, Part I, p. C187–C212. Academia, Prague.

Rosenberg, J. L., and Shombert, D. J. (1961). *J. Phys. Chem.* **65**, 2103.

Saltykov, S. A. (1958). "Stereometric Metallography," 2nd ed. Metallurgizdat, Moscow.

Saltykov, S. A. (1967). *In* "Stereology" (H. Elias, ed.) (*Proc. Int. Congr. Stereol.*, 2nd), p. 163. Springer-Verlag, Berlin and New York.

Scheidegger, A. E. (1974). "The Physics of Flow through Porous Media." 3rd ed. Univ. of Toronto Press, Toronto.

Scheil, E. (1935). *Z. Metallk.* **27**, 199.

Schlogl, R., and Schurig, H. (1961). *Z. Elektrochem.* **65**, 863.

Schlumberger, S. P. (1972). "Log Interpretation. Vol. I—Principles." Schlumberger Ltd., New York.

Schubert, H. (1972). Dr.-Ing. Dissertation, Univ. of Karlsruhe, Germany.

Schwartz, H. A. (1934). *Met. Alloys* **5**, 139.

Scott, G. D. (1962). *Nature* **194**, 956.

Serra, I. (1982). "Mathematical Morphology." Academic Press.

Slider, H. C. (1983). "Slip." Worldwide Practical Petroleum Reservoir Engineering Methods, Pennwell Publishing Co., Tulsa, Oklahoma.

Sneck, T., Kinnunen, L., and Oinonen, H. (1956). The State Inst. Tech. Res. Lab. Build. Technol., Interim Report Otaniemi.

Sohn, H. Y., and Moreland, C. (1968). *Can. J. Chem. Eng.* **46**, 162.

Spektor, A. G. (1950). *Zavod. Lab.* **16**, 173.

Srbek, F., Kunes, K., and Hanukyr, VI. (1973). *In* "Pore Structure and Properties of Materials." (*Proc. Int. Symp. RILEM/UPAC*, Prague, September 18–21) (S. Mohry and M. Svata, eds.), Preliminary Report, Part I, p. B137–B153. Academia, Prague.

Stayton, C. L., and Hart, C. A. (1965). *Forest Prod. J.* **15**, 435.

Stroeven, P. (1974). *In* "Pore Structure and Properties of Materials." (*Proc. Int. Symp. RILEM/UPAC*, Prague, September 18–21, 1973) (S. Modry and M. Svata, eds.), Final Report, Part II, p. C529–C543. Academia, Prague.

Stromberg, R. R. (1955). *J. Res. Nat. Bur. Std.* **54**, 73.

Turner, G. A. (1958). *Chem. Eng. Sci.* **7**, 156.

Turner, G. A. (1959). *Chem. Eng. Sci.* **10**, 14.

Turner, G. A. (1972). "Heat and Concentration Waves." Academic Press, New York.

Underwood, E. E. (1970). "Quantitative Stereology." Addison-Wesley, Reading, Massachusetts.

Underwood, E. E., de Wit, R., and Moore, G. A. (eds.) (1976). *Int. Congr. Stereol.*, 4th. National Bureau of Standards, Special Publ. 431, U.S. Government Printing Office, Washington, D.C.

Wadsworth, J. (1960). *National Research Council of Canada*, Mech. Eng. Rep. MT-41, February (NRC No. 5895).

Wadsworth, J. (1961). *Int. Develop. H.T.*, *ASME Conf.*, Colorado, p. 760.

Wagner, E. F. (1967). *Chemiefasern* **17**, 601.

Wakeham, H., and Spicer, N. (1949). *Textile Res. J.* **19**, 703.

Warren, J. E., and Price, H. S. (1961). *Trans. Soc. Pet. Eng. Am. Inst. Min. Met. Pet. Eng.* **222**, 153.

Weibel, E. R. (1976). *Int. Congr. Stereol.*, 4th (E. E. Underwood, R. de Wit, and G. A. Moore, eds.), pp. 341–350. National Bureau of Standards, Special Publ. 431, U.S. Government Printing Office, Washington, D.C.

Weibel, E. R. (1979). "Stereological Methods." Vol. I. Academic Press.

Weibel, E. R. (1980). "Stereological Methods." Vol. II. Academic Press.

Wicksell, S. D. (1925). *Biometrika* **17**, 84.

Wicksell, S. D. (1926). *Biometrika* **18**, 151.

Winsauer, W. A., Shearin, H. M., Masson, P. H., and Williams, M. (1952). *AAPG Bulletin* **36**, No. 2.

Wyllie, M. R. J. (1963). "The Fundamentals of Well Log Interpretation," 3rd ed. Academic Press.

Wynne-Jones, W. F. K. (1958). *In* "The Structure and Properties of Porous Materials." (D. H. Everett and I. S. Stone, eds.), pp. 35–58. Butterworths, London.

Yanuka, M., Dullien, F. A. L., and Elrick, D. E. (1986). *J. Colloid Interface Sci.* **112**, 24.

Yarnton, D., and Simpson, G. R. (1961). *Powder Met.* **42**.

Young, J. F. (1974). *In* "Pore Structure and Properties of Materials." (S. Modry and M. Svata, eds.) (*Proc. Int. Symp. RILEM/UPAC*, Prague, September 18–21, 1973), Final Report, Part I, p. B197–B203. Academia, Prague.

Zwietering, P., and van Krevelen, D. W. (1954). Fuel **33**, 331.

2 | *Capillarity in Porous Media*

In a "capillary system" the surfaces separating the various bulk phases play a significant part in determining the physical–chemical state of the system. According to this definition, the vast majority of porous media are capillary systems. In the case of capillary systems, mechanical equilibrium (i.e., the absence of a net mechanical force acting on the system) is determined not only by the hydrostatic pressure and gravitational attraction but also by forces associated with surface tension.

In this chapter, unless stated otherwise, the discussion is limited to the study of systems in mechanical equilibrium in which the temperature is uniform. They need not, however, be in a state of thermodynamic equilibrium. For example, porous media systems containing "residual" wetting phase or nonwetting saturations are not, in general, in a state of thermodynamic equilibrium. The question of thermodynamic equilibrium is discussed in greater detail in Sections 2.3 and 2.4.

First, the conditions for mechanical equilibrium are studied, and this leads to the definition of capillary pressure. Subsequently, this important parameter is applied to the characterization of the state of porous media containing two or more fluids. Then, the thermodynamics of quasi-static displacement of a fluid by another immiscible fluid is reviewed. The subject of modeling capillary pressure curves with the help of models of pore structure is discussed next.

2.1. SURFACE BETWEEN TWO FLUIDS

When two fluids, for example, a liquid and a gas, are in contact with each other, they are in general separated by a thin layer called the *interface*, *interphase*, or *surface phase*. From a mechanical standpoint, the system behaves as if it consisted of two homogeneous fluids separated by a uniformly stretched membrane of infinitesimal thickness.

2.1.1. Surface Tension

The existence of a tension in a surface or in an interface has been witnessed daily by most people when making a lather with soap or detergent, when wiping up water with a paper towel, etc. The following discussion of the surface tension follows, more or less, the presentation by Defay and Prigogine (1966).

Possibly the most common direct evidence for the existence of tension in an interface is the tendency of a soap bubble to spontaneously decrease its size. This experiment shows that the surface of the soap bubble is in a state of tension, i.e., it behaves as if it were stretched, and if unopposed by any force, the stress or tension is relieved by a continuous decrease in the size of the bubble. The stress causing this decrease is called *surface* or *interfacial tension*. To define the tension in a surface at a point P, let us imagine an arbitrary curve AB in the surface, passing through this point and dividing the surface into two regions as shown in Fig. 2.1. Across an element $\delta\ell$ of AB, region 2 exerts a force $\sigma\,\delta\ell$ tangential to the surface, where σ is the surface or interfacial tension at this point.

With reference to a bubble or a drop, we can consider a spherical cap whose section is shown in Fig. 2.2, which is subjected to a surface tension σ around the base of the cap and to normal pressures P' and P'' at each point on the surface. As is evident from both the previous discussion and this

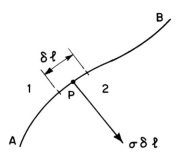

Figure 2.1. Definition of surface tension at a point P on a line AB in the surface (Prigogine and Defay, 1966).

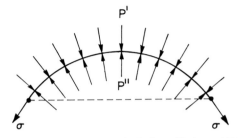

Figure 2.2. Capillary equilibrium of a spherical cap (Defay and Prigogine, 1966).

figure, the effect of the surface tension σ is to reduce the size of the sphere unless it is opposed by a sufficiently great difference between pressures P'' and P'.

The surface forces, as shown in the figure, are exerted by the rest of the sphere on the cap, which they pull downward. An equal and opposite force (not shown in the figure) is exerted by the cap on the rest of the sphere, pulling that upward.

The surface is said to be in a state of *uniform* tension if: (a) at each point, σ is perpendicular to the dividing line and has the same value whatever the direction of this line, and (b) σ has the same value at all points on the surface. In this case, σ can be called the *surface* (or interfacial) *tension* of the surface: It has the dimension of force per unit length and is usually expressed in units of dynes per centimeter or, equivalently, mN/m. A more extensive discussion of the physical meaning of surface tension is given by Defay and Prigogine (1966).

2.1.2. Condition of Mechanical Equilibrium of a Surface, Laplace's Equation

Let us consider an arbitrary smooth surface of which the spherical cap shown in Fig. 2.2 is a special case (see Fig. 2.3). The effect of gravity is neglected in this analysis.

Consider a point P on the surface, and draw on the surface a curve whose distance from P along the surface is a constant ρ. This curve forms the boundary of a cap, whose condition of equilibrium we shall find as ρ tends to zero. Let us draw through P an arbitrary pair of orthogonal lines AB and CD on the surface; let their radii of curvature at P be R_1 and R_2. According to a theorem of Euler (see, e.g., Weatherburn, 1947)

$$(1/r_1) + (1/r_2) = (1/R_1) + (1/R_2), \qquad (2.1.1)$$

where r_1 and r_2 are the principal radii of curvature.

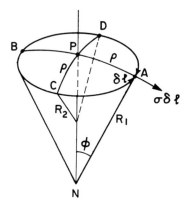

Figure 2.3.   Capillary equilibrium of a nonspherical cap (Defay and Prigogine, 1966).

At point A, say, an element $\delta\ell$ of the boundary line is subjected to a force $\sigma\,\delta\ell$ whose projection along the normal PN is

$$\sigma\,\delta\ell\,\sin\phi = \sigma\phi\,\delta\ell = \sigma(\rho/R_1)\,\delta\ell \qquad (2.1.2)$$

since ϕ is supposed to be very small.

Let us consider four elements $\delta\ell$ of the periphery at A, B, C, and D. These will contribute the following force:

$$\sigma\,\delta\ell\,[2(\rho/R_1) + 2(\rho/R_2)] = 2\rho\sigma\,\delta\ell\,[(1/r_1) + (1/r_2)]. \qquad (2.1.3)$$

Since this expression is independent of the choice of AB and CD, it can be integrated around the whole circumference. As four orthogonal elements have been considered, the integration has to be made over one-quarter of a revolution:

$$\pi\rho^2\sigma[(1/r_1) + (1/r_2)]. \qquad (2.1.4)$$

For mechanical equilibrium of the surface, this force is to be balanced exactly by the pressure forces and, therefore, we must have

$$(P'' - P')\pi\rho^2 = \pi\rho^2\sigma[(1/r_1) + (1/r_2)], \qquad (2.1.5)$$

which reduces immediately to Laplace's equation

$$P'' - P' = \sigma[(1/r_1) + (1/r_2)]. \qquad (2.1.6)$$

It is customary to introduce the mean radius of curvature r_m defined by

$$1/r_m = \tfrac{1}{2}[(1/r_1) + (1/r_2)], \qquad (2.1.7)$$

whence Laplace's equation becomes

$$P'' - P' = 2\sigma/r_m. \qquad (2.1.8)$$

Laplace's equation shows that because of the existence of surface tension an arbitrary surface of mean radius of curvature r_m maintains mechanical equilibrium between two fluids at different pressures P'' and P'. The phase on the concave side of the surface must have a pressure P'' that is greater than the pressure P', on the convex side.

There exist surfaces with a waist where the two principal radii of curvature r_1 and r_2 are of opposite signs. In other words, a straight section along r_1, say, will show fluid at pressure P'', say, on the concave side (and fluid at pressure P' on the convex side), whereas an orthogonal straight section along r_2 will show fluid at pressure P'' on the convex side, and vice versa. According to Eq. (2.1.6) $P'' > P'$ for $r_1 > 0$ and $r_2 < 0$ if the absolute value of r_1 is less than the absolute value of r_2. For the sake of simplicity, in this monograph that side of the surface where the equilibrium pressure is greater will always be referred to as the "concave" side.

The special case of $r_m = \infty$ implies $P'' = P'$. Hence, we see that a plain surface can exist only if the pressures of the fluids on the two sides are equal.

In the absence of gravitational effects, the pressures P' and P'' are uniform throughout the respective phases, while the surface tension has the same value σ at all points in the surface. It follows, therefore, from Eq. (2.1.8) that the mean radius of curvature r_m must be the same at all points in the surface. Hence, the only surfaces that have to be considered in dealing with systems where gravitational effects can be neglected are those of constant mean curvature.

Gravitational forces, however, play an important role in determining the shape of macroscopic surfaces. To include the effects of gravity one simply has to take into account the fact that P'' and P' vary with the height of the point considered. Laplace's equation makes no assumption regarding the variation of P'' and P' with height. It merely relates the values P'', P', r_m, and σ at a given point in the surface. Hence, Laplace's equation is valid at every point in the surface whether or not gravitational forces are negligible, provided only that we may neglect the weight of the surface phase itself.

Because of the variation of P'' and P', gravitational forces cause only a variation of the radius r_m with height. Thus, the surface of a liquid in a wide vessel is plain far from the walls but curves over and has a nonzero curvature near the walls. The shape of a drop resting on a surface or hanging from the end of a capillary tube is also determined by similar considerations. The procedure used to take the effect of gravity into account is discussed at some length by Adamson (1967) in conjunction with a variety of methods for measurement of surface tension, including "capillary rise," "maximum bubble pressure," "drop weight," "ring," "Wilhelmy slide," "sessile drop or bubble," and "pendant drop" methods.

2.2. CAPILLARY EQUILIBRIUM IN THREE-PHASE SYSTEMS

2.2.1. Equilibrium at a Line of Contact, Young's Equation

Capillary systems in porous media always involve a solid phase and at least two fluid phases. In the presence of three fluid phases there are surfaces subjected to surface tension that meet at a point, as shown diagrammatically in Fig. 2.4. This figure represents a normal section of the intersection of the surfaces 1, 2, and 3 along a line through point P and normal to the diagram. Surface 1 separates phases I and III; surface 2 separates I and II; and surface 3 separates II and III. An element of the line through P of length $\delta \ell$ is subjected to the three forces $\sigma_1 \delta \ell, \sigma_2 \delta \ell$, and $\sigma_3 \delta \ell$. If these forces are in equilibrium, the vectorial equation

$$\boldsymbol{\sigma}_1 + \boldsymbol{\sigma}_2 + \boldsymbol{\sigma}_3 = 0 \qquad (2.2.1)$$

must be satisfied. (The forces are drawn consistently so that they are directed toward the interior of the phase where the vector originates. The opposite convection, if used consistently, would be equally satisfactory.) This equation is often called the *law of Neumann's triangle*. Measurements have so far verified the validity of the Neumann triangle law within experimental error. (See Fuchs, 1930 and Miller, 1941.)

Let us consider a drop of liquid placed on a smooth, plain solid surface as shown in Fig. 2.5. Depending on the conditions that will be discussed below, liquid may remain a drop displaying a finite angle of contact θ between the two boundaries: liquid/gas and solid/liquid. The contact angle θ is defined as the angle subtended by the tangent to the liquid/gas boundary constructed at a point on the three-phase line of contact and the tangent to the solid/liquid boundary constructed at the same point. Evidently, if there is equilibrium then Eq. (2.2.1) applies here and may be written also in compo-

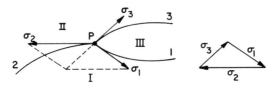

Figure 2.4. Equilibrium at a line of contact: Neumann's triangle (Defay and Prigogine, 1966).

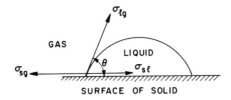

Figure 2.5. Equilibrium at a line of contact (Craig, 1971).

nent form. In particular, for the force components parallel to the solid surface one obtains

$$\sigma_{\ell g} \cos \theta = \sigma_{sg} - \sigma_{s\ell}. \qquad (2.2.2)$$

where $\sigma_{\ell g}$ and σ_{sg} are the surface tensions of liquid and solid, respectively, and $\sigma_{s\ell}$ is the interfacial tension between the liquid and the solid. This equation was first given by Young for a liquid drop resting on a solid surface in air and relates the contact angle with the surface tensions of three interfaces.

The value of the contact angle θ may lie anywhere between 0 and 180°. It is evident from Eq. (2.2.2) that for the range $0° \le \theta° \le 90°$ $\sigma_{\ell g} \cos \theta = \sigma_{sg} - \sigma_{s\ell} \ge 0$, whereas for the range $90° < \theta \le 180°$ $\sigma_{\ell g} \cos \theta = \sigma_{sg} - \sigma_{s\ell} < 0$ must hold. The particular value $\theta = 90°$ requires that $\sigma_{s\ell} = \sigma_{sg}$. The other special case $\theta = 0°$ is of great practical significance because that is the limit where the liquid will start to spread on the solid. The condition for $\theta = 0°$ by Eq. (2.2.2) is $\sigma_{\ell g} = \sigma_{sg} - \sigma_{s\ell}$. As the difference $(\sigma_{sg} - \sigma_{s\ell})$ increases from 0 to $\sigma_{\ell g}$, θ decreases from 90° to 0°. There are also systems where

$$\sigma_{sg} - \sigma_{s\ell} > \sigma_{\ell g} \qquad (2.2.3)$$

to which no equflibrium position of the line of contact can correspond because θ, for physical reasons, cannot be less than 0°. In such instances liquid will spread on the solid, and the driving force of spreading is greater the greater the value of the spreading coefficient $S_{\ell s}$, defined as

$$S_{\ell s} \equiv \sigma_{sg} - \sigma_{s\ell} - \sigma_{\ell g}. \qquad (2.2.4)$$

There is spontaneous spreading of liquid on the surface of the solid whenever $S_{\ell s} \ge 0$.

It is noted that because of the absence of mobility of the surface of a solid, a solid–fluid interface does not assume a curvature in accordance with Laplace's equation as a fluid–fluid interface does. It is not, in general, possible to make direct experimental determination of the solid–fluid interfacial tension. Nevertheless, there are many considerations that lead us to attribute surface tension to solids (see, for example, Defay and Prigogine,

1966), and there are ways to estimate the magnitude of this tension. Evidently, the existence of solid–liquid surface tension has to be invoked when discussing the forces acting on the line of contact between a solid phase and two-fluid phases.

Equations (2.2.2) to (2.2.4) apply equally to situations where there is a film of another liquid (2), which is immiscible with the first liquid, on the solid surface and the drop of liquid (1) rests on this film. In this case the solid s must be replaced by liquid (2) ℓ_2 in all these equations. For a stable film of liquid (2) to exist on the solid surface, the spreading coefficient of liquid (2) on the solid in air, $S_{\ell_2 s}$, must be positive.

The equilibrium contact angle θ of a drop of liquid (1) placed on a film of liquid (2) covering the solid surface may assume any value between 0° and 180°, and liquid (1) may also spread spontaneously on the surface of liquid (2), provided that the spreading coefficient of liquid (2) on the surface of liquid (1) is positive. For example, most mineral oils spread on the surface of water.

Often liquid (1) is contacted with a solid surface covered with bulk liquid (2), as opposed to only a film of liquid (2). In this case Eqs. (2.2.2) to (2.2.4) apply, too, with the substitutions of ℓ_2 for g. In such instances the two liquids compete with each other for contact with the solid surface. After reaching mechanical equilibrium, the liquid through which $0° \leq \theta < 90°$ is usually referred to as the (preferentially) wetting liquid and the other one, through which $90° \leq \theta \leq 180°$, as the nonwetting liquid. Logically, the special case $\theta = 90°$ would correspond to "intermediate or neutral wettability," that is, when neither of the two fluids is either wetting or nonwetting preferentially the solid surface. It will be shown in Section 2.3.3 that "intermediate or neutral wettability" in porous media in reality corresponds to a range of contact angles about 90°, roughly $75° \leq \theta \leq 105°$ (Anderson, 1986). In the case of two liquids in contact with a solid surface, there are also instances when one of the liquids spreads on the solid and denies the other liquid any access to the solid.

If the surface of the solid is rough, Young's equation is applicable to each element of the surface. The observed angle between the liquid–gas surface and the apparent surface of the solid will then vary with the position of the line of contact as shown schematically in Fig. 2.6 (Popiel, 1978). The minimum scale of surface roughness required for this effect to become valid needs investigating. For a rough surface, Wenzel (1936, 1949) has suggested a modified form of Young's equation:

$$\sigma_{\ell g} \cos \theta_a = r(\sigma_{sg} - \sigma_{s\ell}), \tag{2.2.5}$$

where θ_a is the average "apparent" contact angle and r is the ratio of the true-to-apparent area of the solid. Shuttleworth and Bailey (1948) and Good (1952) have given thermodynamic justification of Wenzel's formula.

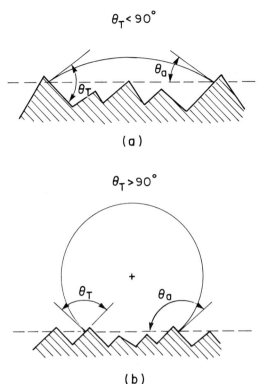

Figure 2.6. Effects of surface roughness on apparent contact angle θ_a. θ_T is the contact angle measured on a smooth, flat surface. (a) The droplet is the preferentially wetting fluid, so $\theta_T < 90°$. (b) The droplet is the nonwetting fluid, so $\theta_T > 90°$. (Popiel, 1978.)

As pointed out by Morrow (1970), among others, Young's equation may not be appropriate in the preceding form when the solid surfaces are deformable or when the solid is partially soluble in either of the fluid phases.

2.2.2. Contact Angle Hysteresis

A rather serious practical problem in conjunction with the concept of contact angle is the fact that the advancing angle (i.e., the contact angle measured through the fluid that is displacing another fluid) is often found to be significantly larger than the receding one (i.e., the contact angle measured through the same fluid when it is being displaced). Therefore, contact angle measurement is very difficult and is fraught with uncertainty. "Hysteresis" is used here to designate dependence of the contact angle on the previous

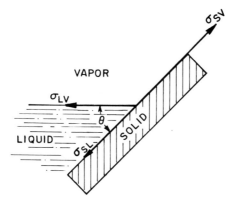

Figure 2.7.   The tilting plate method of measuring contact angle.

history of the system and, in particular, dependence of the contact angle on the direction of displacement of the three-phase line of contact. A sufficient reason for a difference between advancing and receding contact angles is that the properties of the solid surface (e.g., σ_{sg} in Eq. 2.2.2) may be altered by the fluids contacting it.

According to Adamson (1967), there appear to be three types of causes of contact angle hysteresis. The first is contamination of either the liquid or the solid surface. Rigorous cleaning is an absolute must for reliable contact angle determinations. Second, surface roughness causes contact angle hysteresis. The third cause appears to be surface immobility on a macromolecular scale. For example, in the case of a liquid and a solid surface, it is necessary that the adsorbed film of vapor be mobile. Where the liquid contains a surfactant, low mobility can again cause hysteresis.

Some of the methods of contact angle measurement are outlined by Adamson (1967). These include the tilting plate method, illustrated in Fig. 2.7, the use of sessile drops and bubbles (Fig. 2.8), etc.

Craig (1971) describes the contact angle cell used to determine the contact angle in an oil–water–solids system as a function of oil–solid interface age. Flat, polished crystals of the mineral that is predominant in the rock surfaces are immersed in a sample of the "formation water." A drop of "reservoir oil" is placed between two crystals as shown in Fig. 2.9. At first there is a water film between the oil and the crystal, and the initial contact angle measured is, in general, a nonequilibrium value. Only if the equilibrium contact angle measured through water is 0°, or if the spreading coefficient of water on the crystal in oil is positive, will the water film remain indefinitely between the oil and the crystal. Otherwise the water film is under the influence of forces that tend to eliminate it, and in equilibrium there is no water film left. In order to

SESSILE DROPS

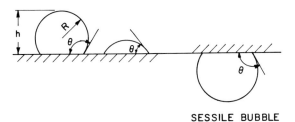

SESSILE BUBBLE

Figure 2.8. Sessile drops and bubble.

accelerate the equilibration process the plates are subsequently displaced relative to each other so that the oil is placed under a stress, forcing it to move from its original location. At those points of the perimeter where the oil drop retreats and the water advances gradually, contact is established between the oil and the crystal.

The contact angle at the surface freshly exposed to water, that is, the "water-advancing contact angle," is measured as a function of time. Figure 2.10 shows that the water-advancing contact angle increases with the age of the oil–solid interface, until finally an equilibrium value of the contact angle is reached. Frequently hundreds or even thousands of hours of aging time are required before equilibrium is attained. The great change in contact angle from apparently water-wet to the actual oil-wet condition is striking and

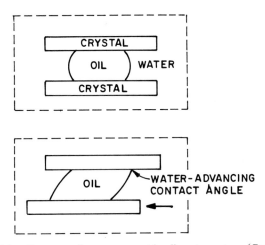

Figure 2.9. Contact angle measurement in oil–water systems (Craig, 1971).

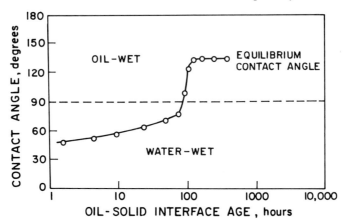

Figure 2.10. Approach to equilibrium contact angle (Craig, 1971).

should be heeded as a reminder that the measured value of the contact angle may depend strongly on the time of exposure of the solid to both fluids. A more detailed description of the contact angle cell method can be found in the work of Wagner and Leach (1959). More recent methods of contact angle hysteresis measurement include the dynamic Wilhelmy plate technique (Anderson *et al.*, 1988).

2.3. CAPILLARY PRESSURE

Capillary pressure is a basic parameter in the study of the behavior of porous media containing two or more immiscible fluid phases. It relates the pressures in the two fluid phases. In this section, first, the definition of capillary pressure is tackled, and then the principles of experimental measurement and the properties of the capillary pressure versus saturation relationship are described.

2.3.1. Definition and Principles of Measurement

Let us first consider a single circular capillary (pore) with variations in its cross section as shown in Fig. 2.11 (see also Fig. 2.15). Let us suppose that initially the capillary was filled with a wetting liquid and subsequently it was connected to a cylinder–piston arrangement containing a nonwetting fluid at atmospheric pressure. Assuming that the pressure inside the wetting liquid is also atmospheric, the nonwetting fluid will not spontaneously penetrate and displace the wetting fluid. In order to achieve any penetration, the pressure

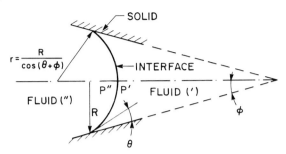

Figure 2.11. Menisci in a conical capillary.

in the nonwetting fluid has to be increased above atmospheric pressure to a value defined by the following equation:

$$P_c = P'' - P' = (2\sigma/R)|\cos(\theta + \phi)|, \qquad (2.3.1)$$

which follows directly from Laplace's equation and the geometry of the menisci shown in Fig. 2.11 ($R/|\cos(\theta + \phi)|$ is the mean radius of curvature of the meniscus). (The effect of the orientation of the solid surfaces, expressed by the angle ϕ, on the radius of curvature $R/|\cos(\theta + \phi)|$ may be visualized by imagining that the solid boundary lines are tilted over a wide range of values of ϕ.) The penetration of a nonwetting phase takes place in the direction of decreasing pore diameter. The pressure difference between the concave and the convex sides of the meniscus is synonymous with the pressure difference between the nonwetting phase and the wetting phase. In this monograph this pressure difference *is defined as the capillary pressure.*

It follows from this that P_c is a positive definite quantity: It is the pressure increment between the concave side and the convex side of a meniscus. In the petroleum literature it is widespread custom to define as capillary pressure the pressure difference between the oil and the water. As a result, in an oil-wet pore the capillary pressure will be negative, according to this definition.

It also follows from Fig. 2.11 that for $\theta + \phi > 90°$ the sign (direction) of curvature of the meniscus will change. Logically, whenever the sign (direction) of curvature of meniscus changes there is also an exchange of roles of the two fluids as regards preferential wettability, which is thus seen to depend also on the orientation of the surface, as well as the intrinsic contact angle.

It is apparent from the diagram that as the applied capillary pressure is gradually increased, farther penetration of the nonwetting phase into the capillary is achieved until the local minimum pore radius, that is, the pore throat, has been reached. The capillary pressure at the throat exceeds the equilibrium value in the subsequent expanding portion of the pore. Therefore, the penetration into the expanding portion of the pore occurs in a nonequilibrium manner; in other words, there is a finite net force present throughout, which drives the wetting phase out. Nonequilibrium penetration will continue until the meniscus reaches a throat that is narrower than the last one. At that point the applied capillary pressure must be increased further in order to penetrate this throat, and so forth. The process in which the nonwetting phase displaces the wetting phase is usually called "drainage," "desaturation," or "de-wetting."

If the displacement process is to be carried out in the reverse direction where now the wetting phase is displacing the nonwetting phase, then the capillary pressure P_c, given by Eq. (2.3.1), must be decreased. If the wetting phase was displaced, as described in the previous paragraph, by increasing the pressure on the nonwetting phase, whereas the wetting phase was kept at constant (e.g., atmospheric) pressure, then in the reverse process, the pressure on the nonwetting phase must be gradually decreased. In this case, with decreasing capillary pressure the meniscus in Fig. 2.11 will move to the left until the locally widest pore cross section, that is, the one with the greatest mean radius R, has been reached. The capillary pressure at this point is a local minimum; the pressure on the nonwetting phase is less than necessary for emptying the convergent narrower portion of the pore lying farther to the left of this point. As a result, this portion will be empty of the nonwetting phase in a nonequilibrium manner and the meniscus will keep moving to the left until an even wider pore cross section is reached. Further emptying of the pore will take place only if the pressure applied on the nonwetting phase P_c is further reduced. This process is usually called "spontaneous imbibition" or wetting.

An alternative way to carry out imbibition is to first fill the pore completely with the nonwetting phase at atmospheric pressure, for example, by evacuating the pore, forcing the nonwetting fluid into it at high pressure, and then exposing both ends of the pore to atmospheric pressure. (It should be born in mind that the nonwetting phase is "nonwetting" only *relative* to certain other fluids that wet the solid surface preferentially. So, for example, mineral oil is normally a nonwetting phase *relative* to water on clean glass or silica surfaces, but it is definitely a wetting phase *relative* to air on the same surfaces.) If the pore filled with the nonwetting fluid at atmospheric pressure is now contacted at the smaller end with a preferentially wetting fluid that is also at atmospheric pressure, a meniscus like the one shown in Fig. 2.11 will be formed; on the right-hand side the pressure will be less than atmospheric and, as a result, the meniscus will advance spontaneously to the left. There will be a pressure drop in the wetting fluid between the right-hand side of the meniscus and the bulk fluid where the pressure is atmospheric. This process is called "free spontaneous" or "uncontrolled" imbibition. In uncontrolled

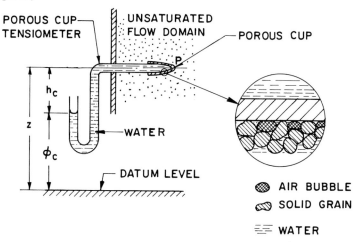

Figure 2.12. Porous cup tensiometer (Bear, 1972).

imbibition the process of displacing the nonwetting fluid from the pore goes to completion without any outside interference by the experimeter.

The capillary pressure is determined experimentally by measuring the difference between the pressures in the two phases at the point of interest. The device called "porous cup tensiometer" has been used (see Fig. 2.12) to measure capillary pressure in soils for a long time. The water-wet porous cup is permeable to water but impermeable to air when the pores are filled with water. The porous cup is an example of the widely used device called "semipermeable membrane," "capillary barrier," or "porous plate," in core analysis. The porous cup tensiometer measures the pressure difference in the wetting phase (water) between the microscopic equilibrium menisci in the soil and a flat bulk water surface in the other branch of the manometer, where the pressure is atmospheric. This pressure difference is equal to the capillary pressure, provided that the nonwetting phase (air bubbles) is also at atmospheric pressure. The reading of the manometer is called the "capillary pressure head," "moisture tension," or "suction head," defined as follows:

$$h_c = P_c/\rho_w g = -(P_w/\rho_w g) \qquad (2.3.2)$$

where P_w is the gauge pressure of water in the soil ($P_w < 0$), ρ_w is the density of water, and g is the gravitational acceleration.

It is customary to define a "piezometric head" or "capillary head" as

$$\phi_c = z + (P_w/\rho_w g) = z - [P_c/\rho_w g] = z - h_c. \qquad (2.3.3)$$

The same principle of capillary pressure measurement can be used in laboratory core samples containing water and oil. In this case two probes are

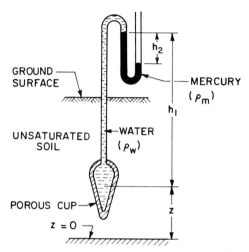

Figure 2.13. Measurement of capillary pressure in the field (Bear, 1972).

inserted into the side of core, one of which is equipped with a water-wet and the other with an oil-wet semipermeable membrane. Pressure transducers are used instead of U-tube manometers to measure the pressure. The probe equipped the oil-wet membrane is filled completely with oil and the other one, with water. In this set-up there is no possibility of the manometer fluid flowing into or out of the sample and thereby altering the original saturation conditions. The capillary pressure is simply the difference between the higher one and the lower one of the two pressure readings.

Often the symbol ψ is used for the capillary pressure head, that is,

$$\psi \equiv h_c. \tag{2.3.4}$$

Sometimes when h_c is very large, a mercury manometer is used with the tensiometer. In the arrangement shown in Fig. 2.13 the difference between the mercury levels in the two branches of the U-tube manometer measures the combined suction exerted by the head of water contained in the straight tube part of the tensiometer and that of the microscopic menisci of water contained in the pores.

2.3.2. Capillary Pressure versus Saturation Relationship—Hysteresis

2.3.2.1. *The "Capillary Pressure Function"*

The capillary pressure in a porous medium is an increasing function of the nonwetting phase saturation or, alternately, a decreasing function of the wetting phase saturation, where the saturation of fluid phase i is defined as follows:

$$S_i = \frac{\text{Volume of fluid phase } i \text{ in the sample}}{\text{Total accessible pore volume in the sample}}. \qquad (2.3.5)$$

It is the custom to start the measurement of the "capillary pressure curves" with the sample 100% saturated with the wetting fluid. Then the nonwetting fluid is injected by very slowly increasing the capillary pressure applied to it.

In this discussion it is assumed that the porous medium is "uniformly wetted." A semipermeable membrane of the same wettability preference as the wetting fluid is used at the exit face of the sample that permits displacement of the wetting fluid but does not permit penetration of the nonwetting fluid up to a certain capillary pressure.

As it is apparent from curve R_0 in Fig. 2.14(a), increasing P_c at first results only in a very small change in saturation. In this initial phase of the experiment the nonwetting phase has penetrated only pores at or near the surface of the sample. The inflection point of curve R_0 corresponds to the first penetration of the sample by the nonwetting phase all the way to the outlet face. This point is called "breakthrough" or "threshold" or "entry" capillary pressure. Evidently, at this point $S_w < 1$. [Another concept used sometimes is the "displacement pressure," which is defined by extrapolating the branch of the curve R_0 to the left of the inflection point to $S_w = 1$ (Leverett, 1941).]

Further increase of P_c results in more penetration of the nonwetting fluid (i.e., displacement of the wetting fluid), but the slope of the P_c versus S_w curve is getting increasingly steep, until the curve appears to be vertical. This behavior has often been interpreted as an indication of "trapped" wetting phase that cannot be displaced by further increase of the capillary pressure, resulting in the term "irreducible" wetting phase saturation. This concept, however, has only limited validity because it has been shown that in sandstone samples the residual wetting phase saturation could be progressively diminished by increasing the externally applied capillary pressure and allowing sufficient time (often weeks) for displacement (Dullien *et al.*, 1986; Melrose, 1987). A wetting phase is not trapped in a sandstone and, probably, in most natural porous media because the grooves, edges, and wedges on the pore surface form a continuous network extending throughout the porous medium. After the nonwetting phase has penetrated the central part of a pore there is always wetting fluid left in the surface grooves, etc. in the form of "thick films." As P_c is increased externally the wetting phase present in these grooves keeps receding, because the excess wetting phase can escape from the sample via the network of surface grooves, etc. An easy way of visualizing this process qualitatively is by thinking of pores with rectangular

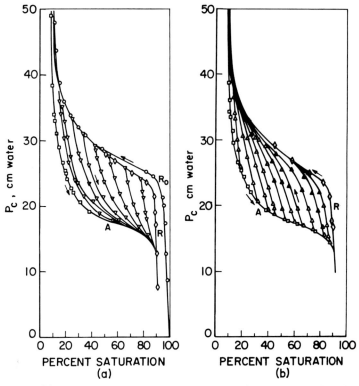

Figure 2.14. (a) Imbibition scanning curves originating from the secondary desaturation curve R (Morrow and Harris, 1965). (b) Desaturation scanning curves originating from the secondary imbibition curve A (Morrow and Harris, 1965).

cross section with the wetting phase present in the corners. Evidently it would require an infinitely great applied capillary pressure to reduce the wetting phase saturation to zero. The role of corners and surface roughness in assuring the hydraulic continuity of the wetting phase at low saturations was first analyzed by Lenormand (1983) and Lenormand and Zarcone (1984).

The correctness of the concept of "irreducible" wetting phase saturation in packs of smooth glass beads was verified by Dullien *et al.* (1989). Here the wetting phase is trapped in the form of pendular rings between touching beads, and there is no hydraulic continuity for reason of lack of corners or surface roughness. In the same work, using packs of beads etched with HF, the residual wetting phase saturation was reduced to 1.4%, at about 4000 Pa capillary pressures, compared to 9% measured in the smooth bead packs.

An additional point of interest is that when the capillary pressure curve is determined by injecting mercury into an evacuated sandstone sample, the same shape curve is obtained as in experiments in which water is displaced by injecting oil. It is obvious, however, that there cannot be an "irreducible" wetting phase saturation when the wetting phase consists of compressible rarified air. The appearance of a fraction of the pore volume that is not penetrated by the mercury even at high applied capillary pressures is probably due, in addition to the presence of surface grooves, etc., to the existence of micropores in the cementing materials (clays, etc.) and silt that require very high pressures to penetrate.

An example of the capillary pressure versus saturation relationship in a pack of smooth glass beads (capillary pressure function) is shown in Fig. 2.14. It is apparent that the relationship between capillary pressure and saturation is not unique but depends on the saturation history of the system. Morrow (1970) lucidly discussed the subject of capillary pressure curves, and his presentation is followed. First, definitions of the main terms are given.

Irreducible saturation S_{wi}: the reduced volume of the wetting phase retained at the highest capillary pressures where the wetting phase saturation appears to be independent of further increases in the externally measured capillary pressure (in porous media other than beds of smooth glass beads the "irreducible saturation" is not a well-defined quantity);

Residual saturation S_{nwr}: the reduced volume of the nonwetting phase that is entrapped when the externally measured capillary pressure is decreased from a high value to zero;

Primary drainage curve (denoted with R_0 in the figure): the relationship characteristic of the displacement of the wetting phase from 100% saturation to the irreducible saturation;

(*Secondary*) *Imbibition curve* (denoted by A in the figure): the relationship characteristic of the displacement of the nonwetting phase from the "irreducible saturation" to the residual saturation;

Secondary drainage curve (denoted by R in the figure): the relationship characteristic of displacement of wetting phase from residual saturation to the "irreducible saturation."

Morrow (1970) points out that most experimental evidence indicates that the "irreducible saturation" obtained by initial drainage is the same as that obtained by secondary drainage. When the residual saturations are the same, the imbibition after secondary drainage will follow exactly the imbibition curve obtained after primary drainage. Thus the secondary drainage curve and the (secondary) imbibition curve constitute a closed and reproducible hysteresis loop (RA in the figure).

Scanning curves within the main hysteresis loop RA can be obtained by reversing the direction of pressure change at some intermediate point along either the secondary drainage or the (secondary) imbibition curve. In the former case, they are called "primary imbibition scanning curves" and in the latter, "primary drainage scanning curves."

In the case of consolidated natural porous media, such as sandstone cores, it has been often found that if, after reaching $P_c = 0$ in imbibition, a positive pressure difference $P_{water} - P_{oil}$ was imposed on the system (this requires exchanging the water-wet porous plate for an oil-wet one, in order to be able to maintain this pressure difference and to prevent flow of water through the core) there was significant additional oil displacement. This process is called "forced" (as opposed to spontaneous) imbibition. Anderson (1987a) in one of his seven review papers on wettability writes "Note that the fact that the water pressure is greater than the oil pressure does not imply that the oil is the wetting fluid at these saturations. If the core were a bundle of cylindrical capillary tubes, then negative capillary pressure would be possible only if the core were oil-wet. However, the interaction of pore structure and wettability allows negative capillary pressures even for strongly water-wet cores (Purcell, 1950)."

As the basic law of capillarity requires that the pressure on the concave side of a nonplanar fluid/fluid interface be greater than on the convex side, the fact that the pressure in the water is greater than in the oil in a sandstone core implies that the water/oil interfaces in the pores under these conditions are curved, with the water on the concave side and the oil on the convex side. The sign of the curvature determines the direction of the capillary driving force, that is, whether water will tend to displace oil or vice versa. Hence, the intrinsic measure of wettability is the sign of curvature of the equilibrium meniscus, or, in which phase the equilibrium pressure is higher. As shown schematically in Fig. 2.11 the effective wettability is determined by both the contact angle θ and the orientation of the solid surface at various points of the three-phase line of contact, represented schematically by the angle ϕ. (In the general three-dimensional case the orientation of the solid surface is given by the direction of the unit normal vector at the particular point on the surface, which needs two angles, the polar angle and the azimuthal angle, for its characterization.) In the case of an irregular surface the equilibrium curvature of the meniscus may be a very complicated function of the pore geometry.

The logical conclusion is that a sign change of the pressure difference, $P_{oil} - P_{water}$, reflects a change from water-wetness to oil-wetness. In such an eventuality it is physically meaningful to change the expression of capillary pressure P_c from $P_{oil} - P_{water}$ to $P_{water} - P_{oil}$, because the equilibrium pressure in a pore is always higher in the nonwetting phase than in the wetting phase. The algebraic convention of allowing both positive and negative capillary pressures is inconsistent with the physical foundations of capillary pressure [Eq. (2.1.8)].

The observation that such a reversal of roles of the two fluids may take place was logically attributed by Killins et al. (1953) to "fractional wetting" (see Fig. 2.20). This means that in certain portions of the pore space of an otherwise strongly water-wet medium, such as a Berea sandstone, the conditions may be such that the oil/water interface is effectively concave on the water side. As shown schematically in Fig. 2.11 this may be due to the effect of the contact angle θ, or the orientation of the pore surface, or both. Evidently, relatively small changes in ϕ will result in wettability reversal when θ is not very different from 90°. Hence, intermediate wettability behavior will be observed in pores such as shown schematically in Fig. 5.16 over a range of contact angles θ on either side of 90°. The characteristic pattern of intermediate wettability is the lack of spontaneous imbibition of either of the two phases. Let the contact angle be $\theta = 70°$. Then in a circular capillary tube the meniscus is as shown in Fig. 5.16 and the liquid will imbibe spontaneously. Now if the pore surface is not cylindrical but conical of

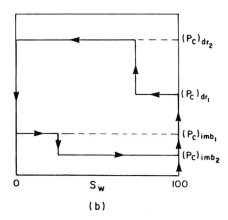

(a)

IN DRAINAGE: AT $(P_c)_{dr_1} = \dfrac{4\sigma\cos\theta}{De_1}$, THE PENETRATION REACHES THE NECK MARKED BY De_2 ;

AT $(P_c)_{dr_2} = \dfrac{4\sigma\cos\theta}{De_2}$, THERE IS COMPLETE PENETRATION

IN IMBIBITION : AT $(P_c)_{imb_1} = \dfrac{4\sigma\cos\theta}{D_1}$, THE PENETRATION REACHES THE POINT MARKED BY "imb1";

AT $(P_c)_{imb_2} = \dfrac{4\sigma\cos\theta}{D_2}$, THERE IS COMPLETE PENETRATION

(b)

Figure 2.15. Example of capillary hysteresis due to pore morphology.

half-cone angle $\phi = 20°$ then the meniscus will be flat as shown in the upper part of Fig. 5.16 and spontaneous imbibition will not take place. Naturally, for $\phi > 20°$ the other fluid will imbibe spontaneously. Imbibition in either direction will stop however, at a point in the pore channel where $\phi = 20°$ that is bound to exist in a convergent–divergent pore channel, provided that $\phi > 20°$ in portions of the pore.

2.3.2.2.   *Capillary Hysteresis, Pore Structure Effects*

Hysteresis is a wide-spread phenomenon in nature, and in some cases hysteresis itself depends either on the history of the system or on the manner in which the experiment leading to the determination of the hysteresis loop is carried out. For example, in some cases, hysteresis can be diminished or eliminated by carrying out the experiment sufficiently slowly. In some other cases, the hysteresis loop exists only in the first traverse of the loop; on second and subsequent traverses no hysteresis loop can be observed.

"Capillary hysteresis" means that the value of the capillary pressure function may have an infinity of different values between an upper and a lower bound, depending on the path along which the particular saturation (or particular capillary pressure) was reached. It has generally two causes: hysteresis of contact angle and pore structure effects.

That type of hysteresis that is independent of the history of the system and is unaffected by waiting even longer in the determination of the experimental points is called "permanent hysteresis" (Everett, 1967).

The theory of hysteresis was developed long ago in conjunction with the magnetic hysteresis by using the concept of "domains." In porous media, "domain" is defined as a group of pores or capillaries that empty at a characteristic drainage pressure and fill at some characteristic imbibition pressure. In general, there are pores or capillaries of different diameters in each domain.

The only experimental measurement of the pore size distribution of domains as a function of the capillary pressure reported in the literature is due to Dullien and Dhawan (1975), who performed their measurements along the drainage branch of the hysteresis loop in sandstone samples. Details of this work are discussed in Chapter 3. It may be pointed out here, however, that in these experiments it was always found that a wide range of different pore sizes were penetrated simultaneously at a given capillary pressure of penetration; that is, the domains had a broad size distribution.

In the strongly simplified version of the domain theory, the so-called independent domain theory, each domain consists of only one type of pore, characterized by a certain drainage pressure P_1 and an imbibition pressure P_2. Thereby it is implied that the pores are independent of each other; in other words, they behave as if all of them were directly connected to the outside of the sample. In the independent domain case, at a given capillary pressure corresponding to the minimum mean radius of curvature of the meniscus in a neck, all bulbs are penetrated by a nonwetting phase. Similarly, bulbs will be emptied from the nonwetting phase at another value of the capillary pressure, corresponding to the maximum mean radius of curvature of the meniscus in a bulb. The necks always empty along with the bulbs.

In the case of interdependent domains, illustrated in Fig. 2.15, however, on increasing the capillary pressure to $(P_c)_{dr1}$ first only the pore between D_{e1} and D_{e2} is penetrated by a nonwetting phase. The rest of the tube is penetrated only at the higher capillary pressure $(P_c)_{dr2}$, even though the neck size in the middle is also D_{e1}.

After the whole system has been filled, on decreasing the capillary pressure to $(P_c)_{imb1}$ only the first pore empties and is filled with the wetting fluid. At the lower capillary $(P_c)_{imb2}$ there is complete displacement.

Everett and coworkers (1952, 1954, 1955, 1967) developed an independent domain theory of permanent hysteresis that has recently been investigated as a model for capillary pressure hysteresis (Poulovassilis, 1962, 1970; Philip, 1964; Topp and Miller, 1966; Poulovassilis and Childs, 1971; Topp, 1969; Talsma, 1970). According to this theory it is possible to predict from the set of experimentally observed imbibition scanning curves (descending scanning curves) the corresponding set of drainage scanning curves (ascending scanning curves), and vice versa, (see Fig. 2.14). The theory is limited to that part of the pore space not occupied by "residual nonwetting phase saturation." Satisfactory predictions were obtained in some tests in beds of sintered glass beads and more recently, by Lai *et al.* (1981), for Berea sandstone; however, the theory was found not to be generally applicable. This negative result is in harmony with our understanding of pore structure, according to which pores of different diameters are connected in series much like in the schematic diagram shown in Fig. 2.15. In most real porous media the pores of capillaries cannot fill or empty independently of each other. In such cases the system is said to exhibit cooperative behavior. In other words, the capillary behavior of any given pore depends on its neighbors. A model of hysteresis that includes cooperative behavior has been investigated by Enderby (1955, 1956).

2.3.2.3. *Methods of Measurement of the Capillary Pressure Function*

The laboratory methods for measuring the relationship $P_c = P_c(S_w)$, where S_w is saturation of wetting phase of porous media may be divided into two main groups:

(1) static methods based on the establishment of successive states of hydrostatic equilibrium; and
(2) dynamic methods based on the establishment of successive states of steady flow.*

---

*Very recently dynamic capillary pressure curves have been measured in the author's lab by *in situ* measurement of capillary pressure and saturation in the course of waterflooding.

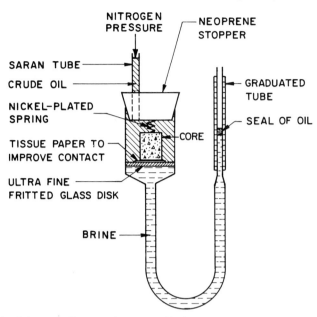

Figure 2.16. Schematic diagram of a porous diaphragm device for capillary pressure determination (Welge and Bruce, 1947 © API).

The latter approach is usually applied in conjunction with relative permeability measurements. It is discussed briefly in Chapter 6.

Some of the equilibrium methods are discussed next. The schematic arrangement used in the "porous diaphragm method," "Welge (restored state) method," or "desaturation method" is shown in Fig. 2.16. Usually the experiment is started with the sample saturated with the wetting fluid. In the example shown in the figure, this is brine. The chamber in which the sample is placed is filled with the nonwetting fluid; in this figure: crude oil. The sample is resting on an ultrafine, fritted glass disk, the purpose of which is to prevent passage of the oil (capillary barrier). The glass disk, however, must conduct the brine between the core and the U-tube. For that reason it is important that good capillary contact is established between the bottom face of the core and the glass disk. This is usually ensured by using paper tissue of $CaCO_3$ paste. In order to prevent breakthrough of the oil across the frit, the latter must have much smaller pores than the core. As the pressure on the oil is raised, the water is displaced from increasingly smaller pores of the sample. The highest attainable capillary pressure is determined by the breakthrough capillary pressure of the fritted glass disk.

The wetting fluid is forced out of the pores of the sample by the nonwetting fluid across the fritted glass disk, into the U-tube where its volume is determined in the graduated portion of the tube. The gas pressure applied is read simultaneously.

The capillary pressure versus saturation curve is determined point by point by displacing the wetting fluid in small increments, at every point waiting until equilibrium is established. The sample saturation is calculated from the initial amount of wetting fluid contained in the pore space and the amount of wetting fluid displaced in each step. Because of the slow rate at which equilibrium is attained, it takes weeks to determine a single capillary pressure curve. (The sample saturation can also be determined by other measurements, e.g., by measuring electrical resistivity, acoustic properties, or x-ray absorption.)

The imbibition capillary pressure curve can be obtained by reversing the above process, that is, decreasing the gas pressure step-by-step. Given good capillary contact between the core sample and the fritted glass, the water will imbibe into the core and displace the nonwetting fluid. A more recent version of the capillary pressure cell is described by Morrow and Mungan (1971) and Dullien *et al.* (1989).

Mainly because of the very long time required for equilibration in liquid–liquid systems, air–liquid capillary pressure saturation data are more often determined instead. In air–liquid systems the time needed for equilibration is significantly shorter than in liquid–liquid systems.

Another technique to determine drainage capillary pressure curves is the well-known method of mercury intrusion porosimetry, discussed in detail in Section 2.5.1.3. As pointed out by Pickell *et al.* (1966), air–mercury capillary pressure data reflect the fluid distribution in water–oil systems only under strong water-wet conditions. Brown (1951) found that gas–oil capillary pressure data can be made to agree with mercury injection capillary pressure data by using an appropriate scaling factor.

Leverett (1941) defined a reduced capillary pressure function, which was subsequently named the Leverett J function by Rose and Bruce (1949) and used for correlating capillary pressure data:

$$J(S_w) = (P_c/\sigma \cos \theta)\sqrt{(k/\phi)} \qquad (2.3.6)$$

where k is the permeability and ϕ is the porosity of the sample.

Brown (1951) found that the J function was successful in correlating pressure data originating from a specific lithologic type within the same formation, but it was not of general applicability.

One of the main limitations of the Leverett J function lies in the fact that the square root of the permeability–porosity ratio is an inadequate scale factor that is incapable of accounting for the individual differences between the pore structures of various samples. The reason for the success of the relationship within a specific lithological type is presumably that within one and the same type of rock the pore structures have similar topology.

Capillary pressure curves have been reported by a large number of workers, for example, Hassler and Brunner (1945), Calhoun *et al.* (1949), Powers and Botset (1949), Rose and Bruce (1949), Stahl and Nielsen (1950), Slobod *et al.* (1951), Dunning *et al.* (1954), Brooks and Corey (1964), Philip (1964), Carey and Taylor (1967).

2.3.3. The Influence of Wettability

The term "wetting" in everyday language means that the liquid spreads over the solid surface, and "nonwetting" means that the liquid tends to ball up and run off the surface. A "preferentially water-set" solid is defined by Jennings (1958) in terms of the advancing contact angle in the solid–water–oil systems. Brown and Fatt (1956) regard "fractional wettability" as the fraction of the total pore surface area that is preferentially oil-wet or water-wet. It is noted that "fractional wettability" can be expected to be a function of both saturation and saturation history of the sample. The concept of a "fractional wettability" takes into account also the heterogeneous mineral composition of most reservoir rocks.

The term "mixed wettability" was introduced probably by Salathiel (1972), who pointed out that in the "mixed wettability condition" the oil-wetted surfaces are distributed in such a way that oil maintains continuity in the relatively large pores and there is permeability to oil, which permits drainage of oil to very low residual saturation. It is noted that "fractional wettability" condition does not imply continuous paths for oil flow.

Melrose and Brandner (1972) state that the contact angle provides the only direct and unambiguous specification of the wettability. Contact angles, however, cannot be measured within the porous medium; the test is conducted with flat, nonporous samples prepared of the material constituting the porous medium.

According to Brown and Fatt (1956) the concept of a contact angle is not representative in the case of reservoir rock because this is often composed of many different minerals, each with a different surface chemistry and a different capacity to adsorb surface active materials from reservoir fluids.

The extent and rate of imbibition have also been used to measure wetting preference (Bobek *et al.*, 1958). In this context, imbibition, sometimes called "free spontaneous imbibition," refers to a process in which a wetting fluid displaces a nonwetting fluid in a porous medium by capillary forces alone, similar to an ink blotter soaking up ink that is expelling air. The rate of imbibition depends on other factors besides wettability (Richardson, 1961; Dullien *et al.*, 1977); these include pore structure, viscosity of the fluids, and interfacial tension. Studies have been conducted by Gatenby and Marsden (1957), Morrow and Mungan (1971), Morrow *et al.* (1972), and Melrose (1965) to correlate contact angles with imbibition rate behavior for porous media.

Amott (1959) proposed combining imbibition and displacement tests to determine wettability indices based on ratios of the displacement volumes determined from the following four measurements, starting with the core sample at residual oil saturation: (1) spontaneous (i.e., free) imbibition of oil after 20 hours; (2) flooding the core sample with oil until no more water is displaced and determining the total volume of water displaced; (3) spontaneous (i.e., free) imbibition of water after 20 hours; and (4) flooding by water.

The results of this test are expressed in terms of two numbers, as follows:

$$\text{Displacement-by-oil ratio} = \frac{\text{Water volume displaced by spontaneous oil imbibition}}{\text{Total water volume displaced by oil imbibition and oil flooding}} \quad (2.3.7)$$

It approaches 1 for strongly oil-wet cores, and it is zero for water-wet and intermediate-wet cores.

$$\text{Displacement-by-water ratio} = \frac{\text{Oil volume displaced by spontaneous water imbibition}}{\text{Total oil volume displaced by water imbibition and water flooding}} \quad (2.3.8)$$

It approaches 1 for strongly water-wet cores, and it is zero for oil-wet and intermediate-wet cores.

More representative results are obtained if imbibition is permitted to proceed until it finally stops, instead of using the arbitrary 20 hours of imbibition time.

In some samples spontaneous imbibition of both oil and water has been observed (e.g., Burkhardt *et al.*, 1958; Mohanty and Salter, 1983; Sharma and Wunderlich, 1985; Dullien *et al.*, 1990). In the last reference the spontaneous imbibition of both Pembina crude and brine into Pembina Cardium cores, saturated initially with brine and Pembina crude, respectively, has been reported. The extent of imbibition in either direction was in the neighborhood of 40% pore volume. Similar results were obtained when styrene monomer was used as the oil phase. After the imbibition was complete the styrene was polymerized and thin sections of the sample were prepared using standard techniques (Yadav *et al.*, 1987). Drawings prepared on the basis of photomicrographs of the thin sections are shown in Fig. 2.17 (a) and (b). The distribution of the two liquids at the end of both imbibition processes is almost identical. This behavior indicates both continuous oil-wet and continuous water-wet paths in the sample. It represents a special case of mixed or fractional wettability.

Both of the Amott wettability indices are positive numbers, approaching 1, in these mixed-wet cores.

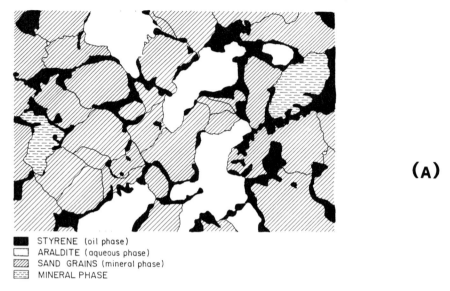

(A)

■ STYRENE (oil phase)
☐ ARALDITE (aqueous phase)
▨ SAND GRAINS (mineral phase)
▤ MINERAL PHASE

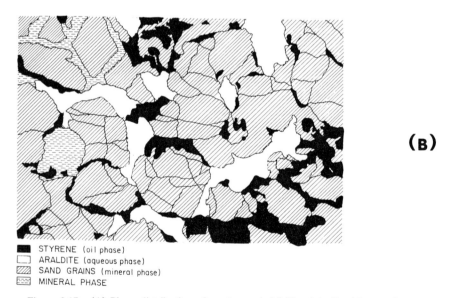

(B)

■ STYRENE (oil phase)
☐ ARALDITE (aqueous phase)
▨ SAND GRAINS (mineral phase)
▤ MINERAL PHASE

Figure 2.17. (A) Phase distribution after styrene imbibition into Pembina sandstone saturated with brine (styrene: black, brine: white) (Dullien *et al.*, 1990). (B) Phase distribution after brine imbibition into Pembina sandstone saturated with styrene (styrene: black, brine: white) (Dullien *et al.*, 1990).

The Amott–Harvey relative displacement test (Boneau and Clampitt, 1977; Trantham and Clampitt, 1977) differs only in some details from the Amott test and uses the index consisting of the difference of the "displacement-by-water ratio" and the "displacement-by-oil ratio." This index varies from $+1$ for complete water wetness to -1 for complete oil wetness.

The USBM Wettability Index, developed by Donaldson *et al.* (1969, 1980, 1981) uses the centrifuge. The test is started at some low water saturation established by centrifugation. Then the cores are centrifuged in brine at incrementally increasing speeds until an applied positive pressure difference of 10 psi (70 kPa) on the brine has been reached (brine drive). In the second step the cores are centrifuged in oil at incrementally increasing speeds until an applied positive pressure difference of 10 psi (70 kPa) on the oil has been reached (oil drive). The areas under the saturation-versus-pressure-difference curves obtained in both steps are calculated and the wettability index is calculated as the logarithm of the ratio of the area under the oil-drive curve to the area under the water-drive curve. Sharma and Wunderlich (1985) have combined the Amott and the USBM wettability indexes. For other methods see Anderson (1986).

A quantitative method by Holbrook and Bernard (1958) has been proposed for determining the relative water wettability of a core sample. This method is based on the observation that a rock surface covered by water and thus, by definition, considered "water wet" will adsorb methylene blue (a water soluble dye) from solution; but an oil covered surface will not. It gives a measure of the fraction of the surface area wetted by water. Shankar (1979) [see also Shankar and Dullien (1979)] found that in a water-wet Berea core nearly the entire pore surface adsorbed dye from a solution injected at 40% water saturation.

In addition to the more direct methods previously discussed, the results of waterflood and relative permeability tests will often provide an indication of the wettability of the oil–water–rock system under study (Craig, 1971). This stems from the major influence that wettability exerts on the flow, as discussed in Chapter 6. There are, in fact, many reservoir engineering applications where flow test data in cores representative of the reservoir and its wettability are the only requirements. Actual knowledge of the reservoir wettability is not required (Treiber *et al.*, 1972). However, the key lies in obtaining representative cores and in maintaining their true reservoir wettability. The possible alteration of the natural wettability of a reservoir rock due to coring, flushing, depressurizing, cooling, preserving, and testing is always open to debate. Mungan (1972) reports that the true natural wettability can be reestablished in an extracted core by contacting it with reservoir fluids at simulated reservoir conditions for a period of days. Dullien *et al.* (1990) have found this only approximately true for Pembina.

An early study of the influence of wettability on oil–water capillary pressure characteristics was performed by Killins *et al.* (1953). Oil–water

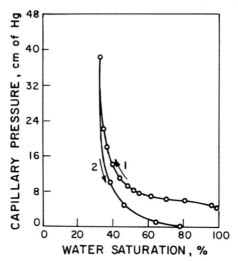

Figure 2.18.   Capillary pressure characteristics, strongly water-wet rock with Venango core VL-2 and k = 28.2 mD. Curve 1, drainage; curve 2, imbibition (Killins *et al.*, 1953).

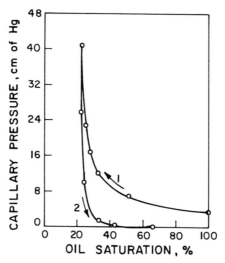

Figure 2.19.   Oil–water capillary pressure characteristics with Tensleep sandstone, oil-wet rock. Curve 1, drainage; curve 2, imbibition (Killins *et al.*, 1953).

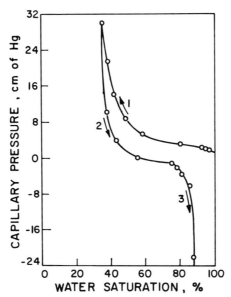

Figure 2.20. Oil–water capillary pressure characteristics with presumed fractional wettability, Berea core 2-MO16-1, and $k = 184.3$ mD. Curve 1, drainage; curve 2, spontaneous imbibition; curve 3, forced imbibition (Killins, Nielsen, and Calhoun, 1953).

capillary pressure curves were determined on consolidated sandstones under both water-wet and oil-wet conditions. As it can be seen from a comparison of Figs. 2.18 and 2.19, there is a qualitative similarity between a strongly water-wet and an oil-wet rock, with water playing the role of oil and vice versa in the case of oil-wet rock. In both figures the capillary pressure is as defined in this monograph.

Even more interesting is the case of the rock shown in Fig. 2.20. When water is displaced with oil along curve 1, the behavior indicates preferential water wetting because of the positive value of the capillary pressure needed for threshold penetration. Also along curve 2 the behavior reflects water wetting because water spontaneously imbibes into the sample along this curve, terminating at zero capillary pressure. At this point, after replacing the water-wet porous plate with an oil-wet one at the exit, positive water pressures was applied relative to the oil, resulting in increasing water saturation, which seems to indicate two things. First, the sign of the radii of curvature of the menisci between the water and the oil remaining in the core at $P_c = 0$ must have been opposite, indicating fractional oil wetness, and from here on, water must be regarded as the nonwetting phase. Second, the fact that more oil could be displaced from the rock indicates that at the end of curve 2 the oil in the sample was still in a continuous form; that is, it was communicating with the outlet face of the core. When the wetting fluid is

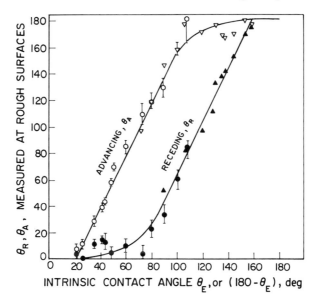

Figure 2.21. Values of advancing (θ_A) and receding (θ_R) contact angles observed at rough surfaces versus intrinsic contact angle θ_E. (Reprinted with permission from Morrow, 1975.)

placed under a higher pressure than the nonwetting fluid, resulting in displacement, there is a case of "forced imbibition."

A great deal of systematic study of the capillary pressure versus saturation relationship in dependence on the contact angle has been performed by Morrow and coworkers (1971, 1976). Porous polytetrafluoroethylene (PTFE) plugs were used that provided systems of fixed solid geometry and of uniform wettability conditions in the entire pore space. Using air with a variety of liquids as the second fluid, a wide range of contact angles could be covered.

In the first of these papers (1971), advancing contact angles, ranging from 0 to 108°, as measured at a plane and smooth PTFE surface were used to correlate the data. It was found that the liquid displacement curves were fairly insensitive to contact angles, measured through the liquid, in the range of 0 to 49° and decreased systematically with the increase of contact angle in the 73–108° range. Once the porous plug was saturated with liquid, suction was needed for displacement of liquid by air even when the contact angle measured through the liquid exceeded 90°.

In his later work, Morrow (1976) used the advancing and receding contact angles, measured at internally roughened tubes of PTFE (Morrow, 1975). He found that these angles differed substantially from the advancing contact angles measured at a smooth surface. Figure 2.21 shows the relationship between the receding and advancing contact angles measured at rough

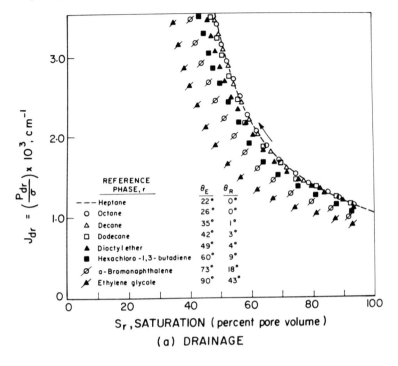

(a) DRAINAGE

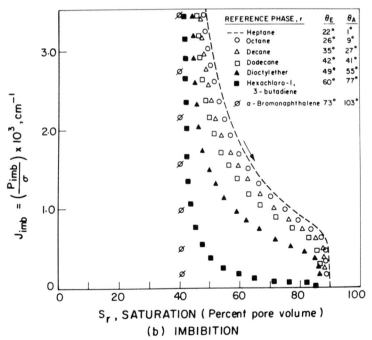

(b) IMBITION

Figure 2.22. Effect of contact angle on capillary pressure, using core No. 3: (a) drainage and (b) imbition. (Reprinted with permission from Morrow, 1976.)

surfaces versus the so-called intrinsic contact angle, that is, the advancing (or the practically indistinguishable receding) contact angle determined at the smooth PTFE surface. It is apparent from this diagram that there can be a very significant contact angle hysteresis at a rough surface. (For low surface energy solids such as PTFE the differences between advancing and receding angles, measured on smooth surfaces, are very small.)

The dependence of capillary pressure on wettability was investigated for six different types of PTFE cores. The total range of intrinsic contact angles was from 0 to 108°, as in the previous study. The results obtained in core No. 3 are shown, as an example, in Fig. 2.22. It can be seen from the diagram that in this case the drainage (i.e., displaced phase is at the lower pressure), and imbibition (i.e., displacing phase is at the lower pressure) displacement curves vary systematically with the receding and advancing contact angle, respectively.

The curves corresponding to different θ diverge in the direction of decreasing S_r; in other words, the ratio $J_{dr}(S_r, \theta_1)/J_{dr}(S_r, \theta_2)$ increases sharply with decreasing S_r and it is not equal to $\cos \theta_1 / \cos \theta_2$ as might be expected on the basis of the Leverett J function correlation (Eq. 2.3.6). Evidently, there is a trend of decreasing trapping of the displaced phase at the lower pressure (generalized drainage) with increasing contact angle.

As expected on the basis of the arguments presented in Section 2.3.1, in the vicinity of $\theta_A = 90°$ there was no, or very little, spontaneous imbibition.

These studies show the difficulty of interpreting correctly the influence exerted by the contact angle in a porous medium on the observed capillary pressure–saturation relationship.

Dumoré and Schols (1974) have correlated quite satisfactorily the widely different drainage capillary pressure functions determined on the same rock material with various fluid–fluid combinations, including mercury–air, using the Leverett J function, but without allowing for existing differences in the contact angle. They explained this apparent insensitivity to contact angle by assuming that the apparent contact angle, effective on the very rough surface of the Bentheim sandstone studied in this work, might have been always zero or nearly zero in all the tests carried out with the various fluid–fluid systems.

An extensive review of the effects of wettability on capillary pressure has been published by Anderson (1987a).

2.3.4. Vertical Distribution of Two Fluids in a Porous Medium under the Action of Gravity

The expected equilibrium distribution of the wetting and the nonwetting fluid in a long vertical column of a porous medium consisting of a pack of smooth glass beads has been described lucidly by Smith (1933).

A long column of beads, if its pores have been initially filled with water and then allowed to drain, retains considerably more liquid than a simple capillary tube subjected to a similar process. After waiting a sufficiently long time, the water retained in the upper parts of the column is found in discrete masses and each mass is a ring of liquid wrapped around the contact point of a pair of adjacent beads, as shown in Fig. 2.23a. Liquid must accumulate in these positions rather than in a uniform layer over the bead surfaces, otherwise the Gibbs free energy of the retained liquid system would not be a minimum. A very thin wetting layer, however, is supposed to exist over the bead grain surfaces. The single liquid rings were called "pendular" rings by Versluys (1917).

As we proceed downward in the column, provided that thermodynamic equilibrium has been established, the rings increase in size until finally they begin to coalesce into more complicated masses; two or more contact points can be imbedded in a common liquid mass. This stage, which was called the "funicular" stage by Versluys and the "transition zone" by Muskat (1949), can be seen in Fig. 2.23b.

As we proceed farther downward along the column, finally a stage is reached where the more complicated masses of the funicular stage merge and form a continuum extending completely across the pore space excluding, of course, those portions where there is trapped residual nonwetting phase present.

2.3.4.1. *Condition of Thermodynamic Equilibrium: The Equation of Capillary Rise in Porous Media*

Smith (1933) claimed that if a sufficiently long time elapsed, the isolated pendular rings would be in thermodynamic equilibrium with the bulk liquid just as if they were connected to it by a pipe, because of mass transfer and consequent equilibration via the vapor phase between the pendular rings and the bulk liquid.

The theoretical basis for this claim is as follows. The well-known barometric formula (i.e., the relationship expressing the change of barometric pressure with elevation at constant temperature) may be applied to the vapor pressures at the elevation of the free liquid surface, where the vapor pressure is P_0, and at a higher elevation, where the vapor pressure of the "pendular" rings is a lower value, P'':

$$\ln(P''/P_0) = -(Mgh/\mathscr{R}T), \qquad (2.3.9)$$

where M is the molecular weight of the vapor, g is the gravitational acceleration, and h is the elevation difference between the pendular rings and the free liquid surface.

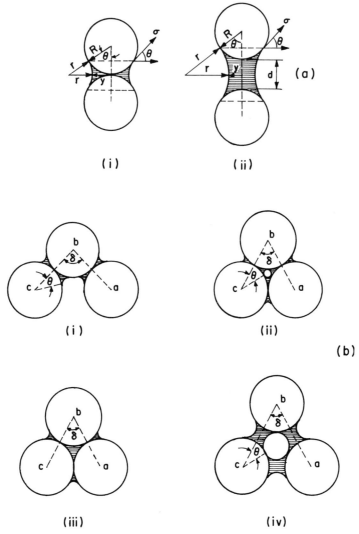

Figure 2.23.   (a) (i) Simple pendular ring of liquid; (ii) simple ring of liquid in spaced hexagonal packing (Smith, 1933). (b) Transformation of single liquid rings: (i) single rings with angle $\delta > 60°$; (ii) single rings about to merge in three-grain pore with $\delta = 60°$; (iii) simple web of liquid existing after coalescence of rings; (iv) rings of liquid in spaced hexagonal packing with $\delta = 60°$. (Reprinted with permission from Smith, 1933.)

On the other hand, Kelvin's equation for the reduction of vapor pressure over a concave liquid surface compared with the value over the free liquid surface is as follows:

$$\ln(P''/P_0) = -(2\sigma/r_m)(V'/\mathscr{R}T), \qquad (2.3.10)$$

where V' is the molar volume of the liquid.

Combining Eqs. (2.3.9) and (2.3.10) we obtain

$$2\sigma/r_m = \rho gh. \qquad (2.3.11)$$

The left-hand side of this equation is equal to $P'' - P'$ [cf., Eq. (2.1.8)]. The right-hand side, however, is equal to the pressure exerted by a liquid column of height h and density ρ.

If the liquid in a pendular ring were connected with the bulk liquid through a capillary tube, the net pressure exerted by the liquid column contained in the tube would be exactly balanced by the suction pressure $\Delta \rho gh$:

$$2\sigma/r_m = P_c = P'' - P' = \Delta \rho gh, \qquad (2.3.12)$$

where $\Delta \rho = \rho$ − density of surrounding medium $\approx \rho$ when the capillary rise occurs in air. Equation (2.3.12) is the equation of capillary rise.

Defay and Prigogine (1966) calculated the influence of curvature on the vapor pressure of water. Their table has been reproduced as Table 2.1. It is evident from these data that the vapor pressure reduction over a concave surface (bubble) is very small compared with the magnitude of the capillary pressure, shown in the last column. This is true for all values of the radius of curvature expected in most porous media.

It may be concluded then that in those parts of the column of beads in which the liquid phase is continuous and connected with the bulk liquid, equilibrium is established relatively readily, whereas in the pendular ring region where the process depends on the vapor pressure reduction, the

TABLE 2.1   Influence of Curvature on the Vapor Pressure of Water[a]

r/cm	Droplet P'/P_0	Bubble P''/P_0	Droplet or bubble $(P'' - P')/10^2$ kPa
∞	1	1	0
10^{-4}	1.001	0.9990	1.46
10^{-5}	1.011	0.9891	14.6
10^{-6}	1.115	0.897	146
10^{-7}	2.968	0.337	1460

[a] From Prigogine and Defay (1966).

establishment of equilibrium is extremely slow and also very tenuous because it can be easily disturbed and upset by disturbances such as small convective currents and the existence of nonuniform temperature (e.g., Morrow, 1970).

Considering the case where the nonwetting fluid is not air but a liquid, the establishment of equilibrium depends on diffusion of the water through the liquid, and thereby the establishment of equilibrium is rendered even slower than in the previous case. For the case when $\Delta \rho \leq 0$ the wetting fluid is weightless or under the influence of a net buoyant force pointing upward and, therefore, the height of capillary rise is theoretically infinite.

The experimental results obtained by King (1897–1898) in drainage columns in which up to $2\frac{1}{2}$ years were allowed for equilibration were used by Smith (1933) to compare with his calculated results predicting a continuous and definite decrease of saturation with increasing capillary pressure. It appears that some of King's results may have been fairly close to equilibrium whereas others were not.

Tschapek et al. (1985) have studied experimentally the amount of "undrainable water" in packs of quartz sand and glass beads by two methods:

(1)   free drainage columns of height up to 100 cm for sands, and
(2)   porous plate up to 1 bar pressure for sands and glass beads.

Different surface tensions and bead sizes ranging from 0.08 to 5.0 mm were used.

The drainage column experiments of 1 week's duration have indicated a constant "undrainable water" saturation past a certain height, which decreased with decreasing surface tension. The porous plate type drainage experiments of about 1 day's duration showed a continuous decrease of "undrainable water" saturation with applied external pressure all the way to the highest pressure of 1 bar used in these experiments for the case of the smallest glass beads used, whereas there was no change with pressure for the largest glass beads. The "undrainable water" saturation was about double in the 0.08-mm bead pack than in the 5.0-mm bead pack, where the "undrainable water" saturation was only about 2.5% as compared with the 6–10% indicated by Morrow (1970).

The conclusion has been reached that "undrainable water" consists both of pendular ring water and film water the thickness of which could be as high as 1 μm, depending on the roughness of the surface.

It is interesting to compare with these results the findings of Dullien et al. (1989), according to which in a smooth bead pack (average bead size about 0.5 mm) an "irreducible wetting phase (Soltrol oil 170) saturation" of 9% was found in a porous diaphragm tensiometer test to be steady over a period of several months.

Tschapek et al. (1985) seem to have used gravimetry throughout their work, but no details of the manipulations and calculations have been disclosed. The mass of experimental evidence against their very low "irreducible

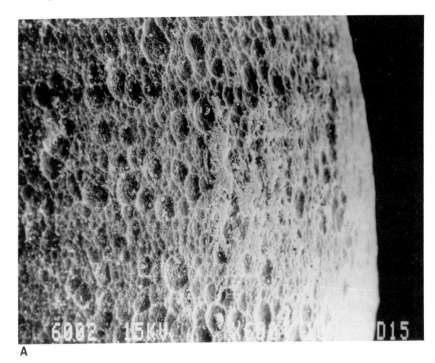

A

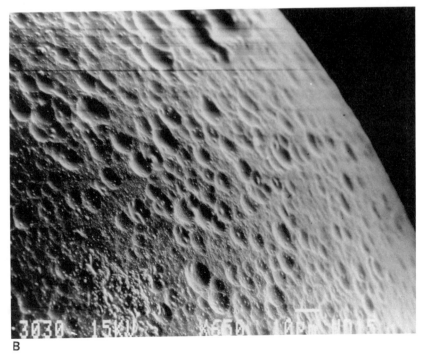

B

Figure 2.24. SEM pictures of etched bead surface (A) unused, (B) at the end of drainage experiment with polymerized wetting phase (Dullien *et al.*, 1989).

saturations" suggests the existence of an undetected systematic error in their measurements and/or calculations.

In the same work Dullien *et al.* (1989) found that when using packs of glass beads whose surface was roughened by etching with HF (see Fig. 2.24) there was no definite "irreducible wetting phase saturation." The drainage experiment was terminated after 5 weeks when the wetting phase saturation at 3600 Pa capillary pressure was about 1.4%, after every preceding step of pressure increment was followed by a decrease in wetting phase saturation. It was concluded that the surface channels created by etching formed a continuous network extending on the beads' surface over the entire pack that made possible transport of the wetting fluid at slow but measurable rates. The film thickness on the surface of smooth beads is not sufficient to permit measurable rate of transport of the wetting fluid.

A

Figure 2.25. Equilibrium heights of continuous wetting phase (molten paraffin wax) for gravity drainage and imbibition in bead packs. (A) Packs of smooth and etched beads and (B) blowup of interfacial region: packs of etched beads. (Dullien *et al.*, 1989.)

B

Figure 2.25. (*Continued*)

Dullien *et al.* (1989) also carried out equilibrium height of capillary rise experiments in both smooth and etched bead packs. Both primary drainage and primary(!) imbibition tests have been used. As shown in Fig. 2.25, as expected, the height of capillary rise was greater in drainage than in imbibition in the case of smooth beads; however, in etched beads the equilibrium heights of rise are identical in both drainage and imbibition. It is apparent from the photograph, however, that the saturation near the wetting phase front is less in imbibition than in drainage because of trapping of air, the nonwetting phase. The models of heights of capillary rise in smooth and etched bead packs in drainage are shown in Fig. 2.26. The role of surface roughness is modeled by a capillary tube permitting equilibration via flow in thick films, much as suggested by Smith (1931) to account for the effect of vapor diffusion.

The phenomenon of "hysteresis of capillary rise" is well established in the literature. It can be due to a capillary surface of heterogeneous wettability, to contact angle hysteresis, or to geometrical factors (see Adamson, 1967). It

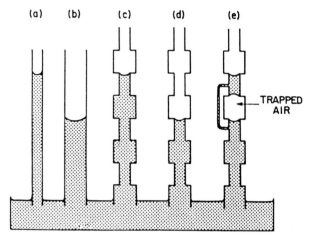

Figure 2.26. Capillary models of drainage and imbibition heights of capillary rise in packs of smooth and etched beads, respectively. (a) Height of capillary rise in a uniform capillary tube of a diameter equivalent to the "window" size. (b) Height of capillary rise in a uniform capillary tube of a diameter equivalent to the "void" size. (c) Capillary model of height of capillary rise in drainage in a pack of smooth beads or in a pack of etched beads of the same size as the smooth beads. (d) Capillary model of height of capillary rise in imbibition in a pack of smooth beads. (e) Capillary model of height of capillary rise in imbibition in a pack of etched beads of same size as the smooth beads modeled in (d). (Dullien *et al.*, 1989.)

appears, however, that both drainage and imbibition are controlled by the pore throats and hysteresis may be due only to trapping of the nonwetting phase. If this can be shown to be the case for real porous media, in general, then the work of Kusakov and Nekrasov (1958, 1960), who have solved the geometrical problem in sinusoidal capillaries and have found several equilibrium heights, depending on the geometry and the depth of immersion of the capillary, is only of theoretical interest.

2.3.5.   Imbibition Mechanisms and Residuals Nonwetting Phase

Pickell *et al.* (1966) investigated and discussed the trapping mechanism of the residual nonwetting phase during an imbibition process. They analyzed the injection and withdrawal of a nonwetting liquid through a restriction in a pore.

Nonwetting phase residual saturations obtained under strong wetting conditions from both quasi-static air–mercury displacement (hysteresis) and dynamic oil–air imbibition were compared with water flood residuals. Sandstones and carbonate rock samples with varied porosity and permeability were included. Figure 2.27 shows good agreement between residual satura-

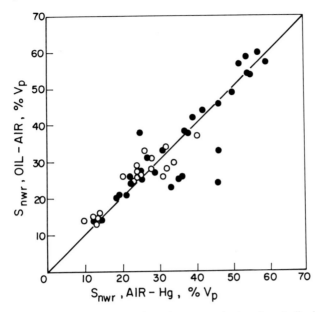

Figure 2.27. Residual saturation comparisons by mercury hysteresis and oil–air systems: O, adjacent samples; •, identical samples (Pickell *et al.*, 1966 © SPE-AIME).

tions obtained by air–mercury and oil–air systems. Figure 2.28 shows the satisfactory comparison of oil–air residual air saturations with residual oil saturations from water floods in water-wet rocks. In both figures equivalent initial saturation conditions were used for the comparisons. It is evident from these data that in these experiments dynamic factors did not play an important role. The residual nonwetting phase saturations cover a wide range of values, extending up to about 60%.

In the context of capillary pressure curves, usually the nonwetting phase saturation at $P_c = 0$ is called "residual." In forced imbibition using a semipermeable membrane to prevent flow of the wetting phase there is no decrease in the residual nonwetting phase saturation, provided that the medium is uniformly water wet. Hence, in this case, the nonwetting phase saturation would appear to be "irreducible." In actual fact, however, under the influence of either long-range, inertial, or viscous forces, the residual saturation can be reduced practically to zero (see Section 6.4.7).

The trapping mechanism of the nonwetting phase is highly complicated and is imperfectly understood at the present (see, e.g., Stegemeir, 1976). So-called "ganglia," "blobs," or "filaments" of various sizes are known to be "trapped," that is, completely surrounded by the wetting phase.

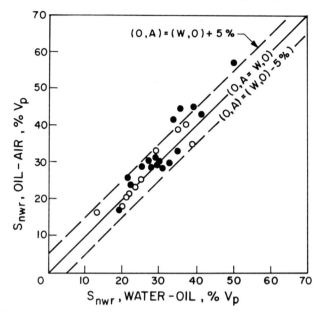

Figure 2.28. Residual saturation comparisons for water-wet samples by water–oil and oil–air techniques: ○, adjacent samples; ●, identical samples (Pickell *et al.*, 1966 © SPE-AIME).

A great deal of information on this subject matter has been gathered recently by Wardlaw and coworkers (1986a, 1986b, 1986c, 1988), following the pioneering observations of Lenormand and coworkers (1983, 1984). Their conclusions are best summarized in terms of Fig. 2.29, showing the various interface types and configurations in imbibition, and Fig. 2.30, summarizing experimental observations made in transparent glass micromodels. The complexity of the displacement mechanisms appears to prevent an accurate description and modeling of the process even if the detailed pore structure of the sample were known. This state of affairs appears to be in contradiction with the observation of Dullien *et al.* (1989) that the height of capillary rise in etched bead packs is simply controlled by the pore throats.

Bacri and coworkers (1985, 1990) have used acoustic technique to determine *in situ* saturation distributions as a function of time along the axis of a sandstone plug and bead packs consisting of varying proportions of glass beads and polymethylmethacrylate beads, obtained in forced imbibition tests run at different constant rates of injection of water, displacing oil (see Fig. 40 in Chapter 5). The tests were started at residual (connate) water saturation. At the lowest injection rate the water saturation profile was very gradual whereas at the fastest flow rate it was steep, approximating a step change. The authors referred to the gradual spreading of water as "hyperdiffusion."

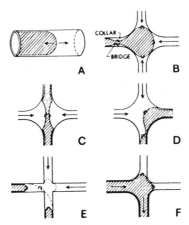

Figure 2.29. Interface types and configurations A–E are for imbibition. Nonwetting phase shaded; (A) convex interface, piston-type advance and retreat; (B) selloidal interface snap-off in throat; (C) selloidal interface, snap-off in pore; (D) selloidal interface, snap-off at pore-throat junction; (E) convex interface, break-off in pore; (F) drainage.

From the point of view of the mechanism of spreading, this phenomenon is not diffusion because it lacks the element of random motion characteristic of typical diffusion processes, it is a manifestation of deterministic transport of the wetting fluid by nonpiston type advance in the pores by convection, controlled by capillarity. Contrary to the authors' opinion, in the 100% glass bead packs the water "pockets" (pendular rings) are well known to be

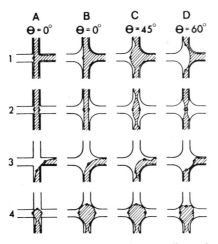

Figure 2.30. Summary of experimental observations on effects of contact angle on interface shape and snap-off sites; nwp connects to pore from three throats, two opposite throats, two adjacent throats, and one throat only 1, 2, 3, and 4, respectively. (Li and Wardlaw, 1986.)

disconnected at irreducible water saturation. (If they were connected then the water saturation could be further reduced.) In forced imbibition, however, water is pumped into the pendular rings lying near the inlet, causing these to swell and coalesce as shown in Fig. 2.23. The increase in water saturation at the inlet is "telegraphed" along the length of the pack by this mechanism. This effect is totally absent in the 100% oil-wet bead packs.

In the gas displacement studies of Keelan and Pugh (1975) by wetting phase imbibition in carbonate reservoir rocks, with gas in place of 80% of pore space, the trapped gas values ranged from 23 to 69% pore space. This wide variation is probably wholly due to differences in the complex pore geometry of the carbonate rocks.

2.3.6. The Centrifuge Method

Capillary pressure curves are often determined by the so-called centrifuge method (Hassler and Brunner, 1945; Slobod *et al.*, 1951). To obtain drainage curves, a small uniform plug is initially saturated with the wetting fluid and is subsequently rotated at a series of increasing angular velocities in a centrifuge, and the quantity of wetting fluid removed at each speed of rotation measured. The sample is held in a cup containing the nonwetting fluid. Normally the wetting fluid (water) has a greater density than the nonwetting fluid (oil), thus causing it to flow out from the sample at the outer radius while it is simultaneously replaced by the nonwetting fluid entering at the inner radius. At a constant rate of rotation, an equilibrium saturation distribution is formed in the sample as is the case in a vertical column in gravity field.

The capillary pressure at the inner radius of rotation r_1 of the sample is equal to

$$P_{c1} = \tfrac{1}{2} \Delta\rho \, \omega^2 \left(r_2^2 - r_1^2\right), \tag{2.3.13}$$

where ω is the angular velocity of rotation and r_2 is the outer radius of rotation of the sample where the pressure has the same value in both phases (i.e., $P_{c2} = 0$). Note that the magnitude of the hydrostatic head varies quadratically with the distance and not linearly, as in the case of gravity field [cf. Eq. (2.3.12)]. Apart from this difference, the situation is analogous to the case of a vertical column of the porous medium saturated with the wetting phase and immersed into the nonwetting phase. The top of the column corresponds to the inner radius of rotation r_1 and the bottom to the outer radius of rotation r_2.

The capillary pressure curve is calculated from a set of capillary pressure versus average saturation data. The capillary pressures are calculated by Eq. (2.3.13) and the average saturations $\bar{S}_w$ are obtained from the volume of displaced wetting phase corresponding to the particular angular velocity. The definition $\bar{S}_w$ is

$$\bar{S}_w = \int_{r_1}^{r_2} S_w \, dr/(r_2 - r_1), \qquad (2.3.14)$$

which, however, has been replaced by the following approximation (Hassler and Brunner, 1945; Slobod *et al.*, 1951):

$$\bar{S}_w \approx \int_0^{P_{c1}} S_w(P_c) \, dP_c/P_{c1}. \qquad (2.3.15)$$

Differentiating this equation with respect to P_{c1} there results

$$\bar{S}_w + P_{c1}(d\bar{S}_w/dP_{c1}) = S_w(P_{c1}). \qquad (2.3.16)$$

From a set of data obtained at a series of different angular velocities, a plot of $\bar{S}_w$ versus P_{c1} is obtained. This makes it possible to evaluate the left-hand side of Eq. (2.3.16), which is equal to the saturation at P_{c1}. It turns out that the correct equation [Eq. (2.3.14)] would yield a result in which S_w on the right-hand side of Eq. (2.3.16) is multiplied by the factor $(r_1 + r_2)/2r_1$. Hence, as long as the value of this ratio is close to 1 (i.e., the ratio r_1/r_2 is close to 1), Eq. (2.3.16) yields a reasonable approximation.

Forced imbibition curves can be obtained by rotating the small core plug at residual (connate) water saturation, surrounded by water in a cup. The displaced oil accumulates on top of the water.

Residual wetting phase saturations have been determined under the action of strong inertial or viscous forces, for example, in a strong centrifugal field. As the work of Dombrowski and Brownell (1954) shows, under these conditions much lower residual nonwetting phase saturations can be obtained than those reported by Morrow (1971b).

A critical discussion of the interpretation of centrifuge capillary pressure data has been presented by Melrose (1988).

2.4. IRREVERSIBLE EFFECTS IN QUASI-STATIC CAPILLARY DISPLACEMENT

Morrow (1970) presented an interesting account of infinitesimally slow displacement of one fluid by another.

2.4.1. Haines Jumps

The so-called Haines jumps result from unstable fluid configurations (Haines, 1930; Melrose, 1965; Miller and Miller, 1956). Haines jumps were demonstrated by Morrow (1970) in the following simple experiments.

In the first experiment, melting point tubes were heated and pulled into long fine capillaries, which were broken into short lengths. These tube segments were then piled in a heap, saturated with a volatile liquid, and the system was observed through a microscope as the wetting phase saturation decreased by evaporation. The tubes were slightly tapered and the air–liquid menisci moved slowly toward the narrower ends of the tubes as the saturation decreased and the capillary pressure increased. Occasionally a meniscus would suddenly recede or jump back toward the wider ends, indicating a sudden pressure change. The meniscus would then continue from its new position until it suddenly kicked back again. In effect, each tapered tube served as a pressure-measuring device that showed that capillary pressure does not rise smoothly and continuously with decrease in saturation.

In a second experiment, the tapered end of a melting point capillary was inserted into the pile of 3-mm spheres, and the liquid in the tube behaved erratically as it followed the pressure changes in the bead pack during the evaporation of a volatile liquid. These pressure fluctuations have also been recorded using a pressure transducer.

These experiments clearly demonstrate that displacement on a microscale is not proceeding reversibly because spontaneous changes in the fluid configuration take place.

2.4.2. Displacement Mechanism

Morrow (1970) considered infinitesimally slow withdrawal of liquid from the capillary model shown in Fig. 2.31. Liquid can be removed reversibly

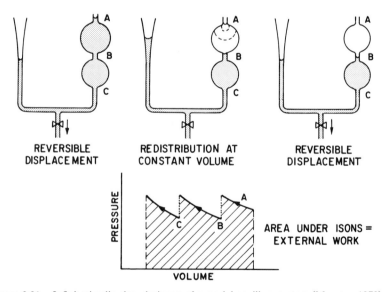

Figure 2.31. Infinitesimally slow drainage of a model capillary system (Morrow, 1970).

from both tubes of the model until the meniscus in the tube on the right meets the enlargement in tube cross section at A. From A the liquid descends rapidly through the upper bulb to a stable position in the vicinity of B. It can be seen from the diagram that spontaneous redistribution of the fluid results in a sudden reduction of the capillary pressure. If the capillary pressure were plotted against the volume of liquid in the model, we would see that displacement under these idealized circumstances consists of smooth reversible changes linked by spontaneous changes in pressure at constant saturations. Morrow suggested the term "ison" for the reversible displacement and "rheon" for the "Haines jump" or irreversible redistribution. No work is done in the constant volume part of the process.

When capillary pressures are measured in a sample of a porous medium, the local discontinuities in pressure are generally too short range to observe and the experimental capillary pressure curves appear to be smooth and continuous.

2.4.3. Efficiency of Displacement

Morrow (1970) determined surface free energy changes directly in various immiscible displacement processes using colored liquid epoxy resins as fluids. After the resin hardened, thin sections were prepared and the relative fractions of space occupied by the "liquid" and the "air" phase as well as the ratios of solid–liquid, solid–air, and liquid–air interfaces were determined by point counting.

The increase in surface free energy over the reference state given by solid overlain by liquid epoxy, due to drainage, was set directly proportional to the sum of the solid–air and liquid–air surfaces. (Note that liquid epoxy spreads on glass.)

The inherent efficiency E_D of conversion of work to the creation of the surface is given by the increase in free energy in the displacement process divided by the work of displacement:

$$E_D = \frac{\sigma_{la}(\Delta S_{sa} + \Delta S_{la})}{V_B \phi \int_{100}^{S_{wl}} P_c \, dS_w},$$

(2.4.1)

where l refers to liquid, a to air, and s to solid surface. S is surface area and V_B is the bulk volume of the porous solid.

Payne (1953) found that the work required to drain a variety of beds of spherical particles was almost equal to the product of the interfacial tension and the total solid surface area of the beads. This is an interesting result because the bead packs were not drained beyond the "irreducible" saturation. Payne's results indicate that the work spent on draining the bed was equal to the amount that would have been required for complete removal of the wetting phase had the process been reversible. (In fact, the process is not reversible and, therefore, the work spent resulted in less than complete removal of the wetting phase. The amount of surface created is less than the total solid surface of the bed.)

Morrow (1970) used Eq. (2.4.1), letting the value of the denominator be 100 units, to evaluate the efficiency of the displacement process. He found that the inherent thermodynamic efficiency of drainage from 100% saturation to the irreducible wetting phase saturation was 79%.

Morrow (1970) also determined the efficiencies of imbibition and secondary drainage. In the imbibition process, surface free energy of the porous system is converted into external work of displacement. The area under the imbibition curve was about 60% of the area under the drainage curve. Hence, 60% is the measure of the work done in the imbibition displacement. The measure of the surface free energy that was available to do the work is the amount present at irreducible wetting phase saturation (i.e., 79% of the total solid surface). However, after imbibition is complete, the system still has surface free energy due to entrapped nonwetting phase. *This value has to be added to the work done.* Morrow (1970) determined this quantity by performing microscopic analysis on samples that were first saturated with liquid epoxy resin, drained, and then imbibed with liquid epoxy, which was allowed to solidify. Microscopy showed that the sum $S_{sa} + S_{la}$, as percent of the total solid surface, was 13%. Adding this to the 60% for recovered work, the

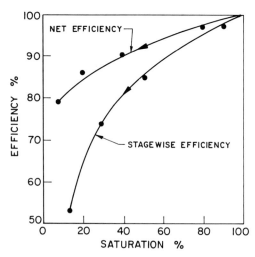

Figure 2.32. Net efficiency and stagewise efficiency versus saturation for initial drainage (Morrow, 1970).

imbibition efficiency was calculated as $73/79 = 92.5\%$. Note that this value is considerably higher than the efficiency of initial drainage.

The efficiency of secondary drainage can be defined as the ratio of the final surface energy (79 units) to the sum of the initial surface free energy (13 units) and the work done on the system, that is, the area under the secondary drainage capillary pressure curve, expressed as a percent of the area under the primary drainage curve (89 units). This gives a value of $79/102 = 77.5\%$.

Using these techniques, Morrow (1970) also determined the displacement efficiencies of initial drainage at intermediate saturations as a function of saturation. The cumulative net efficiency and the differential (stagewise) efficiency for the initial drainage process are shown as functions of saturation in Fig. 2.32. The differential efficiency is defined as

$$(E_D)_{S_w} = \frac{\sigma_{la}\, d(S_{la} + S_{sa})}{V_B \phi P_c\, dS_w}.$$ (2.4.2)

In Fig. 2.32, at 100% saturation, the efficiencies are also 100%. The efficiency of the process is seen to decrease rapidly with decreasing saturation.

The result that the work required for displacement is proportional to the area under the capillary pressure curve has been used in the development of the USBM wettability index.

2.5. PORE STRUCTURE MODELS

Pore structure models may be subdivided into two broad categories. In the first category are models that consist of arrays of (mostly) spherical particles. The second category where the pore structure is modeled by arrays of capillary tubes is usually referred to as bundle of capillary tubes and network models.

In the first category, filling and emptying of the pore spaces can be calculated as a function of the capillary pressure only in the case of some relatively simple arrangements. These models offer ready qualitative explanations for capillary hysteresis, irreducible wetting phase, and residual nonwetting phase saturations. They are suited primarily for capillary pressure calculations in packs of smooth beads. Network modeling has been improved a great deal over the 25 years, and it is now ready for quantitative predictive purposes.

2.5.1. "Bundle of Capillary Tubes" Model and Its Modifications

The simplest possible model of pore structure for capillary pressure purposes consists of a bundle of cylindrical capillary tubes of different diameters and equal lengths. Every capillary has a uniform diameter along its entire length. The drainage capillary pressure curve can be interpreted in terms of this simple model by postulating that at the threshold capillary pressure of penetration of the nonwetting fluid, the largest capillary gets penetrated and filled, and at increasing capillary pressures, increasingly smaller tubes become filled. It is evident that this simple model is incapable of accounting for capillary hysteresis and for the existence of residual wetting and nonwetting phase saturations.

2.5.1.1.   *Mercury Porosimetry*

Drainage capillary pressure curves obtained by mercury intrusion porosimetry are customarily interpreted in terms of the bundle of capillary tubes model.

The so-called pore size distribution of the model can be derived from the capillary pressure curve (Ritter and Drake, 1945) by assigning the volume dV to pores having entry diameters between D_e and $D_e + dD_e$:

$$dV = \alpha(D_e)\, dD_e. \tag{2.5.1}$$

Here $\alpha(D_e)$ is a (volume based) density function of pore diameter distribution. Using Eq. (2.3.1), assuming $\phi = 0$ and letting $R = D_e/2$, in the case of constant and uniform values of the contact angle and the interfacial tension, one obtains by differentiation

$$P_c\, dD_e + D_e\, dP_c = 0. \tag{2.5.2}$$

Eliminating D_e and dD_e from Eqs. (2.3.1), (2.5.1), and (2.5.2) gives

$$dV = -\alpha(D_e)\frac{4\sigma \cos\theta}{P_c^2}\, dP_c = -\alpha(D_e)\frac{D_e}{P_c}\, dP_c. \tag{2.5.3a}$$

Equation (2.5.3a) is often written in the following form:

$$\alpha(D_e) = \frac{P_c}{D_e}\frac{d(V_T - V)}{dP_c}, \tag{2.5.3b}$$

where V_T is the total pore volume of the sample and V is the volume of pores with entry diameters smaller than D_e.

If V_T is interpreted as the drainable volume in the sample, Eq. (2.5.3b) can be used to model a primary drainage capillary pressure curve while taking into account the residual (connate) wetting phase saturation.

It is generally understood that these equations assign the volume of pores to the diameter D_e of the entry pores (necks) through which they are accessible. Drainage capillary pressure curves normally do not give any information on the size of the pores that are accessible through the entry pores.

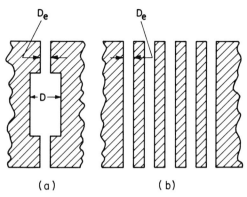

Figure 2.33. Illustration of error introduced by using pore entry diameter in parallel-type capillaric model: (a) actual pore, (b) false pore structure. (Reprinted with permission from Dullien, 1975.)

The type of error committed by using Eqs. (2.5.3) is illustrated in Fig. 2.33, which shows how the pore structure of a sample containing bulges sandwiched between necks is evaluated, erroneously, from the information contained in the drainage capillary pressure curve.

An example of the distortion in the pore size distributions calculated from the drainage capillary pressure curves is shown in Fig. 2.34, where the mercury intrusion porosity pore size distribution curve of sandstone is compared with the pore size distribution obtained by optical means. It is apparent from this figure that mercury porosimetry does not detect the presence of the larger pores and assigns their volume to the pore entry necks. The curves shown in Fig. 2.34 are typical of sandstones. Dullien and coworkers have determined similar curves for a large number of sandstone samples (see Dhawan, 1972; Batra, 1973).

Notwithstanding these experimental verifications of the inadequacy of mercury porosimetry for even approximate characterization of the distribution of pore volume by the appropriate pore dimensions, owing to its simplicity the method is used widely. Some people may believe that the bundle of tubes model approximates the actual pore structure. **In fact, nothing could be farther from the truth.**

In spite of the distortion arising from not giving any information on a wide range of pore sizes and assigning the wrong volumes to the pore diameters that have been detected, the information contained in the drainage capillary pressure curve is very important. This is because the resistance offered by the pore structure to various transport phenomena is controlled by the pore necks, and the sizes of the entry pore necks are deduced from the drainage capillary pressure curves.

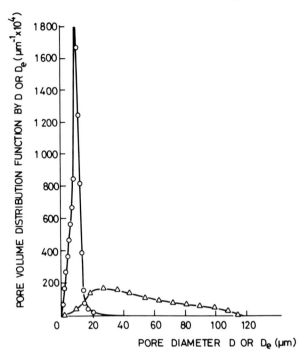

Figure 2.34. Comparison of mercury intrusion pore size distribution with photomicrographic pore size distribution of a sandstone. The points plotted are pore entry diameter distribution D_e ($\bigcirc$) and complete pore size distribution D ($\triangle$). (Reprinted with permission from Dullien, 1975.)

In summary: The pore size distributions calculated from drainage capillary pressure curves should be used with discretion and, preferably, in combination with pore size distributions obtained by optical means. This approach has been emphasized by Dullien and Dhawan (1975) and has resulted in reasonably successful prediction of various transport phenomena, which are discussed in Chapter 3.

Mercury intrusion porosimetry has an extensive literature (see, for example, Modry *et al.*, 1972). As pointed out in Chapter 1, the technique was introduced by Ritter and Drake (1945). A mercury intrusion porosimeter, which has been used by the author and coworkers to determine hysteresis loops and scanning curves (see Fig. 2.14) in sandstones and which permits pore size measurements down to about 1 μm, is shown in Fig. 2.35. There are various special designs for measuring large pores, requiring low capillary pressures (Cameron and Stacey, 1960; Leppard and Spencer, 1968; Baker and Morris, 1971). Mercury intrusion porosimeters have been designed also for very high pressure that can determine pore sizes below 40 Å.

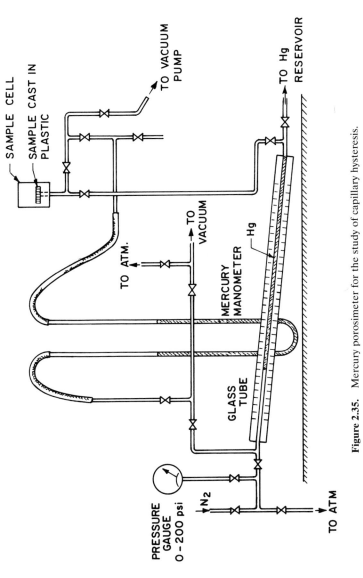

Figure 2.35. Mercury porosimeter for the study of capillary hysteresis.

Mercury intrusion porosimetry has been used to investigate a large number of different materials, such as active carbons, adsorbents, porous catalysts, coal and coke, various rocks, metals, electrodes, ceramics, binding materials (such as cements, plasters, concretes), plastics, paper, textile, leather, wood, etc.

Wardlaw and Taylor (1976) have determined mercury porosimetry curves of reservoir rocks. After mercury was injected into the sample, mercury withdrawal was carried out by gradually decreasing the injection (capillary) pressure. In this manner imbibition and also scanning curves were obtained. For some, but not all, of the different rock samples the withdrawal efficiency was found to increase with higher mercury saturation before withdrawal.

Lai *et al.* (1981) determined primary and secondary drainage, primary imbibition, ascending and descending scanning curves for a Clear Creek sandstone sample, using the mercury porosimeter shown in Fig. 2.35, in order to test the applicability of the independent domain theory of capillary hysteresis. The model was found to apply well to the permanent hysteresis loop (i.e., the region enclosed by the imbibition and the secondary drainage curve), excluding trap hysteresis (i.e., those pores containing "residual" mercury saturation). This limitation of the theory excludes the possibility of determining pore size distribution (including the distribution of both pore body and pore throat sizes) from capillary pressure curves.

2.5.1.2. *Theoretical Analysis of Mercury Intrusion Porosimetry*

Meyer (1953) worked out a computational procedure, based on probability considerations, to correct mercury porosimetry curves. His procedure, however, has not been used widely. No probability consideration is able to determine the unknown sizes of those larger pores entered by the mercury through narrower necks.

Frevel and Kressley (1963) subjected the penetration of mercury into a pack of uniform spheres to theoretical analysis. They pointed out that the characteristic behavior for a pack of particles is an abrupt threshold for mercury penetration. Their model consists of close-packed layers of spheres. The stacking of the layers was assumed to vary between completely close-packed nesting and vertical stacking of particles in layers one above the other. This model gives rise to a range of porosities from 39.54 to 25.95%.

Mayer and Stowe (1965) considered a different model consisting of uniform spheres varying in packing between the two extremes of three-dimensional close packing and three-dimensional cubic packing. The size and shape of the largest openings are identical, as in the model of Frevel and Kressley. In the model of Mayer and Stowe there are six square openings, whereas in the model of Frevel and Kressley there are four square openings and four triangular ones. The model of Mayer and Stowe was defined in terms of a single acute angle, the size of which is determined by the edges of the rhombohedron formed by connecting the centers of a cluster of spheres (see

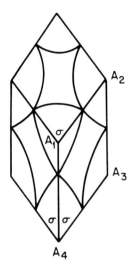

Figure 2.36. Unit cell of packed spheres (Mayer and Stowe, 1965).

Fig. 2.36). This model changes packing equally in all three dimensions. At a given value of the angle, the size and shape of the largest opening is identical for either model; the corresponding porosities, however, differ. Mayer and Stowe calculated the breakthrough pressure in the pack in terms of the porosity and the contact angle. This model is such that only in complete three-dimensional close packing will the square windows split into triangular ones. The mercury penetrating through the window (see Figs. 2.37 and 2.38),

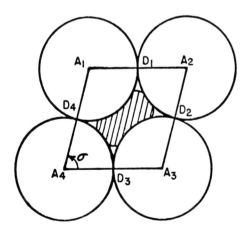

Figure 2.37. Access opening to unit void (Mayer and Stowe, 1965).

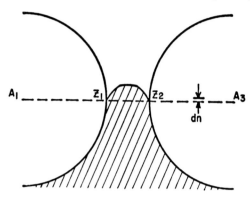

Figure 2.38. Mercury penetration between spheres (Mayer and Stowe, 1965).

however, separates into two branches somewhat before complete three-dimensional close packing is reached, depending on the value of the contact angle. Calculated data, covering the porosity range of 47.64–25.95% and contact angles of 90–180° were presented. The relationships derived by Mayer and Stowe allow the determination of particle radius from standard mercury penetration and porosity data.

In a second paper, Mayer and Stowe (1966) extended their analysis to the filling of the toroidal void volumes (pendular rings) following breakthrough between packed spheres. The solution of their equations gives the toroidal volume as a function of the contact angle and the capillary pressure. They pointed out that the assumption of Frevel and Kressley (1963)—the mercury passes through the access openings in shape of the inscribed circle—leads to a fraction of the space filled following penetration that is lower than the real value. Mayer and Stowe compared their results on the later part of the penetration branch with Kruyer's (1958) results on the early part of the retraction branch of mercury and demonstrated excellent agreement between the results of the two independent methods.

Kruyer (1958) determined the capillary pressure hysteresis loop of mercury in a bed of glass spheres (see Fig. 2.39) and treated the problem of retraction of mercury. He argued that when the capillary pressure is decreased, the size of the pendular rings will increase and finally, at the point where these interfere with each other, the entire pore space will be empty of mercury. This point is determined by the radius of the sphere inscribed in the interstices (see Fig. 2.40). Mercury penetration, on the other hand, will start as soon as a stable surface can pass from one interstice into a neighboring one. The size of the passage is approximately equal to the radius of the circle inscribed in the opening. The radius of this circle is invariably smaller than the radius of the sphere controlling the retraction of mercury from the

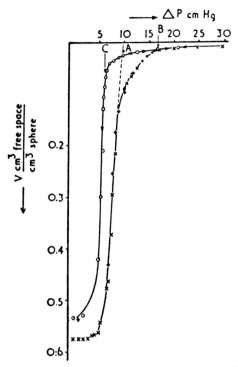

Figure 2.39. Penetration, retraction of mercury in glass spheres 0.42–0.35 mm, and penetration after retraction. (Reprinted with permission from Kruyer, 1958.)

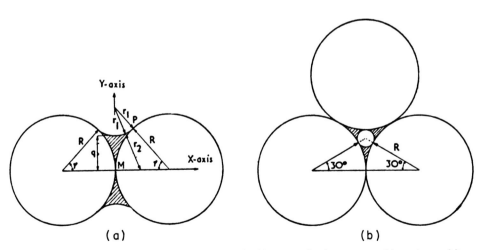

Figure 2.40. Surface of completely wetting liquid penetrating between touching spheres: (a) side view, (b) top view. (Reprinted with permission from Kruyer, 1958.)

interstices; therefore, it takes higher pressure to fill the pack than to empty it, resulting in hysteresis. The trapping of some of the mercury was not explained by Kruyer's treatment.

In the three theoretical treatments of mercury porosimetry just reviewed, "penetration" into the bed was synonymous with "breakthrough" across the entire bed, because the controlling pore necks in the bed were of the same size. A random distribution of pore sizes over the network results in a range of controlling neck sizes, and in such cases "penetration" does not coincide with "breakthrough," but precedes it (Chatzis, 1976, 1980; Chatzis and Dullien, 1977, 1982, 1985).

2.5.1.3. *Adsorption in Porous Media*

Introduction. Adsorption of gases and vapors on solid surface is a highly specialized and important subject on which a vast literature exists, including several treatises (e.g., Brunauer, 1945; Flood, 1967; de Boer, 1958). The interested reader is referred to the literature as well as to texts on surface chemistry (e.g., Defay and Prigogine, 1966; Adamson, 1967) for a more detailed discussion of this subject, where also a good introduction into the subject of adsorption on solid surfaces from solutions can be found. In this section only a brief introduction into this field is given.

Speaking in general terms, adsorption occurs because of a lack of thermodynamic equilibrium between the gas or vapor and the solid surface in contact with it. Equilibrium is achieved by accumulation of the molecules of the gas or vapor, the "adsorbate," at the solid surface, the "adsorbent." As a result of adsorption, an "adsorbed layer" of the adsorbate is formed on the adsorbent. The process of adsorption continues until the adsorbed layer is in thermodynamical equilibrium with the gas or vapor in contact with it. Experience has shown that, for a given pressure below the bulk vapor pressure of the liquid at the existing temperature, a finite amount of adsorption may lead to equilibrium. In thermodynamic terminology it is said that in equilibrium the chemical potential of the adsorbed layer μ_a is equal to the chemical potential of the vapor μ_v''. Before reaching equilibrium, μ_a is less than μ_v''. When dn moles of the adsorbate are transferred from the vapor to the adsorbed layer, at constant temperature and pressure of the vapor phase, the change in the Gibbs free energy is

$$dG_{T,P} = (\mu_a - \mu_v'') \, dn. \qquad (2.5.4)$$

Note that $dG_{T,P}$ is negative, that is, the Gibbs free energy decreases while the system is approaching equilibrium. At equilibrium, transfer of molecules of adsorbate from the vapor to the adsorbed layer or vice versa does not result in any change in the value of the Gibbs free energy (i.e., in equilibrium):

$$dG_{T,P} = 0. \tag{2.5.5}$$

Similar considerations apply in the case of adsorption from a solution, except the chemical potential of the gas is replaced with that of the solute in the solution.

All gases below their critical temperatures tend to adsorb as a result of van der Waals forces between the gas molecules and the solid surface. This is the case of so-called "physical adsorption" as distinct from "chemisorption." Physical adsorption equilibrium is attained very rapidly (except when limited by mass transport rates in the gas phase or within a porous adsorbent) and is reversible. Reversibility means the existence of an adsorption–desorption equilibrium, where "desorption" means negative adsorption.

Chemisorption may be rapid or slow and may occur above or below the critical temperature of the adsorbate. As the name suggests, in chemisorption certain chemical bonding is formed between the adsorbent and the adsorbate molecules. A chemisorbed substance may be difficult to remove from the surface of the adsorbent and there is generally no adsorption–desorption equilibrium in chemisorption. For example, oxygen adsorbed on charcoal at room temperature is held very strongly; pulling vacuum on the charcoal will not remove the adsorbed oxygen from the charcoal. It can be removed by heating the charcoal, but then it comes off in the form of carbon monoxide.

Because of chemical bonding, chemisorption is expected to be limited to a monolayer of molecules on the adsorbent. In contrast, physical adsorption does not stop at monolayer formation but may advance into multilayer adsorption.

There is no sharp dividing line between physical adsorption and chemisorption. There are cases of adsorption possessing characteristics that are intermediate between physical adsorption and chemisorption.

Adsorption Isotherms and Capillary Condensation. In the case of adsorption of gases and vapors on solids, adsorption is usually described phenomenologically in terms of an empirical adsorption function

$$V = f(P'', T), \tag{2.5.6}$$

where V is the amount adsorbed, usually expressed as cubic centimeters at standard T and P per gram of adsorbent. As a matter of experimental convenience one usually determines "adsorption isotherms"

$$V = f_T(P'') \tag{2.5.7}$$

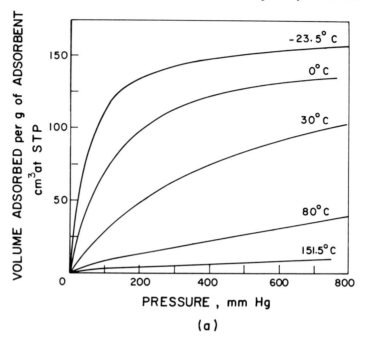

Figure 2.41. (a) Adsorption isotherms for ammonia on charcoal; (b) adsorption isobars of ammonia on charcoal (Brunauer, 1945). (c) Adsorption isosteres of ammonia on charcoal. (From Brunauer, *Physical Adsorption*, Vol. I in *The Adsorption of Gases and Vapors* © 1971 by Princeton University Press, Figs. 3, 11, and 15, pp. 14–25. Reprinted by permission of Princeton University Press.)

at several different temperatures. The data plotted as V versus T at constant P'' are called "adsorption isobars," and data plotted as P'' versus T at constant V are referred to as "adsorption isosteres." An example of the three different representations of data is shown in Fig. 2.41.

A large number of theoretical or semiempirical equations have been developed to fit or predict the adsorption isotherms. One of the best investigated of these equations is due to Langmuir (1918):

$$P''/V = 1/bV_m + P''/V_m,$$ (2.5.8)

where V_m denotes the STP cubic centimeters per gram adsorbed at the complete monolayer point and b is a constant. A plot of P''/V versus P'' should give a straight line, and the two constants V_m and b may be evaluated from the slope and the intercept.

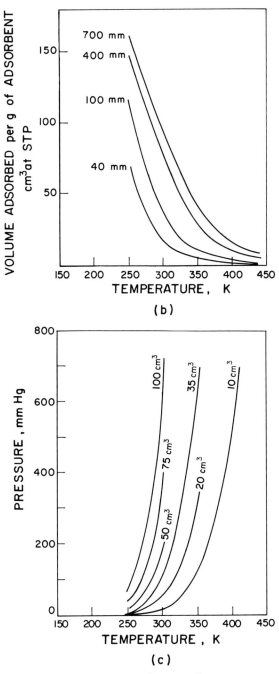

(b)

(c)

Figure 2.41. (*Continued*)

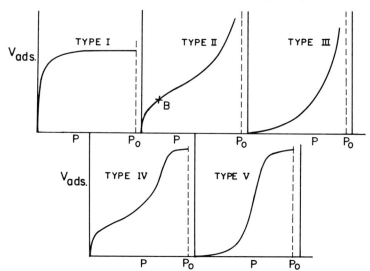

Figure 2.42. Brunauer's five types of adsorption isotherms. (From Brunauer, *Physical Adsorption*, Vol. I in *The Adsorption of Gases and Vapors* © 1971 by Princeton University Press, Fig. 68, p. 150. Reprinted by permission of Princeton University Press.)

Adsorption isotherms are by no means all of the Langmuir type, and Brunauer (1945) considered that there were five principal forms, as shown in Fig. 2.42. Type I is the Langmuir type, characterized by a gradual approach to limiting adsorption that is supposed to correspond to a complete monolayer. Type II is very common in the case of physical adsorption and corresponds to multilayer formation. It used to be the practice to take point B at the knee of the curve, as the point corresponding to complete monolayer. Type III is relatively rare. Types IV and V reflect capillary condensation phenomena because they level off before the saturation pressure is reached. These may show hysteresis effects.

Brunauer *et al.* (1938) extended Langmuir's approach to multilayer adsorption, and their equation has become popular under the name of the "BET equation." The BET equation fits isotherms of types II and III, in addition to type I:

$$x/V(1 - x) = [1/cV_m] + [(c - 1)x/cV_m], \qquad (2.5.9)$$

where $x = P''/P_0$ (P_0 is the vapor pressure of bulk liquid). The values of the constants V_m and c can be obtained from the slope and intercept of the best fitting straight line from a plot of $x/V(1 - x)$ versus x.

Both the Langmuir equation and the BET equation have been used to calculate the specific surface area of the solid S with the aid of the following

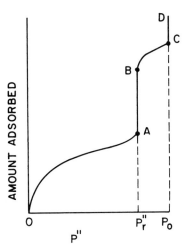

Figure 2.43. Isotherm for adsorption of vapor by an ideal porous body (Defay and Prigogine, 1966).

equation:

$$V_m = V_0 S / N \sigma^o, \tag{2.5.10}$$

where σ^o is the area of a site (i.e., the area occupied by a single adsorbed molecule), V_0 is the STP number (22,400 cc/mole), and N is the Avogadro number.

There are many other equations to predict or fit adsorption isotherms, and many different approaches exist to the problem of adsorption. Their discussion is outside the scope of this monograph, however.

When an evacuated porous solid, in which all the pores are of equal size, is exposed to a gas below its critical temperature (vapor) whose pressure is steadily increased, the phenomena expected to occur are as shown diagrammatically in Fig. 2.43. First, from 0 to A the gas will be adsorbed by the whole solid surface. The shape of this curve may be explained in general terms by the BET theory in which it is supposed that the vapor forms, in succession, several adsorbed layers on the solid. At A, capillary condensation commences, and the amount of liquid adsorbed increases at constant pressure $P'' = P''_r$ (P''_r is the vapor pressure of the liquid present in the pores) up to the point at which the meniscus reaches the mouths of the pores (B). Between B and C further condensation causes the menisci to flatten until at C they are flat, the vapor is saturated, and condensation of the bulk liquid can take place (Defay and Prigogine, 1966).

In the case of real porous solids with distributed pore sizes, small pores will fill first and the largest will not begin to fill until the menisci in the smallest pores have already begun to flatten. Therefore, in real porous solids

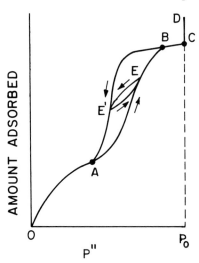

Figure 2.44. Isotherm for adsorption of vapor by real porous body (Defay and Prigogine, 1966).

the ideal vertical section AB in Fig. 2.43 becomes an oblique section AEB shown in Fig. 2.44.

In case the liquid does not wet the surface completely during the filling of the pores, the nonzero value of the contact angle will have the effect of increasing the radius of curvature of the meniscus and of displacing the line AB to the right. If the nonzero contact angle is due to the initial presence on the surface of an adsorbed layer of inert gas, then in the process of filling the pore the adsorbed gas may be displaced from the surface, resulting in a zero contact angle during the reverse process of desorption of the adsorbed liquid from the pores. Consequently, the equilibrium pressures during desorption will lie to the left of the corresponding points on adsorption. The result is a hysteresis loop, shown in Fig. 2.44. Hysteresis is often observed in adsorption isotherms of vapors. Zsigmondy (1911) explained hysteresis of capillary condensation in terms of hysteresis of the contact angle along the lines of this discussion. However, according to this explanation hysteresis should disappear in later experiments, but hysteresis is not in general eliminated, and often the hysteresis loop can be reproduced indefinitely after the first cycle. In other words, hysteresis of capillary condensation usually is a so-called permanent hysteresis.

An alternative explanation of hysteresis is based on the fact that when a liquid spreads on a macroscopic surface it is often found that the advancing and receding contact angles are different. Harkins and coworkers (see Harkins, 1952) have shown, however, that when the surface is clean and smooth the two angles are identical. This is not the case however, when the surface is rough. Bikerman (1958) suggested an explanation of hysteresis of capillary condensation in terms of the dependence of the contact angle on surface roughness.

As discussed in detail later in this section, permanent sorption hysteresis can be accounted for by pore structure considerations.

Pore Size Distribution from Capillary Condensation. Capillary condensation has been used to calculate the distribution of pore sizes, mostly using either the "bundle of capillary tubes" model or the "parallel plate" pore model. The fundamental equation used in relating pore size to vapor pressure is Kelvin's equation [Eq. (2.3.10)].

Wheeler (1946) was apparently the first to point out the necessity of taking into account the thickness of the adsorbed multilayer of gas when applying the Kelvin equation. This is necessary because in very small pores the thickness of the adsorbed multilayer is a significant fraction of the pore radius.

Basic equations and methods: The fundamental equations to calculate pore size distribution from capillary condensation isotherms are as follows:

$$V_s - V = \pi \int_R^\infty (R - t)^2 L(R) \, dR, \qquad (2.5.11)$$

where V_s is the volume of adsorbate at saturation vapor pressure (equal to the total pore volume), V is the volume of adsorbate at intermediate vapor pressure P'', $L(R) \, dR$ is the total length of pores whose radii fall between R and $R + dR$, R is the pore radius, and t is the multilayer thickness that is built up at pressure P''. Equation (2.5.11) merely states the fact that the volume of adsorbate not yet adsorbed at pressure P'' is equal to the volume of pores that has not yet been filled. The Kelvin equation for pores of effective radius $R - t$ is

$$R - t = -[2\sigma V'/\mathscr{R}T \ln(P''/P_0)]. \qquad (2.5.12)$$

In order to be able to evaluate these equations t must be calculated as a function of P''/P_0. This was done first by Shull (1948), using nine published isotherms of nitrogen gas on crystalline materials of large crystal size, by assuming the thickness of one monolayer to be 4.3 Å for nitrogen. In subsequent years the t curve, as it is now called, became an important tool in pore structure analysis.

The physical content of Eq. (2.5.11) is analogous to Eq. (2.5.1).

Shull (1948) assumed that the pore size distribution may be represented by simple analytical forms of the Maxwellian or Gaussian type. These were substituted into Wheeler's equation [Eq. (2.5.11)], permitting this equation to be integrated analytically and resulting in two- or three-parameter expressions for $V_s - V$. Families of standard isotherms were prepared, and the values of the parameters were found by comparing the experimental isotherms with the standard ones. The values of the parameters thus determined the pore size distribution function directly.

Numerous attempts have been made to develop more general solutions to the pore size integral. For more literature references, see, for example, the work of Dullien and Batra (1970).

Barrett *et al.* (1951) worked out a technique that has become popular. The total volume of adsorbate lost in a desorption step has been expressed by them as follows:

$$\Delta V = \Delta V_c + \Delta V_{mt}, \qquad (2.5.13)$$

where ΔV_c is the volume corresponding to capillary desorption and ΔV_{mt} the multilayer desorption. If it is assumed for the moment that ΔV_{mt} may be calculated, then ΔV_c for the desorption step is obtained. This volume is the capillary liquid lost from the porous medium between the beginning and the end of the step between two relative pressures P_1''/P_0 and P_2''/P_0. The actual volume of the pores emptied in the desorption step is

$$\Delta V_p = R_n \, \Delta V_c, \qquad (2.5.14)$$

where

$$R_n = R_p/(\bar{R} + \Delta t)^2, \qquad (2.5.15)$$

with R_p being the actual pore radius, $\bar{R}$ the mean capillary condensate radius for the two relative pressures P_1''/P_0 and P_2''/P_0, and Δt the decrease in multilayer thickness for the desorption step.

The major problem is how to find ΔV_{mt} accurately. The value of ΔV_{mt} for a particular step would appear to be $\Delta t \Sigma S_p$, where ΣS_p is the sum of the surface areas over all steps up to and including the preceding one. However, this assumes the area ΣS_p to be planar, whereas it is actually composed of curved walls of the pores. For each step, S_p is obtained, assuming cylindrical pores, by

$$S_p = 2 \, \Delta V_p/R_p. \qquad (2.5.16)$$

To allow for the curvature, Barrett *et al.* used a factor C equal to

$$C = (R_p - R)/R_p \qquad (2.5.17)$$

and assigned various constant values to this, according to the approximate range of pore sizes expected. This procedure gives reasonably accurate results for pores down to radii of 35 Å. However, for pores of smaller diameter, the value of C changes with pore size too rapidly to be correctly assigned a constant value.

The final equation developed by Barrett *et al.* is

$$\Delta V_p = R_n\left(\Delta V - C \Delta t \sum S_p\right). \qquad (2.5.18)$$

Cranston and Inkley (1957) made the BJH (Barrett, Joyner, and Halenda) method more precise by using a variable C both for the step being calculated and for each of the pore sizes. They multiplied S_p for each pore size range by its individual C, instead of simply summing S_p. While more precise, their method is also far more laborious than the BJH method.

The method of Dollimore and Heal (1964) is more refined mathematically than either the BJH or the Cranston and Inkley method. They used the equation of Halsey (1948).

$$t = 4.3\left[5/\ln(P_0/P'')\right]^{1/3} \qquad (2.5.19)$$

to calculate t as a function of P''/P_0, resulting in variable t from one step to the next. By using this equation there was no need to use a correction factor to allow for the varying curvature of the capillary.

The model of pore structure used in all of the preceding methods consists of straight cylindrical capillaries of uniform cross section, closed at one end. This is a highly idealized model. In the majority of cases, the choice of the adsorption rather than the desorption isotherm would result in very different pore size distribution curves. Because of the crudeness of the model, any increased precision of the calculations can be expected to result in very little real improvement in the accuracy of the pore size distribution obtained. Only by using more realistic pore models can one hope to obtain correct pore size distributions.

The various authors usually compared, as a check, the total surface areas obtained by their respective techniques with the value obtained by the BET method.

Shull (1948) and Barrett *et al.* (1951) used the desorption isotherms and obtained good agreement with BET surface areas in most cases. Cranston and Inkley (1957), however, performed a more comprehensive study of this question by comparing the surface areas calculated from both the adsorption and the desorption branch of the isotherms with the BET areas. This comparison showed, on the average, better agreement in the case of adsorption isotherms. The surface areas calculated by Bernardini *et al.* (1967) were strikingly different from the BET areas. Voight and Thomlinson (1955) used the adsorption branch for porous Vycor glass, where the pores were mostly of the so-called ink bottle type. They obtained poor agreement with the BET surface area. They also demonstrated a considerable variation in the calculated surface area depending on what substance was adsorbed.

It is clear from experimental evidence that it depends on the type of the porous material whether the adsorption or the desorption branch is to be preferred for pore size distribution calculations. It is noted also that agree-

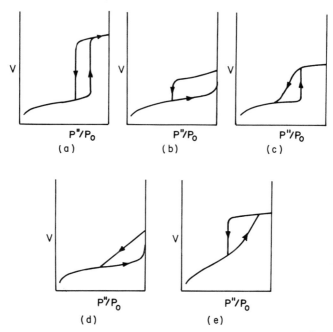

Figure 2.45. The five types of hysteresis loop with at least one steep part (de Boer, 1958).

ment with BET surface areas does not prove that the pore size distribution is accurate.

Sorption Hysteresis and Pore Structure. The problem of a relationship between the type of hysteresis loop and the corresponding capillary shape was approached in great generality by de Boer (1958), who distinguished five types of hysteresis loops (see Fig. 2.45).

Fifteen shape groups of capillaries were considered by de Boer, and the types of hysteresis loop that may result from capillary condensation were analyzed. The reader is referred to the original paper for details of this enlightening treatise: the nature of the problem can be appreciated by considering capillaries of varying widths, open at both ends, as an example (see Fig. 2.46).

One can expect capillary condensation to take place first in the narrow parts at a relative pressure corresponding to the effective radius of curvature r_m of an inscribed cylinder in the narrow part, as given by the Kelvin equation

$$P''/P_0 = \exp[-(2\sigma V'/\mathscr{R}Tr_m)] \qquad (2.5.20)$$

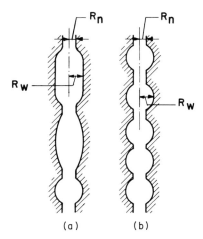

Figure 2.46.   Tubular capillaries with slightly widened parts (de Boer, 1958).

where, for a cylinder of radius R_n,

$$r_m = 2R_n. \tag{2.5.21}$$

As soon as the narrow parts are filled with liquid, however, spherical menisci are formed for which the effective radius of curvature r_m is

$$r_m = R_w, \tag{2.5.22}$$

where R_w is the radius of the wider portions (bulges).

Providing that $R_w \leq 2R_n$, the equilibrium relative pressure corresponding to $r_m = R_w$ is lower than the prevailing pressure and, as a result, the wider parts of the capillary are filled in a nonequilibrium process at a constant relative pressure corresponding to $r_m = 2R_n$.

On the other hand, the entire capillary is emptied in the desorption process at the relative pressure corresponding to the effective radius of curvature $r_m = R_n$. The desorption from the wider part is also a nonequilibrium process because the relative pressure corresponding to $r_m = R_n$ is lower than the value corresponding to $r_m = R_w$. Equations (2.5.21) and (2.5.22) have applications also in imbibition, which has the same mechanism in real porous media (i.e., other than packs of smooth glass beads) as capillary condensation. The transport of wetting phase is provided by means of thick films in surface grooves in one case and vapor phase diffusion, in the other.

De Boer and coworkers used these ideas to analyze hysteresis loops of a variety of substances and published their results in a series of papers starting in 1964. These works represent the most consistent effort made so far to investigate pore structure in depth by analyzing sorption isotherms.

Lippens *et al.* (1964) introduced a modified *t* curve by letting

$$t = (X/S) \times 10^4 \text{ Å},\qquad(2.5.23)$$

where X is the adsorbed volume of liquid adsorbate and S is the specific surface area per gram of adsorbent. They treated the adsorbate consistently as a close-packed liquid and let $S = S_{BET}$. This results in the equation

$$t = 15.47(V_a/S_{BET}) = 3.54(V_a/V_m)\quad\text{Å},\qquad(2.5.24)$$

where V_a is the adsorbed volume of adsorbate in mliter gas STP per gram of adsorbent and V_m is the volume of gas in mliter STP per gram of adsorbent that should be able to cover the whole surface with a unimolecular layer. The value 3.54 Å for the statistical thickness of a monolayer differs from the value 4.3 Å used by Shull and Barrett *et al.*

De Boer and coworkers developed the *t* curve into a powerful tool of pore structure analysis. By using Eq. (2.5.24), *t* was calculated as a function of the relative pressure from experimental results on a variety of different non-porous samples to avoid the influence of capillary condensation (de Boer *et al.*, 1965a). Inasmuch as one can be certain that capillary condensation was absent, the *t* curve appears to be universal; no appreciable influence of the chemical nature of the surface could be observed.

The universal *t* versus P''/P_0 relationship was used by de Boer and coworkers (1965b) and Lippens and de Boer (1965) to investigate pore structure of unknown samples along the following lines. The measured adsorbate volume V_a is converted into a function of *t*, and the V_a versus *t* plot is analyzed. As long as the multilayer is formed unhindered, a straight line is obtained that goes through the origin the slope of which is a measure of the surface area:

$$S_t = 15.47V_a/t.\qquad(2.5.25)$$

One might expect S_t to be equal to S_{BET}; however, this is only approximately true because the *t* curve is an average of various samples. At higher relative pressures, deviation from a straight line may occur that is probably due to capillary condensation.

At a certain pressure, capillary condensation may occur in pores of certain shapes and dimensions; the material takes up more adsorbate than corresponds to the volume of the multilayer; the adsorption branch lies above the *t* curve; and the slope of the V_a versus *t* plot increases.

In pores that are open on all sides (e.g., slit-shaped pores) where a meniscus cannot form until the pores are completely filled by the adsorbed layers on both parallel walls, surface area becomes gradually unavailable for adsorption. At the point in the adsorption branch where this process becomes noticeable, the V_a versus *t* plot will start having a smaller slope, corresponding to the accessible surface area left. Application of these ideas to various samples gives a picture of the pore structure that is consistent with the observed capillary hysteresis behavior.

The case of cylindrical pores open at both ends is quite different from the case of slit-shaped pores open on all sides. In the latter case, the two flat surfaces facing each other are separated by a gap and there is no direct connection between the multilayers adsorbed on them; hence, these layers can build up on both sides until they eventually bridge the gap between them and touch at a point. For the cylindrical pores open at both ends, however, the adsorbed multilayers form a close entity of cylindrical shape; therefore, capillary condensation generally occurs at some point before the capillary is completely filled with multilayers. In fact, the filling of the capillary should occur when the surface of the multilayer starts behaving like a liquid surface.

The problem of finding a quantitative explanation of the phenomenon of capillary condensation in cylindrical capillaries open at both ends attracted considerable attention. Cohan (1944) suggested that capillary condensation takes place in the cylindrical film at the walls of the pores, and consequently, the pores are filled by capillary condensate.

A thermodynamic analysis of this problem was presented by Broekhoff and de Boer (1967). The main idea in their treatment is that the thermodynamic potential of the adsorbed multilayer depends, in addition to the radius of curvature of the capillary, also on the thickness t of the layer. Therefore, for a cylinder with a given diameter, the thermodynamic potential of the adsorbed multilayer will vary continuously with t. Under these conditions the equilibrium condition is as follows:

$$\mathscr{R}T \ln P_0 - [\sigma V'/(R - t_e)] - F(t)_{t-t_e} = \mathscr{R}T \ln P'', \quad (2.5.26)$$

where P_0 is the vapor pressure over the flat surface of the bulk liquid at temperature T and P'' is the equilibrium vapor pressure over the cylindrical layer of a mean radius $R - t_e$ of thickness t_e. The function $F(t) \geq 0$ represents the effect of the surface forces of the solid walls on the thermodynamic potential of the adsorbed multilayer. Its value goes to 0 for very large t and it increases with decreasing values of t. The function $F(t)$ is given by the t curve determined on a flat surface, which was expressed in analytical form by curve fitting the data points (Broekhoff and de Boer, 1968a). The physical meaning of Eq. (2.5.26) is that the vapor pressure P'', in equilibrium with an adsorbed multilayer in a capillary is lower than P_0, the vapor pressure of the bulk liquid with a flat surface, partly because the surface of the adsorbate is curved and partly because of the proximity of the solid walls of the adsorbent.

By using the analytical expressions for $F(t)$, Eq. (2.5.26) was solved for t_e as a function of P''/P_0 and R, with nitrogen as adsorbate, and the results were tabulated. For a given P''/P_0, t_e increased with decreasing R, a fact that is evident from the physical content of Eq. (2.5.26).

The left side of Eq. (2.5.26) may be regarded as μ_a, the chemical potential of the adsorbed layer, whereas the right side is μ_v'', the chemical potential of the vapor. Therefore, when dn moles of the adsorbate are transferred from the vapor to the adsorbed layer, at constant temperature T and pressure P'' of the vapor phase, the change in the Gibbs free energy is given by Eq. (2.5.4).

While the radius of curvature is changing, the pressure in the condensed phase cannot simultaneously remain constant; however, the effect of this pressure change on G can be neglected. Equation (2.5.26) is a special case of Eq. (2.5.4) when the system is in stable equilibrium; that is, Eq. (2.5.5) holds.

For a fixed value of R there are an infinity of equilibrium situations, each characterized by a pair of values $(P''/P_0, t_e)$.

The second derivative of G is

$$d^2G/dn^2 \geq 0 \qquad (2.5.27)$$

for stable equilibrium. Under the conditions of an adsorption experiment, it is justified to assume that the pressure P'' in the vapor phase remains constant while dn moles of vapor are transferred from the vapor to the adsorbed phase. Therefore, the condition of stable equilibrium from Eqs. (2.5.27) and (2.5.26) becomes

$$-[dF(t)/dt] - \sigma V'/(R-t)^2 \geq 0, \qquad (2.5.28)$$

where the substitution

$$dn = -(dS/V')(R-t) = (2\pi L/V')(R-t)\,dt \qquad (2.5.29)$$

has been made, with dS the change in surface area accompanying the adsorption of dn moles. With the aid of an analytical expression for $F(t)$, Eq. (2.5.28) can be evaluated for various values of t_e, for a fixed value of R. The particular value of t_e for which the left side of Eq. (2.5.28) is exactly equal to zero is the "critical thickness" t_{cr}, corresponding to incipient metastable equilibrium. This value of the equilibrium thickness and the corresponding value of P_a''/P_0 define the point where, for a given R, capillary condensation takes place and the whole pore spontaneously fills with capillary condensate.

Equation (2.5.28) was solved, giving a relationship between t_{cr} and R. At the same Eq. (2.5.26) provides a relation among t_{cr}, R, and P_a''/P_0. By eliminating t_{cr} between these two equations, R has been calculated as a function of P_a''/P_0. For each P_a''/P_0, a larger value of R was obtained than when Cohan's equation was used. The physical meaning of this result is that surface forces help to bring about capillary condensation.

The preceding treatment lends itself to pore size distribution calculation from the adsorption branch of the isotherm by a technique similar to that used by Barrett *et al.* (1951). An important difference is that t now depends on R as well as on P''/P_0. An exact computational method as well as a simplified technique were developed by Broekhoff and de Boer (1968a).

This thermodynamic analysis was extended to ink-bottle type pores with essentially spherical shapes (Broekhoff and de Boer, 1968b).

The desorption condition from cylindrical capillaries was also developed by Broekhoff and de Boer (1967, 1968c), using two different approaches. The first and somewhat indirect approach consisted of integrating Eq. (2.5.4) from equilibrium thickness t_e of the multilayer to $t = R$, the pore radius, and letting $\Delta G_{T,p}$ equal zero, resulting in a relationship between t_e, R, and P_0. Simultaneous solution of this equation with the equilibrium condition, Eq. (2.5.26), gave the desired relation between R and P_d'', the desorption pressure. This approach is indirect, however, because it approaches the desorption from the opposite direction (i.e., adsorption). Indeed, when Eq. (2.5.4) is integrated at $P'' = P_d''$ from $t_e \leq t \leq R$, ΔG has a hump between $t = t_e$ and $t = R$, which represents the thermodynamic reason for the absence of capillary condensation (complete filling) at the pressure P_d''. This will only occur at a higher pressure P_a'', as previously discussed. The argument, that from the desorption direction the hump in ΔG does not exist, is intuitive.

In the more direct approach (Broekhoff and de Boer, 1968d), the authors analyzed the change in shape of the meniscus during evaporation from a cylindrical capillary that was originally completely filled with condensate. The geometrical condition of desorption is that the meniscus is tangential to the adsorbed multilayer. By using the thermodynamic analysis just presented, an expression was obtained for the slope of the meniscus at a distance t from the pore wall. Introducing the geometrical condition of desorption into this equation, the same relationship among R, t_e, and P''/P_0 was obtained as by the adsorption process. A similar equation was derived earlier by Deryagin (1940, 1957), by making use of the concept of "disjoining" pressure. His equations have not found practical application in the calculation of pore size distribution from sorption isotherms.

Interesting consequences of the analysis of Broekhoff and de Boer are that the shape of the meniscus in a pore may not be hemispherical but rather conical and that the emptying of the pores is an intrinsically continuous process rather than the classical discontinuous evaporation step.

Using the analytical expressions (Broekhoff and de Boer, 1968a) obtained for $F(t)$, the pore radius R was calculated (Broekhoff and de Boer, 1968b) as a function of P_d''/P_0. For any given P_d''/P_0, a greater value of R was obtained than by the form of the Kelvin equation given by Eq. (2.5.12). The physical meaning of this result is that the forces emanating from the walls of the absorbent have the effect of holding the capillary condensate in the pore.

Using the theoretical apparatus developed by Broekhoff and de Boer, it was an easy matter to calculate, for open-ended cylindrical pores, the theoretical width of the hysteresis loop $P_a''/P_0 - P_d''/P_0$ as a function of P_a''/P_0. The calculated results were compared with the experimental ones for a variety of adsorbents (1968b). In many cases good agreement was obtained, which was taken as an indication of the type of pore structure of the adsorbents.

Pore size distributions were calculated (1968b) from the desorption branch by a technique completely analogous to the one given in (1968a) and compared with the curves obtained from the adsorption branch of the isotherms. The good agreement obtained for samples where the shape of the hysteresis loop indicated a pore structure consisting approximately of open-ended cylinders was a major achievement reached by these authors.

Brunauer *et al.* (1967) have developed a method for micropore analysis (MP method), using the de Boer t curve method. They plotted the adsorbed volume (as liquid) versus t from $t = 35$ Å units upward. The slope of the curve was accepted as a measure of the available surface area of the sample. From the surface area of the pores filled and the corresponding value of t, the size as well as the volume of micropores was calculated.

According to Dubinin *et al.* (1968), however, it is meaningless to talk about layerwise adsorption on a micropore surface and it is not even possible to determine the surface areas of such pores. The MP method was criticized also by Marsh and Rand (1970, 1972). Details of this controversy are outside the scope of this text and the reader is referred to the original literature [see, e.g., Brunauer *et al.* (1973) and Dubinin (1973)].

2.5.2.   Network Models

The pore structure of the vast majority of porous media consists of an interconnected three-dimensional network of voids, pores, or capillaries. The network usually has irregular geometry, and capillary segments of various different shapes and sizes are distributed over the network in some irregular fashion.

Hence, it is logical to model capillary pressure curves and other transport properties of porous media with the help of network models of the pore structure. Were the pore geometry of a small macroscopic sample of a porous medium known in sufficient detail (the question what represents "sufficient" detail depends on both the phenomenon to be modeled and the accuracy required) then with the help of high powered computers eventually it will be possible to calculate quasi-static capillary displacement in the sample deterministically. At the present time there is still a great deal to be learned about the mechanism of the process, particularly in imbibition, and there is also much to be gained by using network models that are representative of the sample in a statistical sense.

2.5.2.1.   *Introduction*

A network of straight lines and/or points is an abstract mathematical concept called a space lattice. The most important examples of space lattices consist of periodic arrays of a unit cell. As a matter of emphasis, it is convenient to distinguish between networks of straight lines, on the one hand, and of points, on the other. In a network of straight lines, the lines form "bonds" that meet at points, called "nodes," whereas in a network of points these represent centers of spheres, the "sites." The points of contact between adjacent spheres may be interpreted as "bonds" in a network of points. In the present paper the words "bond," "capillary," "pore neck," and "pore throat" have the same meaning. The same holds true for "sites," "nodes," and "pore bodies." The word "pore" may be used to mean either a bond or a site.

A network model may be used to represent symbolically the pore structure of a porous medium, as illustrated in Fig. 2.47, for the simple case of an artificial two-dimensional pore structure. Figure 2.47(a) shows the artificial

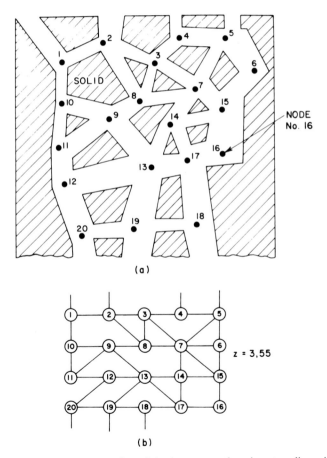

Figure 2.47. Symbolic representation of the interconnections in a two-dimensional porous medium (Chatzis and Dullien, 1985).

pores whose intersections, the nodes, are numbered. Underneath, in Fig. 2.47(b), the symbolic representation of this pore structure is shown in the form of a network of bonds and nodes. The network is characterized by the same coordination number of z as the pore structure in Fig. 2.47(a). The coordination number is defined as the average number of bonds meeting at each node, that is,

$$z = \sum f_i z_i \tag{2.5.30}$$

Here, z_i is the number of bonds meeting at a node of type i, and f_i is the frequency of the nodes of type i.

The network in Fig. 2.47(b) is representative only of the manner of interconnectedness (i.e., the topology) of the pore structure shown in Fig. 2.47(a). Other aspects of the pore structure, which will be called "geometry," such as the dimensions and the orientations of the pores, are not modeled in Fig. 2.47(b), in which the length and the orientations of the bonds were chosen completely arbitrarily.

One can distinguish one-, two-, and three-dimensional "networks." A one-dimensional "network" is a chain, that is, not a network in the true sense of the word. The only difference between the bundle of capillary tubes model and a one-dimensional "network" is that in the former model each tube has a uniform diameter, whereas in the latter the diameter changes along the axis of the tube in some regular or irregular manner.

A two-dimensional network is a lattice. A few examples are shown in Fig. 2.48. In these examples all networks have some kind of symmetry. Naturally, a network need not be regular, but it is convenient to study the properties of regular rather than irregular networks. It is noted that there are no dead-end capillaries in these networks. Experiments have indicated that blind pore volume is not much greater than 1% in sandstone or sintered glass (Russell *et al.*, 1947; Everett *et al.*, 1950; Mysels and Stigter, 1953).

Originally, networks were studied extensively in electrical circuitry, where the "junctures" or "nodes" can be regarded as mathematical points without size or resistance. Often the networks studied for the purpose of modeling pore structure are assumed to be like electrical networks. In actual fact, however, networks obtained by joining capillary segments will have "nodes," entities where two or more capillaries intersect, of a very definite characteristic shape, size, volume, and resistance.

In any network, the capillary connecting two adjacent nodes may have a uniform diameter, or it may consist of any number of capillary segments with different diameters.

The difference between a two- and a one-dimensional network is probably very clear to every reader and does not need any explaining. It may not be equally clear, however, that there are similarly great differences between three- and two-dimensional networks, too. Some of these differences are

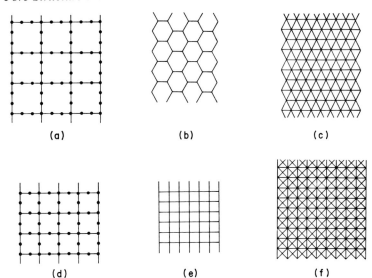

Figure 2.48. Two-dimensional networks investigated by Chatzis and Dullien (1977): (a) Tritetragonal ($Z = 2.66$); (b) hexagonal ($Z = 3$); (c) hexatriangular ($Z = 6$); (d) ditetragonal ($Z = 3$); (e) square ($Z = 4$); (f) tetratriangular ($Z = 6.66$).

discussed in greater detail below. At this point, let us emphasize only one basic difference: only in three-dimensional networks is it possible to have two continuous phases present simultaneously. In two-dimensional networks only one phase can be continuous—the second phase must be discontinuous. In the terminology of topology this fact is expressed as follows: "Bicontinua cannot exist in two dimensions." The term "continuous phase" or "continuum" means that one can go from any point within that phase to any other point within the same phase without ever going outside the phase.

It is the generally accepted view that in actual porous media there is a range of saturations where at least a portion of each phase is continuous. Hence, two-dimensional networks cannot be used to accurately model this "transition stage" of immiscible fluid–porous media systems. Notwithstanding this, much of the work with network models has been limited to the use of various two-dimensional networks. The reason for this preference lies in the greatly increased complexity of the calculations when resorting to three-dimensional models.

The use of network models for the purpose of capillary pressure characterization studies appears to have been introduced by Fatt (1956). He used two-dimensional networks of the electrical network type (see Fig. 2.48), where two adjacent nodes were always connected by a capillary of uniform diameter.

Fatt assumed various number distributions of tube radii and distributed the tube radii over the network in a random manner using tables of random numbers. He assumed that initially the network was completely saturated with the wetting fluid. Then he postulated that the network was completely surrounded by the nonwetting fluid and assumed that the pressure difference between the fluid phases (i.e., the capillary pressure) was increased at first until its value was equal to the entry pressure of the largest tube in the network. At that point all the largest tubes in the network located on the edge of the network were penetrated, and if there were any additional largest tubes in direct contact with any one of those on the edge of the network, these were penetrated too. Upon increasing the capillary pressure to that required to enter tubes of the second largest radius, all such tubes that were accessible—either because they were situated on the edge of the network or because they were connected to tubes of the same or of larger size such that were already penetrated—were also penetrated by the nonwetting phase.

Using the above principles, Fatt continued the penetration into the network, step by step, until even the narrowest tubes were penetrated. He calculated the capillary pressure by Eq. (2.3.1), assuming both angles ϕ and θ to be zero. The wetting phase saturation at each stage of the penetration process was calculated by the following formula:

$$S_{wk} = 100\left[1 - \left(\sum_{i=1}^{k} \mu_{ik}\pi R_i^2 \ell_i \middle/ \sum_{i=1}^{n} N_i\pi R_i^2 \ell_i\right)\right], \qquad (2.5.31)$$

where N_i is the number of tubes of size i in the network, μ_{ik} is the number of penetrated tubes of size i when the smallest penetrated tube is of size k, ℓ_i is the tube length, and R_i is the tube radius. Fatt also assumed that

$$\ell_i = C/R_i, \qquad (2.5.32)$$

where C is a constant independent of i.

The reason for this assumption was that it gave closer agreement between network flow properties and those observed in real porous media than any other relationship between tube length and radius. This relationship, however, does not agree with the results of pore structure investigations, and it is topologically impossible to build a random network in the porosity range of typical sandstone of tubes the lengths of which are subject to this relationship (Chatzis and Dullien, 1977). Using Eq. (2.5.32), Eq. (2.5.31) becomes

$$S_{wk} = \left[1 - \left(\sum_{i=1}^{k} \mu_{ik} R_i \middle/ \sum_{i=1}^{n} N_i R_i\right)\right], \qquad (2.5.33)$$

Using this procedure with a variety of tube radius distributions and two-dimensional networks, Fatt found that in general the curves resembled typically capillary pressure curves obtained with sandstone samples.

Fatt's ingeneous initiative provoked some immediate discussion of his work. Rose, Carpenter, and Witherspoon (loc. cit., Discussion) and Rose (1957) criticized Fatt's work because of the assumptions he made regarding the escape of the wetting phase during the penetration of the nonwetting phase. These authors permitted the water different escape routes and, depending on the type of escape route used, different wetting phase saturations were obtained by them.

Dodd and Kiel (1959) modified Fatt's procedure. The displacing fluid was allowed to enter on three sides of the model and the displaced fluid to exit on the fourth side. The displaced fluid was trapped whenever no continuous path to the effluent end was available. An attempt was made also to include the effect of intermediate wettability by equating the probability of entrance of the displacing fluid into a pore filled with the fluid to be displaced to a number less than unity.

Dodds and Lloyd (1971/1972) applied the network model to predict capillary pressure curves in multicomponent sphere packs. The void structure was represented by a regular two-dimensional network of capillary tubes with spherical void spaces at the junctions. The sizes of the tubes and the spherical voids were calculated from a tetrahedral model of packing for unequal spheres (Wise, 1952, 1960). After making a correction, using the hydraulic radius model, reasonably good agreement with experimental capillary pressure curves was obtained.

2.5.2.2. *Generalized Capillary Pressure Functions Obtained with the Help of Network Models*

The work of Dodds and Lloyd (1971/1972) represented an improvement over previous efforts in at least two respects. First, they assigned a volume of its own to each node. Second, they estimated the dimensions of both the nodes and the tubes connecting the nodes from geometrical considerations on sphere packs. Indeed, a pore structure model can be said to have passed scrutiny only if it can predict the capillary pressure function using data that are independent of the information contained in the capillary pressure function.

In this section the recent work by Chatzis (1976, 1980), Chatzis and Dullien (1977, 1982, 1985) and Diaz *et al.* (1987) is reviewed, along with the fundamental properties of capillary network models and some pertinent results of percolation theory.

In the same time period when Fatt introduced his network model the powerful mathematical theory of percolation was formulated chiefly with the purpose of understanding the conductivity properties of random mixtures of electrically conducting and nonconducting objects (Broadbent and Hammersley, 1957; Shante and Kirkpatrick, 1971; Frisch and Hammersley, 1962, 1963; Vissotsky *et al.*, 1961; Kirkpatrick, 1973; Sykes and Essam, 1963;

Domb and Sykes, 1961). It was realized that the same principles apply to a pore network in which a proportion of the pores is occupied, in a random manner, by fluid I, whereas the rest of the pores are filled with fluid II. Important for the applicability of percolation theory to the modeling of capillary pressure curves was the assumption that the capillary displacement process is random, that is, that pores of different sizes are randomly distributed in the pore network.

Calculation of the penetration of the network in a way that is independent of the density function $f_b(D_b)$ of the sample and depends only on the type of the network considered, its dimensionality, and its coordination number (Chatzis and Dullien, 1977) has certain advantages with respect to economy of computations. In this calculation, a fraction p_b of the bonds in the network is assumed to be open, whereas the remaining fraction $(1 - p_b)$ is blocked. As every bond has the same probability p_b of being open, it follows that the bonds which are open are distributed randomly over the network. As far as the nodes or sites are concerned, whenever a bond is open, both nodes situated at the two ends of that bond are also assumed to be open.

A bond or a node is penetrated by the nonwetting fluid if both of the following conditions are fulfilled. (1) The bond or the node is open; (2) the bond or the node is connected to the injection face of the sample by an uninterrupted chain of bonds and nodes that are also open (i.e., is accessible from the outside).

According to this procedure, known as "bond percolation," the value of p_b is increased, step-by-step, from zero to one and, at each value of p_b, the number of bonds and nodes that have been penetrated by the nonwetting fluid is determined.

Some details of these calculations are described below as well as the determination of the mercury porosimetry curve from the results.

2.5.2.3. *Results of Bond Percolation*

The various two-dimensional lattices investigated by Chatzis and Dullien (1977) had widths of 20–40 pore segments and depths ranging from 15 to 80 segments. As shown in Fig. 2.49, the nonwetting phase was permitted to penetrate into the network through one of the faces; the two sides of the network perpendicular to the penetration face were assumed to have impervious boundaries; and the fourth face was left open.

The pores were assumed to be evacuated initially, as is the practice in mercury intrusion porosimetry tests. Hence, the question of trapping or escape of the wetting phase did not arise in these studies.

The pore diameters were distributed according to an arbitrary distribution function and were allocated to the various bonds of the network in a random manner, similarly as was done also by Fatt (1956). The bonds in the network

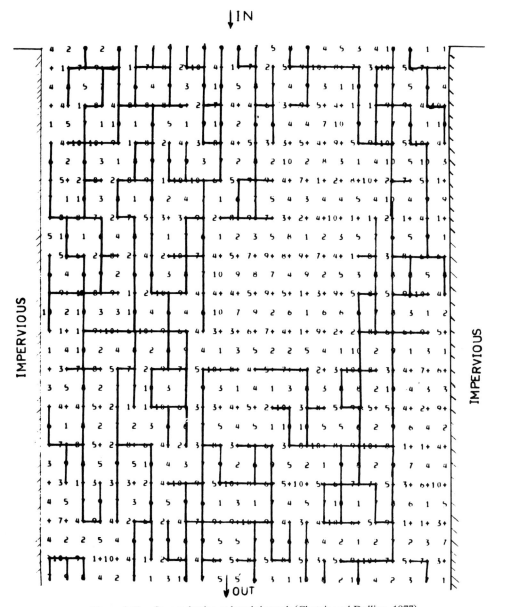

Figure 2.49. Square lattice at breakthrough (Chatzis and Dullien, 1977).

were numbered in a consecutive manner from 1 to N, N being the total number of bonds in the network, starting from the top left-hand corner of the network and terminating at the bond of the lower right-hand corner of the network. Using uniformly distributed floating-point integers of some integer scale (e.g., $1, 2, 3, \ldots, 10$) a vector of N elements was generated on the computer with the help of a pseudorandom number generating routine. The elements of the one-dimensional random vector were assigned, one by one, as a pore size to the bonds of the network, in the same sequence in which the bonds were numbered from 1 to N.

In the first step of the stagewise penetration process, the largest pores ($k = 1$) were opened, and only those pores of this size that were accessible through the penetration face of the network were actually penetrated. In the next step, the second largest pores ($k = 2$) were opened, and so forth. For every value of the fraction of "open" bonds, p_{bk}, both the number fraction of all the bonds that were actually penetrated, Y_k, and the number fraction of "open" bonds that were penetrated, y_k, were determined. At a certain value of the cumulative pore size distribution, corresponding to some intermediate pore size, the penetration reached the opposite face of the network. When at a particular step the penetration reaches the opposite face of the network, this condition is termed "breakthrough." Figure 2.49 shows a network at breakthrough. Bonds indexed by the integer 10 in this figure represent $k = 1$ type pores; those indexed by the integer 9 represent $k = 2$ type pores, etc.

"True" breakthrough persists even for a network of infinite size. The cumulative probability p_{cr} at true breakthrough is called "critical percolation probability." At true breakthrough, infinitely long channels containing the nonwetting phase and capable of conducting flow, or electric current, are established for the first time. Before "true" breakthrough, the penetrated pores cannot conduct flow or current between the injection and the exit face of the sample and, of macroscopic media, the corresponding saturation with respect to the nonwetting phase is negligibly small; that is, the penetration is a "boundary" or "end effect."

The depth of penetration is plotted in Fig. 2.50 for the various two-dimensional lattices as a function of p. The exponential increase of the depth of penetration makes it possible to obtain a good estimate of the critical probability p_{cr} for breakthrough by extrapolation. It was found that a network size of about 40×40 pore segments is the minimum to represent the true value of p_{cr} to within ± 0.01 probability units. The value of p_{cr} is independent of the form of the distribution function of capillary diameters.

Inspection of Fig. 2.49 shows several features:

(1) The saturation at breakthrough is finite.
(2) Only the nonwetting phase forms continuous paths over the whole extent of the network—the wetting phase does not.

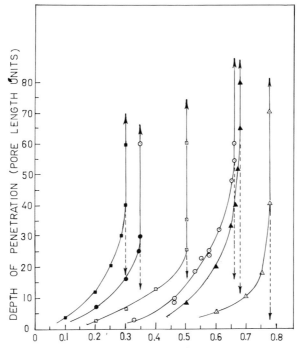

Figure 2.50. Depth of penetration in two-dimensional networks. The curves shown are for the following networks: tetratriangular (■); hexatriangular (●); square (▫); hexagonal (○); ditetragonal (▲); tritetragonal (▵) (Chatzis and Dullien, 1977).

(3) Much of the nonwetting phase is still nonconducting because it is contained in pores ("dendritic" pores) from which it cannot exit at the breakthrough capillary pressure, which is not high enough for the penetration of the relatively small pores blocking exits.

(4) There are only two conducting capillaries at the exit face.

Number-based capillary pressure curves of the various two-dimensional networks investigated are shown in Fig. 2.51. It is noted that whereas the critical probability at breakthrough p_{cr} can be estimated to a high precision, the corresponding number-based saturation Y_c is somewhat uncertain because of the great error in Y_k caused by even a very small error in p_{cr} at or near the point of breakthrough.

Using the product $p_b Z$ as the abscissa instead of the probability p, all of the various different two-dimensional networks show breakthrough at the same value of the $p_b Z = 2$, as shown in Fig. 2.52.

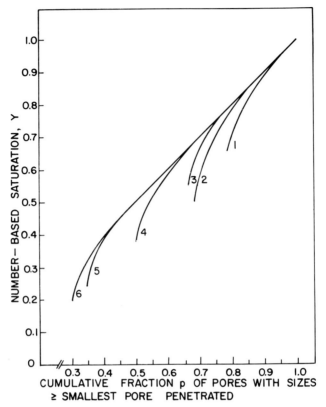

Figure 2.51. Saturation curves for two-dimensional networks after breakthrough: curve 1, tritetragonal ($Z = 2.66$); 2, hexagonal ($Z = 3$); 3, ditetragonal ($Z = 3$); 4, square ($Z = 4$); 5, hexatriangular ($Z = 6$); 6, tetratriangular ($Z = 6.66$) (Chatzis and Dullien, 1977).

Z is the coordination number of the network, defined by the following equations:

$$Z = \sum_r Z_r f_r \qquad\qquad (2.5.34a)$$

and

$$Z_r = \frac{1}{2}\left(\sum m\right)_r + 1 \qquad\qquad (2.5.34b)$$

where $(\sum m)_r$ is the number of pores connected to a pore of type r at both ends and f_r, the relatively frequency of such pores in the lattice.

Because of the similarity of the bond percolation problem to the penetration process in a network of capillary tubes, bond percolation results in three-dimensional lattices can serve as data for the corresponding three-

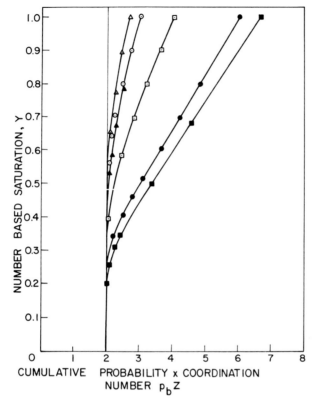

Figure 2.52. Generalized saturation curves for two-dimensional networks after break-through: ⊙, hexagonal ($Z = 3$); ▲, 2-tetragonal ($Z = 3$); △, 3-tetragonal ($Z = 2.6$); ⊡, square ($Z = 4$); ●, hexatriangular ($Z = 6$); ■, tetratriangular ($Z = 6.6$) (Chatzis and Dullien, 1977).

dimensional network of randomly distributed pores. The three-dimensional bond percolation probability data obtained by Frisch *et al.* (1962) have been translated to the terminology of pore networks, and the results are shown in Fig. 2.53.

2.5.2.4. *Prediction of the Mercury Intrusion Capillary Pressure Curve*

Chatzis and Dullien (1977) realized that the number-based capillary pressure curves obtained as a result of the network analysis and shown in Figs. 2.51 and 2.53, obtained with the help of arbitrary probability distributions of capillary diameters, can be transformed into the usual capillary pressure curve of the actual porous sample with the help of the photomicrographically obtained "total" pore size distribution of the sample. This is why the

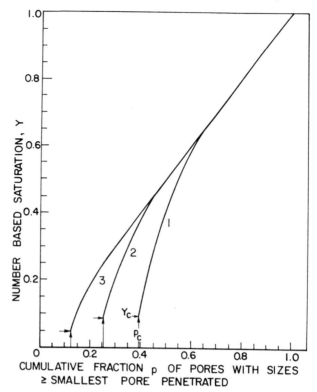

Figure 2.53. Saturation curves for three-dimensional networks after breakthrough: Curve 1, tetrahedral ($Z = 4$); 2, simple cubic ($Z = 6$); 3, face-centered cubic ($Z = 12$) (Chatzis and Dullien, 1977).

number-based saturation versus probability distribution functions have been named "number-based generalized capillary pressure functions."

The abscissa of the number-based generalized capillary pressure curve can be transformed into capillary pressure with the help of the following relations. The integrated number density distribution of pores sizes $f(D)$ has been equated to p_{bk}:

$$p_{bk} = \int_{D_k}^{D_{max}} f(D) \, dD; \qquad D_{min} \leq D_k \leq D_{max} \qquad (2.5.35)$$

to find the value of the smallest capillary diameter D_k in the sample penetrated at p_{bk}. Using D_k obtained in this manner, the corresponding capillary pressure is calculated by Laplace's equation

$$(P_c)_k = 4\sigma \cos \theta / D_k. \qquad (2.5.36)$$

As a result of these operations the dimensionless "capillary pressure" p_{bk} used in the network analysis is transformed into the corresponding value of the actual capillary pressure $(P_c)_k$.

The ordinate of the generalized number-based capillary pressure curve is converted to the volume-based saturation of the real sample under consideration with the help of the following relations:

$$S_k = \sum_{j=1}^{k} y_j s_j = y_k S(D_k) \tag{2.5.37a}$$

and

$$S(D_k) = \frac{\int_{D_k}^{D_{max}} V(D) f(D)\, dD}{\int_{D_{min}}^{D_{max}} V(D) f(D)\, dD}, \tag{2.5.37b}$$

where s_j is the volume fraction of pores of size D_j in the sample, $S(D_k)$ is the (cumulative) volume fraction of pores of size $D \geq D_k$ in the sample, $V(D)$ is the volume of a pore characterized by diameter D at both ends, and S_k is the volume-based saturation predicted by the network analysis for the sample when the smallest capillary permitted to be penetrated had diameter D_k.

Equation (2.5.37a) follows because at any stage of the penetration process the saturations of all pore sizes $j \leq k$ are identical (i.e., $y_1 = y_2 = \cdots = y_k$); in the network analysis the different penetrated pore size categories are completely equivalent.

Various assumptions have been tested (Chatzis and Dullien, 1977) regarding the relationship between the volume and the diameter of a capillary in the network. In the cases (1)–(3), discussed below, the original distribution $f(D)$, given in terms of equivalent sphere diameters, was transformed into $f'(D)$ by using the condition $f(D)V(D) = f'(D)V'(D)$, where $V(D)$ is the volume of a pore of size D in the original distribution and $V'(D)$ is the volume of the corresponding cylindrical capillary in the transformed distribution. The new distribution $f'(D)$ was used, after normalization, to calculate D_k by Eq. (2.5.35).

Photomicrographic studies have shown that the length of a pore is of the same order of magnitude as its diameter. Mathematical models of permeability, formation resistivity factor, and rate of capillary rise gave good agreement with experiment for 14 widely different sandstones, using this assumption regarding the geometrical proportions of pores (El-Sayed and Dullien, 1977).

(1) $\ell/D = $ constant: $V'(D)$ is proportional to D^3 unchanged from the distribution of equivalent spheres. The capillary pressure curves predicted by both two- and three-dimensional networks are shown for the special case

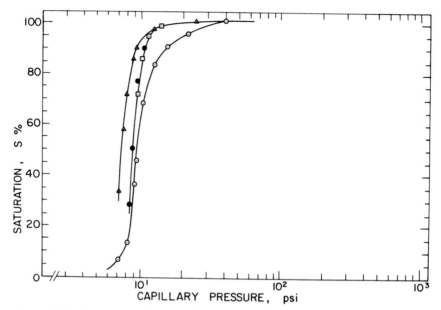

Figure 2.54. Comparison of the capillary pressure curve predicted using the assumption ℓ/D = constant with the experimental curve of Berea sandstone (◯). The computed data are shown for the following networks: three-dimensional cubic (△); three-dimensional tetrahedral (●); two-dimensional square (◻) (Chatzis and Dullien, 1977).

$\ell = D$ in Fig. 2.54. The predicted saturations are too high, but the tetrahedral network gave correct breakthrough capillary pressure and capillary diameter.

(2) $\ell =$ constant: $V'(D)$ is proportional to D^2. The predicted breakthrough saturations are slightly lower; however, the breakthrough capillary diameter now is too high.

(3) $\ell D =$ constant: $V'(D)$ is proportional to D. This is the relationship used by Fatt (1956). It does tend to predict a little lower saturation at breakthrough than those given previously, but the diameter at breakthrough is shifted in the direction of larger values, farther away from the experimental one.

(4) Almost perfect prediction was obtained with the tetrahedral network using the assumption $V'(D) \propto D$ and calculating $S(D_k)$ from the original photomicrographic $f(D)$ by Eq. (2.5.37b).

It is obviously very important to check whether any of the assumed relationships between ℓ_j and D_j are physically possible. It can be shown (Chatzis, 1976) that it is topologically impossible to have a porous medium where the capillary lengths are either directly or inversely proportional to the diameters, provided that the diameters are distributed randomly and that the porosity is in the range from 0.2 to 0.4. For the case of constant ℓ with cubes of edge length a forming the points of connection, the porosity ϕ is given by

$$\phi = \left(r_1^3 + \tfrac{3}{4}\pi r_2\right)/\left(r_1 + r_2\right)^3, \qquad (2.5.38)$$

where $r_1 = a/D$ and $r_2 = \ell/D$. It is found that for ϕ to be in the range from 0.2 to 0.4, the value of r_1 must be in the range from 1 to 1.5 and that of r_2 in the range from 1 to 2.

It may be concluded that the type of network models introduced by Fatt (1956) and studied also by Chatzis and Dullien (1977) are oversimplified and are not representative of mercury intrusion porosimetry curves of sandstones. One possible reason for this is that in these all the volume is assigned to the bonds that also control penetration of mercury and, hence, the value of the capillary pressure. In reality, most of the volume belongs to those geometrical entities (bulges) that do not have any relationship to the capillary pressure. A model consisting of alternating throats and pore bodies can be expected to be more representative. In a network model, pore bodies occupy quite naturally the positions of nodes connected by pore necks. Hence, the pore bodies will be the "sites" and the pore necks the "bonds."

2.5.2.5. Bond-Correlated Site Percolation

According to site percolation, a fraction p_s of the nodes of the network is selected at random and is imagined to be open, whereas the remaining fraction $(1 - p_s)$ remains blocked. (Each node has the same probability of being open.) In classical site percolation all the bonds are assumed to be open at every stage of the process. The assumptions have been made by Chatzis (1980) that the nodes of different sizes are distributed randomly over the network and whenever a node is closed all the bonds that meet at the node are also closed; that is, a bond is open if both nodes situated at its two ends are open. According to this scheme, to each bond an index equal to the larger index of the two nodes at its two ends (corresponding to the smaller node, physically speaking) was assigned. As the frequency of nodes of different indices was uniform, the number of possible combinations of adjacent nodes characterized by a given larger index k increased statistically linearly with k. Hence the frequency of bonds increased statistically linearly with increasing index k. Thus, the cumulative frequencies of nodes p_s, and of bonds p_b, having indices $j = 1, 2, \ldots, k$, follow linear and quadratic relationships, respectively, or

$$p_b = p_s^2 \qquad (2.5.39)$$

As a result of this relationship the site percolation threshold of, for example, the simple cubic lattice $p_{sc} \approx 0.33$ has been reduced in the "bond-correlated site percolation" scheme to $p_{bc} = (0.33)^2 = 0.10$, which deter-

FACE OF INJECTION

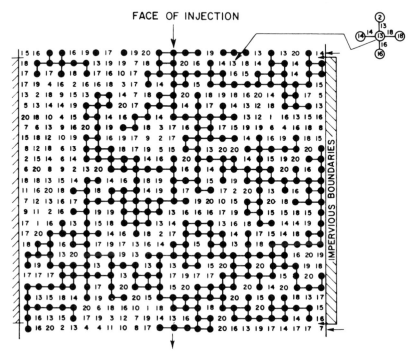

Figure 2.55. Breakthrough for a 25 × 25 square network (site percolation) (Chatzis and Dullien, 1985).

mines the breakthrough (threshold) capillary pressure in primary drainage, while keeping the advantage of site percolation over bond percolation for modeling drainage, consisting of a much more gradual penetration of different size sites that represent most of the pore volume.

The calculation was carried out by increasing the value of p_s, in small increments, from zero to one and counting for each value of p_s the number of accessible (i.e., penetrated) nodes and bonds. As illustration see Fig. 2.55, in which the nodes are numbered. In this example, there are 20 different numbers (or classes), starting at 1, denoting those nodes picked to be open in the first step of the calculation, and ending at 20, denoting those nodes picked to be open in the last step of the calculation. This picture corresponds to the stage of the process in which all the nodes denoted by the numbers ranging from 1 to 12 are open. Those nodes that have been penetrated by the invading nonwetting phase are shown as filled-in circles, whereas the penetrated bonds are indicated by full lines. The numbers originally assigned to those sites that have been penetrated are missing from the illustration, because they have been covered up by the filled-in circles. The details of the calculations, as well as the transformation of the results made in order to obtain the mercury porosimetry curve, are described below in detail.

Although the connectivity of the pore structure of a porous medium can be represented by a regular network, this representation does not cover the problems of pore size and pore shape. Hence, in order to be able to calculate various pore structure properties of porous media, additional assumptions must be made regarding the size and shape of pores. The well-known structure of packs of monometric beads is characterized by an ensemble of voids (nodes) connected to each other by windows (bonds). Following this pattern, the pores in porous media are assumed to be represented by two sets of objects, the set of nodes and the set of bonds. Each node and each bond is assigned a "diameter," D_s and D_b, and a volume, $V_b(D_b)$ and $V_s(D_s)$, respectively. The diameters follow certain distribution functions, $f_b(D_b)$ and $f_s(D_s)$, respectively. The meaning of "diameter" must be clearly defined in every case, because for irregular pore shapes the value of pore diameter depends also on the pore shape and the method of measurement used in the determination of pore diameters. Last, but not least, it must be borne in mind that, in addition to the main pore network extending throughout the porous medium, secondary systems of pores (i.e., "dual porosity") may exist. This is the case for sandstones where, in addition to the network of larger pores constituted by the interstitial spaces between the sand grains, there are also smaller pores inside the cementing materials (clays) ordinarily present in sandstones.

2.5.2.6. *Results of Bond-Correlated Site Percolation*

The calculation was started by letting $k = 1$ open and then the continuous paths formed by bonds and nodes characterized by the index 1 and connected to the injection face were traced in order to establish which bonds and nodes were penetrated by the nonwetting fluid. At that point the penetrated bonds and nodes were counted and the depth of penetration, expressed in mesh-point units, was also determined. Then, letting $k = 2$ open, the continuous paths, consisting of bonds and nodes characterized by both indices 1 and 2 and connected to the injection face, were traced and the counts described under step $k = 1$ were made. The same procedure was repeated for $k = 3, 4, 5, \ldots$, all the way to $k = c$, at which point the continuous path consisting of penetrated bonds and nodes reached, for the first time, the face situated at the opposite extremity of the network. This point in the penetration process is called the "breakthrough" which corresponds, in the case of an infinitely large network, to an exact value $p_{bk} = p_{bc}$ (or $p_{sk} = p_{sc}$), termed the "percolation threshold" in percolation theory. In Fig. 2.55 a square network is shown at the point of breakthrough according to the modified site percolation problem.

After reaching the point of breakthrough the procedure was continued by letting $k = c + 1, c + 2, \ldots, n$ open. In this phase of the calculation a few additional quantities were determined at each step. For example, the number of nodes contained in ganglia, formed by the nonwetting phase, with no exit was counted. These structures, which have been called alternately "pseudo dead-end pores," "nonconductive saturation," and "dendritic pores" in the literature, are clearly visible in Fig. 2.55. More details of these calculations can be found in Chatzis (1980).

Using the results of the counts made in each step of the procedure, the following other important quantities were calculated:

(1) The cumulative proportions of all the bonds and all the nodes that were penetrated, Y_{bk} and Y_{sk}, respectively.
(2) The cumulative proportions of the open bonds and nodes, y_{bk} and y_{sk}, respectively, that were penetrated in the kth step.
(3) The cumulative proportions of the penetrated bonds and nodes, q_{bk} and q_{sk}, respectively, that are contained in ganglia with no exit, many examples of which can be seen in Fig. 2.55.

The proportions y_{bk}, y_{sk}, q_{bk}, and q_{sk} are observed to be independent of j. This is because all of the bonds and nodes of index j, such that $j = 1, 2, 3, \ldots, k$, are open and have the same probability of being penetrated.

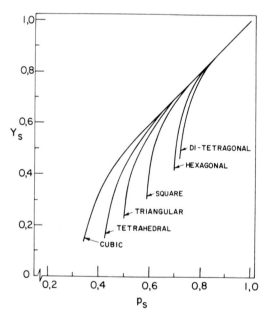

Figure 2.56. Curves of Y_s versus p_s for various networks (Chatzis and Dullien, 1985).

TABLE 2.2 Comparison of Some Networks at Breakthrough

			I. Bond percolation						II. Site percolation					
			p_{bc}			Zp_{bc}			p_{sc}			zp_{sc}		
Network	Z	z	(1)	(2)	(3)	(1)	(2)	(3)	(1)	(2)	(3)	(1)	(2)	(3)
Tri-tetragonal	2.66	2.44	0.78	—	—	2.07	—	—	—	—	—	—	—	—
Di-tetragonal	3	2.66	0.68	—	—	2.04	—	—	0.72	—	—	1.92	—	—
Hexagonal	3	3	0.66	0.653	—	1.98	1.958	—	0.70	0.70	0.697	2.10	2.10	2.091
Square	4	4	0.50	0.500	0.499	2.0	2.0	1.966	0.59	0.590	0.591	2.36	2.36	2.364
Triangular	6	6	0.345	0.347	0.349	2.07	2.08	2.094	0.50	0.50	0.500	3.0	3.0	3.0
Tetra-triangular	6.66	6	0.30	—	—	2.0	—	—	—	Ref. (4)	—	—	—	—
Tetrahedral	4	4	—	0.388	0.390	—	1.55	1.56	—	0.425	—	—	1.7	—
Simple cubic	6	6	0.27	0.257	0.254	1.62	1.54	1.52	0.34	0.307	0.32	2.04	1.842	1.92
Face-centered cubic	12	12	—	0.119	0.125	—	1.43	1.50	—	0.195	0.208	—	2.34	2.496

(1) Chatzis (1980), (2) Sykes and Essam (1963), (3) Dean (1983) and Dean and Bird (1967), (4) Domb and Sykes (1961).

Hence, $y_{bjk} = y_{bk}$ and $y_{sjk} = y_{sk}$. Similarly, all penetrated bonds and nodes have the same probability of belonging to ganglia with no exit, regardless of the value of j ($j = 1, 2, 3, \ldots, k$). Thus, $q_{bjk} = q_{bk}$ and $q_{sjk} = q_{sk}$.

Another important point to note is that the results obtained in this study depend only on p_{bk} (or p_{sk}) and are independent of the size of the fractional increments

$$\frac{N_{bj}}{\sum\limits_{j=1}^{N} N_{bj}},$$

which were chosen arbitrarily. Naturally, by increasing p_b in smaller increments (i.e., making n greater), more points $k = 1, 2, 3, \ldots, n$ will result, because the simulation is carried out in greater detail.

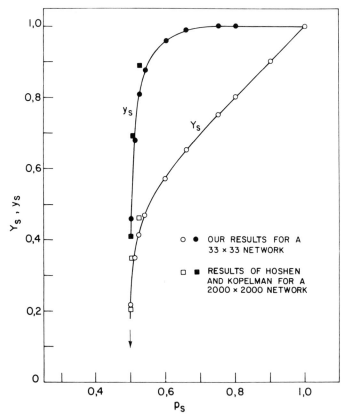

Figure 2.57. Comparison between results obtained in this work and those of Hoshen and Kopelman (1976) for a triangular network (site percolation) (Chatzis and Dullien, 1985).

The variation of Y_s calculated in this work for various, different networks is shown in Fig. 2.56. Each curve starts at the point of breakthrough in these figures, because below p_c, Y_s is negligibly small in the case of a network that is large enough to be of comparable size with a real porous medium. The percolation thresholds p_c obtained in this work are compared with those calculated by other authors in Table 2.2, where the very good agreement between our results and those obtained by others is apparent.

Usually only those results representative of very large samples are of interest. Hence, the size of the networks plays an important role in determining whether the results obtained in this work are meaningful. For the case of two-dimensional networks the minimum size necessary for the obtainment of good results are found to be about 33×33 nodes, as is shown in Fig. 2.57.

2.5.2.7. *Calculation of the Mercury Porosimetry Curve of Sandstones*

In this section the method of transformation, used for obtaining the capillary pressure curve of a sample of the porous medium from the curves giving the variation of Y_b and Y_s with p_b and p_s is described.

D_{bk} and D_{sk}, corresponding to p_{bk} and p_{sk}, respectively, are calculated with $f_b(D_b)$ and $f_s(D_s)$, respectively, as already indicated by Eq. (2.5.35). p_b and p_s are, respectively, the number fractions of bonds and nodes characterized by the diameters $D_b \geq D_{bk}$ and $D_s \geq D_{sk}$; $f_b(D_b)$ and $f_s(D_s)$ are, respectively, the densities of different bond and node diameters in the sample of the porous medium. It is assumed that $f_b(D_b)$ and $f_s(D_s)$ are known and the integral equations can be solved for D_{bk} and D_{sk}, respectively, once the values of p_{bk} and p_{sk} have been specified. Then, the capillary pressure P_{ck} corresponding to P_{bk} can be calculated by the relationship:

$$P_{ck} = \frac{4\sigma \cos \theta}{D_{bk}}. \tag{2.5.40}$$

The most striking characteristic of the functions Y versus p is their behavior at breakthrough, which corresponds to the percolation threshold of an infinite network. According to the results obtained in this work, at the point of breakthrough, Y increases suddenly from near zero to a finite and significant value. In order to be able to interpret correctly these results, which were obtained on networks of a finite size, it is necessary to examine carefully the conditions at the percolation threshold.

The following relationship exists for values $p > p_c$ in the neighborhood of the percolation threshold (De Gennes and Guyon, 1978):

$$Y \propto (p - p_c)^\beta, \tag{2.5.41}$$

where β is a constant exponent that depends only on the dimensionality of the network ($\beta \simeq 0.40$ for the three-dimensional case).

Consider what happens as p approaches p_c. It is important to recall the physical meaning of p (i.e., a change dp corresponds to opening bonds, or nodes, of diameters between D and $D + dD$), present in the network with finite probability density $f(D)$, given by the expression

$$f(D) = -\frac{dN(D)}{N\,dD}, \qquad (2.5.42)$$

where $dN(D)$ is the number of bonds (nodes) in the diameter range $(D, D + dD)$ and N is the total number of bonds (nodes) present in the network. It is important to note that dN is an integer ($dN \geq 1$).

Consider now an infinite and statistically homogeneous network, from which representative samples, characterized by the same probability density function $f(D)$ as that of the entire network, are taken. As in each sample the quantities $f(D)$, $dN(D)$, and N are all finite, it follows from Eq. (2.5.42) that dD must also be finite. It also follows from this equation that $N \to \infty$ implies $dN(D)/dD \to \infty$, for which two different possibilities can be distinguished: either $dN =$ constant, independent of N, and

$$dD \to 0 \qquad (2.5.43a)$$

or

$$\left\{\begin{array}{l} dD = \\ dN(D)/N = \end{array}\right\} \text{constant, independent of } N. \qquad (2.5.43b)$$

Since the capillary pressure P_c cannot be increased in an experiment in a truly continuous fashion, corresponding to a sequence of infinitesimally small increments, Eq. (2.5.40) indicates that dD cannot be infinitesimally small either. As the size of the measurable increment whereby the value of P_c can be augmented (i.e., the precision of the measurement of P_c) is evidently independent of the size of the sample, dP_c and also dD are both independent of N. Thus, the conclusion may be reached that the condition expressed by Eq. (2.5.43a) is not realistic and that the correct condition is expressed by Eq. (2.5.43b).

Finally, as $dp/dD = f(D)$, it follows that dp cannot be infinitesimally small either. Hence, the limit $p \to p_c$ in Eq. (2.5.41) is physically impossible to satisfy and, consequently, the threshold value of Y at the breakthrough Y_c is not zero, but finite, even in an infinitely large sample, for the physical reasons explained above.

In this work the percolation threshold could be determined to within $\pm 1\%$. However, owing to the rapid change in Y at the percolation threshold, the value of Y_c could not be determined accurately.

An important trend is evident from Fig. 2.58. After passing the percolation threshold, there is a region of p_b in which only a fraction of the open bonds and sites is penetrated ($Y < p$; $y < 1$), because in this phase of the process

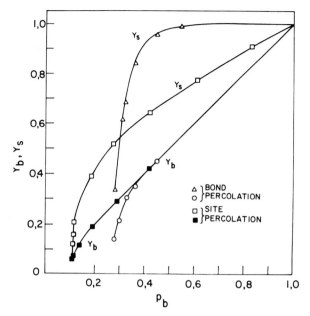

Figure 2.58. Comparison of Y_b and Y_s versus p_b curves obtained in the bond problem and the site problem, respectively (18 × 18 × 12 cubic network) (Chatzis and Dullien, 1985).

the access to certain open bonds and sites *i* blocked by bonds and nodes that are not open yet. In a later phase of the penetration process, however, long before all the bonds (or nodes) are open, all the open bonds (or nodes) are penetrated $(Y = p)$.

Comparing, in Fig. 2.58, the curves corresponding to the bond problem with those of the site problem, the following observations can be made:

(1) There is a considerable difference between the two percolation thresholds (bond percolation: $p_{bc} \approx 0.27$; site percolation: $p_{bc} \approx 0.12$).

(2) Y_b tends to become linear soon after breakthrough, both for the site percolation and for the bond percolation.

(3) After breakthrough, Y_s increases and approaches unity, much faster in the case of bond percolation than in site percolation. It is this behavior that makes site percolation suitable for modeling the properties of the porous media studied in this work.

As the networks studied here represent statistically homogeneous media, the quantities Y and q are expected to be uniformly distributed over the network.

The curves in Fig. 2.59 show that, apart from the region of the network near the injection face, Y has an approximate statistical uniformity in a

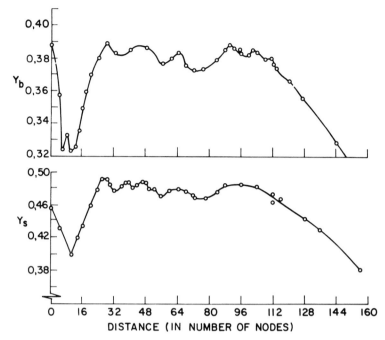

Figure 2.59. Change of *Y* with distance measured from the injection face (Chatzis and Dullien, 1985).

two-dimensional network up to a depth of about 100 mesh points. Starting at the point where the depth of penetration exceeds the width of the network, which is equal to 80 nodes in this case, *Y* starts to decrease gradually. Therefore, in order to make sure that there is a statistical uniformity of the quantities *Y* and *q*, only such networks were used in this study whose depth never exceeded their width.

Before transforming $Y_b(p_{bk})$ and $Y_s(p_{bk})$ to the volume-based saturation S_{nwk} of the nonwetting fluid, some assumptions must be made on the relationship between the volume of nonwetting liquid, $V_{b\ell k}^*$ or $V_{s\ell k}^*$, contained in a bond or in a node ℓ ($\ell = 1, 2, 3, \ldots, k$) and the capillary pressure P_{ck}.

The following assumptions were made (Chatzis and Dullien, 1985).

(1) The bonds were assumed to be capillary tubes of uniform, but angular, cross section and of a volume $V_b(D_b)$. A "corner" of a pore is shown schematically in Fig. 2.60.

(2) The nodes were assumed to be reservoirs of volume $V_s(D_s)$ and of angular cross section (see Fig. 2.60 for a typical corner).

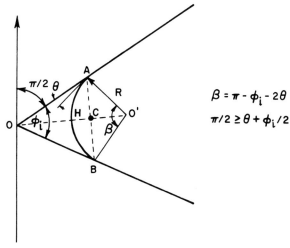

$$\beta = \pi - \phi_i - 2\theta$$
$$\pi/2 \geq \theta + \phi_i/2$$

Figure 2.60.   Geometry of a meniscus in a corner (Chatzis and Dullien, 1982, 1985).

In contrast with the case of a network of cylindrical capillary tubes where $V^*_{b\ell k} = V_s(D_s)$, the bonds and nodes of angular cross section are completely filled with the nonwetting fluid at infinite pressure only. The fraction of their volume filled in dependence of the P_{ck} must therefore be calculated. The unfilled area $(A_v)_i$ in the "corner" of a pore, shown in Fig. 2.60, is given by the following expression:

$$(A_v)_i = R^2 \left[\frac{\cos\theta \cos\left(\theta + \dfrac{\phi_i}{2}\right)}{\sin\dfrac{\phi_i}{2}} - \frac{1}{2}(\pi - \phi_i - 2\theta) \right],$$

where R is the radius of curvature of the meniscus. Hence, the unfilled fraction S_v of a pore cross section is as follows:

$$S_v = \frac{R^2 \displaystyle\sum_{i=1}^{m} \left[\dfrac{\cos\theta \cos\left(\theta + \dfrac{\phi_i}{2}\right)}{\sin\dfrac{\phi_i}{2}} - \dfrac{1}{2}(\pi - \phi_i - 2\theta) \right]}{A_t}, \quad (2.5.44)$$

where A_t is the cross-sectional area of the pore.

The volume $V_{b\ell k}^*$ as a function of D_{bk} follows immediately from Eq. (2.5.44):

$$V_{b\ell k}^* = \left(1 - S_{vb}(\theta)\left(\frac{D_{bk}}{D_{b\ell}}\right)^2\right)V_b(D_b), \qquad (2.5.45a)$$

where $D_{b\ell}$ is the diameter of the capillary (bond) ℓ ($\ell = 1, 2, 3, \ldots, k$). Consistency with the customary definition of pore size used in mercury porosimetry [i.e., with Eq. (2.5.40)] is assured by defining the capillary diameter as follows:

$$D_{b\ell} = 2r_{b\ell} \cos \theta,$$

where $r_{b\ell}$ is the radius of curvature of the meniscus of the nonwetting phase at the least capillary pressure necessary to force it into the capillary ℓ, which is assumed to be removed from the network.* $D_{b\ell}$ is calculated with the aid of Eq. (2.5.36) by substituting the index ℓ for k. $S_{vb}(\theta)$ is the unfilled fraction of the cross section of the capillary at this minimum capillary pressure. Finally,

$$D_{bk} = 2r_{bk} \cos \theta,$$

where r_{bk} is the radius of curvature of the meniscus at the prevailing capillary pressure. (Naturally, D_{bk} is also the diameter of capillary k).

The calculation of $V_{b\ell k}^*$ starts at the percolation threshold, that is, at $k = c$. For values such that $p_{bk} < p_{bc}$ there follows $Y \to 0$ whenever $N \to \infty$ (i.e., in the case of macroscopic porous media, which are the only ones of importance in this monograph).

In a similar fashion, as shown above for the case of the bonds V_{sk}^*, the volume of the nonwetting liquid contained in a node ℓ at the capillary pressure P_{ck} is found from the expression

$$V_{s\ell k}^* = \left(1 - S_{vs}(\theta)\left(\frac{D_{bk}}{D_{s\ell}}\right)^2\right)V_s(D_s), \qquad (2.5.45b)$$

where the diameter of node ℓ, $D_{s\ell} = 2R_{s\ell} \cos \theta$ was defined analogously as D_b. R_s is the radius of curvature of the meniscus at the least capillary pressure necessary to make the nonwetting fluid penetrate the node ℓ. The latter is assumed to be removed from the network and cut into two halves so that the penetration may start in the center of the node. In this case, "penetration" is specified as the point in the process of intrusion into the

---

*Lenormand (1981) pointed out that whereas the capillary pressure of penetration into a capillary of rectangular cross section of edge lengths $2x$ and $2y$ is $P_c = \sigma \cos \theta(1/x + 1/y)$, approximately, the radius of curvature in the plane of normal cross section, r, is given by $P_c = \sigma \cos \theta(1/r + 1/\infty) = (\sigma \cos \theta)/r$. [E.g., for $x = y, r = x/2$.]

node when more than one meniscus is formed in the various "corners" of the node, for the first time.

The diameters $D_{s\ell}$ are calculated with the aid of expressions analogous to Eq. (2.5.36) by substituting the index ℓ for k. Eq. (2.5.45) may be noted to contain the diameter D_{bk} (instead of D_{sk}), because the curvature of the meniscus in the nodes at the prevailing capillary pressure is the same value as in the bonds and is related to D_{bk} rather than D_{sk}.

If all bonds of a size $D_b \geq D_{bk}$ were penetrated at the capillary pressure p_{ck}, the fraction of the total bond volume in the network S_{bk}^*, filled with the nowetting fluid at that pressure, would be given by the following expression:

$$S_{bk}^* = \frac{\int_{D_{bk}}^{D_{bl}} V_{b\ell k}^* f_b(D_b)\, dD_b}{\int_{D_{bn}}^{D_{bl}} V_b(D_b) f_b(D_b)\, dD_b}. \qquad (2.5.46a)$$

Analogously, if all the nodes of a size $D_s \geq D_{sk}$ were penetrated at the capillary pressure P_{ck}, the fraction of the total node volume in the network S_{sk}^*, filled with the nonwetting fluid at that pressure, would be

$$S_{sk}^* = \frac{\int_{D_{sk}}^{D_{sl}} V_{s\ell k}^* f_s(D_s)\, dD_s}{\int_{D_{sn}}^{D_{sl}} V_s(D_s) f_s(D_s)\, dD_s}. \qquad (2.5.46b)$$

The value of D_{sk} corresponding to each D_{bk} is obtained by using Eqs. (2.5.39) and (2.5.35) for D_{bk} and D_{sk}. For the special case of pores of circular cross section the fractions S_{bk}^* and S_{sk}^* are denoted by S_{bk}^0 and S_{sk}^0, respectively, in Table 2.3.

As long as only a proportion y_{bk} of the bonds is penetrated, however, the volume-based saturation of the bonds is given by

$$S_{bk} = y_{bk} S_{bk}^* \qquad (2.5.47a)$$

and that of the nodes by

$$S_{sk} = y_{sk} S_{sk}^*, \qquad (2.5.47b)$$

where y_{sk} is the penetrated proportion of the nodes. Hence, the following expression is obtained for the volume-based saturation of the nonwetting liquid of the network.

$$S_{nwk} = X_b S_{bk} + (1 - X_b) S_{sk}. \qquad (2.5.48)$$

Here, X_b is the fraction of the total volume of the network contributed by the bonds. In Table 2.3, S_{nwk} is superscripted as S_{nwk}^* and S_{nwk}^0 for pores with angular cross section and circular cross section, respectively.

TABLE 2.3 Calculation of the Mercury Porosimetry Curve of the Berea BE-1 Sandstone Sample[a]

k	P_{ck}^{*}	D_{bk}	$D_{bk}^{2} f_{b}$	S_{bk}^{0}	S_{bk}^{*}	y_{bk}	S_{bk}	$X_{b} S_{bk}$	D_{sk}	$D_{sk}^{2} f_{s}$	S_{sk}^{0}	S_{sk}^{*}	y_{sk}	S_{sk}	$(1 - X_{b})S_{k}$	S_{nmk}^{*}	S_{nmk}^{0}
7	1.00	29.5	102.3	0.311	0.154	0.46	0.071	0.036	44.5	848.0	0.506	0.394	0.39	0.154	0.077	0.113	0.170
8	1.02	28.8	104.9	0.340	0.173	0.73	0.126	0.063	44.0	843.8	0.524	0.412	0.49	0.202	0.101	0.164	0.252
9	1.09	27.0	109.5	0.418	0.224	0.88	0.197	0.099	42.5	826.3	0.577	0.466	0.80	0.373	0.187	0.286	0.415
10	1.15	25.5	110.6	0.484	0.270	0.96	0.259	0.130	41.0	802.1	0.629	0.516	0.91	0.470	0.235	0.365	0.519
11	1.23	24.0	109.4	0.552	0.320	0.99	0.317	0.159	39.5	771.8	0.679	0.570	0.96	0.547	0.274	0.443	0.599
12	1.32	22.4	105.6	0.621	0.379	1.00	0.379	0.190	38.0	735.9	0.727	0.623	0.98	0.611	0.306	0.496	0.667
13	1.43	20.7	99.1	0.691	0.442	1.00	0.442	0.221	37.0	709.2	0.758	0.661	1.00	0.661	0.331	0.552	0.724
14	1.54	19.2	91.4	0.749	0.497	1.00	0.497	0.249	35.5	665.3	0.802	0.712	1.00	0.712	0.356	0.605	0.776
15	1.68	17.6	81.6	0.807	0.557	1.00	0.557	0.279	34.0	617.0	0.842	0.760	1.00	0.760	0.380	0.659	0.824
16	1.84	16.0	70.6	0.856	0.622	1.00	0.622	0.311	33.0	582.6	0.868	0.794	1.00	0.794	0.397	0.708	0.862
17	2.06	14.3	58.1	0.900	0.688	1.00	0.688	0.344	31.5	527.6	0.903	0.840	1.00	0.340	0.420	0.764	0.902
18	2.36	12.5	44.7	0.938	0.753	1.00	0.753	0.377	30.0	468.3	0.935	0.882	1.00	0.882	0.441	0.818	0.937
19	2.78	10.6	31.1	0.967	0.817	1.00	0.817	0.409	28.5	403.1	0.963	0.920	1.00	0.920	0.460	0.869	0.965
20	3.47	8.5	17.8	0.987	0.880	1.00	0.880	0.440	27.0	327.5	0.986	0.956	1.00	0.956	0.478	0.918	0.987
21	5.90	5.0	0	1.000	0.953	1.00	0.953	0.477	25.0	0	1.000	0.990	1.00	0.990	0.495	0.972	1.000
	10.0	—	0	1.000	0.982	1.00	0.982	0.491	—	0	1.000	0.996	1.00	0.996	0.498	0.989	1.000

[a]Chatzis and Dullien (1982, 1985).

Since the ratio of the number of bonds to the number of sites is equal to $Z/2$ in all regular networks (Chatzis, 1980), X_b was calculated as follows:

$$X_b = \frac{Z\langle V_b(D_b)\rangle}{Z\langle V_b(D_b)\rangle + 2\langle V_s(D_s)\rangle}. \qquad (2.5.49)$$

Here the symbol $\langle\rangle$ designates the average value. For sandstones the following relationships were assumed:

$$V_b(D_b) \propto D_b^2, \qquad V_s(D_s) \propto D_s^2.$$

The density functions $f_b(D_b)$ and $f_s(D_s)$ were assumed to be given by the beta function:

$$f_b(D_b) \propto \left(\frac{D_b - D_{b\,min}}{D_{b\,max} - D_{b\,min}}\right)^{0.5}\left(1 - \frac{D_b - D_{b\,min}}{D_{b\,max} - D_{b\,min}}\right)^2, \quad (2.5.50a)$$

$$5 > D_b > 45;$$

$$f_s(D_s) \propto \left(\frac{D_s - D_{s\,min}}{D_{s\,max} - D_{s\,min}}\right)^{0.25}\left(1 - \frac{D_s - D_{s\,min}}{D_{s\,max} - D_{s\,min}}\right), \quad (2.5.50b)$$

$$25 > D_s > 65.$$

$f_s(D_s)$ and $f_b(D_b)$ are shown in Fig. 2.61.

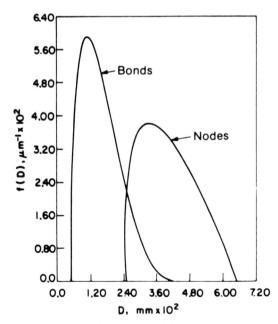

Figure 2.61. Probability density distributions of bond and node diameters (Chatzis and Dullien, 1985).

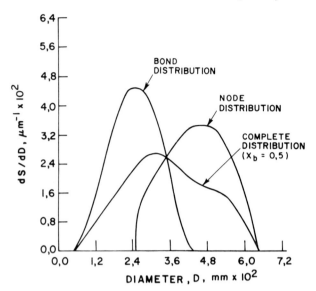

Figure 2.62.   Volume-based diameter-distribution densities (Chatzis and Dullien, 1985).

The density functions $f_b(D_b)$ and $f_s(D_s)$ were determined to be consistent with existing knowledge of pore structure of Berea BE-1 sample.

The volume-based pore-size distributions, obtained from $f_b(D_b)$ and $f_s(D_s)$ as follows,

$$f_b(D_b)V_b(D_b)$$

and

$$f_s(D_s)V_s(D_s),$$

are shown in Fig. 2.62, along with the complete pore-size distribution of the sample, that is,

$$X_b f_b(D_b)V_b(D_b) + (1 - X_b)f_s(D_s)V_s(D_s)$$

for a simple cubic network with $X_b = 0.5$. This distribution is seen to have a form similar to the photomicrographic pore-size distribution determined by Dhawan (1972) for two sandstone samples (see Fig. 1.28).

Equations (2.5.39) and (2.5.35) (for D_{bk} and D_{sk}) were used in conjunctions with Eqs. (2.5.50a) and (2.5.50b), to establish the correlation between D_{bk} and D_{sk} shown in Table 2.4. Thus, the indices assigned to bonds and nodes in the network could be replaced with the diameter of each bond and each node.

To calculate the mercury porosimetry curve, three different values of $S_{vb}(\theta)$ and $S_{vs}(\theta)$ were used in Eqs. (2.5.45a) and (2.5.45b): 0.65, 0.46, and 0.

TABLE 2.4   Correlation, for Berea BE-1 Sandstone Sample, between
Bond Diameter and the Diameter of the Smaller One of the
Two Nodes Situated at The Ends of the Bond[a]

k	p_{sk}	p_{bk}	D_{sk} (μm)	D_{bk} (μm)
2	0.100	0.010	54.0	38.2
3	0.150	0.023	51.5	36.0
4	0.200	0.040	49.6	34.0
5	0.250	0.063	47.5	32.2
6	0.300	0.090	45.5	30.5
7	0.330	0.100	44.5	29.5
8	0.350	0.123	44.0	28.8
9	0.400	0.160	42.5	27.0
10	0.450	0.200	41.0	25.5
11	0.500	0.250	39.5	24.0
12	0.550	0.303	38.0	22.4
13	0.600	0.360	37.0	20.7
14	0.650	0.423	35.5	19.2
15	0.700	0.490	34.0	17.6
16	0.750	0.563	33.0	16.0
17	0.800	0.640	31.5	14.3
18	0.850	0.723	30.0	12.5
19	0.900	0.810	28.5	10.6
20	0.950	0.092	27.0	8.5
21	1.000	1.000	25.0	5.0

[a]Chatzis and Dullien (1982, 1985).

The latter corresponds to the special case of pores of circular cross section, which are filled completely once they are penetrated by the nonwetting phase. As apparent from Fig. 2.63, the first two choices for $S_v(\theta)$ gave practically the same results. The simple cubic network ($X_b = 0.5$) and the tetrahedral network ($X_b = 0.4$) were found to give practically indistinguishable results for the mercury porosimetry curve. Only for the simple cubic network were the quantities Y_b and Y_s calculated in this work. For tetrahedral network they were obtained from the work of Frisch *et al.* (1962).

The mercury porosimetry curves calculated in the manner described above, by using the simple cubic network as the model, are compared in Fig. 2.63 with the experimental curve representing 11 sandstone samples of different origin (Batra, 1973). In this diagram the capillary pressure is expressed in a nondimensional form P_c^* as the multiple of the breakthrough capillary pressure: $P_c^* = P_c/P_{cb}$. In addition, the curve was normalized arbitrarily for $S_{nw} = 100$ at $P_c^* = 25$ for all the samples. The curve starts at the point characterized by the coordinates $\{P_c^* = 1, S_{nw} \approx 10\%\}$ because the experiments of Chatzis (1980) performed on large sandstone samples showed that,

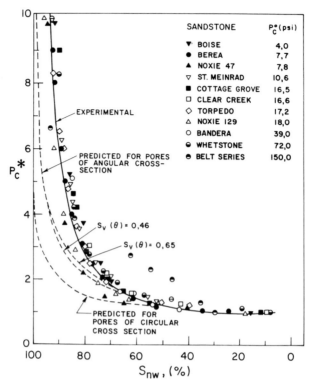

Figure 2.63. Dimensionless mercury-porosimetry curve of sandstone samples (Chatzis and Dullien, 1985).

on the one hand, for $P_c^* < 1$, $S_{nw} \to 0$ and, on the other hand, at $P_c^* = 1$ the lowest value of S_{nw} is approximately 10%. Some intermediate results of the calculations leading to the prediction of the mercury porosimetry curve are listed in Table 2.3. It should be noticed that at $P_c^* = 5.9$ (i.e., for $k = 21$) the whole network is invaded ($p_b = p_s = 1$, as can be seen from Table 2.4), but the filling of the corners (edges or wedges) of pores continues indefinitely as the value of P_c^* is increased. This phenomenon is termed late-filling.

Inspection of the curves in Fig. 2.63 leads to the following observations.

(1) The experimental data on mercury porosimetry for all of the sandstone samples, with the exception of one (Belt Series), are well fitted by a single curve (solid line), but especially well at both (almost asymptotic) ends of the curve. It may be concluded from this result that the pore structure of the 11 sandstone samples is geometrically similar.

(2) One measured quantity, the breakthrough capillary pressure P_c^0, is sufficient for calculation of the mercury porosimetry curve of most sandstones from the solid curve in Fig. 2.63.

(3) The curves calculated by using the modified site-percolation method on a cubic or a tetrahedral network and pores of angular cross section were found to be in agreement with the experimental data up to values of S_{nw} in the range of 70 to 80%

(4) Above 70 to 80% of S_{nw} the calculated values are too high. The largest discrepancy is $+8\%$ for a given value of P_c^*.

(5) The porosimetry curve calculated by assuming pores of circular cross section is in agreement with the experimental curve only up to $S_{nw} \approx 50\text{–}60\%$, but beyond these values it is in excess of the experimental S_{nw} data by a margin of at least 18%.

The agreement between the experimental data and the predictions up to $S_{nw} \approx 70\text{–}80\%$ may be regarded as a strong support for the claim that the network model used in this study provides a sound representation of the main pore network in sandstones, consisting of the passages between sand grains. The lack of agreement at higher nonwetting phase saturations makes very good sense, too, in view of the presence in sandstones of micropores, both inside the cementing materials and also between sand grains, in such instances when the void has been narrowed down by deposits to a very narrow gap. These micropores are characterized by a higher displacement capillary pressure than the smallest pores accounted for in this study and are, therefore, penetrated only at higher values of P_c^* than the pores considered in this work. The topology of the Belt Series sample is probably different from the other samples, as a result of high degree of cementation. Micropores (dual porosity) were not considered in this work because of the lack of the necessary data on the pore structure.

The successful prediction of the behavior of 10 very different sandstone samples may appear surprising at first sight, considering that the range of pore diameters, D_{bk} and D_{sk}, used is not representative of all these different materials. The explanation lies partly in the presumed geometrical similarity of the principal pore structures of many different sandstones. Also, it should be pointed out that the absolute magnitudes of the pore sizes do not determine the results of these calculations. If the nondimensional pore diameters $x = D_{bk}/D_{bc}$ and $y = D_{sk}/D_{bc}$, where D_{bc} is the bond diameter corresponding to P_{cb} by Eq. (2.5.40), are introduced into Eqs. (2.5.35), (2.5.50), and (2.5.46), these equations and the results of the calculations remain unchanged. Hence, the dimensionless pore diameters, expressed as multiples of the breakthrough bond diameter, rather than the absolute magnitudes of the pore diameters, are determining in these calculations.

2.5.2.8. *Calculation of Saturations of Ganglia with No Exit*

Since the fractions q_{bk} and q_{sk} have the same value for all the pores j, such that $j = 1, 2, 3, \ldots, k$, they are transformed to a volume basis simply by

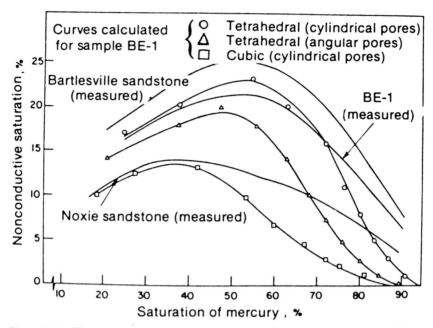

Figure 2.64. The "nonconductive saturation" function of some sandstone samples (Chatzis and Dullien, 1985).

letting

$$\tilde{S}_{nwk} = X_b q_{bk} S_{bk} + (1 - X_b) q_{sk} S_{sk}, \qquad (2.5.51)$$

where $\tilde{S}_{nw}$ is termed nonconductive saturation. This portion of the saturation does not contribute either to the electric or the hydraulic conductivity, but does give the fraction of the pore volume filled with the nonwetting fluid in the form of ganglia with no exit (see Fig. 2.55 for many examples of such ganglia). Thus, the conductive fraction of the nonwetting phase saturation S'_{nwk} is given by

$$S'_{nwk} = S_{nwk} - \tilde{S}_{nwk} \qquad (2.5.52)$$

In this work the nonconducting saturation of a sandstone was calculated by using the same input as for the calculation of the mercury porosimetry curve. The curves shown in Fig. 2.64 reveal a significant difference between the results obtained for the simple cubic network and the tetrahedral network models. The difference between the results obtained by assuming pores to have circular or angular cross section, respectively, is also seen to be significant in this type of calculation. In addition to the experimental curve representing the Berea BE-1 sample, measured by El-Sayed and Dullien (1977) by displacement of mercury, using radioactive mercury (see Chapter

1), two more experimental curves obtained on two sandstones of different origin were included in the plot in order to demonstrate that the different sandstone samples exhibit significant individual characteristics from the point of view of the nonconductive saturation, and that their behavior cannot be represented by the same curve.

Some differences between the pore structures of various sandstones have probably eluded us so far, and these differences may have an important effect on the formation of ganglia with no exit. Qualitatively speaking, however, the predicted curves have the same shape as the experimental ones. All the curves have a maximum in the saturation interval between 35% and 55%. Both curves calculated by means of the tetrahedral network are in reasonably good agreement with the curve measured for the Berea BE-1 sample with the exception of the high saturation end of the diagram where all the experimentally determined curves decrease more gradually than the calculated ones. Such behavior indicates that at high nonwetting phase saturations the amount of mercury that could not be miscibly displaced by tagged mercury was greater in the sandstones than in the network models used in these calculations. The true cause, or causes, of this discrepancy are difficult to discern, partly because the experimental data scatter badly at high nonwetting phase saturations. However, the presence of systems of very fine pores in the sandstones, which were not accounted for in our model, would result in additional amounts of mercury trapped in the form of ganglia with no exit.

2.5.2.9.   *Modeling of Drainage and Imbibition Capillary Pressure Curves Including Trapping*

The bond-correlated site percolation introduced by Chatzis (1980) was later applied by Diaz (1984) and Diaz *et al.* (1987) to the simulation of oil–water capillary pressure curves of Berea sandstone. The same procedure described in Section 2.5.2.6 was used to calculate the number-based saturation (Y_s versus p_s and Y_b versus p_b) and accessibility (y_s versus p_s and y_b versus p_b) curves for primary drainage, with the important difference that trapping of the wetting phase by the displacing nonwetting phase was also kept track of. Trapping of a cluster of wetting phase was defined by the condition that all the nodes in actual contact with the cluster have been occupied by the nonwetting phase. The fact that the wetting phase so trapped may "leak" past the surrounding nonwetting phase in the form of thick films through surface grooves and edges was not considered.

The penetration history has four characteristic values of p_s (see Fig. 2.65):

(1)   The percolation threshold value p_{sc}, the lowest value at which the ganglia of the penetrating fluid have infinite extent in an infinitely large lattice. This corresponds to breakthrough at the discharge face in a lattice of finite but sufficiently large size. For the simple cubic lattice $p_{sc} \approx 0.33$; below this value $y_s = y_b \approx 0$.

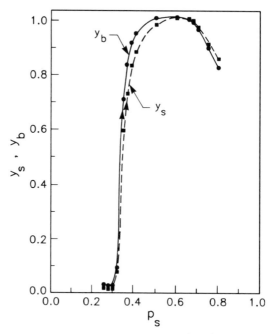

Figure 2.65. Penetrated number fraction of allowed (open) nodes and bonds in primary drainage with trapping as a function of allowed (open) number fraction of all the nodes in the network (Diaz 1984).

(2)   The values at which all open nodes and bonds first become accessible to the penetrating fluid and are, therefore, all penetrated (i.e., $y_s = 1$ and $y_b = 1$, respectively). These were found to be $p_s = 0.55$ and $p_s = 0.60$, respectively.

(3)   The values at which some open nodes and bonds start to become inaccessible because they are occupied by trapped ganglia of the wetting fluid. It has been found that this occurs also at the value $p_s = 0.55$ and $p_s = 0.60$, respectively (i.e., y_s and y_b exhibit maxima at this point).

(4)   The value beyond which there is no more penetration because all the nodes and bonds left to be made "open" past this value were occupied by trapped ganglia. This is also the value of p_s at which all the wetting phase becomes discontinuous; this happened at $p_s \approx 0.68$. The corresponding number-based nonwetting phase saturation is also $Y_s \approx 0.68$.

The information contained in Fig. 2.65 has been used to calculate the primary drainage capillary pressure curve of an actual sample by the same procedure described in Section 2.5.2.7, except that effect of angular pore cross section was ignored. Instead, the probability density distributions of pore body and pore throat diameters, $f(D_s)$ and $f(D_b)$, were varied but were always kept consistent with photomicrographic evidence of pore structure.

Secondary Imbibition. During secondary imbibition, the wetting phase displaces the nonwetting phase, starting at "irreducible" wetting phase saturation. In the simulation of secondary imbibition the nodes and bonds have been opened up for penetration by the wetting fluid in the order of index numbers that is the reverse than in drainage. The model of imbibition was based on the assumption that the wetting phase displaces the nonwetting phase in a pistonlike manner from each and every pore. According to Laplace's equation of capillarity, at the highest capillary pressure only the smallest pores may be penetrated by the wetting phase; at somewhat lower capillary pressures, somewhat larger pores may also be penetrated; and so forth, always subject to the condition that the pores in question are accessible to the wetting fluid in contact with the inlet face of the sample. Supposing that the mechanism of displacement is pistonlike, the only way a node or a bond can be accessible to the wetting phase is if it is in communication with the inlet face (= discharge face in drainage) either directly or by the agency of an uninterrupted sequence of open nodes and/or bonds. Therefore, this was the condition of accessibility used in the model of imbibition, analogously to the case of drainage, notwithstanding the fact that it has been shown in a variety of experiments (Dullien *et al.*, 1986; Dullien *et al.*, 1989; Wardlaw and LiYu, 1988; Lenormand *et al.*, 1983) that the wetting fluid can reach pores by advancing on the solid surface in grooves and wedges.

As a result of this type of mobility, pores of a certain small size that are not connected to the inlet face by the agency of an uninterrupted sequence of other pores that are just as small as or smaller than the pores in question, are also accessible to the wetting fluid. As a result, in imbibition into a sandstone in principle, all pores are accessible and may all be penetrated by the wetting fluid at an appropriate capillary pressure related to their size according to Laplace's equation, unless they already contain trapped nonwetting phase.

Decreasing p_s from 1, the wetting phase became continuous at $p_{sc} = 0.68$ and the nonwetting phase became discontinuous at $p_s = 0.32$, which is tantamount to running the primary discharge simulation in the reverse order. The trapped residual nonwetting phase saturation on a number basis is $Y_s = 0.32$, which is also equal to the number-based irreducible wetting phase saturation found in the simulation of primary drainage. The Y_s versus p_s curves of primary drainage and secondary imbibition (including also secondary drainage, to be discussed next) are shown in Fig. 2.66. It is apparent that on a number basis there is no hysteresis.

The number-based secondary imbibition displacement curve has been converted to the corresponding capillary pressure curve by calculating the capillary pressure P_c, using D_{sk} instead of D_{bk} in Eq. (2.5.36), and the calculation of saturation was the same as in primary drainage.

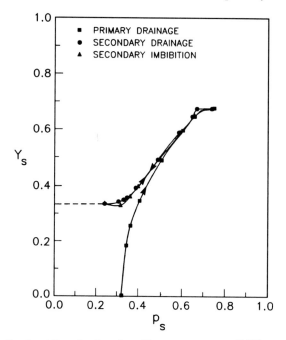

Figure 2.66. Simulated "number-based capillary pressure curves" (Diaz *et al.*, 1987).

Secondary Drainage.   Secondary drainage is a repetition of primary drainage, starting at residual nonwetting phase saturation. The results of simulation in terms of the Y_s versus p_s curve are shown in Fig. 2.66.

Discussion: *The Simulated Capillary Pressure Curves.*   The simulated capillary pressure curves are compared with experiment on Berea sandstone in Fig. 2.67 in the form of nonwetting phase saturation versus reduced capillary pressure P_c^*.

Inspection of Fig. 2.67 shows the following:

(1)   The simulated primary drainage curve is in good agreement with experiment except for underestimating the "irreducible" wetting phase saturation. The reason for this discrepancy lies in the network model representing only the macropores in the sandstone. The micropores, which were unaccounted for in the model but are actually present in the cementing materials in the sandstone, trap additional wetting phase.

(2)   The simulated secondary imbibition curve predicts much greater capillary hysteresis than the measured one, because the predicted capillary pressure corresponding to the breakthrough of the wetting phase is very low. Further, the change from "irreducible" wetting phase saturation to residual nonwetting phase saturation takes place over a fractional change in the

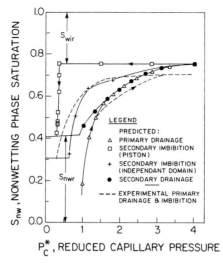

Figure 2.67. Comparison of simulated and measured capillary pressure curves of sandstone samples (adapted from Diaz *et al.*, 1987).

capillary pressure that is far too small. Both of these discrepancies are due to the assumption made in the model that sites control imbibition.

The simulated secondary imbibition curve has been recalculated by the author, using the same accessibility data obtained by Diaz (1984), by assuming that every bond and every site is penetrated by the wetting phase at a value of $P_c^* = 28 \ \mu m/D$, where 28 μm is the "breakthrough" bond diameter in primary drainage and D is the diameter of the bond, or site, in question expressed in μm units. Bonds and sites occupied by trapped nonwetting phase were excluded. It is apparent that there has been a great improvement as a result of changing the assumption of piston-type imbibition displacement to an assumption of independent domain.

In summary, it may be concluded that network simulation of drainage capillary pressure curves when one phase has a strong wetting preference shows a strong promise of becoming a useful predictive tool, approaching the accuracy that can be expected from core analysis. In addition to fluid and wettability properties a good reconstruction of pore structure is required.

The situation in the case of secondary imbibition is more difficult. The hierarchy of displacement has been demonstrated in 2-D micromodels to be a very complicated function of pore aspect ratios and other details of pore geometry, in particular, how pores and throats of different sizes are distributed in space relative to each other. In addition, it is well-known, that the capillary pressure at which various pores are penetrated in imbibition is a different function of the pore size, depending on the fluid topology (see Chapter 5). The simple-minded approach leading to reasonably good predictions in the case of the imbibition curve in Fig. 2.67, in which it was assumed that every pore throat and pore body is filled with the wetting phase in secondary imbibition as if it were directly accessible to the wetting fluid, much like in the case of the independent domain theory, will need critical testing with other 3-D porous media.

It is of interest to note in this connection that Lapidus *et al.* (1985) obtained excellent agreement between experimental mercury intrusion and withdrawal curves of Aerosil 0X50 and a controlled-pore glass, on the one hand, and network simulation using a cubic network of pores and throats, on the other. Two size distributions, one for the pores and another for the throats, were assumed and fitted optimally. The sizes were randomly distributed over the pores and the throats of the network. The throats were assumed to have no volume. Intrusion of mercury into the pores was simulated to take place through all six faces of the cube; and it was assumed to be controlled by the throats in the order of decreasing throat sizes, subject to the condition of accessibility of the throats to mercury, in the same way as was done by Chatzis and Dullien (1982, 1985). Mercury withdrawal also was assumed to take place across all six faces of the cube; but it was assumed also that mercury ruptures at throats of a size below a specified value, thus creating mercury/gas interfaces that permit the withdrawal of mercury from pores all over the network in the order of increasing pore size. Also in this treatment there has been a great deal of arbitrariness but, at the same time, it has been demonstrated that both the intrusion and the withdrawal curves can be simulated consistently by using the same two size distribution curves and assuming that in the imbibition (i.e., withdrawal) process each pore empties at the equilibrium capillary pressure corresponding to its size, regardless of its position in the network.

Detailed mechanisms of imbibition have been studied mostly in conjunction with dynamic (rather than quasistatic) displacement; this subject is more appropriately discussed in Chapter 5.

REFERENCES

Adamson, A. W. (1967). "Physical Chemistry of Surfaces." Wiley (Interscience), New York.
Amott, E. (1959), *Trans. AIME* **216**, 156.
Andersen, M. A., Thomas, D. C., and Teeters, D. C. (1988). *SPE/DOE* 17368.
Anderson, W. G. (1986), *JPT* (October), 1125, (November), 1246.
Anderson, W. G. (1987), *JPT* (October), 1283.
Bacri, J. C., Leygnac, C., and Salin, D. (1985). *J. Physique Lett.* **46**, L467.
Bacri, J. C., Rosen, M., and Salin, D. (1990). *Europhys. Lett.* **11**(2), 127.
Baker, D. J., and Morris, J. B. (1971), *Carbon* **9**, 687.
Barrett, E. P., Joyner, L. G., and Halenda, P. P. (1951). *J. Am. Chem. Soc.* **73**, 373.
Batra, B. J. (1973). Ph.D. Thesis, Univ. of Waterloo, Waterloo, Ontario.
Bear, J. (1972), "Dynamics of Fluids in Porous Media." American Elsevier, New York.

Bernardini, F., Collepardi, M., and Armisi, I. (1967). *Chim. Ind.* (Milan) **49**, 366.

Bikerman, J. J. (1958). "Surface Chemistry." Academic Press, New York.

Bobek, J. E., Mattax, C. C., and Denekas, M. O. (1958). *Trans. AIME* **213**, 155.

Boneau, D. F., and Clampitt, R. L. (1977). JPT (May), 501.

Broadbent, S. R., and Hammersley, J. M. (1957). *Camb. Phil. Soc.* **53**(3), 629.

Broekhoff, J. C. P., and de Boer, J. H. (1967). *J. Catal.* **9**, 8.

Broekhoff, J. C. P., and de Boer, J. H. (1968a). *J. Catal.* **10**, 153.

Broekhoff, J. C. P., and de Boer, J. H. (1968b), *J. Catal.* **10**, 377.

Broekhoff, J. C. P., and de Boer, J. H. (1968c), *J. Catal.* **10**, 391.

Broekhoff, J. C. P., and de Boer, J. H. (1968d). *J. Catal.* **10**, 368.

Brooks, R. H., and Corey, A. T. (1964). "Hydrology Papers." Colorado State Univ., Fort Collins, Colorado.

Brown, H. W. (1951). *Trans. AIME* **192**, 67.

Brown, R. J. S., and Fatt, I. (1956). *Trans. AIME* **207**, 262.

Brunauer, S. (1945). "The Adsorption of Gases and Vapors," Vol. I. Princeton Univ. Press, Princeton, New Jersey.

Brunauer, S., Mikhail, R. Sh., and Bodor, E. F. (1967). *J. Colloid Interface Sci.* **24**, 451.

Brunauer, S., Skalny, J., and Odler, I. (1973). *In* "Pore Structure and Properties of Materials." (S. Modry and M. Svata, eds.), Part I, pp. C3–C26. Academia, Prague.

Brunauer, S., Emmett, P. H., and Teller, E. (1938). *J. Am. Chem. Soc.* **60**, 309.

Burkhardt, J. A., Ward, M. B., and McLean, R. H. (1958). Paper SPE 1139-G presented at the 1958 SPE Annual Meeting, Houston, Oct. 5–8.

Calhoun, J. C., Jr., Lewis, M., Jr., and Newman, R. C. (1949). *Trans. AIME* **186**, 189.

Cameron, A., and Stacy, W. O. (1960). *Chem. Ind.* **9**, 222.

Carey, J. W., and Taylor, S. A. (1967). *Monogr. Am. Soc. Agron.* **11**, 13.

Chatzis, I. (1976). M.A.Sc. Dissertation, Univ. of Waterloo, Waterloo, Ontario.

Chatzis, I. (1980). Ph.D. Thesis, Univ. of Waterloo, Waterloo, Ontario.

Chatzis, I., and Dullien, F. A. L. (1977). *J. Can. Pet. Technol.* **16**, 97.

Chatzis, I., and Dullien, F. A. L. (1982). *Revue de L'Institut Francais du Petrole*, March/April.

Chatzis, I., and Dullien, F. A. L. (1985). *ICE* **25**, 47.

Cohan, L. H. (1944), *J. Am. Chem. Soc.* **66**, 98.

Craig, F. F., Jr. (1971). "The Reservoir Engineering Aspects of Water-flooding." Society of Petroleum Engineers of AIME, Monograph, Vol. **3**, Dallas, Texas.

Cranston, R. W., and Inkley, F. A. (1957). *Adv. Catal.* **9**, 143.

de Boer, J. H. (1958). *In* "The Structure and Properties of Porous Materials" (D. H. Everett and F. S. Stone, eds.), Vol. 10, p. 68. Butterworths, London.

de Boer, J. H., Linsen, B. G., and Osinga, T. L. (1965a). *J. Catal.* **4**, 403.

de Boer, J. H., van de Plas, Th., and Zondervan, G. J. (1965b). *J. Catal.* **4**, 649.

Dean, P. (1963). *Proc. Camb. Phil. Soc.* **59**, 397.

Dean, P., and Bird, N. F. (1967). *Proc. Camb. Phil. Soc.* **63**, 447.

Defay, R., and Prigogine, I., (with collaboration of Bellemans, A.) (1966). "Surface Tension and Adsorption" (translated by D. H. Everett). Longmans, Green and Co. Ltd., London.

De Gennes, P. G., and Guyon, E. (1978). *J. Mecanique*, **17**, 403.

Deryagin, B. V. (1940). *Acta Phys.-Chim. USSR* **12**, 181.

Deryagin, B. V. (1957). *Proc. Int. Congr. Surface Activ.*, 2nd, London **2**, 153.

Dhawan, G. K. (1972). Ph.D. Thesis, Univ. of Waterloo, Waterloo, Ontario.

Diaz, C. E. (1984). M.A.Sc. Thesis, Univ. of Warerloo, Waterloo, Ontario.

Diaz, C. E., Chatzis , I., and Dullien, F. A. L. (1987). *Transport in Porous Media* **2**, 215.

Dodd, C. G., and Kiel, O. G. (1959), *J. Am. Chem. Soc.* **63**, 1646.

Dodds, J. A., and Lloyd, P. J. (1971/1972). *Power Technol.* **5**, 69.

Dollimore, D., and Heal, G. R. (1964). *J. Appl. Chem.* **14**, 109.

Domb, C., and Sykes, M. F. (1961). *Phys. Rev.* **122**, 77.

Dombrowski, H. S., and Brownell, L. E. (1954). *Ind. Eng. Chem.* **46**, 1207.

Donaldson, E. C., Thomas, R. D., and Lorenz, P. B. (1969). *SPEJ* **9**, 13.

Donaldson, E. C. , *et al*. (1980). Report DOE/BETC/IC-79/5, Bartlesville Energy Technology Center, U. S. DOE.

Donaldson, E. C. (1981). *Preprints, American Chemical Soc., Div. of Petroleum Chemistry* **26**, No. 1, 110.

Dubinin, M. M. (1973). *In* "Pore Structure and Properties of Materials" (S. Modry and M. Svata, eds.), Part I, pp. C27–C59. Academia, Prague.

Dubinin, M. M., Lezin, Y. S., Kadlec, O., and Zukai, A. (1968). *Russ. J. Phys. Chem.* **42**, 500.

Dullien, F. A. L., Allsop, H., and Macdonald, I. F. (1990). *JCPT* **29**, 63.

Dullien, F. A. L., and Batra, V. K. (1970). *ISEC.* **62**, 25.

Dullien, F. A. L., and Dhawan, C. K. (1975). *J. Interface Colloid Sci.* **52**, 129.

Dullien, F. A. L. (1975). *AIChE.* **21**, 299.

Dullien, F. A. L., El-Sayed, M. S., and Batra, V. K. (1977). *J. Interface Colloid Sci.* **60**, 497.

Dullien, F. A. L., and Batra, V. K. (1970). *In* "Flow through Porous Media." 2, American Chemical Society. Washington, D.C.

Dullien, F. A. L., Lai, F. S. Y., and Macdonald, I. R. (1986). *J. Colloid Interface Sci.* **109**, 201.

Dullien, F. A. L., Zarcone, C., Macdonald, I. F., Collins, A., and Bochard, R. D. E. (1989). *J. Colloid. Interface Sci.* **127**, 362.

Dumoré, J. M., and Schols, R. S. (1974). *Soc. Pet. Eng. J.* **14**, 437.

Dunning, H. N., Hsiao, L., and Johansen, R. T. (1954). U.S. Bureau of MinesReport Investigations No. 5020, December 1953; *Pet. Eng.* **26**(1), 882 (1954); *Oil and Gas J.* **53**(15), 139 (1954).

El-Sayed, M. S., and Dullien, F. A. L. (1977). "Investigations of Transport Phenomena and Pore Structure of Sandstone Samples," 28th Technical Meeting of Petroleum Society CIM, Edmonton, Alberta. Preprint No. 7722.

Enderby, J. A. (1956). *Trans. Faraday Soc.* **51**, 835.

Enderby, J. A. (1956).*Trans. Faraday Soc.* **52**, 106.

Everett, D. H. (1954). *Trans. Faraday Soc.* **50**, 1077.

Everett, D. H. (1955). *Trans. Faraday Soc.* **51**, 1551.

Everett, D. H. (1967). *In* "The Solid-Gas Interface" (E. Alison Flood, ed.), Vol. II, pp. 1055–1010. Dekker, New York.

Everett, D. H., and Smith, F. W. (1954). *Trans. Faraday Soc.* **50**, 187.

Everett, D. H., and Whitton, W. I. (1952). *Trans. Faraday Soc.* **48**, 749.

Everett, J. R., Gooch, F. N., and Calhoun, J. C. (1950). *Trans. AIME* **189**, 215.

Fatt, I. (1956). *Trans. AIME Pet. Div.* **207**, 144, 160, 164.

Flood, E. S. (ed.) (1967). "The Solid-Gas Interface," Vols. 1–2. Dekker, New York.

Frevel, L. K., and Kressley, L. J. (1963). *J. Anal. Chem.* **35**, 1492.

Frisch, H. L., Hammersley, J. H., and Wells, D. J. A. (1962). *Phys. Rev.* **127**, 949.

Frisch, H. L., and Hammersley, J. M. (1963). *Journal Soc. Indust. Appl. Math.* **11**, 894.

Fuchs, N. (1930). *Kolloid Z.*, **52**, 262.

Gatenby, W. A., and Marsden, S. S. (1957). *Prod. Monthly* **22**, 5.

Good, R. J. (1952). *J. Am. Chem. Soc.* **74**, 5041.

Haines, W. B. (1930). *J. Agr. Sci.* **20**, 97.

Halsey, C. D. (1948). *J. Chem. Phys.* **16**, 931.

Harkins, W. D. (1952). "The Physical Chemistry of Surface Films." Van Nstrand-Reinhold, Princeton, New Jersey.

Hassler, G. L., and Brunner, E. (1945). *AIME T.P.* 1817.

Holbrook, O. C., and Bernard, G. G. (1958). *Trans. AIME* **213**, 261.

Hoshen, J., and Kopelman, R. (1976). *Phys. Rev.* B, **14**, 3438.

Jennings, H. Y. (1958). *Prod. Monthly* **22**, 26.

Keelan, D. K., and Pugh, V. J. (1975). *Soc. Pet. Eng. J.* **15**, 149.

Killins, C. R., Nielson, R. F., and Calhoun, J. C., Jr. (1953). *Producers Mon.* **18**(2), 30.

King, F. H. (1897–1898). *Ann. Rep. U.S.G.S.*, 19th Part II, 59–294.

Kirkpatrick, S. (1973). *Rev. Mod. Phys.* **45**, 574.

Kruyer, S. (1958). *Trans. Faraday Soc.* **54**, 1758.

Kusakov, M. M., and Nekrasov, D. N. (1958). *Dokl. Akad Nawk SSSR* **119**, 107.

Kusakov, M. M., and Nekrasov, D. N. (1960). *J. Phys. Chim.* **34**, 1602.

Lai, F., Macdonald, I. F., Dullien, F. A. L., and Chatzis, I. (1981). *J. Colloid Interface Sci.* **84**, 362.

Langmuir, I. (1918). *J. Am. Chem. Soc.* **40**, 1361.

Lapidus, G. R., Lane, A. M., Ng, K. M., and Corner, W. C. (1985). *Chem. Eng. Commun.* **38**, 33.

Lenormand, R. (1981). Thèse. L'Institut National Polytechnique de Toulouse, France.

Lenormand, R. (1983). *CR Acad. Sci. Paris Sér. II* **297**, 437.

Lenormand, R., Cherbuin, C., and Zarcone, C. (1983). *CR Acad. Sci. Paris. Sér. II* **297**, 637.

Lenormand, R., and Zarcone, C. (1983). *J. Fluid Mech.* **135**, 337.

Lenormand, R., and Zarcone, C. (1984). *Soc. Pet. Eng.* **13**, 264.

Lenormand, R., Zarcone, C., and Sarr, A. (1983). *J. Fluid Mech.* **135**, 351.

Leppard, C. J., and Spencer, D. H. T. (1968). *J. Sci. Instrum. J. Phys. E. Ser.* **35**, 573.

Leverett, M. C. (1941). *Trans. AIME* **142**, 152.

Li, Y., and Wardlaw, N. C. (1986). *J. Colloid Interface Sci.* **109**, 461.

Lippens, B. C., and de Boer, J. H. (1965). *J. Catal.* **4**, 319.

Lippens, B. C., Linsen, B. G., and de Boer, J. H. (1964). *J. Catal.* **3**, 32.

Marsh, H., and Rand, B. (1970). *J. Colloid and Interface Sci.* **33**, 478.

Marsh, H., and Rand, B. (1972). *J. Colloid and Interface Sci.* **40**, 121.

Mayer, R. P., and Stowe, R. A. (1965). *J. Colloid Sci.* **20**, 893.

Mayer, R. P., and Stowe, R. A. (1966). *J. Phys. Chem.* **70**, No. 12, 3867.

Melrose, J. C. (1965). *Soc. Pet. Eng. J.* **5**, 259.

Melrose, J. C. (1987). "Characterization of Petroleum Reservoir Rocks by Capillary Pressure Techniques." Presented at IUPAC Symposium, Bad Solen, West Germany, April 26–29.

Melrose, J. C. (1988). *Log Analyst* **29**, No. 1 (February).

Melrose, J. C. and Bradner, C. F. (1972). *Ann. AIChE Meeting*, 65th, New York. Paper 33a.

Meyer, H. I. (1953). *J. Appl. Phys.* **24**, 510.

Miller, E. E., and Miller, R. D. (1956). *J. Appl. Phys.* **27**, 324.

Miller, N. F. (1941). *J. Phys. Chem.* **45**, 1025.

Modry, S., Svata, M., and Jindra, J. (1972). "Bibliography on Mercury Porosimetry." House of Technology, Prague.

Mohanty, K. K., and Salter, S. J. (1983). Paper SPE 12127 presented at the 1983 SPE Annual Technical Conference and Exhibition, San Francisco, Oct. 5–8.

Morrow, N. R. (1970). *Ind. Eng. Chem.* **62**, 32. *Preprinted as a chapter in* "Flow through Porous Media." American Chemical Society, Washington, D.C.

Morrow, N. R. (1971a). *J. Can. Pet. Technol.* **10**, 38, 47.

Morrow, N. R. (1971b). *Bull. A.A.P.G.* **55**, 42.

Morrow, N. R. (1975). *J. Can. Pet. Technol.* **14**, 42.

Morrow, N. R. (1976). *J. Can. Pet. Technol.* **15**, 49.

Morrow, N. R., Cram, P. J., and McCaffery, F. G. (1972). Preprint SPE 3993 for the Ann. SPE Fall Meeting, 47th, San Antonio, October 8–11, 1972. Displacement Studies in Dolomite with Wettability Control by Octanoic Acid.

Morrow, N. R., and Harris, C. C. (1965). *SPE Jour.* **5**, 15.

Morrow, N. R., and Mungan, N. (1971). *Revue IFP* **26**, 629.

Mungan, N. (1972). *Soc. Petr. Eng. J.* **12**, 398.

Muskat, M. (1949). "Physical Principles of Oil Production." McGraw-Hill, New York.

Mysels, K. J., and Stigter, D. (1953). *J. Phys. Chem.* **57**, 104.

Payne, D. (1953). *Nature* (London) **172**, 261.

Philip, J. R. (1964). *J. Geophys. Res.* **69**(8).

Pickell, J. J., Swanson, B. F., and Hickman, W. B. (1966). *Soc. Pet. Eng. J.* **6**, 55.

Popiel, W. J. (1978). "Introduction to Colloid Science." Exposition Press, Hicksville, New York.

Poulovassilis, A. (1962). *Soil Sci.* **93**, 405.

Poulovassilis, A. (1970). *Soil Sci.* **109**, 154.

Poulovassilis, A., and Childs, E. C. (1971). *Soil Sci.* **112**, 301.

Powers, E. L., and Botset, H. G. (1949). *Producers Mon.* **13**, 15.

Purcell, W. R. (1950). *Trans. AIME* **189**, 369.

Richardson, J. G. (1961). *In* "Handbook of Fluid Dynamics" (V. L. Streeter, ed.). McGraw-Hill, New York.

Ritter, H. L., and Drake, L. C. (1945). *Ind. Eng. Chem.* **17**, 782.

Rose, W. (1957). "Studies of Waterflood Performance," Vol. III, Use of Network Models. Illinois State Geology Survey, Circ. No. 237, Urbana, Illinois.

Rose, W. D., and Bruce, W. A. (1949). *Trans. AIME* **186**, 127.

Russell, R. G., Morgan, F., and Muskat, M. (1947). *Trans. AIME* **170**, 51.

Salathiel, R. A. (1973). Preprint SPE 4104 for the Ann. SPE Fall Meeting, 47th , San Antonio, October 8–11, 1972. Oil Recovery by Surface Film Drainage in Mixed-Wettability Rocks.

Shankar, P. K. (1979). Ph.D. Thesis, University of Waterloo.

Shankar, P. K., and Dullien, F. A. L. (1979). *In* "Surface Phenomena in Enhanced Oil Recovery." (D. O. Shah, ed.). Plenum Press, 453.

Shante, V. K. S., and Kirkpatrick, S. (1971). *Adv. Phys.* **42**, 395.

Sharma, M. M., and Wunderlich, R. W. (1985). Paper SPE 14302 presented at the 1985 *SPE Annual Technical Conference and Exhibition*, Las Vegas, Sept. 22–25.

Shull, C. G. (1948). *J. Am. Chem. Soc.* **70**, 1405.

Shuttleworth, R., and Bailey, G. L. J. (1948). *Disc. Faraday Soc.* **3**, 16.

Slobod, R. L., Chambers, A., and Prehn, W. L., Jr. (1951). *Trans. AIME* **192**, 127.

Smith, W. O. (1933). *Physics* **4**, 425.

Stahl, C. D., and Nielsen, R. F. (1950). *Producers Mon.* **14**(3), 19.

Stegemeir, G. L. (1976). Mechanism of entrapment and mobilization of oil in porous medium. Preprint for the AIChE Meeting, 81st, Missouri, April.

Sykes, M. F., and Essam, J. W. (1963). *Phys. Rev. Lett.* **10**, 3.

Talsma, T. (1970). *Water Resources Res.* **6**(3), 964.

Topp, G. C. (1969). Contribution No. 297. Soil Research Institute, Canada Dept. of Agriculture, Ottawa. Presented before Div. S-1, Soil Science Society of America, New Orleans, Louisiana.

Topp, G. C., and Miller, E. E. (1966). *Soil Sci. Soc. Am. Proc.* **30**, 156.

Trantham, J. C. and Clampitt, R. L. (1977). *JPT* (May), 491.

Treiber, L. E., Archer, D. L., and Owens, W. W. (1972). *Soc. Petr. Eng. J.* **12**, 531.

Tschapek, M., Falaska, S., Wasowski, C. (1985). *Powder Tech.* **42**, 175.

Versluys, J. (1917). *Int. Mitt. Bodenk.* **7**, 117.

Vissotsky, V. A., Gordon, S. B., Frisch. H. L., and Hammersley, J. H. (1961). *Phys. Rev.* **123**, 5.

Voigt, E. M., and Tomlinson, R. H. (1955). *Can. J. Chem.* **33**, 215.

Wagner, O. R., and Leach, R. O. (1959). *Trans. AIME* **213**, 155.

Wardlaw, N., and Taylor, R. P. (1976). *Bull. Can. Pet. Geol.* **24**, 225.

Wardlaw, N., and Li, Y. (1986a). *J. Colloid. Interface Sci.* **109**, 461.

Wardlaw, N., and Li, Y. (1986b). *J. Colloid Interface Sci.* **109**, 473.

Wardlaw, N., Li, Y., and Laidlaw, W. G. (1986c). *Advances in Colloid and Interface Sci.* **26**, 1.

Wardlaw, N., and Li, Y. (1988). *Transport in Porous Media* **3**, 17.

Weatherburn, C. E. (1947). "Differential Geometry," Vol. I, p. 72. Cambridge Univ. Press, London.

Welge, H. J., and Bruce, W. A. (1947). "Drilling and Production Practices," *Amer. Petrol. Inst.*, p. 166.

Wenzel, R. N. (1936). *Ind. Eng. Chem.* **28**, 988; (1949). *J. Phys. Coll. Chem.* **53**, 1466.

Wheeler, A. (1946). *In* "Presentations at Catalysis Symposia" (Am. Assoc. Adv. Sci. Conf., Gibson Island, June 1945, 1946).

Wise, M. E. (1952). *Philips Res. Rep.* **7**, 321.

Wise, M. E. (1960). *Philips Res. Rep.* **15**, 101.

Yadav, G. D., Dullien, F. A. L., Chatzis, I., and Macdonald, I. F. (1987). *SPE Reservoir Engineering* **2**, 137.

Zsigmondy, R. (1911). *Z. Anong. Allgem. Chem.* **71**, 356.

3 | *Single-Phase Transport Phenomena in Porous Media*

3.1. INTRODUCTION

The conductivity of a porous medium for single-phase fluids is customarily expressed, by analogy with electrical conduction, by giving the specific permeability k (or "permeability," for short) of the medium defined by Darcy's law written in differential form

$$\mathbf{v} = -(k/\mu)\nabla\mathscr{P} = -(k/\mu)(\nabla P - \rho\mathbf{g}) \qquad (3.1.1)$$

where $\mathscr{P}$ is defined by the following equation:

$$\mathscr{P} = P + \rho g z; \qquad (3.1.2)$$

z is the distance measured vertically upward from an arbitrarily chosen datum level, P the hydrostatic pressure, ρ the fluid density, and g the acceleration due to gravity. $\mathscr{P}$ is measured by a pipe, called the "piezometer" (see Fig. 3.1) and is indicated as the "piezometric head" ϕ (dimension of length):

$$\phi = \mathscr{P}/\rho g = (P/\rho g) + z, \qquad (3.1.3a)$$

which is the sum of the "elevation head" z and the "pressure head" $P/\rho g$.

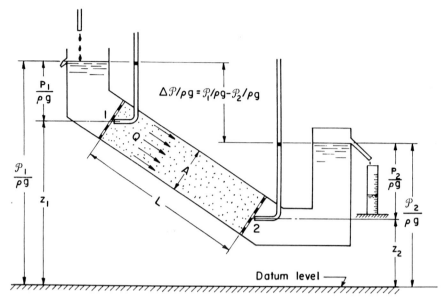

Figure 3.1. Illustration of "piezometric head," "elevation head," and "pressure head" (Bear, 1972).

For a compressible fluid the pressure head is defined by

$$\int_{P_0}^{P} \frac{dP}{\rho(P)g},$$ (3.1.3b)

where P_0 is the hydrostatic pressure at the datum level.

As it is apparent from Fig. 3.1, the difference in $\mathscr{P}$ is equal to the pressure change in the fluid flowing through the porous sample. When the liquid is at rest, $\mathscr{P}$ is constant everywhere.

$\mathbf{v} = (\delta Q/\delta A)\mathbf{n}$ is the "filter" or Darcy velocity or "specific discharge," and $\mathbf{n}$ is the unit normal vector of the surface of the area δA through which there is a volume flow at the rate δQ. The rationale behind Eq. (3.1.1) is as follows: The porus medium is imagined to be subdivided into a network of small blocks, and Darcy's law is applied to each block. The size of each block must be small enough to approximate $\mathbf{v}$, $\mathscr{P}$, k, ρ, and μ with constant values within each block; but the size of each block must also be large enough for Darcy's law in its macroscopic form to apply in the block. These conditions appear to be satisfied to an acceptable degree in most practical situations.

In ground water hydrology and soil mechanics, the only fluid of interest is water and, therefore, the so-called "hydraulic conductivity" k_H defined as

$$k_H = k\rho g/\mu$$ (3.1.4)

is used. Darcy's law can then be written as

$$\mathbf{v} = -k_{\mathrm{H}}\nabla\phi. \tag{3.1.5}$$

For the case of beds of particles or fibers a different way of expressing the resistance of the porous medium to flow is with the help of the friction factor f_{p}, defined by the equation

$$f_{\mathrm{p}} = \overline{D}_{\mathrm{p}}\,\Delta\mathscr{P}/\rho v^2 L \tag{3.1.6}$$

as a function of the "superficial" or "particle" Reynolds number Re_{p}:

$$\mathrm{Re}_{\mathrm{p}} = \overline{D}_{\mathrm{p}} v\rho/\mu, \tag{3.1.7}$$

where $\overline{D}_{\mathrm{p}}$ is some effective average particle or fiber diameter, ρ is the fluid density, and L is the length of the bed in the macroscopic flow direction.

Many different model approaches for the treatment of single-phase flow have been tried, which may be categorized in a number of different ways depending on one's point of view and preference. One can distinguish two fundamentally different approaches: in one, the flow *inside* conduits is analyzed; in the other, the flow *around* solid objects immersed in the fluid is considered. For low and intermediate porosities the conduit flow approach is more appropriate, whereas for very high porosities only the second approach is suitable. In between, however, there is a gray area where neither approach seems to have a clear advantage over the other.

Empiricism, often aided by dimensional analysis and theoretical considerations, is one way of approaching the problem. The correlations obtained as a result of such procedures are called "phenomenological models" in this book. This approach is independent of considerations pertaining either to conduit flow or to flow around submerged objects.

Phenomenological models have proved to be particularly useful in the case of packs of fairly uniform and isometric particles or of fibers. They relate the transport coefficients of the porous media to grain or fiber properties and packing structure.

Within the conduit flow approach it is useful to distinguish "capillaric" and "statistical" models. The simplest kind of capillaric model consists of a bundle of straight cylindrical capillaries of uniform cross section. Empirical "capillaric" models have resulted in excellent correlations. Knowledge of pore structure should make capillary network models the ultimate answer. In the statistical models the random nature of the interconnectedness and/or orientation of the pores has been emphasized.

Channel flow has been treated mostly in the approximation that neglects all but one velocity component, resulting in Hagen–Poiseuille type flow equations. There have been reported a few contributions in which the complete Navier–Stokes equations have been solved for special channel geometries. These calculations are also reviewed in this chapter.

The continuum approach to modeling porous media does not distinguish between conduit flow or flow around submerged objects. This is essentially a deterministic approach because certain basic physical laws (the continuity equation for mass and the momentum equation) are assumed and these equations are averaged over the volume under consideration.

The approaches considering flow around solid objects, sometimes also called "drag theories," are variants, extensions, or generalizations of Stokes' law.

It is important to note that the models just listed are not to be considered mutually exclusive approaches to the problem on hand, but merely represent the main viewpoints emphasized by various workers. Each approach has a certain validity, and the better the various models will be reconciled with each other, the more adequate will become our understanding of flow through porous media. The various model approaches are, nevertheless, dealt with separately below, because this method of discussing the subject is thought to help the reader to gain a better appreciation of and insight into the great variety of problems of flow through porous media.

3.2. PHENOMENOLOGICAL FLOW MODELS

Phenomenological permeability models have been in use since the end of the nineteenth century (Slichter, 1899), but only relatively recently has there been published a more-or-less general dimensional analysis of the problem (Rumpf and Gupte, 1971). The phenomenological approach has been used almost entirely for porous media consisting of packs of particles or fibers.

3.2.1. Beds of Particles

According to the analysis of Rumpf and Gupte (1971), there is the following relationship among the dimensionless parameters:

$$\Delta \mathscr{P} / \rho v^2 = f\left(v\overline{D}_{\mathrm{p}}/\nu, \, L/\overline{D}_{\mathrm{p}}, \, \phi, \, q_i, \, \psi_i, \text{structure}\right), \qquad (3.2.1)$$

where $\Delta \mathscr{P}$ is pressure drop over the distance L, ν is kinematic viscosity, ϕ is bulk porosity, q_i are particle size distribution parameters, and ψ_i are particle shape parameters.

The various parameters that characterize particle shape and packing structure may be very difficult to evaluate in the general case. The complications may even become unsurmountable if there are several very different particle shapes present in the pack. Neglected in the dimensional analysis were the compressibility or expandability of the pack, the effects of interfacial energy, suspended matter, dissolved gases, and, in the case of gaseous flow, the mean free path of the molecules. Rumpf and Gupte (1971) studied the special case of uniformly random packs consisting of various distributions of spherical particles over the relatively wide range of porosities, $0.35 \leq \phi \leq 0.7$, and the range of Reynolds numbers $10^{-2} \leq \text{Re}_p \leq 10^2$. A narrow distribution and two broader distributions extending over a range of particle diameters, $D_{pmax}/D_{pmin} \simeq 7$ were tested by them.

As it may be seen in Fig. 3.2, showing a log-log plot of $f_p \phi^{5.5}$ versus Re_p, all the results could be represented by two parallel lines, the gap between them amounting only to 5%. It was justified, therefore, to write $f_p \phi^{5.5} = Kf(\text{Re}_p)$, where K is a different constant for the narrow distribution and the two wider distributions, respectively. The straight-line portion of the curves could be expressed analytically as

$$f_p = K(5.6/\text{Re}_p)\phi^{-5.5} \tag{3.2.2}$$

with $K = 1.00$ and 1.05 for the narrow distribution and the two wider distributions, respectively.

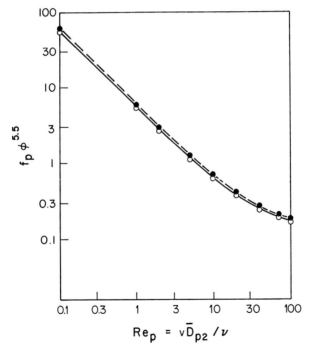

Figure 3.2. Friction factor × porosity function for uniformly random bead packs: ●, narrow distribution; ○, broad distribution. (Adapted with permission from Rumpf and Gupte, 1971.)

It is important to realize that an empirical relationship of the form

$$f_p = \text{const}/\text{Re}_p \tag{3.2.3}$$

implies Darcy's law with

$$k = \overline{D}_p^2/\text{const.} \tag{3.2.4}$$

The relationship [Eq. (3.2.2)] found by Rumpf and Gupte (1971) becomes, after introducing the definitions of f_p and Re_p,

$$(\Delta \mathscr{P}/L)(1/\mu v) = 5.6 K \phi^{-5.5}/\overline{D}_{p2}^2, \tag{3.2.5}$$

which, on comparison with Darcy's law, results in the following expression for the permeabilities:

$$k = \overline{D}_{p2}^2 \phi^{5.5}/5.6 K. \tag{3.2.6}$$

($\overline{D}_{p2}$ is "surface average" sphere diameter.)

Much effort has been expended in order to find the best $\overline{D}_p$ and porosity function $f(\phi)$ so as to be able to represent all data points by the same constant. Had these attempts been successful, there would exist now a general formula for the permeability. The work of Rumpf and Gupte (1971) suggests that such a general formula does not exist because the value of K depends on the particle size distribution parameters, and K, as well as $f(\phi)$, may be expected to depend on the particle shape and the packing structure.

The different forms that have been suggested for the porosity function $f(\phi)$ by the various authors are shown in Table 3.1. Rumpf and Gupte compared the fits obtained by their data when using their $f(\phi)$ and that of Blake and Kozeny. As is evident from Fig. 3.3, the form $\phi^{5.5}$ is the better

TABLE 3.1   Different Porosity Functions for Low Reynolds Number Flow[a]

$1/f(\phi)$	Author
$(1 - \phi)^2/\phi^3$	Blake (1922), Kozeny (1927), Carman (1937)
$(1 - \phi)^2/\phi$	Zunker (1920)
$[(1 - \phi)^{1.3}/(\phi - 0.13)]^2$	Terzaghi (1925)
$[1.115(1 - \phi)/\phi^{1.5}][(1 - \phi)^2 + 0.018]$	Rapier (1949)
$69.43 - \phi$	Hulbert and Feben (1933)
$\phi^{-3.3}$	Slichter (1898)
$\phi^{-1.0}$	Krüger (1918)
$\phi^{-6.0}$	Hatch (1934), Mavis and Wilsey (1936)
$\phi^{-4.0}$	Fehling (1939)
$\phi^{-4.1}$	Rose (1945)
$\phi^{-5.5}$	Rumpf and Gupte (1971)

[a]Adapted with permission from Rumpf and Gupte (1971).

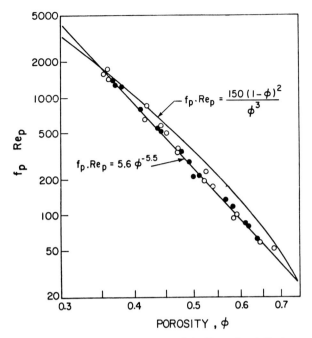

Figure 3.3. Comparison of the Blake–Kozeny and the Rumpf and Gupte porosity functions for the data of Rumpf and Gupte. (Adapted with permission from Rumpf and Gupte, 1971).

one. Whether this function is better also for other types of packs, however, cannot be said with certainty.

The value of the constant in the various equations relating f_p, Re_p, and ϕ naturally depends on the form of $f(\phi)$ used, but there are differences in the literature even for the same form of $f(\phi)$. For example, the so-called Blake–Kozeny equation (e.g., Bird *et al.*, 1960), valid for Re < 1, may be written as

$$f_p\left[\phi^3/(1-\phi)^2\right] = 150/Re_p \qquad (3.2.7)$$

from which

$$k = \left(\bar{D}_{p2}^2/150\right)\left[\phi^3/(1-\phi)^2\right]; \qquad (3.2.8)$$

whereas according to the Carman–Kozeny equation (see Section 3.3.1)

$$k_{CK} = \left(\bar{D}_{p2}^2/180\right)\left[\phi^3/(1-\phi)^2\right]. \qquad (3.2.9)$$

The two "best values" of the constant are seen to differ by as much as 20%.

The value of $\bar{D}_p$ used in Eqs. (3.1.2) and (3.1.3) is of considerable significance. It has been found (e.g., Macdonald *et al.*, 1979 and Rumpf and

Gupte, 1971) that the most consistent data is obtained if $\overline{D}_p = \overline{D}_{p2}$ is used, where $\overline{D}_{pr}$ is defined as follows:

$$\overline{D}_{pr} = \frac{\left[\int_0^\infty D_p D_p^r N(D_p)\, dD_p\right]}{\left[\int_0^\infty D_p^r N(D_p)\, dD_p\right]}, \tag{3.2.10}$$

where $N(D_p)$ is the density function of distribution of (spherical) particle diameters, $r = 0$ corresponds to the number average, $r = 2$ to the "surface average," and $r = 3$ to the "volume average" diameter. In the case of nonspherical particles the surface average equivalent sphere diameter is defined as

$$\overline{D}_{p2} \equiv 6V_p/S_p, \tag{3.2.11}$$

where V_p/S_p is the volume-to-surface ratio of the particulate system.

Batel (1959) introduced the group $(\overline{D}_{pm}^2/S_0)^{1/3}$ which is proportional to $(\overline{D}_{p3}^2 \overline{D}_{p2})^{1/3}$, where $\overline{D}_{pm}$ is the average particle diameter by weight, S_0 is the mean specific surface of the particles, and $\overline{D}_{p3}$ is the "volume average" particle size. Keulegan (1954), Rumer (1962), Harleman et al. (1963), and Berg (1970) used the median diameter of the particle size distribution by weight. The last author used, as an additional parameter, the percentile deviation (i.e., a measure of sorting or spread of the distribution).

In the case of nonspherical particles, the various methods of particle size analysis often do not yield $\overline{D}_{p2}$ as defined in Eq. (3.2.11). Macdonald et al. (1979) converted the results of sieve tests reported in the literature by assuming that for more-or-less isometric, but not spherical, particles, the sieve test measures D_v, the diameter of a sphere, having the same volume as the actual particle. They corrected this as

$$\overline{D}_{2p} = \phi_s D_v, \tag{3.2.12}$$

where ϕ_s is the "sphericity," that is, the ratio of surface area of the hypothetical sphere of the same volume as the particle to the actual surface area of the particle.

Empirical permeability correlations with porosity and grain size distribution have been given for reservoir sands and sandstones by Morrow et al. (1969), Berg (1970), and Pryor (1971).

3.2.2. Nonlinearity of the Friction Factor–Reynolds Number Relation

The transition from the linear portion of the f_p versus Re_p relation (range of validity of Darcy's law) to the nonlinear one is gradual. Nevertheless it is customary to talk about a critical Reynolds number in analogy with transition from laminar to turbulent flow used in pipe flow.

It was pointed out by Scheidegger (1960) that for various porous media the value of the Reynolds number above which Darcy's law is no longer valid has been found to range between 0.1 and 75. This uncertainty of a factor of 750 of the "critical" Re numbers is partly due to differences of pore structure, including "surface roughness," of the various materials, but it also depends on the criteria used for the validity of Darcy's law. For most of the data analyzed by Macdonald *et al.* (1979), the deviation from linearity starts to become noticeable in the range $Re_p/(1 - \phi) \approx 1\text{--}10$.

Some light has been shed on the effect of pore geometry on the deviations from Darcy's law by the results of the numerical solution of the Navier–Stokes equations (Azzam, 1975; Azzam and Dullien, 1977) for various simple pore geometries, as discussed in Section 3.3.5.1, and of the experiments in which (Dullien and Azzam, 1972) flow rates were determined as a function of pressure gradient for artificial capillaries with periodical step changes in diameter. The diameter ratio as well as the length of a segment were varied in order to see the influence of the geometrical parameters on the extent of the deviation from "Poiseuille permeability." The results indicated greatly different relative deviations, depending on the channel geometry. The smaller the length-to-diameter ratio of the narrow segment, the stronger is the relative deviation, whereas as a function of small-to-large segment diameter ratio the relative deviation has a maximum at a certain value of the diameter ratio.

Another issue, already pointed out by Scheidegger (1960) and emphasized by Happel and Brenner (1965) is the physical interpretation of the breakdown of Darcy's law. The latter authors maintain that it is a serious misinterpretation of the phenomenon to attribute the failure of Darcy's law to turbulence: "Failure of Darcy's law results when the distortion that occurs in the streamlines owing to changes in direction of motion is great enough that inertial forces become significant compared with viscous forces. The incidence of turbulence will occur at much higher Reynolds numbers (if indeed such a phenomenon occurs at all)." There is considerable evidence, however, to indicate that turbulence does occur in porous media (e.g., Kyle and Perrine, 1971; Dudgeon, 1966).

At flow rates outside the range of validity of Darcy's law, the two best known relationships are (see, e.g., Bird *et al.*, 1960), for intermediate values of Re_p up to about 10^3, the Ergun equation

$$f_p\big[\phi^3/(1 - \phi)\big] = (150/Re_p)(1 - \phi) + 1.75, \qquad (3.2.13)$$

and, for higher values of Re_p, the Burke–Plummer equation

$$f_p\big[\phi^3/(1 - \phi)\big] = 1.75. \qquad (3.2.14)$$

The subdivision into three ranges of Re_p is by no means precise and the transition from one region to the next one is a continuous and gradual one.

These relationships were obtained for packed columns with uniform packing and no "channeling."

Macdonald *et al.* (1979) have tested the Ergun equation, using much more data than was ever used before by others. The following relations were found to give the best fit to all of the data:

Smooth particles: $f_p[\phi^3/(1 - \phi)] = (180/\mathrm{Re}_p)(1 - \phi) + 1.8$,   (3.2.15)

Roughest particles: $f_p[\phi^3/(1 - \phi)] = (180/\mathrm{Re}_p)(1 - \phi) + 4.0$.   (3.2.16)

For intermediate surface roughness, the value of the second term on the right (see Section 3.4.3) lies between 1.8 and 4.0.

The data used by Macdonald *et al.* (1979) in the preceding test included narrow size distributions of spherical glass beads packed in a "uniformly random" manner (Gupte, 1970); beds of cylindrical fibers (Kyan *et al.*, 1970); spherical marble mixtures, sand and gravel mixtures, and ground Blue Metal mixtures (Dudgeon, 1966); consolidated media (Fancher and Lewis, 1933); a variety of cylindrical packings (Pahl, 1975); and a wide variety of different materials (Doering, 1955; Matthies, 1956; Luther *et al.*, 1971).

Equations (3.2.15) and (3.2.16) can be expected to predict experimental data for a wide variety of unconsolidated porous media with a maximum error of $\pm 50\%$. For sandstones the equations may underpredict the results by as much as 300%.

Gupte's results for uniform smooth spherical particles are better fitted by values of both parameters that are lower by about 27% than those in Eqs. (3.2.15) and (3.2.16). Gupte, however, created media with higher porosity than are normally encountered with spherical particles. Coulson (1949), Fischer (1967), Brauer (1971), and Barthels (1972) obtained results for more normal range of porosities from 0.34 to 0.48, which give values of 157 to 180, with an average of 169, and 1.49–1.77, with an average of 1.63, for the two coefficients.

The Ergun equation is a special form of the Forchheimer equation (Forchheimer, 1901; Green and Duwez, 1951; Brownell *et al.*, 1950; Cornell and Katz, 1953; Greenberg and Weger, 1960, etc.):

$$\Delta \mathscr{P}/L = \alpha \mu v + \beta \rho v^2 \qquad (3.2.17)$$

where α is reciprocal permeability and β the so-called inertia parameter, with α and β expressed in terms of pore structure parameters as follows:

$$\alpha = A\left[(1 - \phi)^2/\phi^3 \overline{D}_{p2}^2\right] \qquad (3.2.18)$$

and

$$\beta = B\left[(1 - \phi)/\phi^3 \overline{D}_{p2}\right], \qquad (3.2.19)$$

where A and B are the "constants," 180 and 1.8 (or 4.0), in Eqs. (3.2.15) and (3.2.16).

A great deal of work dealing with the presentation of the pressure gradient–flow rate relationship at high flow rates has been published. A detailed critical review of this field has been published by Scheidegger (1974).

The basic issue is whether or not the Forchheimer equation is the correct flow equation over the entire practical range of flow rates. There have been suggestions to add a cubic third term in the velocity. Polubarinova-Kochina (1952) and, later, Irmay (1958) have suggested the following form:

$$\Delta \mathscr{P}/L = \alpha\mu v + \beta\rho v^2 + \gamma\rho\, \partial v/\partial t; \qquad (3.2.20)$$

but a mass of experimental data have been correlated by Macdonald *et al.* (1979) by Eq. (3.2.17), within experimental error.

Ahmed and Sunada (1969), following previous work by Green Duwez (1951) and Ergun (1952), rearranged the Forchheimer equation in the form:

$$\frac{(\Delta \mathscr{P}/L)}{(\beta\rho v^2)} = \left(\frac{\alpha\mu}{\beta\rho v}\right) + 1 \qquad (3.2.21)$$

and suggested a plot on a log-log scale

$$\frac{(\Delta \mathscr{P}/L)}{(\beta\rho v^2)} \text{ versus } \frac{\beta\rho v}{\alpha\mu}$$

as a generalized friction factor–Reynolds number correlation. Macdonald *et al.* (1979) have tested this form, using a mass of literature data, and have found the correlation satisfactory, as did Geertsma (1974), Ahmed and Sunada (1969), and Green and Duwez (1951) before them. Therefore, the conclusion can be reached that the physical model underlying the Forchheimer equation is adequate.

Geertsma (1974) has pointed out that an alternate way of writing the Forchheimer equation is as follows

$$\text{Da} = 1 + \text{Re}, \qquad (3.2.22)$$

where

$$\text{Da} = \Delta \mathscr{P}/\alpha\mu v L \qquad (3.2.23)$$

is the "Darcy number" and

$$\text{Re} = \beta\rho v/\alpha\mu \qquad (3.2.24)$$

a Reynolds number.

In the absence of slip flow, the Forchheimer equation has been written for gases as follows (Geertsma, 1974):

$$-(P_2^2 - P_1^2)/[2L(\mathscr{R}T/M)\mu G_m] = \alpha + \beta(G_m/\mu), \qquad (3.2.25)$$

where G_m is mass flow rate per unit area and M is molecular weight of gas.

It has been customary in the literature to refer to a "turbulent" range whenever the inertia term in the Forchheimer equation is predominant. According to Scheidegger (1974) there appear to be two critical Reynolds numbers at which the flow regime changes: the first indicating the conditions when inertia effects in laminar flow become important and the second when true turbulence sets in. The data extending over a very wide range of flow rates, do note display any jumps when plotted in the Ahmed–Sunada form (see e.g., Fig. 3.4). As a matter of fact, there are no measured data that would indicate that the factors in the Forchheimer equation suddenly assume new values at any point when turbulence sets in. The available evidence suggests that in porous media flow the transition from laminar to turbulent flow is gradual.

Macdonald *et al.* (1979) found the following correlations, of limited validity, for Gupte's (1970) and Dudgeon's (1966) data, for A:

$$A = 118.2 + 107.5\overline{D}_{p2}, \tag{3.2.26}$$

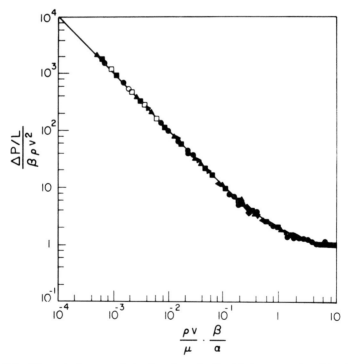

Figure 3.4. Dimensionless correlation based on the Forchheimer equation. (Reprinted with permission from Ahmed and Sunada, 1969.)

with $\overline{D}_{p2}$ in cm (for any ϕ), and for Gupte's data (1970) and the sand data of Dudgeon (1966), at an average value $\overline{D}_{p2} = 0.07$ cm:

$$A = 214.25 + 151.72\phi. \tag{3.2.27}$$

Stanek and Szekely (1974) have written the "Ergun equation" in vectorial form

$$-\nabla\mathscr{P} = \mathbf{v}\left\{\left[150\mu(1-\phi)^2/\overline{D}_p^2\phi^3\right] + \left[1.75\rho v(1-\phi)/\overline{D}_p\phi^3\right]\right\}, \tag{3.2.28}$$

which they applied to the analysis of three-dimensional flow in nonuniform packed beds, using a numerical technique of solving the differential equations.

Beavers and Sparrow (1971) analyzed the problem of compressible gas flow through porous media, including unsteady state as well as effects of inertia. The analysis used some of the results of Emanuel and Jones (1968). Their flow equation is

$$-\frac{dP}{dx} = \frac{\mu v}{k} + \frac{c\rho v^2}{\sqrt{k}} + \rho\left(\frac{v}{\phi}\right)\frac{d(v/\phi)}{dx}, \tag{3.2.29}$$

which they recast in the following dimensionless form:

$$\gamma M^2 \frac{dv}{v} + \frac{dP}{P} + \gamma M^2\phi^2\left(\frac{dx}{\sqrt{k}}\right)\left(\frac{1}{\mathrm{Re}} + c\right) = 0, \tag{3.2.30}$$

where

$$\mathrm{Re} = \rho v\sqrt{k}/\mu,$$

$$M = (v/\phi)/a = \text{Mach number} \qquad (a \text{ is the speed of sound}),$$

and

$$\gamma = C_p/C_v = \text{ratio of specific heats}. \tag{3.2.31}$$

They point out that only when $1/\mathrm{Re} \gg c$ are the irreversibilities limited to the viscous losses of Darcy flow. To date c values varying from 0.075 to 0.5 have been measured (Schwartz and Probstein, 1969; Beavers and Sparrow, 1971; Ward, 1964). Therefore, in some cases significant departures from Darcy's law occur at "Reynolds numbers" as low as 0.04. It is very important to note, however, that this Reynolds number is different from the one defined in Eq. (3.1.7). For porosities in the range from 0.2 to 0.3 the latter is greater by a factor of about 100–50, whereby the above limit of validity of Darcy's law is shifted to Re_p in the range from 2 to 4.

3.2.3.   Fibrous Beds

Pressure drop versus flow rate data are fairly satisfactorily correlated in many fibrous beds by the "Ergun equation" (see Macdonald *et al.*, 1979). Nevertheless, fibrous beds have received special attention for a number of reasons, for example, their practical importance as gas filters and the fact that they can form stable structures of very high porosity (e.g., $\phi = 0.99$). Under such conditions, however, the dependence of the friction factor on porosity deviates sharply from the form given in Eq. (3.2.13). Most of the data in fibrous beds have been obtained with gases.

Various empirical correlations for the permeability of fibrous beds have been proposed in the literature. These include the following equations, applicable to $Re_f < 1 (Re_f = D_f v / \nu)$, due to Davies (1952):

$$k = \frac{D_f^2}{64(1 - \phi)^{3/2}\{1 + 56(1 - \phi)^3\}} \tag{3.2.32}$$

and for $\phi > 0.98$,

$$k = \frac{D_f^2}{70(1 - \phi)^{3/2}\{1 + 52(1 - \phi)^{3/2}\}} \tag{3.2.33}$$

where the subscript f refers to "fiber."

Another expression is due to Chen (1955):

$$k = \frac{\pi D_f^2 \ln\{k_5/(1 - \phi)^2\}}{4k_4} \frac{\phi}{1 - \phi}, \tag{3.2.34}$$

where $k_4 = 6.1$ and $k_5 = 0.64$.

More recently Kyan *et al.* (1970) have proposed a new friction factor versus Reynolds number correlation based on the idea that the total pressure drop across the bed is the sum of the pressure drops due to viscous losses, form drag, and deflection of fibers.

Based on a geometric model of the fibrous bed they have assumed that in a significant portion of the pore space the fluid is not flowing. An "effective" porosity ϕ_e has been defined as

$$\phi_e = \frac{\text{volume of flow region}}{\text{total volume}}. \tag{3.2.35}$$

A relationship between the porosity ϕ and the effective porosity ϕ_e has been derived:

$$\phi_e = N_e^2(1 - \phi)(0.5/\pi), \tag{3.2.36}$$

where N_e is the "effective pore number"

$$N_e = [\pi/(1 - \phi)0.5]^{1/2} - 2.5. \qquad (3.2.37)$$

It is easy to ascertain that ϕ_e is much less than ϕ for the whole practical range.

The pressure drop due to viscous flow has been expressed by Kyan *et al.*, as follows:

$$(\Delta P)/L)_{\text{flow}} = k_1[\mu v/D_f^2(1 - \phi)N_e^4], \qquad (3.2.38)$$

where k_1 is a constant.

The pressure drop due to the total drag force has been referred to as "form drag" (FD):

$$\left(\frac{\Delta P}{L}\right)_{\text{FD}} = k_2\left(\frac{\rho v^2}{D_f N_e^6}\right)\left(\frac{\mu}{D_f v \rho}\right)\frac{1}{(1 - \phi)^2}, \qquad (3.2.39)$$

where k_2 is a second constant. It is evident that, after cancellation of terms, Eq. (3.2.39) reduces to $(\Delta P/L) = k_2[\mu v/D_f^2 N_e^6(1 - \phi)^2]$, which is identical to Eq. (3.2.38) apart from the different dependence on porosity. The total drag includes the effects of viscous shear and, therefore, Eq. (3.2.39) appears to be redundant.

Finally,

$$\left(\frac{\Delta P}{L}\right)_d = k_3\left(\frac{\mu}{D_f v \rho}\right)^2 \frac{\rho^2 v^4}{ED_f N_e^{10}(1 - \phi)^{5.5}}, \qquad (3.2.40)$$

where k_3 is a third constant, E the modulus of elasticity of the fiber, and subscript d stands for deflection.

The final form of the expression obtained by these authors is

$$f_p f_1(\phi) = [(1 - \phi)/\text{Re}_f] + N_d f_2(\phi), \qquad (3.2.41)$$

where the two porosity functions, $f_1(\phi)$ and $f_2(\phi)$, are given by

$$f_1(\phi) = \frac{N_e^4(1 - \phi)^2}{k_1 + [k_2/N_e^2(1 - \phi)]} \qquad (3.2.42)$$

and

$$f_2(\phi) = \frac{k_3}{\{k_1 + [k_2/N_e^2(1 - \phi)]\}N_e^6(1 - \phi)^{3.5}}, \qquad (3.2.43)$$

and N_d is "deflection number"

$$N_d = \mu^2/ED_f^2\rho. \qquad (3.2.44)$$

(The Reynolds number N_{Re} used by these authors is actually

$$N_{Re} = D_f v \rho / \mu (1 - \phi) = Re_f / (1 - \phi), \qquad (3.2.45)$$

which, although often used, is physically meaningless at high porosities as it increases without limit as $\phi \to 1$.)

Experimental data were obtained with glass, Dacron, and Nylon fibers of 8 to 28 μm in diameter, with water and aqueous glycerol solutions. The reported porosity range was 0.682–0.919. The constants, k_1 and k_2, were evaluated from the slopes of the best straight lines fitted to the experimental data of the two runs with $\phi = 0.682$ and 0.919. Instead of obtaining values of the modulus of elasticity E for the various fibers used, the values k_3/E were found empirically from the measured data.

In Fig. 3.5, Eq. (3.2.41) was plotted for various values assigned to $f_d \equiv N_d f_2(\phi)$, as shown in the legend, together with the experimental data for selected runs. At low values of N_{Re} all data points seem to cluster about the

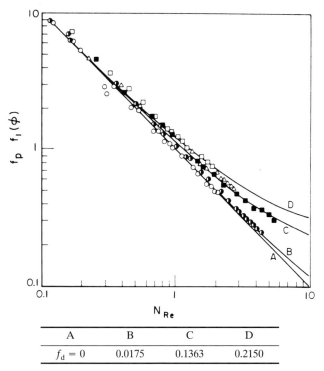

	A	B	C	D
$f_d = 0$		0.0175	0.1363	0.2150

Figure 3.5. Normalized friction factor plot for randomly packed fibrous beds. (Reprinted with permission from Kyan *et al.*, 1970 © by American Chemical Society.)

same line. The different behavior displayed by various samples at higher values of N_{Re} has been attributed to the different values of N_{d} for the various fibers, as hydrodynamic nonlinearity was ruled out at the low Reynolds numbers used. It should be noted, however, that the customary drag coefficient C_{D} starts deviating noticeably from linearity above $\mathrm{Re}_{\mathrm{p}} > 1$ because of wake formation. Therefore, the claim made by the authors that the nonlinear behavior is due to "deflection" has not been proven, particularly since they failed to provide an independent check on values that might have been obtained for the elastic modulus E from their data.

Kyan *et al.* have derived the following expression for the so-called Kozeny constant k' [see Eq. (3.3.7)], based on a comparison of Eq. (3.2.41) and the Carman–Kozeny type of friction factor versus Reynolds number relationship, i.e., $f_{\mathrm{p}} = 36k'(1 - \phi)^2/\mathrm{Re}_{\mathrm{f}} \, \phi^3$:

$$k' = \frac{\left[62.3N_{\mathrm{e}}^2(1 - \phi) + 107.4\right]\phi^3\left[1 + N_{\mathrm{d}}f_2(\phi)\right]}{16N_{\mathrm{e}}^6(1 - \phi)^4} \qquad (3.2.46)$$

and have compared the values calculated by this expression with some published values in Fig. 3.6. In Eq. (3.2.46), $k_1 = 62.3$, $k_2 = 107.4$. The sharp

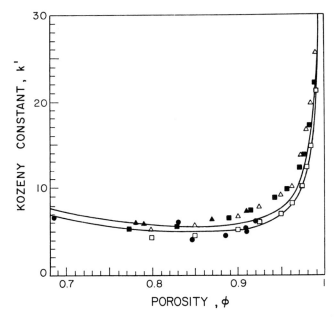

Figure 3.6. Variation of the Kozeny constant with porosity in fibrous beds (Kyan *et al.*, 1970).

increase in the value of the "Kozeny constant" starting around $\phi = 0.95$ is noted. The two curves in the figure are for $f_d = 0$ and $f_d N_{Re} = 0.1$.

3.3. PERMEABILITY MODELS BASED ON CONDUIT FLOW

The flow of a fluid in many porous media can be imagined with equal justification to occur either in a network of closed conduits or around solid particles forming a spatial array. Both of these approaches have been pursued extensively by various researchers.

The simplest approaches based on the idea of conduit flow do not pay any attention to the fact that different pores are interconnected with each other. These are called "capillaric permeability models" in this monograph. All "capillaric" models are inherently one-dimensional.

3.3.1. Capillaric Permeability Models

Among the different capillaric permeability models one enjoys much greater popularity than the rest: the so-called Carman–Kozeny model.

3.3.1.1. *Carman–Kozeny or Mean Hydraulic Diameter Model*

The Carman–Kozeny approach, often called the "hydraulic radius theory," is sometimes regarded as a phenomenological approach. In fact, it is closer to being a capillaric model because certain assumptions on pore structure have been made in its development, whereas a purely phenomenological theory, such as that of Rumpf and Gupte (1971), avoids making any assumption of that nature. The fact that the Carman–Kozeny equation is similar to some phenomenological equations is irrelevant when the classification is based on the type of approach rather than the result obtained.

In the Carman–Kozeny theory (Carman, 1937, 1938, 1956; Kozeny, 1927), the porous medium was assumed to be equivalent to a conduit, the cross section of which has an extremely complicated shape but, on the average, a constant area. In analogy with the established practice in hydraulics, the channel diameter D_H governing the flow rate through the conduit was assumed to be four times the hydraulic radius, defined as the flow cross-sectional area divided by the wetted perimeter, that is,

$$D_H = \frac{4 \times \text{void volume of medium}}{\text{surface area of channels in medium}}.$$ (3.3.1)

Sometimes the pore diameter calculated from the breakthrough capillary pressure on the drainage curve is also called "hydraulic diameter" (e.g., Macmullin and Muccini, 1956), although this value is not in general, related to the volume-to-surface ratio of porous media. Only if the flow channels were all equal and of uniform diameter would there be a close correspondence between the two quantities (see Table 1.3).

As the Carman–Kozeny theory is aimed at laminar (creeping) flow, whereas the hydraulic radius concept is known in hydraulics to be a good approximation of reality only under turbulent flow conditions, the use of the hydraulic radius represents an assumption that can find confirmation only in flow through porous media experiments and cannot claim my justification by appealing to analogy with hydraulics.

A Hagen–Poiseuille type equation is assumed to give the average pore, interstitial, or seepage velocity v_p in the flow channels:

$$v_p = (\Delta \mathcal{P}/L_e)(D_H^2/k_0 16\mu), \tag{3.3.2}$$

where L_e is the average path length of flow and k_0 is a "shape factor." The pore velocity v_p and the filter velocity v in Darcy's law,

$$v = (k/\mu)(\Delta \mathcal{P}/L), \tag{3.3.3}$$

are assumed to be related as follows:

$$v_p = (v/\phi)(L_e/L) = v_{DF}(L_e/L). \tag{3.3.4}$$

The division of v by ϕ is the "Dupuit–Forchheimer assumption," which is very often used to define an average interstitial velocity. The multiplication of v by L_e/L in Eq. (3.3.4) is due to Carman. It corrects for the fact that a hypothetical fluid particle used in the macroscopic flow equations and flowing with velocity v covers a path length L in the same time as an actual fluid particle, flowing with velocity v_p, covers an average effective path length L_e, as illustrated schematically in Fig. 3.7. Unfortunately, the accurate value of L_e is seldom, if ever, known and, therefore, the true value of v_p is also uncertain in most porous media. Combination of the preceding three equations gives the result:

$$k_{CK} = \phi D_H^2/16k_0(L_e/L)^2. \tag{3.3.5}$$

This is the basic form of all capillaric models, differing only in the method of calculating the mean square diameter and in the value used for $k_0(L_e/L)^2$, which is a function of pore geometry.

The "hydraulic diameter" can be expressed as follows:

$$D_H = 4\phi/S_0(1 - \phi), \tag{3.3.6}$$

where S_0 is the specific surface area based on the solid's volume. By combining Eqs. (3.3.5) and (3.3.6), the usual form of the Carman–Kozeny

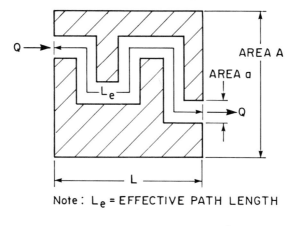

$$Q = vA = v_p a \quad , \quad \phi = \frac{a L_e}{AL}$$

$$\therefore \; v_p = v/(a/A) = (v/\phi)(L_e/L)$$

Figure 3.7. Illustration of the physical meaning of Eq. (3.3.4) (Dullien, 1990a).

equation for permeability is obtained:

$$k_{CK} = \phi^3/k_0(L_e/L)^2(1 - \phi)^2 S_0^2, \qquad (3.3.7)$$

where $(L_e/L)^2$ is usually called hydraulic "tortuosity factor" (Υ). The quantities L_e/L and $(L/L_e)^2$ have also been called "tortuosity" by various authors (e.g., Bear, 1972). According to Carman, the best value of the combined factor $k' = k_0(L_e/L)^2$ to fit most experimental data on packed beds is equal to 5, where k' is the so-called Kozeny constant. Defining the mean particle diameter $\overline{D}_{p2}$ as the diameter of the hypothetical sphere with the same S_0 as the particles [cf. Eq. (3.2.11)], that is,

$$\overline{D}_{p2} = 6/S_0, \qquad (3.3.8)$$

Eq. (3.2.9) can be obtained immediately.

The hydraulic "tortuosity factor" $\Upsilon \equiv (L_e/L)^2$ is not a property of the porous medium, but it is a parameter of the *one-dimensional model* of the medium. As illustrated schematically in Fig. 3.8, if the sample length in the direction of macroscopic flow is L, the porosity of the sample is ϕ, the pressure drop is ΔP and the flow rate is Q, the capillaric model consists of tubes of length L whereas the effective length of pores in the medium is $L_e > L$. Therefore, in the medium the pore velocity for the given ΔP is

(L/L_e) times less, and also number of pores in parallel is (L/L_e) times less because the model's porosity must match ϕ, the porosity of the medium. As a result, a tube length L_e gives, for the same values of ϕ, L and ΔP, $(L/L_e)^2$ times lower flow rate (i.e., $(L/L_e)^2$ times lower permeability than the choice of tubes of length L).

All along it has been tacitly assumed in this argument that the capillary diameter D of the model is representative of the effective pore diameter of the medium in the sense that equal lengths of model capillary and pores of the medium have the same hydraulic conductivity. Unless this condition is met Υ has no meaning at all. For this reason, often the "tortuosity factor" Υ is simply interpreted as a "fudge factor" for any permeability model (i.e., $\Upsilon \equiv k_{model}/k_{measured}$). It is important to stress the point that the very concept of Υ is limited to one-dimensional models because the tortuosness of the pore network is intrinsically incorporated into three-dimensional models.

The Carman–Kozeny equation is of approximate validity. It has been found particularly useful for measuring surface areas of some powders. In the case of particles that deviate strongly from the spherical shape, broad particle size distributions, and consolidated media, the Carman–Kozeny equation is often not valid, and, therefore, it should always be applied with great caution.

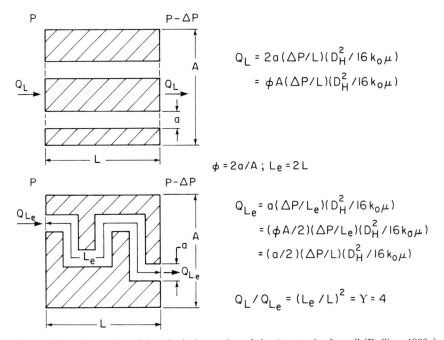

$$Q_L = 2a(\Delta P/L)(D_H^2/16k_0\mu)$$
$$= \phi A(\Delta P/L)(D_H^2/16k_0\mu)$$

$$\phi = 2a/A \; ; \; L_e = 2L$$

$$Q_{L_e} = a(\Delta P/L_e)(D_H^2/16k_0\mu)$$
$$= (\phi A/2)(\Delta P/L_e)(D_H^2/16k_0\mu)$$
$$= (a/2)(\Delta P/L)(D_H^2/16k_0\mu)$$

$$Q_L/Q_{L_e} = (L_e/L)^2 = \Upsilon = 4$$

Figure 3.8. Illustration of the physical meaning of the "tortuosity factor" (Dullien, 1990a).

The Carman–Kozeny equation was tested extensively by Wyllie and coworkers (1952, 1955), who sometimes obtained values for k' very much in excess of the 5.0 recommended by Carman. The shape factor k_0 was found to lie between 2.0 and 3.0, as suggested originally by Carman.

The frequent claim that the main reason for disagreement between permeabilities predicted by the Carman–Kozeny equation and experimental values lies in (anomalously) high sample tortuosities is unfounded. There are a number of other possible reasons for this: (1) the type of dependence of permeability on porosity; (2) parallel-type pore nonuniformities; (3) serial-type pore nonuniformities (Dullien, 1975). These factors are discussed briefly below.

Dependence of Permeability on Porosity.   The Carman–Kozeny model is based on the assumption of conduit flow. However, at very high porosities this concept breaks down and the experimental facts are better described by the "flow around submerged objects" approach, which is discussed in Section 3.6.

As discussed in Section 3.2.3, for fibrous structures of porosities $\phi = 0.95$ and higher, rapidly increasing values of k' are required to obtain agreement with experiments. This shows that at high porosities the friction drag mechanism represents an increasingly higher portion of the resistance to flow as compared with the viscous shear mechanism that determines the friction in conduit flow. By Eqs. (3.3.5) and (3.3.6), $k_{CK} \propto (1 - \phi)^{-2}$ as $\phi \to 1$. In reality, $k \propto (1 - \phi)^{-1}$. Therefore, $k' \to \infty$ as $\phi \to 1$.

It is evident from Eq. (3.3.7) that for a fixed $k_0(L_e/L)^2$ the value of k_{CK} is uniquely determined by the porosity ϕ and the specific surface area S_0. It is easy to show, however, that in reality the permeability depends also on the size distribution and the topographical arrangement of capillaries.

Analogously, as is the case in electrical circuits, capillaries of different conductivities also may be either in parallel or in serial connection.

Parallel-type Pore Nonuniformities.   The reason for the existence in the sample of continuous channels that are controlled by relatively large pore necks may be understood by probability considerations. The capillaries forming a network are interconnected with each other in a disordered random fashion. There is always a choice for the fluid particles at a juncture to enter a narrower or a larger pore. If at every juncture the largest available pore is entered, then the fluid particle travels through a continuous flow channel of the largest possible conductivity in the sample. There are channels of smaller conductivity in parallel with portions of such a main channel. Channels of different conductivities in parallel connection are called "parallel-type pore nonuniformities."

The effect of parallel-type nonuniformities on the sample permeability can be illustrated easily if, for simplicity, a bundle of cylindrical capillaries, consisting only of two different diameters D_ℓ and D_s, is considered (ℓ for large and s for small). By using the Hagen–Poiseuille equation for each capillary, it can be readily shown that the permeability k of the bundle is given by the following expression:

$$k = k_{CK}(1 + yx^4)(1 + yx)^2 / (1 + yx^2)^3, \qquad (3.3.9)$$

where $x = D_\ell/D_s$ and $y = \nu_\ell/\nu_s$ = the number ratio of large-to-small capillaries. Equation (3.3.9) yields, for example, for $x = 3$ and $y = 0.1$, $k = 2.24k_{CK}$. It is apparent that the permeability is greater than it would be if all capillaries had the same diameter D_H.

Serial-Type Pore Nonuniformities. The effective cross section of every channel in a porous medium varies in a quasi-periodical manner. In other words, a bulge is followed by a neck, which in turn is followed by a bulge, and so forth. This kind of sequential variation in the effective cross section of flow channels tends to result in a smaller permeability than the one calculated from the hydraulic radius even if excess viscous dissipation due to the resulting convergent–divergent flow is not considered (see Section 3.3.5.1).

The effect of serial-type nonuniformities can be illustrated readily by considering a bundle of identical straight cylindrical capillaries each of which consists of an alternating sequence of segments of two different diameters, D_ℓ and D_s. By using the Hagen–Poiseuille equation for each segment, neglecting expansion and contraction losses, it can be readily shown that the bundle permeability k varies as follows:

$$k = k_{CK}(1 + y)^2(1 + yx)^2 / (1 + yx^2)^3(1 + y/x^4), \qquad (3.3.10)$$

with $x = D_\ell/D_s$ and $y = \ell_\ell/\ell_s$, where ℓ_ℓ and ℓ_s are the aggregate lengths of the large and narrow segments, respectively. For example, with $x = y = 2$, Eq. (3.3.10) yields $k = 0.27k_{CK}$, where k_{CK} is calculated by Eq. (3.3.5). It is noted that the factor on the right of Eq. (3.3.10) contributes an additional decrease in k, due to the quasi-periodic contractions and expansions of the capillaries, which is not taken into account by the factor L_e/L.

It is evident that the serial-type nonuniformity affects the permeability in the opposite sense than the parallel-type nonuniformity and, therefore, there is a possibility for some mutual cancellation of the two effects.

3.3.1.2. *Bundle of Capillary Tubes Models*

These models consist of systems of parallel capillary tubes of well-defined geometry. The tubes can be uniform and identical, uniform but with distributed diameters (organpipes), periodically constricted and identical, or periodically constricted and different.

Bundles of Identical Capillaries. Assuming a cube shaped sample of edge length L, the model consists of n parallel capillaries of length L and diameter D, satisfying the condition of porosity ϕ of the sample:

$$\phi = \frac{nD^2\pi}{4L^2} \tag{3.3.11}$$

and the condition of flow rate Q under the influence of the pressure drop ΔP:

$$Q = n\frac{D^4\pi}{128\mu}\frac{\Delta P}{L}, \tag{3.3.12}$$

given by the Hagen–Poiseuille equation, which is expressed also be Darcy's law

$$Q = \frac{k_1}{\mu}L^2\frac{\Delta P}{L}. \tag{3.3.13}$$

Combination of Eqs. (3.3.11) to (3.3.13) gives for the permeability k_1 of this one-dimensional model the following expression

$$k_1 = \phi\frac{D^2}{32}. \tag{3.3.14}$$

A *pseudo three-dimensional* capillaric model would arrange the n capillaries in such a way that $n/3$ capillaries would be parallel to the x-axis, $n/3$ parallel to the y-axis, and $n/3$ parallel to the z-axis. In this more realistic, isotropic model the permeability k for the same porosity ϕ is $1/3$ of the permeability k_1 of the one-dimensional model, that is,

$$k = \phi\frac{D^2}{96} \tag{3.3.15}$$

It is evident that merely by the artifice of orienting pores with equal probability in the three space coordinate directions a hydrodynamic "tortuosity factor" $\Upsilon = 3$ has been introduced into the model. These model equations, however, can be written in a form that does not contain explicitly either ϕ or Υ, as follows.

With reference to Fig. 3.9, let us introduce the "conductance" g_1 of a capillary segment of length ℓ_1 ("lattice constant"), equal to the spacing of the capillaries in the model $[n = (L/\ell_1)^2]$:

$$g_1 = \frac{D^4\pi}{128\mu\ell_1}. \tag{3.3.16}$$

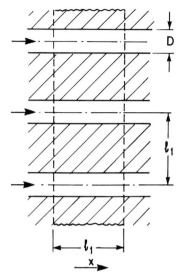

Figure 3.9. Illustration of the physical basis of capillaric models of permeability (Dullien, 1990a).

Combining Eqs. (3.3.16), (3.3.12), and (3.3.13) results in the relationship

$$ng_1\ell_1\frac{\Delta P}{L} = \frac{k_1}{\mu}L^2\frac{\Delta P}{L}, \qquad (3.3.17)$$

which with $n = (L/\ell_1)^2$ yields

$$\frac{k_1}{\mu} = \frac{g_1}{\ell_1}. \qquad (3.3.18)$$

For the pseudo three-dimensional model an analogous procedure gives the result

$$\frac{k}{\mu} = \frac{g_3}{\ell_3}\left(= \frac{1}{3}\frac{k_1}{\mu}\right) \qquad (3.3.19)$$

with $\ell_3 = \ell_1\sqrt{3}$ and $g_3 = g_1/\sqrt{3}$. The importance of Eq. (3.3.19) is that it has the same form as a true (i.e., interconnected) three-dimensional model.

Bundles of Capillaries of Different Diameters in Parallel. Each capillary is assumed to have a uniform cross section and the frequency of each diameter D is given by a (volume-based) pore size density $\alpha_p(D)$. There is flow in $1/3$ of the capillaries lying in the direction of the macroscopic flow in this model.

The permeability of this type of capillary model is as follows:

$$k = (\phi/96) \int_0^\infty D^2 \alpha_p(D) \, dD, \qquad (3.3.20)$$

where the subscript p refers to capillaries of different diameters in parallel.

It is evident that the value calculated for $\overline{D^2}$ by the integral in Eq. (3.3.20) is extremely sensitive to errors at the extremity of the pore size distribution function corresponding to the largest pore sizes. In addition if, as customary, the pore entry size distribution is used, the model is very different from the actual pore structure. It is not surprising, therefore, that the parallel-type capillaric model has not resulted in good permeability predictions (Henderson, 1949; Scheidegger, 1974).

Bundles of Periodically Constricted Tubes. In this "serial-type" capillaric model each channel is assumed to consist of segments of different diameters distributed according to some (volume-based) pore size distribution density $v_s(D)$, where the subscript s refers to tube segment of different diameters in series.

The permeability of this type of model is

$$k = \frac{\phi}{96} \frac{\left[\int \alpha_s(D) \, dD/D^2 \right]^2}{\int \alpha_s(D) \, dD/D^6}. \qquad (3.3.21)$$

The factor 96 was again introduced on the grounds that $1/3$ of the capillaries are pointing in each spatial direction.

The serial-type model tends to underestimate permeabilities and is very sensitive to uncertainties at the extremity of the pore size distribution function corresponding to the smallest pores present in the sample.

Bundles of Different Periodically Constricted Tubes. Both the models consisting of bundles of capillaries of different diameters in parallel and those consisting of bundles of identical periodically constricted tubes are incapable of modeling the permeability of most porous media. The physical reason for this is that in most porous media both types of spatial variations in pore size are present. In other words, (1) progressing through the medium inside any pore channel the fluid particles will pass through an alternating sequence of pores and pore throats, and (2) along certain pore channels the pore throats are bigger than along others. Therefore a realistic permeability model must incorporate both the serial type and the parallel type of spatial nonuniformities of pore size. Such a "serial–parallel" type model was introduced by Dullien (1975a). The unit cell of the model, in one dimension, is shown schematically in Fig. 3.10 where it is apparent that the model consists of

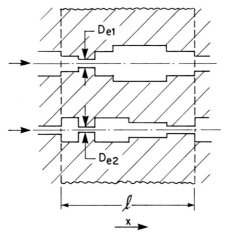

Figure 3.10. Illustration of the physical basis of the capillaric model of permeability expressed by Eqs. (3.3.22) and (3.3.23) (Dullien, 1990a).

periodically repeating tube sections of different kinds. Two of these are shown in Fig. 3.10, made of tube segments of different diameters. Of special importance are the throats, of diameters D_{e1} and D_{e2} in the figure.

For each subset of capillaries of the model, characterized by a certain range of throat diameters between D_e and $D_e + dD_e$, an expression analogous to Eq. (3.3.21) is obtained, giving the contribution to the total permeability k by those tubes controlled by throat diameters lying in this range, that is,

$$k(D_e)\,dD_e = \frac{\phi(D_e)}{96} \frac{\left\{ \int_{De}^{\infty} \left[\beta_{p,s}(D_e, D)\,dD/D^2 \right] \right\}}{\int_{De}^{\infty} \left[\beta_{p,s}(D_e, D)\,dD/D^6 \right]}\,dD_e \quad (3.3.22)$$

where $\beta_{p,s}(D_e, D)\,dD_e\,dD$ is the volume-based bivariate distribution of capillary diameters, giving the pore volume fractions consisting of pores in the diameter range D to $D + dD$ the passage through which is controlled by throats in the diameter range D_e to $D_e + dD_e$. Hence all the pores of different diameters D, characterized by the same D_e, are in series. The pores characterized by different values of D_e, however, are in parallel. $\phi(D_e)\,dD_e$ is the contribution to the model porosity by capillaries characterized by throats in the diameter range D_e to $D_e + dD_e$.

The permeability k of this model is obtained, finally, by integrating Eq. (3.3.22) over all values of D_e:

$$k = \int_0^\infty k(D_e)\, dD_e. \tag{3.3.23}$$

Quite generally, the capillary model permeability may be written as follows:

$$k = \frac{\phi}{32\Upsilon} \overline{D^2} \tag{3.3.24}$$

or, equivalently, as

$$k/\mu = g/\ell, \tag{3.3.25}$$

where ℓ is the "lattice constant" (see Fig. 3.9) and g is the conductance of an element of length ℓ; g is proportional to $\overline{D^4}$. It is important to note that neither $\overline{D^2}$ nor $(\overline{D^4})^{1/2}$ is compatible with Eq. (3.3.11).

Equations (3.3.22) and (3.3.23) have been used by El-Sayed (1978) to predict the permeabilities of numerous sandstone samples. In these calculations the bivariate pore size distributions (BPSD) $\beta_{p,s}(D, D_e)$, determined by the procedures discussed in Chapter 1 on Wood's metal porosimetry were used where D_e is the pore entry diameter calculated by Eq. (2.3.1) with $\phi = 0$. The pore diameters D were determined photomicrographically.

In 1975 there was no accurate information available on the breakthrough capillary pressures of the various samples. Only after El-Sayed (1978) measured the breakthrough pressures accurately did it become possible to use the previously published BPSDs for permeability predictions by assigning a more accurate entry pore diameter to those pores that were penetrated at the lowest capillary pressures.

Because of the time-consuming nature of the experimental work involved in measuring the bivariate distribution data, such measurements have been performed only on two different sandstone samples (Berea BE1 and Bartlesville). A technique has been devised to estimate the BPSD for samples where only the two pore size distribution curves, exemplified in Fig. 2.34, were available.

The calculation of the BPSD was done as follows: The area under the mercury porosimetry curve was divided into three parts of equal area. For very narrow mercury porosimetry peaks two subdivisions were used, and for the broadest peaks four subdivisions were used. The lower boundary of each zone represents the diameter of the smallest pore penetrated in the pore volume represented by the entire area under the curve to the right of the boundary. For each zone a curve (not a good assumption) to the experimental photomicrographic pore size distribution curve was drawn such that the area under the curve was proportional to the pore volume penetrated. From

	j i	1	2	3	4	5	6	7	8	$\frac{D_i}{\mu m}$
V_{ij} %	1	0.8	1.0	3.9	9.2	8.1	6.4	4.3	1.4	4.25
	2	0	0.3	3.4	9.3	8.6	6.6	3.1	1.0	13.0
	3	0	0	1.5	10	8.6	6.0	4.3	1.9	15.5
$\frac{D_j}{\mu m}$		4.25	13.0	15.5	30	50	70	90	110	

Figure 3.11. Estimated bivariate pore size distribution (BPSD) of Big Clifty sandstone (El-Sayed, 1978).

the curves, the BPSD was calculated by subdividing the whole diameter range into suitable increments and calculating the areas of the zones lying between adjacent curves within these increments.

The calculated BPSDs were presented in the form shown in Fig. 3.11, where i denotes pore of entry and j denotes any pore; V_{ij} is the percent volume in the sample of pores with mean diameter D_j such that can be penetrated from the outside of the sample through pore necks of mean diameter D_i. Increasing values of both i and j correspond to increasing pore sizes.

El-Sayed (1978) corrected the BPSDs calculated on the basis of measured entry pore diameters to take into account the fact that flow through the dendritic pores is controlled by exit pore diameters because the controlling pore throats for the dendritic pores are the exit pore throats. (see Chapter 1). El-Sayed (1978) subtracted the volume of the nonconductive dendritic pores from the bottom row of the histogram in Fig. 3.11 and added it to the next row above it, and so forth, until each row contained the appropriate percent conductive pore volume in agreement with his experiments (see Fig. 1.31).

The effect of excess viscous dissipation due to convergent–divergent flow in the capillaries was estimated on the basis of the results of Azzam (1975) and Azzam and Dullien (1976, 1977), who solved the Navier–Stokes equations numerically for periodically constricted tubes of the type shown in Fig. 3.12 and found an average decrease of predicted permeability due to such effects by a factor of $1/1.15$.

The permeabilities predicted by Eqs. (3.3.22) and (3.3.23), using the BPSDs corrected for nonconductive dendritic pore fractions, were generally smaller than the measured values. The mean deviation between prediction and experiment was minimum when a "tortuosity factor" $T = 1.725$ was used in Eq. (3.3.22) *instead of* value $T = 3$ incorporated in the factor of 96 ($= 3 \times$

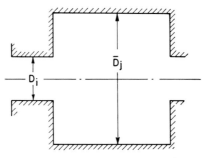

Figure 3.12.   One period of a periodically constricted tube. (Reprinted with permission from Azzam and Dullien, 1976 © by American Chemical Society.)

32). Hence, the pore network sandstone samples was considerably less tortuous than a cubic network of straight lines or set of randomly oriented lines, for both of which $\Upsilon = 3$ (Wiggs, 1958; Johnson and Stewart, 1965; Satterfield, 1970; Haring and Greenkorn, 1970; Dullien, 1975a). The reason for this discrepancy is that pores have a finite diameter-to-length ratio which increases, on the average, with increasing porosity. As a result, the orientation of streamlines in the pore network is not random even though the orientation of the pore axes may be random, because the streamlines will tend to follow the paths of least resistance and "cut (pore) corners" as shown schematically in Fig. 6.31. The results of permeability predictions are shown in Table 3.2.

TABLE 3.2   Comparison of Experimental and Calculated Permeabilities[a]

Sandstone	ϕ	k_{exp}	k_{calc}	k_{exp}/k_{calc}
Boise	0.232	5370	5459	0.984
Bartlesville	0.270	1307.5	1299.5	1.006
Berea 108	0.178	613	634.4	0.966
Berea BE1	0.220	400	453	0.882
Noxie 47	0.250	298	300.4	0.992
St. Meinrad	0.247	210	235.5	0.892
Torpedo	0.246	184	190.2	0.965
Big Clifty	0.192	164	156.4	1.049
Noxie 129	0.238	147.5	154.0	0.958
Cottage Grove	0.216	116	110.0	1.051
Clear Creek	0.191	94.5	94.54	1.032
Bandera	0.191	8.6	8.92	0.964
Whetstone	0.134	0.75	0.73	1.027
Belt Series	0.083	0.36	0.306	1.18

[a]Value of "hydraulic tortuosity factor," for all the samples, $\Upsilon = 1.725$.

3.3.2. Statistical Permeability Models

Whenever a probability law is used in the permeability model, the term "statistical model" seems appropriate, regardless of the other assumptions that may also be used in the model.

3.3.2.1. *Cut-and-Random-Rejoin-Type Models*

The cut-and-random-rejoin-type model was introduced by the work of Childs and Collis-George (1950) and was subsequently modified by Marshall (1958) and Millington and Quirk (1961). More recently, Brutsaert (1967, 1968) has written reviews on this subject.

In all of these treatments, the existence of capillary hysteresis is ignored and the drainage (or suction) capillary pressure curve is used. Therefore, this model also suffers from the shortcoming of assigning all the pore volume to entry pores.

It has been assumed (erroneously) that $\alpha_s(D_e)\,dD_e$, determined from drainage capillary pressure curve, gives the fraction of total pore volume occupied by pores of diameters between $D_e \to D_e + dD_e$, where D_e is the pore entry diameter. Assuming that the areal porosity is equal to the bulk porosity ϕ, and that $\alpha_s(D_e)\,dD_e$ gives also the fraction of the cross section of the pore space consisting of two-dimensional features of diameters between D_e and $D_e + dD_e$, there follows that $\phi\alpha_s(D_e)\,dD_e$ is the fraction of the area of a section occupied by pore openings with sizes in the range $D_e \to D_e + dD_e$.

It is imagined that the sample is sectioned into two by a plane perpendicular to the direction of flow, and the two parts are joined together again in a random fashion. If the pores in the solid matrix are assumed to be distributed at random, their chance of overlapping is governed by random probability. It follows that the fractional area in the rejoined section, which consists of an overlap between pore openings in the range between $D_e \to D_e + dD_e$ in one of the two surfaces and $D_e' \to D_e' + dD_e'$ in the other surface, is equal to $\phi^2\alpha_s(D_e)\alpha_s(D_e')\,dD_e\,dD_e'$. The second assumption used here would be valid only if the pores were uniform capillary tubes and were sectioned with a plane perpendicular to the tube's axis.

The assumption was made in this model also that the flow rate in the capillaries is given by a Hagen–Poiseuille-type relationship. Thus, the average velocity $\bar{u}$ in a pore with diameter D_e is

$$\bar{u} = -(D_e^2/32\mu)(\partial\mathscr{P}/\partial z),\qquad(3.3.26)$$

where z is the coordinate of the direction of macroscopic flow through the system. The tacit assumption made when introducing this equation is that the pore velocity is in the direction of the macroscopic flow.

The volumetric flow rate $q(D_e, D_e')\,dD_e\,dD_e'$ across the fraction of the rejoined cross section, which is occupied by overlapping pores in the range

$D_e \to D_e + dD_e$ in one of the rejoined surfaces and $D'_e \to D'_e + dD'_e$ in the other one, is obtained by multiplying the fractional overlapping area into the average pore velocity in the smaller pore (e.g., D'_e):

$$q(D_e, D'_e)\, dD_e\, dD'_e = -\left(\left(D'^2_e/32\mu\right)\partial\mathscr{P}/\partial z\right)\phi^2\alpha_s(D_e)\alpha_s(D'_e)\, dD_e\, dD'_e.$$

(3.3.27)

The total flow per unit area of the rejoined section is given by the integral

$$v = -\frac{\phi^2}{32\mu}\frac{\partial\mathscr{P}}{\partial z}\left[\int_0^\infty \alpha_s(D_e)\, dD_e \int_0^{De} D'^2_e\alpha_s(D'_e)\, dD'_e\right.$$

$$\left. + \int_0^\infty D^2_e\alpha_s(D_e)\, dD_e \int_{De}^\infty \alpha_s(D'_e)\, dD'_e\right].$$ (3.3.28)

Note that this equation is based on a fictitious areal porosity ϕ^2 rather than the actual value ϕ.

Comparing this result with Darcy's law,

$$v = -(k/\mu)(\partial\mathscr{P}/\partial z)$$ (3.3.29)

results in the following expression for the permeability k:

$$k = \frac{\phi^2}{32}\left\{\int_0^\infty \alpha_s(D_e)\left[\int_0^{De} D'^2_e\alpha_s(D'_e)\, dD'_e\right]dD_e\right.$$

$$\left. + \int_0^\infty D^2_e\alpha_s(D_e)\left[\int_{De}^\infty \alpha_s(D'_e)\, dD'_e\right]dD_e\right\}.$$ (3.3.30)

This expression predicts that the permeability is proportional to ϕ^2. This has been interpreted occasionally to mean that the "tortuosity factor" Υ varies as $1/\phi$, which is unlikely to be true. Hence Eq. (3.3.30) should thereby underpredict k. On the other hand, by using the drainage capillary pressure curve for $\alpha_s(D_e)\, dD_e$, the volume of pores of diameter D_e is grossly overestimated because the volume of larger pores that are penetrated through entry pores of diameter D_e is included in $\alpha_s(D_e)$. Hence, there is a possibility for cancellation of errors. The expression in curly braces is $\overline{D^2}$, the mean-square pore diameter obtained according to this model. A sample calculation given by Marshall (1958) shows that, surprisingly, the various pore diameters contribute nearly equal amounts to $\overline{D^2}$.

Millington and Quirk (1961) obtained the following expression for the permeability:

$$k = \frac{2\phi^{4/3}}{32}\int_0^\infty \alpha_s(D_e)\, dD_e \int_0^{De} D'^2_e\alpha_s(D'_e)\, dD'_e.$$ (3.3.31)

It can be shown that the two terms in the curly braces in Eq. (3.3.30) are equal and, therefore, use of only one of the terms multiplied by 2 is correct. However, the porosity functions in Eqs. (3.3.30) and (3.3.31) are different. Marshall (1958) also used ϕ^2. Each of these authors, using different porosity functions, claimed to have predicted permeabilities in good agreement with the experimental values.

Klock *et al.* (1969) applied the equation of Millington and Quirk to glass beads and quartz sand samples using mercury intrusion porosimetry. The calculated permeabilities were too small, on the average, by about a factor of 1.7.

The cut-and-random-rejoin models have been used also to predict "phase permeabilities" as a function of saturation. This subject is discussed in Chapter 5.

The cut-and-random-rejoin model is an interesting mathematical artifice devised for the calculation of permeabilities.

3.3.2.2. *The Model of Haring and Greenkorn*

Haring and Greenkorn (1970) modeled the pore structure by a large number of randomly oriented, straight, cylindrical pores. There are four independent variables to begin with: the pore length ℓ, the pore radius R, and the two angles, θ and ϕ, which define the orientation of the pore in three-dimensional space. Haring and Greenkorn let

$$0 \leq \theta \leq \pi/2, \qquad 0 \leq \phi \leq 2\pi;$$

$$0 \leq \ell^* \leq 1, \qquad \ell^* = \ell/\ell_{max}$$

$$0 \leq r^* \leq 1, \qquad r^* = R/R_{max}$$

where ℓ_{max} and R_{max} are the greatest values of ℓ and R in the sample. The angles θ and ϕ are random variables, that is to say, they can have any value with equal probability in the ranges stated, and ℓ^* and r^* are assumed to be distributed according to the beta function. Arbitrarily setting the length of all pores of different radii r^* equal to $\overline{\ell^*}$ (i.e., the "average length" of the ensemble of pores) the differential volume of the nonwetting fluid that is invading the pore space is expressed as

$$V_p \, dS = -n\pi R_{max}^2 \ell_{max} \, r^* \overline{\ell^*} g(r^*) \, dr^*, \tag{3.3.32}$$

where V_p is the pore volume of the sample, S is the saturation with respect to the nonwetting fluid, n is the total number of pores in the model, and $g(r^*) \, dr^*$ is the number fraction of pores of radii between r^* and $r^* + dr^*$. The penetration starts with the largest value $r^* = 1$. Hence, $dr^* < 0$, whereas $dS > 0$. Algebraic manipulations followed by integration between the limits

$r^* = 0$, $S = 1$ and r^*, S give

$$1 - S = \int_0^{r^*} \frac{(\alpha + \beta + 3)!}{(\alpha + 2)!\beta!}(r^*)^{\alpha+2}(1 - r^*)^\beta \, dr^*, \qquad (3.3.33)$$

where α and β are parameters of the beta function. The limit $r^* = 0$, $S = 1$ is physically not realizable when using water as the wetting phase, as Haring and Greenkorn have done.

In the model of Haring and Greenkorn all pores are directly accessible from the outside of the sample (i.e., all pores are entry pores and independent domains). Hence, this model also is incompatible with the existence of capillary hysteresis. Moreover, if the entry to no pore is blocked by a narrower pore, then all pore segments in series connection with each other must have the same diameter. Since all pores of different diameters are also assumed to have the same length, the model of Haring and Greenkorn is a special case of the capillaric model consisting of uniform tubes of different diameters where the orientations of all tubes are random.

After introducing the capillary pressure $P_c(r)$, the pore radius r^* is replaced with P_c^* as follows:

$$r^* = P_c(R_{\max})/P_c(r) \equiv 1/P_c^*(r), \qquad (3.3.34)$$

giving

$$1 - S = \int_0^{1/P_c^*} \frac{(\alpha + \beta + 3)!}{(\alpha + 2)!\beta!} \left(\frac{1}{P_c^*}\right)^{\alpha+2}\left(1 - \frac{1}{P_c^*}\right)^\beta d\left(\frac{1}{P_c^*}\right)$$

$$= B(x; p_1, p_2), \qquad (3.3.35)$$

where $B(x; p_1, p_2)$ is the incomplete beta function. The parameters α and β are determined from tabulated values $B(x; p_1, p_2)$, where $x = 1/P_c^*$, $p_1 = \alpha + 2$, and $p_2 = \beta$, using the threshold capillary pressure of penetration corresponding to $S = 0$. The parameters permitted a good fit to the experimental points up to $S = 0.85$. The deviations between predictions and experimental data above this value are understandable because the water saturation in bead packs has a limiting, irreducible value, whereas the calculated values go, by the definition of beta function, all the way to $S = 1$.

The parameters α and β are used to estimate permeabilities. The average velocity $\bar{u}$ in each pore is assumed to be independent of the velocity in other pores, which is consistent with the tacit assumption of uniform cross section of each flow capillary. As usual, $\bar{u}$ is expressed by the Hagen–Poiseuille equation:

$$\bar{u} = -(R_{\max}^2/8\mu)r^{*2}(\partial \mathscr{P}/\partial z)\cos\theta, \qquad (3.3.36)$$

where the macroscopic flow is in the z direction.

The average velocity $\langle \bar{u}_z \rangle$ in the z direction (i.e., the "pore velocity" v_p) is calculated as follows:

$$\langle \bar{u}_z \rangle = \int \bar{u} \cos \theta \, g(r^*) \, dr^* \sin \theta \, d\theta$$

$$= -\frac{R_{max}^2}{24\mu} \frac{\partial \mathcal{P}}{\partial z} \frac{(\alpha + 1)(\alpha + 2)}{(\alpha + \beta + 2)(\alpha + \beta + 3)}. \qquad (3.3.37)$$

Here the authors weight the individual pore velocities with the function $g(r^*) \, dr^*$, which is the number fraction of the tubes with radius r^* in the sample. As pointed out by Guin *et al.* (1971a, b), this is incorrect because the velocities must be weighted also with the pore cross sections.

Haring and Greenkorn relate $\langle \bar{u}_z \rangle$ to the filter velocity v by using the Dupuit–Forchheimer assumption and Darcy's law, that is,

$$v = \langle \bar{u}_z \rangle \phi = -(k/\mu)(\partial \mathcal{P}/\partial z). \qquad (3.3.38)$$

The final expression for the permeability is

$$k = \frac{R_{max}^2}{24} \frac{(\alpha + 1)(\alpha + 2)}{(\alpha + \beta + 2)(\alpha + \beta + 3)} \phi. \qquad (3.3.39)$$

This expression is of a similar form as all other capillaric and statistical models. The random orientation results in a "tortuosity factor" $\Upsilon = 3$.

The fact that good agreement was obtained with measured permeabilities for random bead packs suggests that the channels in the packs were fairly uniform (Pakula and Greenkorn, 1971).

The formula given by Guin *et al.* (1971a, b) for the filter velocity is

$$v = \int \bar{u} A \ell f(A, \ell, \alpha) \, dA \, d\ell \, d\alpha, \qquad (3.3.40)$$

where $f(A, \ell, \alpha) \, dA \, d\ell \, d\alpha$ is the number of cylindrical pores per unit volume having cross-sectional area $A \to A + dA$, length $\ell \to \ell + d\ell$, and orientation $\alpha \to \alpha + d\alpha$.

The conceptual picture behind Eq. (3.3.40) is that the volumetric flow rate in a capillary of cross-sectional area A is multiplied with the number of pores that intersect unit area of the porous medium, and the result is integrated over all the different pores that intersect with the unit area under consideration. Guin *et al.* depart from a generally valid model, however, when they calculate $\bar{u}$ essentially by Eq. (3.3.36). Applying Eq. (3.3.36) to the case when two capillaries of the same orientation but different radii R_1 and R_2 are in series connection with each other, we find that this equation predicts, for the ratio of the two pore velocities, $v_{p1}/v_{p2} = (R_1/R_2)^2$, whereas in actual fact it is the inverse relation, $v_{p1}/v_{p2} = (R_2/R_1)^2$, that holds.

Hence we can see that in the model of Guin *et al.* (1971a, b) the pores characterized by different values of the parameters A and α do not interact with each other, but we have here a sophisticated model in which both the orientations and the lengths of the capillaries are distributed over a range of values. Some of the conclusions reached by Guin *et al.* are interesting. For example, they showed that the permeability of their model depends on $\overline{A^2}/\overline{A}$ rather than $\overline{A}$. It can be shown that in models that take serial-type connections into account, the permeability depends on $\overline{A^3}/\overline{A^2}$.

3.3.3.   Empirical Permeability Models

Some of the most successful permeability models are based on pure empiricism—even though attempts are often made to justify the good empirical correlation by a theoretical model.

3.3.3.1.   *Permeability Correlation with Breakthrough (Bubbling) Pressure and Porosity*

The physical basis of this model is that the permeability of a tube with step changes in diameter, of given length ℓ and volume, can be matched by the permeability of a bundle of n uniform tubes of diameter D_x, the same length ℓ and same total volume if there is the following relationship between

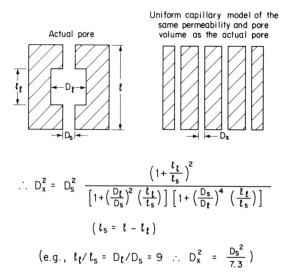

$$\therefore \; D_x^2 = D_s^2 \; \frac{\left(1+\frac{\ell_\ell}{\ell_s}\right)^2}{\left[1+\left(\frac{D_\ell}{D_s}\right)^2\left(\frac{\ell_\ell}{\ell_s}\right)\right]\left[1+\left(\frac{D_s}{D_\ell}\right)^4\left(\frac{\ell_\ell}{\ell_s}\right)\right]}$$

$$(\ell_s = \ell - \ell_\ell)$$

$$\left(\text{e.g.,} \; \ell_\ell/\ell_s = D_\ell/D_s = 9 \; \therefore \; D_x^2 = \frac{D_s^2}{7.3}\right)$$

Figure 3.13. Bundle of uniform capillary tubes model of identical permeability as the periodically constricted tube model (Dullien, 1990).

geometrical dimensions of the tubes (see Fig. 3.13):

$$D_x = D_s \frac{(1 + \ell_\ell/\ell_s)^2}{\left[1 + (D_\ell/D_s)^2(\ell_\ell/\ell_s)\right]\left[1 + (D_s/D_\ell)^4(\ell_\ell/\ell_s)\right]} . \qquad (3.3.41)$$

It is interesting that we have always $D_x < D_s$. For example, in the case of sandstone samples Dullien (1990) has found that $D_x = D_b/3.5$, where D_b is the "breakthrough" diameter (see Table 1.1). Hence, the calculated permeability k_c could be expressed in the conventional form, that is,

$$k_c = (D_b/3.5)^2(\phi/32) \times 10^3 \quad (mD) \qquad (3.3.42)$$

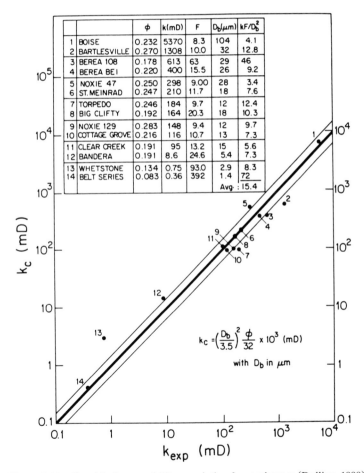

Figure 3.14. Empirical permeability correlation for sandstones (Dullien, 1990).

where D_b is in μm. The predicted and calculated values are compared in Fig. 3.14, where also some other pertinent information is included. As D_b, ϕ, and k could not be measured on identical pieces of a sample and there is a natural scatter in each of these values within the same sample, the agreement between k_c and k_{exp} in Fig. 3.14 is every bit as good as the agreement between k_{exp} measured in different pieces of the same core sample.

Another correlation of similar accuracy has been found by Chatzis (1980):

$$k = (85.63/P_{cb})^{2.71} \qquad (3.3.43)$$

where P_{cb} is the breakthrough capillary pressure in psi units, measured in the Hg/air system.

Macmullin and Muccini (1956) presented a correlation among bubbling pressure, formation factor F (see Section 3.7), and permeability, with a probable error of single observation of about 18%. The materials tested have included glass beads, glass frits, sand, sandstone, procelain, alumina, carbon, graphite, and PVC sheet. Chatzis (1980) has tested this correlation for sandstones and found a $\pm 60\%$ scatter and a very different value of the constant c in the correlation [see Eq. (3.3.44)]. Exactly the same form of correlation has been proposed recently by Katz and Thompson (1986) by making appeal to percolation theory and claiming general validity for this correlation, that is,

$$k = cD_b^2/F \text{ (millidarcy)} \qquad (3.3.44)$$

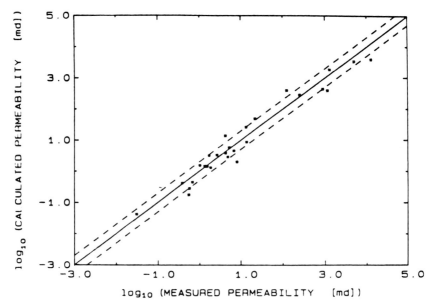

Figure 3.15.   Permeability correlation based on Eq. (3.3.44) (Katz and Thomson, 1986).

There are n fine capillaries for every large capillary

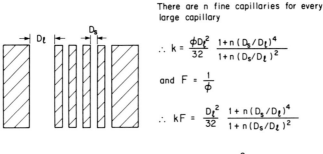

$$\therefore k = \frac{\phi D_\ell^2}{32} \frac{1+n(D_s/D_\ell)^4}{1+n(D_s/D_\ell)^2}$$

and $F = \frac{1}{\phi}$

$$\therefore kF = \frac{D_\ell^2}{32} \frac{1+n(D_s/D_\ell)^4}{1+n(D_s/D_\ell)^2}$$

(e.g., $D_s/D_\ell = 0.5$, $n = 100$ $\therefore$ $32 \, kF/D_\ell^2 = 0.28$)

Figure 3.16. Demonstration of the effect of pore size distribution on the value of c in Eq. (3.3.49) in the case of parallel-type pore nonuniformities (Dullien, 1990).

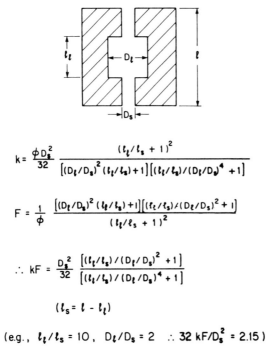

$$k = \frac{\phi D_s^2}{32} \frac{(\ell_\ell/\ell_s + 1)^2}{\left[(D_\ell/D_s)^2(\ell_\ell/\ell_s)+1\right]\left[(\ell_\ell/\ell_s)/(D_\ell/D_s)^4 +1\right]}$$

$$F = \frac{1}{\phi} \frac{\left[(D_\ell/D_s)^2(\ell_\ell/\ell_s)+1\right]\left[(\ell_\ell/\ell_s)/(D_\ell/D_s)^2 +1\right]}{(\ell_\ell/\ell_s + 1)^2}$$

$$\therefore kF = \frac{D_s^2}{32} \frac{\left[(\ell_\ell/\ell_s)/(D_\ell/D_s)^2 +1\right]}{\left[(\ell_\ell/\ell_s)/(D_\ell/D_s)^4 +1\right]}$$

$(\ell_s = \ell - \ell_\ell)$

(e.g., $\ell_\ell/\ell_s = 10$, $D_\ell/D_s = 2$ $\therefore$ $32 \, kF/D_s^2 = 2.15$)

Figure 3.17. Demonstration of the effect of pore size distribution on the value of c in Eq. (3.3.49) in the case of serial-type pore nonuniformities (Dullien, 1990).

This correlation is shown in Fig. 3.15. The value of the constant c was proposed to be $1/226$ (in consistent units?). The great variation of this "constant" for 14 sandstone samples is shown in Fig. 3.14. They range, in consistent units, from $1/313$ (Noxie 47) to $1/14$ (Belt Series).

The two very simple examples illustrated in Figs. 3.16 and 3.17 show that the value of the group kF/D_b^2 depends on both the aspect ratios and the spatial arrangement of pores relative to each other. In Fig. 3.16 the model consists of capillaries of two different diameters in parallel. In this case $D_b = D_\ell$. In Fig. 3.17, however, the model is made of capillaries of two different diameters in series. In this example $D_b = D_s$. It is evident that for a bundle of identical uniform capillaries $kF/D^2 = 1/32$ and that parallel and serial type pore uniformities make the value of this group vary, depending on pore structure. In the numerical examples chosen in Figs. 3.16 and 3.17 the values of kF/D_b^2 are $1/114$ and $1/15$, respectively. It is evident that in real porous media the effects of parallel and serial type of pore nonuniformities will tend to cancel each other to some extent. Comparing these values with those in Fig. 3.14, one sees that for most of the sandstones the effect of parallel-type pore nonuniformities is stronger than that of serial-type pore nonuniformities (the "neutral" value $1/32$ corresponds to 31, in Fig. 3.14). This may be due to the presence of many fine pores in pore throats in the cementing materials, in parallel with the main pore throat.

3.3.4. Network Models of Permeability

Models consisting of three sets of capillary tubes aligned with the three coordinate axis x, y, and z are not true network models, unless the tubes intersect with each other. Models of pore structure, consisting of three-dimensional networks where the pore sizes are distributed over the bonds and/or the nodes of the network in a disordered manner, come to resemble closest the structure of real porous media (Ng and Payatakes, 1985). Partly because of the lack of accurate information on the details of pore structure, such network models have not yet been used successfully to predict permeabilities.

Koplik (1982) carried out calculations on two-dimensional regular networks consisting of circular junctions (pores) centered at the intersections of straight channels (necks) and found linear relations of the Hagen–Poiseuille type between the flow rate and the pressure drop in both the pore and the neck. Hence the problem of calculating the network conductance could be reduced to the analogous electrical problem. The following regular networks were treated: square, hexagonal, kagomé, trigonal, and crossed square, shown in Fig. 3.18. The conductances g were distributed according to some probability density function (pdf) and were assigned randomly to the "bonds" of the network. For the special case of a uniform (flat top) conductance pdf, ranging

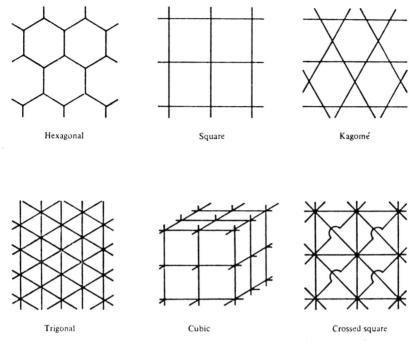

Figure 3.18. Various network models of permeability (Koplik, 1982).

from 1 to 5000 by increments of one, the effective "bond" conductance g_m was calculated, using effective medium theory, which when substituted for all the "bond" conductances will result in the same network conductance as obtained for the case of the distributed "bond" conductances.

As it is apparent from Table 3.3 the effective "bond" conductance g_m was found to be a slowly (almost logarithmically) increasing function of the

TABLE 3.3 Effective Medium Theory Conductances $g_m(z)$ and Geometrical Factors t for Various Lattices[a]

Lattice	z	$g_m(z)$	t
Hexagonal	3	1752	$1/\sqrt{3}$
Kagomé	4	1991	$\sqrt{3/2}$
Square	4	1991	1
Trigonal	6	2186	$\sqrt{3}$
Cubic	6	2186	1
Crossed square	8	2272	3

[a]After Koplik (1982).

network coordination number z. As pointed out by Koplik (1982),

$$g_m(2) = \langle g^{-1}\rangle^{-1} \le g_m(z) \le \langle g\rangle = g_m(\infty) \qquad (3.3.45)$$

where $g_m(2)$ corresponds to conductors (bonds) in series, whereas $g_m(\infty)$ corresponds to conductors (bonds) in parallel.

Equation (3.3.45) applies also for $g = g_m$ (i.e., when all "bonds" have the same conductance). In this case there follows that $g_m(z)$ is the same value g for any coordination number z. The tortuosity, however, is different for different networks, hence, $g_m(z)$ is not the analog of tortuosity.

Koplik (1982) also expressed the network conductivities (k_2/μ) in terms of $g_m(z)$ and a geometrical factor t_2 as follows:

$$(k_2/\mu) = g_m(z)t_2 \qquad (3.3.46)$$

for 2-D networks, where k_2 has the dimensions (length)^3. For the square network, $t_2 = 1$ and the values of t_2 for the other 2-D networks are also listed in Table 3.3. The factor t arises quite naturally when comparing the Hagen–Poiseuille type flow through the network with Darcy's law, the same way as it was done in the case of the capillaric models [cf. Eqs. (3.3.11) to (3.3.19)]. Its value increases with the network coordination number z. The value t_2 (in three dimensions, t_3/ℓ) is the sum of the projected lengths of the bonds onto the direction of the pressure gradient, divided by the cross section, perpendicular to this direction, intersecting the bonds (see Fig. 3.19). The more bonds intersect with unit area and the greater their projected lengths, the higher is the value of t.

The effect of network tortuosity on permeability is built into t, and the effect of the porosity on the permeability is built into g_m.

Koplik (1982) also gave a straightforward extension of Eq. (3.3.46) to three-dimensional networks:

$$(k/\mu) = g_m(z)t_3/\ell_3. \qquad (3.3.47)$$

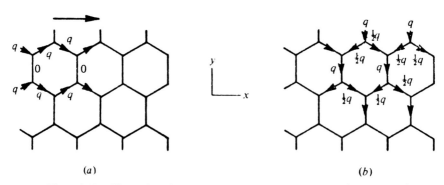

(a)   (b)

Figure 3.19.   Illustration of calculating t_2 in a hexagonal network (Koplik, 1982).

TABLE 3.4   Geometrical Factors for Three-Dimensional Networks[a]

Lattice	σ	t_3
Diamond	4	$1/4\sqrt{3}$
Simple cubic	6	1
Body-centered cubic	8	$\sqrt{3}$
Face-centered cubic	12	$2\sqrt{2}$
Hexagonal close-packed	12	$2\sqrt{2}$
Filled body-centered cubic[b]	16	$3/2\sqrt{3} \simeq 2.60$

[a]After Koplik (1982)
[b]This refers to a lattice formed by superposing simple and body-centered cubic, or including the cube axes in the body-centered structure.

For the cubic network $t_3 = 1$. Hence, in this case Eq. (3.3.47) has exactly the same form as Eq. (3.3.19). For the diamond (tetrahedral) lattice ($z = 4$) $t_3 = (\frac{1}{4})\sqrt{3}$. More values of t_3 for other lattices are listed in Table 3.4. Koplik (1983) extended his calculations to random networks in which the pore centers are randomly distributed in space and where there is also a pdf of coordination numbers.

Network models of permeability require as input a (simplified) reconstruction of the actual pore network of a sample of the porous medium. Such a reconstruction has been based, up to the present, on "serial sections" of a sample (Lin and Cohen, 1982; Lin and Hamasaki, 1982; Lin and Perry, 1982; Lin, 1983; Kaufmann *et al.*, 1983; Macdonald *et al.*, 1986; Kwiecien, 1987; Kwiecien *et al.*, 1990). The reconstructed pore structure has included node-to-node (pore-to-pore) distances, the coordination number of every node, the shape of pores and necks, pore and pore neck "diameters" and, in some cases, the orientations of the lines connecting two neighbor nodes. Indeed, all this information was used by Koplik *et al.* (1984) in an attempt to model the permeability k and the formation factor F (see Section 3.7) of a Masillon sandstone, using the computational technique developed by Koplik (1982, 1983) which has been discussed above. Serial sections about 10 μm apart were obtained. In the absence of details of processing the pore structure data it is impossible to form a clear picture of what was done and what the reconstructed pore structure looked like. Of particular interest would be to know how the conductances g were calculated. The reconstructed flow paths were twisting stacks of elliptical cylinders, each cylinder spanning the gap between two consecutive serial sections. How the problems (e.g., of overlap of the faces of neighboring stacks of cylinders and the branching from one-to-several cylinders, and vice versa) were handled is not explained in the paper.

The permeability k of 27.4 D, predicted by the model, was about 10 times greater than the value of 2.5 D measured on a larger sample and the predicted formation factor F of 5.7 was about one-half of the measured value of 11.8. It is quite possible, as the authors also suggest, that their sample of 0.4 mm thickness in the direction of flow was not representative of the sandstone. It is quite likely that had the permeability and the formation factor of the thin sample been measured it would have been close to the predicted values. The presence of a few "holes" in a sample would have a much greater effect on the permeability than on the formation factor.

3.3.5. Deterministic Permeability Models

Models that use either an explicit or an indicated solution of the Navier–Stokes equations in the pore space are called "deterministic" in this monograph. Explicit solutions have been obtained in various "periodically constricted tubes" by Payatakes *et al.* (1973), Azzam (1975), and Azzam and Dullien (1976, 1977) with interesting results on the magnitude of the effects of convergent–divergent flow and circulation patterns in "Turner structures" (Turner, 1958, 1959). All these effects were found to increase the predicted permeability by not more than 30% and mostly by much less than that.

3.3.5.1. *Solution of the Navier–Stokes Equations in Periodically Constricted Tubes*

In the models discussed in Sections 3.3.1 to 3.3.4, only the z velocity component has been considered, resulting in Hagen–Poiseuille-type flow equations. It is evident that in convergent–divergent type of flow the radial velocity component may not always be negligible. Solution of the complete Navier–Stokes equation yields the correct conductivity to fluids of a periodically constricted tube. This may be compared with the result obtained with the help of the Hagan–Poiseuille equation, which is applied to capillary segments of approximately constant diameter, assuming that contraction and expansion losses which occur when the fluid passes between neighboring segments are negligible.

The stream function–vorticity method consists of expressing the Navier–Stokes equation in terms of the stream function as the dependent variable and results in a fourth-order partial differential equation which, in turn, is replaced with a system of two partial differential equations, each of second order, with the stream function and the nonvanishing component of the vorticity as the dependent variables (Payatakes *et al.*, 1973; Christiansen *et al.*, 1972; Runchal *et al.*, 1969).

This technique has been used by Payatakes *et al.* for periodically constricted tubes of a geometry shown in Fig. 3.20. The pressure drop and the

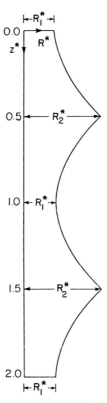

Figure 3.20.   Dimensionless unit cell of a periodically constricted tube. (Reprinted with permission from Payatakes *et al.*, 1973.)

velocity components have been obtained in dimensionless form for various values of the Reynolds number

$$\text{Re}_\lambda = \mu \bar{u}_0 / \nu, \tag{3.3.48}$$

where λ is the wavelength of a periodically constricted tube, $\bar{u}_0$ is the mean velocity at the entrance to the tube, and ν is the kinematic viscosity. Various values of the Reynolds number are assumed and the dimensionless pressure $\mathscr{P}^*$ and velocity (u_z^*, u_r^*)

$$\mathscr{P}^* = (P + \rho\, gz)/\rho\bar{u}_0^2, \qquad u_z^* = u_z/\bar{u}_0, \qquad u_r^* = u_r/\bar{u}_0 \tag{3.3.49}$$

are calculated. By integrating the dimensionless pressure over the length of a period of the periodically constricted tube, the pressure drop is obtained. The friction factor f_v of the periodically constricted tube is defined as

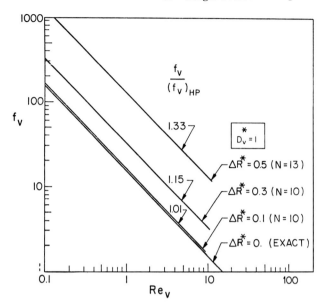

Figure 3.21. Effect of the variation of the amplitude on the dimensionless pressure drop for constant volumetric diameter. (Reprinted with permission from Payatakes *et al.*, 1973.) The values of $f_v/(f_v)_{HP}$ are due to Azzam (1975).

follows:

$$f_v = -\left(D_v^{*5}/32 R_1^{*4}\right) \Delta \mathcal{P}^*, \tag{3.3.50}$$

where D_v^* is the dimensionless volume average diameter

$$D_v^* = D_v/\lambda \tag{3.3.51}$$

with D_v the volume average diameter and R_1^* the dimensionless constriction radius

$$R_1^* = R_1/\lambda. \tag{3.3.52}$$

Using these definitions, it may be shown that

$$f_v = \left(\Delta \mathcal{P} D_v/2\lambda \rho \bar{u}^2\right), \tag{3.3.53}$$

where $\Delta \mathcal{P}$ is the pressure drop and

$$\bar{u} = 4Q/\pi D_v^2, \tag{3.3.54}$$

with Q the volumetric flow rate. Some of the results obtained by Payatakes *et al.* are shown in Fig. 3.21 in the form of f_v versus the Reynolds number

$$\mathrm{Re}_v = D_v \bar{u}/\nu \tag{3.3.55}$$

with D_v^* and ΔR^* as two geometrical parameters.

$$\Delta R^* = R_2^* - R_1^*, \tag{3.3.56}$$

that is, the difference between the largest and the smallest dimensionless radius of the periodically constricted tube.

Interpretation of these plots presents some problems, however, as they do not show by how much the friction factor is increased due to excess viscous dissipation because of convergent–divergent flow.

Azzam (1975) calculated the friction factors $(f_V)_{HP}$ for the periodically constricted tubes considered by Payatakes *et al.* by dividing the tubes into small increments perpendicular to the tube axis of symmetry and using the Hagen–Poiseuille equation for each increment. Increments of 100, 1000, and 10,000 gave identical results. The friction factor ratios $f_v/(f_v)_{HP}$, shown in Fig. 3.21, indicate that the friction factor calculated by solving the complete Navier–Stokes equation is greater, in the worst case, only by a factor of about 1.3 than the value obtained by the Hagen–Poiseuille equation.

Azzam (1975) (see also Azzam and Dullien, 1976, 1977) solved the complete Navier–Stokes equation for tubes with periodical step changes in diameter. The mathematical formulation has been based on the stream function–vorticity approach. The numerical technique that has been used, however, is what is known as the "upwind" ("upstream" or "donor cell") finite difference scheme. The results, expressed as permeability ratios, for three different values of D_s/λ are shown in Fig. 3.22 as functions of D_s/D_ℓ. The results indicate minimum-type curves, much like those obtained experimentally (Dullien and Azzam, 1972) for much smaller values of D_s/λ. For these the calculations predict very much smaller deviations from unity that those found experimentally. It is possible that the experiments were in error, probably because of the presence of unnoticed air bubbles.

Azzam's calculations indicate that end effects in the narrow capillary segments (necks) become significant only in the case of small length-to-diameter ratios. End effects in the large segments (bulges) always contribute negligibly to the excess losses because most of the viscous dissipation takes place in the narrow segments. Azzam has applied the results of his calculations to the permeability model of Dullien (1975a) (see Section 3.3.1.2) and found an average decrease of about 15% in the permeabilities predicted by the model (Azzam and Dullien, 1976).

3.3.5.2. *Volume-averaged Form of the Navier–Stokes Equations in Porous Media, Deterministic Model*

Volume averaging or homogenization of the equations of motion is considered to lie outside the scope of this monograph. Only occasional reference to the results obtained by this approach is made whenever a discussion of these results is deemed appropriate for the purpose of this text.

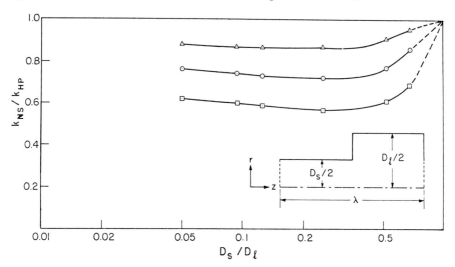

Figure 3.22. Calculated permeability decrease found by solving the Navier–Stokes equations for tubes with periodic step changes in diameter. Values shown are for $D_s/\lambda = 0.215$ ($\triangle$), 0.43 ($\odot$), and 0.86 ($\square$). (Reprinted with permission from Azzam and Dullien, 1977.)

Slattery (1972) and Whitaker (1969) considered steady, incompressible, laminar flow of a constant viscosity Newtonian fluid in a rigid porous medium.

For any point function ψ (e.g., velocity, density, pressure, etc.) associated with the fluid, Slattery (1972) derived the following important theorem between gradients of averages and averages of gradients for functions ψ defined in both the solid and the void phase and suffering a jump discontinuity at the phase interface:

$$\langle \nabla \psi \rangle = \nabla \langle \psi \rangle + \frac{1}{V} \int_{A_i} \psi \mathbf{n} \, dA, \tag{3.3.57}$$

where the volume average of ψ is defined as

$$\langle \psi \rangle = \frac{1}{V} \int_{V_\mathrm{f}} \psi \, dV, \tag{3.3.58}$$

with A the surface bounding the volume V and A_i the area of the solid–fluid interface contained within V. In deriving Eq. (3.3.57) the restriction has been made that the average volume V is constant and the orientation relative to some inertial frame remains unchanged.

Using Slattery's averaging theorem, a general volume-averaged flow equation can be obtained (e.g., Dullien and Azzam, 1972):

$$\nabla \langle P \rangle = \mu v \left\{ \frac{1}{D^2} \frac{1}{V} \int_{Vf} \nabla *^2 \mathbf{u} * \, dV \right\}$$

$$- \rho v^2 \left\{ \frac{1}{D} \frac{1}{V} \int_{Vf} \mathbf{u} * \cdot \nabla * \mathbf{u} * \, dV + \frac{1}{D} \frac{1}{V} \int_{Ai} \mathscr{P} * \mathbf{n} \, dA \right\}, \quad (3.3.59)$$

where $\mathbf{u} * = \mathbf{u}/v$; $\mathscr{P} * = \mathscr{P}/\rho v^2$; $\nabla * = D\nabla$; $\nabla *^2 = D^2 \nabla^2$.

Ahmed and Sunada (1969) have presented a derivation of the Forchheimer equation [Eq. (3.2.17)] by averaging the Navier–Stokes equation and accounting for "turbulent fluctuations." Based on heuristic arguments regarding alleged turbulence effects and the distribution of pressure, the authors arrived at expressions for the permeability and the inertial coefficient in the Forchheimer equation. Comparing Eq. (3.3.59) with Eq. (3.2.17) one can conclude that a sufficient condition for the constancy of the permeability is that velocity $\mathbf{u}$ at any fixed point is proportional to v. This idea was put forward by Heller (1972). The same condition, however, is sufficient for the constancy of β only if the second term in the expression in curly braces of Eq. (3.3.59) is negligible.

3.3.6. Flow Models Based on Flow Around Submerged Objects

The subject of flow around submerged objects at low Reynolds numbers is discussed in great detail in the treatise by Happel and Brenner (1965) who have paid attention also to important applications in flow through porous media. Using the concentric spheres cell model, they have calculated the velocity through the bed compared with the Stokes' law velocity, as a function of porosity. Their treatment is limited to creeping flow where the inertial term in the Navier–Stokes equation may be neglected. Their result provides theoretical support for the validity of Darcy's law but does not attempt to answer the question of how the solids' concentration (i.e., the porosity of the bed affects deviations from Darcy's law at increasing Reynolds numbers. For a very dilute system (i.e., high porosity) these deviations are known to parallel those from Stokes' law. At lower porosities, however, the situation can be expected to be different, because of the fact that the proximity of other spheres influences wake formation downstream from every sphere.

Happel and Brenner show the variation of the Kozeny "constant" with porosity for the case of cell model calculations involving four different geometries: flow parallel to cylinders, flow perpendicular to cylinders, flow through random orientation of cylinders, and flow through assemblages of spheres. In every case they predict a steady increase in the value of the Kozeny constant with porosity, which becomes very rapid as the porosity approaches unity. For 0.99 porosity, the value of the "Kozeny constant" predicted by them ranges from 31.1 in the case of flow parallel to cylinders to 71.63 in the case of flow through assemblages of spheres. The predicted variations of the "Kozeny constant" with porosity in flow relative to circular cylinders due to various authors have also been compared. All theories predict a *steady* increase of Kozeny constant with porosity, which is in contradiction with the experimental results of Rumpf and Gupte (1971) showing a minimum of $k' = 3.4$ at 0.55 porosity. The model of Kyan *et al.* (1970) for fibrous beds also gives a minimum for the Kozeny constant but at a much higher porosity of about 0.87. It is likely that flow-around-submerged-objects models lose their validity increasingly with decreasing porosities.

The simplified treatment due to Brinkman, who reasoned that the force on a particle situated in a pack of particles can be calculated as if the particle were imbedded in a porous mass, has been reviewed and generalized by Lundgren (1972) (see also Scheidegger, 1974). The relationship derived by Brinkman for the permeability is as follows:

$$k = \frac{D_p^2}{72}\left[3 + \frac{4}{1-\phi} - 3\left(\frac{8}{1-\phi}\right)^{1/2} - 3\right]. \qquad (3.3.60)$$

It has been pointed out by Happel and Brenner that for $\phi = 1/3$ this equation gives $k = 0$, which certainly disqualifies it from being used at relatively low porosities.

A detailed discussion of Lundgren's highly mathematical work is outside the scope of the present monograph. He proposed that the Brinkman equation should be modified as

$$k = \frac{D_p^2}{72 M(\phi)}\left[3 + \frac{4}{1-\phi} - 3\left(\frac{8}{1-\phi}\right)^{1/2} - 3\right], \qquad (3.3.61)$$

which differs from the original in the "effective viscosity" $M(\phi)$:

$$M(\phi) = \frac{4\pi}{3}\frac{R^2\alpha^2}{(1-\phi)F(\alpha^2 R^2, \alpha R)}, \qquad (3.3.62)$$

where $R = D_p/2$ and αR is the following function of the porosity:

$$\alpha R = \frac{3}{4}\frac{3 - [8/(1-\phi) - 3]^{1/2}}{[1/(1-\phi)] - \frac{3}{2}}, \qquad (3.3.63)$$

and $F(\alpha^2 R^2, \alpha R)$ is a complicated expression, involving Bessel functions and Legendre polynomials.

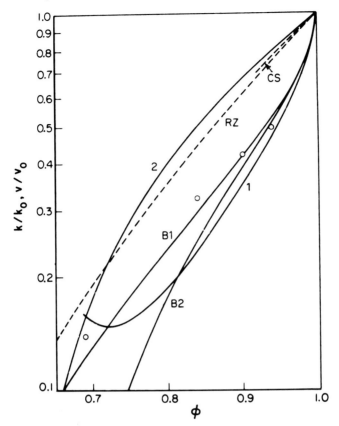

Figure 3.23.   Reduced permeability k/k_0 or settling velocity v/v_0 in a random bed of spheres versus void fraction ϕ: B1, Brinkman (1949); B2, modification of the Brinkman formula due to Lundgren (1972); 1, Eq. (3.3.61); 2, settling velocity calculated by Eq. (5.40) of Lundgren; ●, experiments of Happel and Epstein (1954); RZ, sedimentation experiments of Richardson and Zaki (1954); CS, ultracentrifuge experiments of Cheng and Schachman (1955). (Reprinted with permission from Lundgren, 1972 © Cambridge University Press.)

The interesting aspect of the "effective viscosity" obtained by Lundgren is that it is not always greater than 1 but in fact it decreases rapidly for porosities less than 0.7. The various permeability predictions together with the data of Happel and Epstein (1954) are shown in Fig. 3.23. In the figure k_0 is the permeability of the infinitely dilute bed (i.e., $\phi \to 1$) calculated from the Stokes drag $[k_0 = 2D_p^2/36(1 - \phi)]$. It is evident that the Brinkman relationship as modified by Lundgren shows a behavior below porosities of about 0.7 that differs radically from other predictions. Lundgren thinks this is due to the approximation made by him in which the inpenetrability of the particles was neglected. This criticism appears to be in order but, on the other hand, the change in trend introduced by Lundgren's modification appears to be correct, and further improvements along these lines may be worth pursuing.

The Navier–Stokes equation or, if the inertia term is neglected, the Stokes equation, may be solved also for the case of flow around objects arranged in a regular array (see Section 3.3.5.1) if certain simplifying assumptions are made (e.g., Philip, 1972; Brenner, 1979).

3.4. FLOW OF GASES AND DIFFUSION IN POROUS MEDIA

Flow of gases through porous media is treated in the literature mostly together with pore diffusion. This field has an extensive literature which has been reviewed, among others, in the books of Satterfield (1970), Petersen (1965), and Geankoplis (1972) and in the review by Youngquist (1970).

Flow of gases in porous media has certain special characteristics that distinguish it from the flow of liquids.

3.4.1. "Slip" and Compressible Flow

An important difference between flow of liquids and gases in porous media is that in the latter case the velocity at the solid walls cannot, in general, be considered zero, but a "slip" or "drift" velocity at the wall must be taken into account. This effect becomes significant when the mean-free path of the gas molecules is of magnitude comparable to the pore size. When the mean-free path is much less than the pore size, the slip velocity becomes negligibly small. As in liquids the mean-free path of molecules is of the order of the molecular diameter, so the no-slip condition always applies in liquid flow.

The physical meaning of slip velocity is easily understood if only the mechanism of momentum transport in a gas is considered. Gas molecules travel a certain distance between two consecutive collisions and also between the *last* collision with another gas molecule and the wall collision. "Wall velocity" means the average flow velocity of the molecules at a distance equal to the mean-free path from the wall. In gas flow through a capillary, the molecules at a distance equal to the mean-free path from the wall have, on the average, a nonzero velocity in the direction of flow. As the mean-free path becomes an increasingly greater fraction of the capillary diameter, the "wall velocity" increases in significance relative to the average velocity. The elementary quantitative treatment of this phenomenon may be found, for example, in the book by Present (1958), and it has also been discussed by Hewitt (1967).

The limiting situation in which the mean-free path of the gas molecules is greater than either the diameter of the capillary or its length is called "molecular streaming" or Knudsen flow. Under these circumstances only collisions between gas molecules and the walls of the tube arise. Therefore, Knudsen flow takes place by diffusive, in contrast to viscous, mechanism. As the mean-free paths of gas molecules are, at room temperatures and pressures, roughly in the range of 0.01 to 0.1 μm, flow in pores under 100 Å diameter is usually of the Knudsen type. Under conditions of high vacuum, molecular streaming assumes significance also relative to macroscopic systems (e.g., Nocilla, 1961).

The Hagen–Poiseuille equation, written for gas flow in a capillary tube, is as follows:

$$G = (D^3\pi/64\mathcal{R}T)\left[\pi\bar{u}_m + (D\bar{P}/2\mu)\right](\Delta P/L), \qquad (3.4.1)$$

where G is molar flow rate, $\bar{u}_m$ is the mean molecular speed ($\bar{u}_m = (8\mathcal{R}T/\pi M)^{1/2}$), $\bar{P} = (P_1 + P_2)/2$ and M is molecular weight. The effect of "slip" in porous media gas flow is expressed by the Klinkenberg equation (Eq. 1.1.7).

Thus, a plot of molar flow rate divided by the pressure gradient versus $\bar{P}$ is expected to give a straight line. In fact, at low enough values of the mean pressure, the straight line starts curving upward and a minimum of the function has been found at about $D/\lambda = 0.4$ (λ = mean-free path) (Knudsen, 1909).

At $\bar{P} = 0$ the experimental curve intercepts the ordinate axis at the specific flow rate corresponding to pure Knudsen flow. The transition region has been analyzed by Weber (1954) and, independently, by Scott and Dullien (1962a), in terns of three flow components: a Poiseuille term, a slip term, and a Knudsen flow term. In the treatment of Scott and Dullien the molecules were divided into two groups: those that on the average do not collide with other molecules between two consecutive wall collisions and those that do. The fractional number of molecules in each group was estimated on the basis of collision probability considerations. The two flows were calculated and were assumed to be additive. The flow of the molecules in the first group was calculated by applying the known expression for molecular streaming, of Knudsen flow, and those in the second group by using Eq. (3.4.1). All parameters were calculated from the kinetic gas theory (i.e., no adjustable parameters were used):

$$\frac{G}{\Delta P/L} = \left[1 - e^{-\sinh^{-1}\left(\frac{2R}{\lambda}\right)}\right]\left(\frac{\pi R^3}{M\bar{u}_m} + \frac{\pi R^4\bar{P}}{8\mu\mathcal{R}T}\right)$$

$$+ e^{-\sinh^{-1}\left(\frac{2R}{\lambda}\right)}\frac{16R^3}{3M\bar{u}_m} \qquad (3.4.2)$$

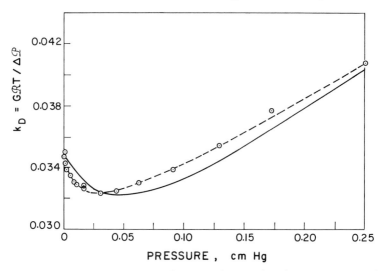

Figure 3.24. Comparison of calculated (solid curve) by Eq. (3.4.2) and experimental (dashed curve) flow rates for carbon dioxide in a glass capillary, low pressure range. The experimental points are from Knudsen. (Reprinted with permission from Scott and Dullien, 1962.)

As it can be seen in Fig. 3.24, where $G\mathscr{R}T/\Delta P$ has been plotted versus $\bar{P}$, Eq. (3.4.2) has been able to predict Knudsen's results very closely.

Later, Simons (1967) derived a general solution of the Boltzmann equation for Poiseuille gas flow in the transition region. However, in order to obtain numerical results, considerable simplifications were required. In the end, he obtained numerical answers that are no closer to experiment than those of the simple Scott–Dullien treatment, which has been applied to porous media by Wakao, Otani, and Smith (1965), and Otani, Wakao, and Smith (1965). Whether the minimum-type behavior in the transition region discussed above also exists in the case of flow through porous media is a controversial issue (Wicke and Vollmer, 1952). Grove and Ford (1958) have reported finding the minimum-type behavior in some samples and not in others (see Fig. 3.25). They have suggested that the pore size distribution may be the factor that determined whether or not a minimum was observed. This is a reasonable explanation, as for the case of fairly uniform pore size the medium can be expected to behave like a bundle of capillary tubes of the same diameter, whereas for increasingly nonuniform pore sizes the conditions responsible for the occurrence of the minimum are different for the different flow channels, resulting in blurring and, eventually, disappearance of the minimum.

The problem of calculating gas flow in the transition region in a model of a porous medium consisting of randomly distributed spheres in space was

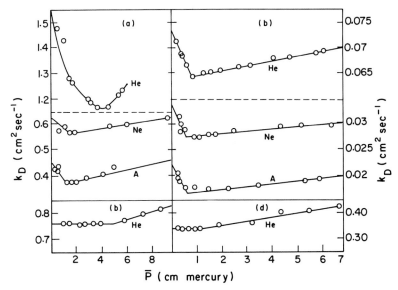

Figure 3.25. Permeability coefficients for flow of inert gases of 25°C through (a) porous ceramics and (b) graphites. (Reprinted with permission from Grove and Ford, 1958.)

approached by solving the Boltzmann equation by Derjagin and Bakanov (1957). Their treatment also showed the existence of a minimum. The model used by these authors is often called the "dusty gas model."

It is to be noted that in gas permeametry a "permeability" k_D, defined by the equation

$$vP = k_D(P_1 - P_2)/L, \qquad (3.4.3)$$

is often used, where k_D has the same dimensions [area/time] as the diffusion coefficient. For example, in Eq. (3.4.1) we have $(D/16)[\pi \bar{u}_m + (D\bar{P}/2\mu)] = k_D$.

The "time lag" is an important parameter that can be used to calculate diffusion coefficients of gases. Ash *et al.* (1963) published a theoretical analysis of transient slip flow, leading to a time lag in long capillary tubes with or without adsorption.

Goodknight and Fatt (1961) investigated the effect of dead-end pore volume on time lag. It was established by them that the time lag is influenced only by the total pore volume (flow channels plus dead-end pores), but the time required to reach the asymptotic conditions increases with increasing resistance between the flow channels and the dead-end pores.

3.4.2.   Gas Diffusion

Diffusion of gases in porous media has traditionally been discussed along with gas flow. Therefore, it is discussed here, even though it involves diffusional mixing just like "hydrodynamic dispersion," which is the subject of Chapter 6.

One-dimensional diffusion of two components, A and B, is described by the following phenomenological equation (e.g., deGroot and Mazur, 1969):

$$N_A = \mathscr{D}_{AB}(dC_A/dx) + y_A(N_A + N_B), \qquad (3.4.4)$$

where C_A is concentration of A in moles per unit volume, $\mathscr{D}_{AB}$ is the mutual diffusion coefficient, and N_A and N_B are the molar fluxes of A and B. y_A is the mole fraction of A. Equation (3.4.4) can be integrated for the steady-state case at constant pressure P to give

$$N_A = (\mathscr{D}_{AB}P/\mathscr{R}TL\alpha) \ln[(1 - \alpha y_{AL})/(1 - \alpha y_{A0})], \qquad (3.4.5)$$

where

$$\alpha \equiv 1 + (N_B/N_A) \qquad (3.4.6)$$

with y_{A0} and y_{AL} the mole fractions of A at $x = 0$ and $x = L$, respectively. In an important fundamental contribution, Hoogschagen (1955) showed experimentally that at constant total pressure $-N_B/N_A = (M_A/M_B)^{1/2}$. He thereby extended this relationship, which was known to hold only to Knudsen flow, to ordinary diffusion of gases in porous media.

Often ordinary diffusion and Knudsen flow (also called "Knudsen diffusion") take place simultaneously in porous media. In the case of mixed ordinary and Knudsen diffusion, the two mechanisms must be properly averaged to obtain an accurate expression for the net flux. This has been done by Evans *et al.* (1961) using the "dusty gas" model, Scott and Dullien (1962b), Rothfeld (1963), and Hewitt and Sharratt (1963) in terms of momentum transfer to the walls of the capillary, and by Spiegler (1966) using friction factor analysis, with the following result:

$$N_A = \frac{L}{[(1 - \alpha y_A)/\mathscr{D}_{AB}] + [(1/\mathscr{D}_{KA})]} \frac{dC_A}{dx} \qquad (3.4.7)$$

or, after integrating at constant temperature and pressure,

$$N_A = \frac{\mathscr{D}_{AB}P}{\mathscr{R}TL\alpha} \ln\left[\frac{1 - \alpha y_{AL} + (\mathscr{D}_{AB}/\mathscr{D}_{KA})}{1 - \alpha y_{A0} + (\mathscr{D}_{AB}/\mathscr{D}_{KA})}\right]. \qquad (3.4.8)$$

In the case of diffusion in a porous medium with a distribution of pore sizes, the simplifying assumption is usually made that Eq. (3.4.8) is valid in the individual pores. The calculation of the diffusion flux in a porous medium is performed with the aid of various different pore structure models in much the same way as is done in the case of fluid flow in a porous medium. The models used to predict pore diffusion include the parallel-type capillary model and the cut-and-random-rejoin model. The capillary model consisting of a bundle of capillaries of different diameters results in the following expression for N_{AS}, the diffusion flux based on the cross section of the sample (solids and pores):

$$N_{AS} = (1/\Upsilon) \int_0^\infty N_A(R) f(R) \, dR, \qquad (3.4.9)$$

where $f(R) \, dR$ is a volume-based pore size distribution normalized such that

$$\int_0^\infty f(R) \, dR = \phi. \qquad (3.4.10)$$

By this definition, $f(R) \, dR$ represents the contribution to the pore volume of pore segments of size between R and $R + dR$, *regardless of their position occupied in a pore network*. In Eq. (3.4.9), however, it is assumed that all pore segments *of the same R* are arranged *in series* to form uniform tubes of the same length, which are arranged *in parallel* with each other *and* with the rest of the tube *of different R*. It is noted that $N_A(R)$ in Eq. (3.4.9) is a function of the pore size R through the Knudsen diffusion coefficient $\mathcal{D}_{KA}$, which is given by the following formula:

$$\mathcal{D}_{KA} = 2/3R(8\mathcal{R}T/\pi M_A)^{1/2}. \qquad (3.4.11)$$

Effects of pore constructions (i.e., serial-type pore uniformities) may cause severe deviations from Eq. (3.4.9) as was pointed out by Petersen (1958), Michaels (1959), and Dullien (1975b), among others, who have introduced a "constriction factor" to allow for this effect.

Haynes and Brown (1971) analyzed theoretically the effect of pressure in the transition region on the departure from the parallel pore model in the case of pore diffusion though constricted pores.

The cut-and-random-rejoin model was introduced to pore diffusion by Wakao and Smith (1962) who applied it to the special case of bimodal pore size distribution such as often exists in catalyst pellets. Cunningham and Geankoplis (1968) extended this approach to trimodal pore size distributions. The cut-and-random-rejoin model does not contain an adjustable tortuosity factor Υ; however, it implies that $\Upsilon = 1/\phi$, which is not generally true.

The pore structure model proposed by Dullien (1975a, 1975b) appears to represent a step in the right direction because it takes both parallel- and serial-type pore uniformities into account. This model is presented in Section 3.7 "Formation Resistivity Factors," along with the definition of the "electrical tortuosity factor" X.

Experimental data are commonly reduced to effective diffusivities $\mathscr{D}_{\text{eff}}$, defined as

$$\mathscr{D}_{\text{eff}} = N_{\text{AS}}\mathscr{R}TL/P(y_{\text{AL}} - y_{\text{A0}}), \tag{3.4.12}$$

and, to facilitate comparison, the predictions of the various models are also brought in this form. It is shown in Section 6.3.4.1 that, in the case of unsteady-state pore diffusion experiments, $X = \mathscr{D}/\mathscr{D}_{\text{eff}}$.

3.4.3.   Combined Diffusion and Flow

In the presence of both a concentration and a pressure gradient both diffusion and flow contribute to the flux. It has been shown by Evans *et al.* (1961) that Eq. (3.4.8) still applies under these conditions. The total molar flux N_{T} is written as follows:

$$N_{\text{T}} = N_{\text{A}} + N_{\text{B}} + F, \tag{3.4.13}$$

where N_{A} and N_{B} are diffusive fluxes, and F is (Scott and Dullien, 1962a)

$$F = \left[\frac{R^2 P}{8\mu} + \frac{R\mathscr{R}T}{M\bar{u}_{\text{m}}} \right] \frac{1}{\mathscr{R}T} \frac{dP}{dx} \tag{3.4.14}$$

with (Wakao *et al.* 1965) the following expression of $M\bar{u}_{\text{m}}$ for a binary gas:

$$M\bar{u}_{\text{m}} = M_A\bar{u}_{\text{MA}}y_{\text{A}} + M_B\bar{u}_{\text{mB}}y_{\text{B}}. \tag{3.4.15}$$

The total molar flux N_{T} is found by simultaneous numerical integration of the various flux equations (Wakao *et al.*, 1965; Otani *et al.*, 1965).

An interesting effect, usually called the "Kirkendall effect," is observed if two closed vessels, containing two different gases (different molecular weights), but at the same initial pressure and temperature, are connected through a capillary or a porous medium ("membrane"). The molecules of the lighter gas will diffuse relatively more rapidly across the "membrane." This gives rise to a pressure difference between the two sides of the membrane, which causes a compensating bulk flow (e.g., Suetin and Volobuev, 1964, 1965; McCarty and Mason, 1960; Miller and Carman, 1960, 1961).

Allawi and Gun (1987) have derived an equation for total diffusional *and* viscous flux of A:

$$(N_A)_T = \frac{(1-\alpha)\mathscr{D}_{AB}k\bar{P}^2}{\mathscr{R}TL(\mathscr{D}_{KB} - \mathscr{D}_{KA})} \left[\frac{\mathscr{D}_{KB}/\mu_A - \mathscr{D}_{KA}/\mu_B}{\mathscr{D}_{KA} + k\bar{P}/\mu_A - \alpha(\mathscr{D}_{KA} + k\bar{P}/\mu_B)} \right]$$

$$\cdot \ln \left[\frac{(\mathscr{D}_{KB} - \mathscr{D}_{KA})y_{AL} + \mathscr{D}_{KA} + \mathscr{D}_{AB}}{(\mathscr{D}_{KB} - \mathscr{D}_{KA})y_{A0} + \mathscr{D}_{KA} + \mathscr{D}_{AB}} \right]$$

$$+ \frac{(1-\alpha)(\mathscr{D}_{KA} + k\bar{P}/\mu_A)(\mathscr{D}_{KB} + k\bar{P}/\mu_B)(P_1 - P_2)}{\mathscr{R}TL\left[(1-\alpha)/(\mathscr{D}_{KB} + k\bar{P}/\mu_B) - (\mathscr{D}_{KA} - k\bar{P}/\mu_A) \right]} \quad (3.4.16)$$

Equation (3.4.16) has been tested in diffusion and flow experiments with good results. Among other things, "electrical" (or diffusional) tortuosity factors X of catalyst pellets have also been calculated from the measured data. The values of X ranged from about 5 to about 9.

For isobaric diffusion, $P_1 - P_2 = 0$, $1 - \alpha = \mathscr{D}_{KA}/\mathscr{D}_{KB}$ and Eq. (3.4.16) reduces to Eq. (3.4.8).

3.4.4. Surface Flow

Gases tend to adsorb on the pore surface and the apparent "permeabilities" measured with highly adsorbing gases greatly exceed those determined with nonadsorbing gases or with liquids. The excess permeability has been attributed to surface flow or surface diffusion in the absorbed phase. This subject has been under intensive study and its literature is big enough to warrant devoting a whole chapter or even a monograph to it. It has been reviewed, for example, by Dacey (1965) and by Barrer (1967), whose presentation is followed here.

According to Barrer, in thick films and for capillary condensate, the rate of movement of sorbed fluid should be related to the gradient of capillary pressure of "hydrostatic stress intensity." The main resistance to movement will be determined by viscosity in the fluid and by frictional drag between the fluid and the stationary substrate along which it moves. Thus, the transport can be considered in macroscopic hydrodynamic terms. When the adsorbed film is no more than a monolayer in thickness, it has been customary to apply the concept of two-dimensional surface or spreading pressure as defined in treatments of the thermodynamics of interfaces. For monolayers and more dilute adsorbed films, Barrer and Gabor (1960) proposed the following picture of "surface" migration. Activated surface hops of absorbed molecules for distances of a few angstroms alternate with evaporative flights across crevices over longer and variable distances. A surface-pressure concept has been used by Babbitt (1950), Gilliland *et al.* (1958), and Ash *et al.* (1963) for correlating surface-flow measurements. More recent contributions in this area include Roybal and Sandler (1972), Horiguchi *et al.* (1971), etc.

Flood (1958) and Flood *et al.* (1952) formulated surface flow in terms of the gradient dP'/dx, where P' is sometimes referred to as the "hydrostatic stress intensity" in the absorbed phase. One can visualize capillary conden-sate to be present in relatively small pores or in "surface pores" (i.e., capillary channels or crevices on the pore surface). The condition of thermo-dynamic equilibrium in the coexisting capillary condensate phase (') and the vapor phase (") (e.g., Defay *et al.*, 1966) is

$$\mu'' = \mu', \tag{3.4.17}$$

where μ'' and μ' are the chemical potentials.

Between two neighboring equilibrium states this results in the condition

$$d\mu'' = d\mu'. \tag{3.4.18}$$

Hence, by using the isothermal Gibbs–Duhem equations

$$-V'\,dP' + d\mu' = 0, \qquad -V''\,dP'' + d\mu'' = 0, \tag{3.4.19}$$

we obtain

$$dP'/dP'' = V''/V', \tag{3.4.20}$$

where V'' and V' are the molar volumes in the two equilibrium phases.

In a good approximation

$$V'' = \mathscr{R}T/P''. \tag{3.4.21}$$

Hence we can write Eq. (3.4.20) as follows:

$$dP'/dP'' = \mathscr{R}T/P''V'. \tag{3.4.22}$$

It is evident from this equation that at low values of P'' the pressure drop in capillary condensate phase can be very much greater than in the equilib-rium vapor phase. If the pressure in the gas phase is maintained at P_1'' and P_2'' at the two ends of the sample, then the corresponding pressure drop $\Delta P'$ in the capillary condensate phase is

$$\Delta P' \equiv P_2' - P_1' = (\mathscr{R}T/V')\ln(P_2''/P_1''). \tag{3.4.23}$$

This equation presupposes the physical existence of a capillary condensate phase, which is subject to the condition

$$P'' - P' = [2\sigma\cos\theta]/R. \tag{2.3.1}$$

Hence,

$$dP'' - dP' = d(2\sigma\cos\theta/R) \approx -dP'. \tag{3.4.24}$$

Combining this result with Eq. (3.4.23) gives

$$P'_1 - P'_2 = \Delta(2\sigma \cos\theta/R) = -(\mathscr{R}T/V')\ln(P''_2/P''_1). \quad (3.4.25)$$

In the physical situation considered, the highest possible value of P''_1 is P_0, the vapor pressure of the bulk liquid, when $2\sigma \cos\theta/R = 0$.

The value of R rapidly becomes very small at decreasing values of P''/P_0 because of the smallness of the coefficient $2\sigma \cos\theta(V'/\mathscr{R}T)$. The macroscopic thermodynamic formulas cannot be depended upon when R approaches molecular dimensions. For higher values of R, however, when the value of P''/P_0 is not less than about 0.2–0.4, Eq. (3.4.25) can explain the great increase in the observed "permeability" observed with adsorbable gases.

The mechanism consisting of a combination of surface hops and evaporative flights proposed first by Barrer and coworkers is, in the opinion of this author, in need of a rigorous thermodynamical justification. Considering the nature of the process, it appears to be amenable to treatment by the techniques of the theory of fluctuations.

Barrer (1967) has gone to considerable lengths to emphasize the formal analogy between Fick's law of diffusion and Darcy's law, regardless of the fundamental differences between diffusive and viscous mechanisms of mass transport.

Surface (film) flow of a wetting liquid under the influence of capillary forces has recently become of importance in immiscible displacement processes and it is discussed also in Chapter 5.

3.5. NON-NEWTONIAN FLOW IN POROUS MEDIA

This important field has been reviewed by Savins (1970) very thoroughly (185 references). In the present review only a short discussion is given of some of the aspects of this field.

The interpretation of many observed effects is complicated by the occurrence of spurious phenomena such as adsorption of the polymers on the capillary walls and partial pore blockage. Often some authors interpret the observations in terms of partial pore blockage, whereas others emphasize non-Newtonian effects as the explanation. There is little doubt that under certain conditions partial pore blockage does occur, but the mechanism of this phenomenon is also controversial.

In the absence of spurious effects the flow behavior in bead packs is described by a generalization of the Carman–Kozeny treatment (see Section 3.3.1.1) as shown in Fig. 3.26. The functions $F(\tau_w)$ and $F(\tau_\phi)$ are defined as follows:

$$F(\tau_w) = \frac{2\bar{u}}{D} = \frac{1}{\tau_w^3}\int_0^{\tau_w} \tau^2 \Gamma(\tau)\, d\tau \quad (3.5.1)$$

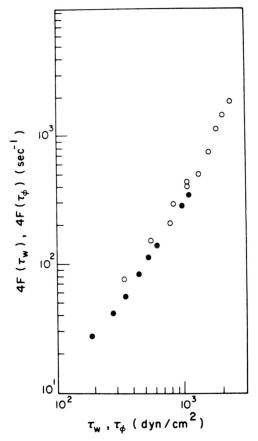

Figure 3.26. Flow behavior of 2% carboxymethylcellulose solution in a bead pack (●), with $k = 445$ darcy, $\phi = 0.37$, and $C' = 2.5$, Eqs. (3.5.2) and (3.5.4) compared with capillary flow behavior (○) Eqs. (3.5.1) and (3.5.3) (Savins, 1970).

and

$$F(\tau_\phi) \equiv \frac{2v}{\phi D_{\mathrm{H}}} = \frac{1}{\tau_\phi^3} \int_0^{\tau_\phi} \tau^2 \Gamma(\tau)\, d\tau, \qquad (3.5.2)$$

where D is capillary diameter, D_{H} the hydraulic diameter of the pack, Γ the shear rate, τ_{w} the wall shear stress in the capillary:

$$\tau_{\mathrm{w}} = (D/4)(\Delta \mathscr{P}/L), \qquad (3.5.3)$$

and τ_ϕ the corresponding quantity in the pack defined as

$$\tau_\phi \equiv (D_{\mathrm{H}}/4C')(\Delta \mathscr{P}/L), \qquad (3.5.4)$$

where C' is the "shift factor" used to bring the curve obtained for the porous medium in line with the capillary flow data. It is easy to show that for Newtonian flow the function $4F(\tau_w)$ versus τ_w simplifies to the straight line relationship

$$4F(\tau_w) = \tau_w/\mu, \qquad (3.5.5)$$

and the relationship $4F(\tau_\phi)$ versus τ_ϕ becomes

$$v = (\phi D_H^2/32C'\mu)(\Delta\mathscr{P}/L), \qquad (3.5.6)$$

which shows that C' is identical with the "tortuosity factor" Υ. For non-Newtonian fluids, both functions deviate from the straight line behavior. Since the treatment was based on a comparison with cylindrical capillaries, the value of the shape factor in the Carman–Kozeny equation, k_0, is equal to 2 in this case. Two different values for the "shift factor" C' have been used. One is the value 2.5, corresponding to the customary value of 5 of the Kozeny constant k' and to 180 in the corresponding friction factor versus Reynolds number relationship. The other value used for C' is 25/12, which corresponds to the constant of 150 in the Blake–Kozeny equation.

In many cases, deviations from the simple behavior shown in Fig. 3.26 have been reported. In the majority of the cases, the deviations indicate real or apparent shear thickening behavior in packed beds when no such behavior could be found in capillary flow experiments performed with the same solutions. As example is shown in Fig. 3.27 where the "Darcy viscosity" defined as

$$\eta_\phi \equiv \frac{\tau_\phi}{4F(\tau_\phi)} = \frac{\phi}{32C'} \frac{D_H^2}{v} \frac{\Delta\mathscr{P}}{L} \qquad (3.5.7)$$

increases sharply with increasing shear rate, whereas the apparent viscosity in Poiseuille flow η, defined analogously as

$$\eta \equiv \tau_w/4F(\tau_w), \qquad (3.5.8)$$

is either constant or actually decreases. The fact that with the higher molecular weight polymer two distinct curves were obtained for the two different packs is interesting because it indicates the effect of pore structure on the observed phenomenon.

Another way of representing the data is in the form of friction factor versus Reynolds number plots. An example is shown in Fig. 3.28 where the friction factor f_ϕ is defined as

$$f_\phi = (\beta\,\Delta\mathscr{P}\phi^2/\rho v^2 L), \qquad (3.5.9)$$

where β is one of the two parameters of a second-order memory fluid and

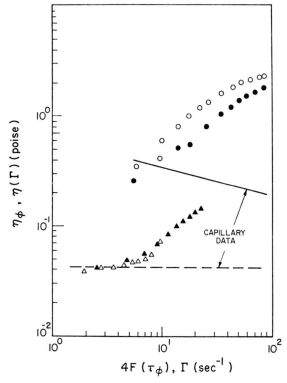

Figure 3.27.   Flow behavior of polyethylene solutions in bead packs compared with capillary flow experiments with $C' = 2.5$ (Savins, 1970).

| | Packed bead data | | Polymer | |
Symbol	k(darcy)	ϕ	Conc. (wt.%)	Mol. wt. ($\times 10^{-6}$)
○	16.03	0.364	0.199	4.0
●	18.0	0.347	0.199	4.0
△	16.0	0.365	0.585	0.2
▲	2.43	0.381	0.585	0.2

the Reynolds number N'_{Re} is defined such that

$$f_\phi N'_{Re} = 1, \qquad (3.5.10)$$

giving

$$N'_{Re} = D_p v^{2-n} \rho / 72 C' (1 - \phi) \psi, \qquad (3.5.11)$$

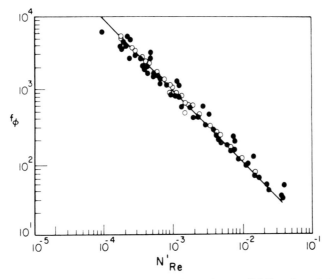

Figure 3.28. Reynolds number versus friction factor for parallel flow through bead packs, using the data of Sadowski (1963). The values shown are for variable polyvinyl alcohol and polyethylene glycol concentrations (○) and variable hydroxyethylcellulose concentrations and molecular weight grades (●) (Savins, 1970).

where ψ is the "viscosity level parameter" in the equation

$$v^n = (k\,\Delta\mathscr{P}/\psi L).\qquad(3.5.12)$$

It is evident from Fig. 3.28 that Eq. (3.5.10) is obeyed for the systems covered in this case. There are exceptions, however, such as those shown in Fig. 3.29 where the solutions used show either constant viscosity or shear thinning behavior in viscometric experiments. It needs emphasizing that this kind of real or apparent shear thickening behavior is quite common, and the interested reader is referred to the original literature for more details.

A disturbing fact is the observation made by several researchers [e.g., Sadowski (1963), Gogarty (1967), and Ershaghi (1972)] that reproducible results could be obtained only in constant flow rate experiments because in constant pressure drop runs, the velocity kept decreasing. This kind of observation indicates deposition of polymers on the solid surface by one mechanism or another. At constant flow rate the increased pressure drop seems to provide the necessary force to keep a reproducible portion of the flow channels open.

Some researchers who considered the role of progressive plugging and adsorption of polymer aggregates concluded that these aberrations were absent from their experiments (Marshall and Metzner, 1967; Dauben and

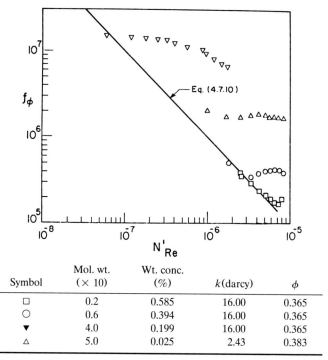

Symbol	Mol. wt. ($\times$ 10)	Wt. conc. (%)	k(darcy)	ϕ
□	0.2	0.585	16.00	0.365
○	0.6	0.394	16.00	0.365
▼	4.0	0.199	16.00	0.365
△	5.0	0.025	2.43	0.383

Figure 3.29.   Reynolds number versus friction factor for non-Newtonian flow through bead packs for polyethylene oxide solutions (Savins, 1970).

Menzie, 1967). They explained the observed shear thickening by postulating viscoelastic effects. Others (e.g., Gogarty, 1967) have postulated retention by the solid surface. According to Gogarty, permeability is reduced because of selective pore blockage of the smaller pore openings. He appeared to regard the retention of polymer units by the porous medium to be similar to the mechanism whereby sieves retain solid particles.

Szabo (1975a, b) and Dominguez and Willhite (1977) have show that under dynamic conditions a "mechanical entrapment" or "hydrodynamic retention" of polymers from partially hydrolyzed polyacrylamide solutions occurred in sandstone as well as synthetic porous Teflon plugs. For a given set of conditions the retention seems to level off at a constant value, which suggests a mechanism different from ordinary filtration of "plugging." The observation that flushing the sample with brine does not result in elution of the entrapped polymer indicates that the entrapped phase does not diffuse into the following solvent. The fact that retention was found to *increase* with increasing flow rates and to *decrease* with decreasing flow rates (Dominguez and Willhite, 1977) shows a certain reversibility in the entrapment mechanism with respect to the flow rate. There is no evidence to indicate, however, that hydrodynamic retention would be completely reversible if the flow rate were made vanishingly small.

Hydrodynamic retention possibly depends on flow through quasi-periodically constricted channels, where the polymer builds up immediately upstream from the pore necks because highly branched polymer molecules get attached to the capillary surface and then capture, by entanglement, other polymer molecules from the flowing solution. This build-up is counteracted by backdiffusion into the stream against the pressure exerted by the flowing field.

3.6. ANISOTROPIC PERMEABILITY

This field has been reviewed by Rice *et al.* (1970).

Anisotropy is the dependence on the direction of measurement of the magnitude of certain properties of a substance. In the case of crystals and fluids anisotropy is an intrinsic property.

It is well known that porous media are often anisotropic, as reflected in different values of the permeability and the formation factor, depending on the direction of measurement. Sedimentary porous media, such as sandstones, have a layered structure and the permeability parallel to the layers is mostly greater than in the perpendicular direction. The anisotropy of sandstones has been investigated in detail by Greenkorn *et al.* (1964). Intrinsic anisotropy is caused by grain orientation, whereas a great deal of anisotropy of another type is due to the presence of layers of low permeability.

As a result of anisotropy, at a given point in the medium the direction of the pressure gradient vector is, in general, different from that of the filter velocity vector. Therefore, for a complete description of the flow phenomenon it is necessary to specify both the pressure gradient and the filter velocity vector fields. Let us suppose, for example, that the medium has an arbitrary orientation with respect to the coordinate system and let the pressure gradient point in the x-coordinate direction. In the general case of an anisotropic medium, there will result three different flow rates in each of the x, y, and z directions, whereas in an isotropic medium the flow is in the x direction. It was apparently derived first by Ferrandon (1948) that for anisotropic media Darcy's law should be written as follows:

$$v_i = -\frac{1}{\mu}\left(k_{i1}\frac{\partial \mathscr{P}}{\partial x_1} + k_{i2}\frac{\partial \mathscr{P}}{\partial x_2} + k_{i3}\frac{\partial \mathscr{P}}{\partial x_3}\right) \qquad (i = 1, 2, 3), \quad (3.6.1a)$$

where 1, 2, and 3 represent the x, y, and z coordinates and k_{ij} form the elements of a second-order tensor, the values of which depend on the orientation of the medium with respect to the coordinate system.

Alternatively, in vector notation, Eq. (3.6.1a) becomes

$$\mathbf{v} = -\left(\bar{\bar{k}}/\mu\right)\nabla\mathscr{P}. \tag{3.6.1b}$$

A great deal of effort has been expended on proving that the permeability tensor $\bar{\bar{k}}$ is symmetric (Whitaker, 1969; Guin et al., 1971b; Szabo, 1968; Case and Cochran, 1972), that is,

$$\bar{\bar{k}} = \begin{pmatrix} k_{11} & k_{12} & k_{13} \\ k_{12} & k_{22} & k_{23} \\ k_{13} & k_{23} & k_{33} \end{pmatrix}. \tag{3.6.2}$$

Usually it has been assumed that anisotropic porous media are "ortho-tropic" (i.e., they have three mutually orthogonal principal axes). For or-thotropic media the tensor is symmetric, and rotation of the coordinate system will produce a diagonal matrix when the three coordinate axes are aligned with the principal axes of the medium. For this particular orientation of the medium, the pressure gradient and the velocity have the same direction and, therefore, in this case Darcy's law becomes

$$v_i = -(k_i/\mu)(\partial\mathscr{P}/\partial x_i) \qquad (i = 1, 2, 3), \tag{3.6.3}$$

where the three different values of k_i are still, in general, not equal.

In the usual simple laboratory measurements, so-called directional perme-abilities rather than the components k_{ij} of the permeability tensor are measured. Two kinds of simple permeability measurements have been per-formed on anisotropic sandstones (Johnson and Hughes, 1948; Johnson and Breston, 1951). One of the methods used consists of cutting plugs at various angles from well core of fluid-bearing strata. The permeabilities of the plugs are then measured in an ordinary permeameter. In the second method, a hole was drilled down the center of a cylindrical piece of a well core whose faces had been parallel. The equipment used consisted of a system of clamps, mounted with bearings that allowed the hollow porous cylinder to be rotated to any position for flow measurement while fluid was continuously being flowed from the center to the outside of the cylinder. A collecting head clamped to the cylinder was used to collect the fluid which flowed from that portion of the cylinder. The directional permeability was then calculated according to Darcy's law from the volume of fluid flowing in the given direction and collected by the head.

Scheidegger (1954) investigated the problem of relating the directional premeabilities measured by Johnston et al. to the components k_{ij} of the permeability tensor. He distinguished two mathematically equally possible cases.

Case 1: The filter velocity is measured directly and the component of the pressure gradient in the direction of the filter velocity must be used in Darcy's law.

Case 2: The component of the filter velocity taken in the direction of the pressure gradient is measured directly and used in Darcy's law.

In the first case, letting the direction of the filter velocity be given by the unit vector $\mathbf{n}$, we have for the component $\nabla\mathscr{P}_n$ of the pressure gradient parallel to $\mathbf{n}$

$$\nabla\mathscr{P} = \mathbf{n} \cdot \nabla\mathscr{P} = -\mu\mathbf{n} \cdot \overline{\overline{k}}^{-1} \cdot \mathbf{n}v, \qquad (3.6.4)$$

where Eq. (3.6.1b) has been used and $\overline{\overline{k}}^{-1}$ is the inverse permeability tensor. The directional permeability k'_n is defined by the expression

$$v = -(k'_n/\mu)\nabla\mathscr{P}_n. \qquad (3.6.5)$$

Combining Eqs. (3.6.4) and (3.6.5) gives the result

$$k'_n = 1/\left(\mathbf{n} \cdot \overline{\overline{k}}^{-1} \cdot \mathbf{n}\right). \qquad (3.6.6)$$

If the principal axes of the permeability tensor are chosen as coordinates axes, the permeability tensor becomes diagonal. Denoting the "principal" permeabilities as k_1, k_2, and k_3 and the angles enclosed by $\mathbf{n}$ with the principal axes as α, β, and γ, Eq. (3.6.6) becomes

$$\frac{1}{k'_n} = \frac{\cos^2\alpha}{k_1} + \frac{\cos^2\beta}{k_2} + \frac{\cos^2\gamma}{k_3}. \qquad (3.6.7)$$

If $r = (k'_n)^{1/2}$ is measured from the origin and plotted corresponding to the various directions of $\mathbf{n}$, Eq. (3.6.7) predicts that an ellipsoid must be obtained.

In the second case, however, Darcy's law is applied to the pressure gradient, whose direction is given by the unit vector $\mathbf{n}$ and the velocity component v_n taken in the same direction:

$$v_n = (k''_n/\mu)|\nabla\mathscr{P}|, \qquad (3.6.8)$$

where k''_n is the directional permeability defined by Eq. (3.6.8) and $|\nabla\mathscr{P}|$ the absolute value of $\nabla\mathscr{P}$.

By Eq. (3.6.1b) we can write

$$v_n = \mathbf{n} \cdot \mathbf{v} = (1/\mu)(\mathbf{n} \cdot \overline{\overline{k}}) \cdot (|\nabla\mathscr{P}|\mathbf{n}). \qquad (3.6.9)$$

Comparison of Eqs. (3.6.8) and (3.6.9) yields the result

$$k''_n = \mathbf{n} \cdot \overline{\overline{k}} \cdot \mathbf{n}. \qquad (3.6.10)$$

In his original paper Scheidegger (1954) concluded that $k'_n = k''_n$, but later (Scheidegger, 1956) he corrected this mistake. Indeed, if the principal axes of the permeability tensor are chosen as coordinate axes, a similar treatment as that used in case 1 now results in the following expression:

$$k''_n = k_n \cos^2 \alpha + k_2 \cos^2 \beta + k_3 \cos^2 \gamma \qquad (3.6.11)$$

which is the central equation of an ellipsoid if $r = 1/(k''_n)^{1/2}$ is plotted for various directions of **n**.

Scheidegger (1956) showed that the relative difference between k'_n and k''_n is small in practice.

The fact that k'_n and k''_n are not equal is readily understood on physical grounds if it is considered that in the case of k'_n the direction specified by **n** is parallel to the direction of the filter velocity **v**, whereas in the case of k''_n it is not. It is not surprising that the directional permeability has different values for different orientations of the filter velocity vector.

According to Scheidegger, the method of measuring directional permeabilities on plug-shaped samples yields k'_n (i.e., **n** gives the direction of both the axis of the plug and of the filter velocity, whereas the second method used by Johnson *et al.* gives k''_n).

Interesting measurements have been performed on anisotropic packs by Latini (1967) and Fontugne (1969). Various sand and crushed limestone samples, as well as one mica sample and two activated carbon samples have been investigated. The permeabilities were measured in two directions: parallel to the bedding plane and perpendicular to it. For the sand and limestone samples, the horizontal-to-vertical permeability ratios varied from 0.8 to 1.36. For the mica and the activated carbon samples, however, the permeability ratios were around two. Equivalent particle diameters were calculated by the Carman–Kozeny formula using the vertical permeabilities. These were smaller in every case than the mean diameters obtained from sieve analysis, even for those cases where the permeability ratio was less than one. These differences were particularly dramatic for mica and the activated carbons. In the former case, the Carman–Kozeny formula gave an equivalent diameter of 2 μm compared with 334 μm obtained from sieve analysis. For the activated carbon the corresponding data were about 16 μm compared with 322 μm by sieve analysis. Quantitative interpretation of these data in terms of pore structure has not been attempted.

3.7. RESISTIVITY (FORMATION) FACTOR

An important, clearly defined, and easily measurable property of porous media is the so-called formation factor F defined in Eq. (1.1.10). The formation factor is a dimensionless quantity whose value is always greater than one. It is defined only for porous matrices of negligible electrical conductivity. In the absence of surface conduction its value is uniquely determined by pore geometry. The influence of pore structure on the electrical conductivity (and the diffusion coefficient) may be divided into two contributions: the reduction of the cross section available for conduction and the orientation and topography of the conducting pores.

For isotropic disordered media, the ratio of the cross section available for conduction to the bulk cross section is equal to the bulk porosity ϕ (i.e., F is inversely related to ϕ). In fact, it has been found empirically that F varies more than just in inverse proportion to ϕ. The first such known relationship was suggested by Archie (1942):

$$F = \phi^{-m}, \tag{3.7.1}$$

where m is the "cementation exponent" (m was supposed to have values between 1.3 and 2.5 for various types of rocks). For random suspensions of spheres, cylinders, and sand in aqueous solutions of zinc bromide of approximately the same densities as the particles, De La Rue and Tobias (1959) found Eq. (3.7.1) to hold with $m = 1.5$ with a high degree of precision over a range of ϕ extending from about 0.25 to 0.55. For consolidated specimens, however, Archie's equation could not be generally confirmed.

Maxwell (1881) derived the following expression for uniform nonconducting spheres imbedded in a fluid continuum:

$$F = (3 - \phi)/2\phi. \tag{3.7.2}$$

Wyllie (1957) introduced the following relationship:

$$F \equiv X/\phi, \tag{3.7.3a}$$

where X is "electric tortuosity" (Schopper, 1966). Equation (3.7.3a) is identical, in form, to the ratio of bulk molecular diffusivity to the effective molecular diffusivity measured in steady-state pore diffusion experiment, that is,

$$\mathcal{D}/\mathcal{D}_{\text{eff}} = X/\phi. \tag{3.7.3b}$$

The physical basis of Eq. (3.7.3) in the case of uniform capillaries, may be readily grasped with reference to Fig. 3.8, where now

$$R_{\text{w}} = \rho_{\text{w}} L/A \tag{3.7.4}$$

and, for the tortuous path,

$$R_0 = \rho_{\text{w}} L_{\text{e}}/a, \tag{3.7.5}$$

where ρ_{w} is the resistivity of the electrolyte. This results, with

$$\phi = \frac{L_{\text{e}}}{L} \frac{a}{A} \tag{3.7.6}$$

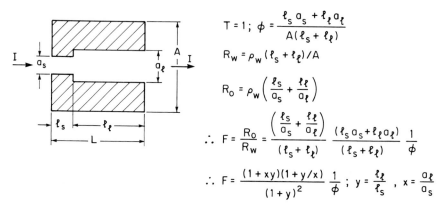

$$T = 1; \quad \phi = \frac{\ell_s a_s + \ell_\ell a_\ell}{A(\ell_s + \ell_\ell)}$$

$$R_w = \rho_w (\ell_s + \ell_\ell)/A$$

$$R_0 = \rho_w \left(\frac{\ell_s}{a_s} + \frac{\ell_\ell}{a_\ell} \right)$$

$$\therefore F = \frac{R_0}{R_w} = \frac{\left(\dfrac{\ell_s}{a_s} + \dfrac{\ell_\ell}{a_\ell} \right)}{(\ell_s + \ell_\ell)} \frac{(\ell_s a_s + \ell_\ell a_\ell)}{(\ell_s + \ell_\ell)} \frac{1}{\phi}$$

$$\therefore F = \frac{(1 + xy)(1 + y/x)}{(1 + y)^2} \frac{1}{\phi} \; ; \; y = \frac{\ell_\ell}{\ell_s} \;, \; x = \frac{a_\ell}{a_s}$$

Figure 3.30.   Illustration of the physical basis of Eq. (3.7.3) (Dullien, 1990a).

in the following expression for F

$$F = \frac{R_0}{R_w} = \frac{L_e}{L} \frac{A}{a} = \frac{1}{\phi} \left(\frac{L_e}{L} \right)^2 = \frac{\Upsilon}{\phi}. \tag{3.7.7}$$

In this case $X = \Upsilon$. For the case of capillaries with variations in the cross section, however, $X > \Upsilon = 1$, as illustrated in Fig. 3.30.

Hence, it is convenient to write

$$X = \Upsilon \cdot S' \tag{3.7.8}$$

where $S' \geq 1$ is the "constriction factor". For example, for $y = 2$ and $x = 4$, $S' = 1.5$.

X may be calculated from measured values of F and ϕ as illustrated in Table 3.5.

The model suggested by Dullien (1975a) for the prediction of permeabilities, and discussed in Section 3.3.1.2, was also applied to the prediction of formation factors (Dullien, 1975b). For each subset of capillaries of the model, characterized by a range of throat diameters between D_e and $D_e + dD_e$, the following expression is obtained for the contribution to the total formation factor F by those tubes controlled by throat diameters lying in this range

$$\frac{dD_e}{F(D_e)} = \frac{\phi(D_e)}{3} \frac{\left\{ \int_{De}^{\infty} [\beta_{p,s}(D_e, D) \, dD/D^2] \right\}^2}{\int_{De}^{\infty} [\beta_{p,s}(D_e, D) \, dD/D^4]} dD_e. \tag{3.7.9}$$

TABLE 3.5 Comparison of Measured and Calculated Formation Factors[a]

Sandstone	F_m	$X = \Upsilon \cdot S'$	F_{calc}	F_{calc}/F_m	Fractional volume of micropores
Boise	8.31	1.92	7.92	0.95	0.18
Bartlesville	10.02	2.71	10.29	1.03	—
Berea 108	62.63	11.15	39.68	0.63	0.16
Berea BE1	15.48	3.41	15.11	0.98	—
Noxie 47	8.99	2.25	8.66	0.96	0.20
St. Meinrad	11.71	2.90	10.70	0.91	0.18
Torpedo	9.74	2.39	9.74	1.00	0.24
Big Clifty	20.28	3.90	21.35	1.05	0.19
Noxie 129	9.42	2.66	9.00	0.96	0.18
Cottage Grove	10.65	2.30	11.06	1.04	0.21
Clear Creek	13.21	2.51	15.84	1.20	0.14
Bandera	24.64	4.68	22.71	0.92	0.34
Whetstone	92.73	12.43	93.07	1.00	0.23
Belt Series	392.22	32.55	395.1	1.01	0.68

[a]Value of "hydraulic tortuosity factor," for all the samples, $\Upsilon = 1.5$.

The formation factor F is obtained by integrating Eq. (3.7.9) over all D_e, that is,

$$F = 1 \bigg/ \int_0^\infty dD_e/F(D_e). \qquad (3.7.10)$$

Straightforward application of the pore structure data contained in "flow control distribution functions" (FCDF), such as shown in Fig. 3.11, resulted in the prediction of $F(F_{cal})$ that were considerably greater than the measured values (F_m) in almost every case. It was realized (El-Sayed, 1978) that this discrepancy was caused by the presence in the sandstones of a considerable amount of pore volume, consisting of very small pores, so called micropores, which was mostly undetected by photomicrography but was established by mercury porosimetry at high pressures.

The bulk of the micropores is between particles of the cementing materials (clays, etc.) contained in the sandstones. Cementing materials are also present in the capillaries between the sand grains, probably especially in the neck portions of these. The fact of the matter is that the micropores in the cementing materials compared with the much larger capillaries between the sand grains contribute only negligibly to the permeability of the medium. Therefore, the fact that they had not been considered in the BPSD used in Section 3.3.1.2 the FCDF, did not have a noticeable effect on the predicted permeabilities. On the other hand, the electric conductivity of a unit cube of a porous matrix containing capillaries that are uniform along their entire lengths is independent of the pore diameters and is the same value for two samples, containing uniform capillaries of different diameters in random orientations if the samples have the same porosities.

The reason for this behavior is that the electric conductance of a uniform conductor i varies with A_i, the cross-sectional area of the conductor, whereas the hydraulic conductance in laminar flow varies as A_i^2. It is instructive to consider a hypothetical sample consisting of n identical parallel capillaries with a total constant cross-sectional area equal to A. The cross-sectional area of each capillary is $A_i = A/n$. Now let us imagine that n is increased beyond any limit, while A is kept constant, thus leaving the electric conductance of the sample unchanged. The hydraulic conductance, however, is proportional to $nA_i^2 = n(A/n)^2 = A^2/n$, which becomes vanishingly small as n is increased beyond any limit.

These considerations led El-Sayed (1979) to correct the BPSD for the purpose of prediction of formation factors by taking into account the volume of very small pores not included in the permeability predictions. The volume of small pores was distributed among the various component networks in proportion to the relative volume of each network. It was assumed that the small pores are in parallel with the capillaries of the networks. Arbitrarily, pore volume was first added to the controlling pore neck and a new "equivalent" neck diameter was calculated. If the resulting ratio of equivalent bulge diameter to equivalent neck diameter was less than two, a portion of the pore volume of the small pores was transferred to the bulge, so as to obtain a ratio equal to two. The formation factor predicted by Eqs. (3.7.9) and (3.7.10) have been compared with the measured values in Table 3.5. The fractional volumes of small pores are also shown in the table. For the best fit to all the measured values F_m the factor 3 in Eq. (3.7.9) had to be replaced with $\Upsilon = 1.5$.

The concept of "tortuosity factor," Υ or X, is limited to one-dimensional models of the transport process. In network models of flow or conduction the effects of both the tortuosity and the porosity are built into the factors $g_m(z)$ and t_3/ℓ in Eq. (3.3.47). The electrical conductivity σ of a three-dimensional network may be written as follows (Koplik *et al.*, 1984)

$$\sigma = g'_m(z)\, t_3/\ell_3 \qquad\qquad (3.7.11)$$

where $g'_m(z)$ is the single effective medium value to be used instead of the random "bond" conductances g', where

$$g' = \sigma_w(a/\ell) \qquad\qquad (3.7.12)$$

where σ_w is the conductivity of the electrolyte with which the "bonds" are filled. The t_3 are the same as in the permeability case. Hence, according to the network model,

$$F = \frac{R_0}{R_{\mathrm{w}}} = \frac{\sigma_{\mathrm{w}}}{\sigma} = \frac{\sigma_{\mathrm{w}}}{g'_{\mathrm{m}}(z)(t_3/\ell_3)}. \tag{3.7.13}$$

As pointed out in Section 3.3.4., this predictive equation has not been tested yet on a sample, using satisfactory pore structure data.

3.8. THE TORTUOSITY TENSOR

It has been shown that the tortuosity factors Υ and X introduced in the models of permeability and formation resistivity factor are not true pore structure parameters, but rather parameters associated with one-dimensional models of pore structure. Notwithstanding this, the tortuosity of a porous medium is a fundamental property of the streamlines, or lines of flux, in the conducting capillaries. It measures the deviation from the macroscopic flow direction of the fluid at every point.

Bear and Bachmat (1966, 1967) (see also: Bear, 1972) have introduced the tensors

$$\Upsilon_{ij}^* = \Upsilon_{ij}(d\sigma/ds)^2$$

and

$$\Upsilon_{ij} = (d\xi_i/d\sigma)(d\xi_j/d\sigma),$$

where σ is the length measured along the streamline, s is the length measured along axis of channel, and ξ_i are the local coordinates in the fixed coordinate system x_i (e.g., Spiegel, 1959), with $(d\xi_i/d\sigma)(d\xi_j/d\sigma)$ representing a matrix whose nine elements (for i, $j = 1, 2, 3$) are products of cosines of angles between the direction of a streamline at a point and the coordinate axis. It can be shown that Γi_j is a symmetrical, second-rank matrix. The coefficient $(d\sigma/ds)^2$ takes into account the effect of the convergence-divergence of streamlines within individual channels. If all channels have constant cross section, then $(d\sigma/ds)^2 = 1$ and $\Upsilon_{ij} = \Upsilon_{ij}^*$.

Bear and Bachmat (1966, 1967) introduced the assumption that the force **R** resisting the motion of a particle at a point inside a channel is

$$\mathbf{R} = -(\mu/B)\mathbf{v}^*, \tag{3.8.1}$$

where $\mathbf{v}^*$ is the particle's mass-averaged velocity at that point and B, the (hydraulic) conductance of the channel at the point considered, is a function of the shape of the channel's cross section and of the location of the point under consideration.

Introducing Eq. (3.8.1) into the equation of motion of a fluid particle in the channel, the resulting expression is averaged, first over the channel cross section and then over the representative elementary volume of the porous medium. As a result, the coefficient $\overline{B\Upsilon_{ij}^*}$ appears in the averaged equation.

The difficulty with this procedure is that volume-averaging B does not result in the correct conductance of a network of channels or even of a single channel whose cross section varies along the axis of the channel. The same holds true also for $B\Upsilon_{ij}^*$ of which B is a special case (i.e., when $\Upsilon_{ij}^* = 1$). Indeed, the problem of modeling permeabilities is, to a large extent, one of finding the correct average channel conductance (i.e., $\overline{B}$), which differs from the mean-square channel diameter, denoted by $\overline{D^2}$ in the present work, by only a constant numerical factor.

Bear and Bachmat (1966, 1967) defined permeability by the relation

$$k_{ij} = \overline{B\Upsilon_{ij}^*} \tag{3.8.2}$$

from which it is apparent that the tortuosity Υ_{ij}^* is defined in the inverse sense (L/L_e, instead of L_e/L) compared with the more common definition.

A further step in the development of Bear and Bachmat (1966, 1967) is to write

$$\overline{B\Upsilon_{ij}^*} = \overline{B}\,\overline{\Upsilon_{ij}^*}, \tag{3.8.3}$$

where $\overline{\Upsilon_{ij}^*}$ is interpreted as the porous medium's tortuosity.

One difficulty with this step again is that as simple volume averaging will, in general, not give correct results, there is no way of evaluating $\overline{B}$ in the general case.

In the special case when all channels have the same conductance $\overline{B} = B$, we are left only with the problem of evaluating $\overline{\Upsilon_{ij}^*}$.

It has been shown by Bear and Bachmat (1966) that for isotropic media the tortuosity tensor reduces to a single scalar $\overline{\Upsilon} - \overline{\Upsilon}_{11} = \overline{\Upsilon}_{22} = 1/3$. This was shown, however, only for the case of a model consisting of straight circular channels of constant radius. The averaging procedure consisted of the usual evaluation of $\overline{\cos^2\theta}$ for uniform (random) distribution of directions. It is evident that volume averaging would give the same answer in this case.

Thus, for the case of randomly oriented uniform tubes, the procedure of Bear and Bachmat is clearly defined; however, it is difficult to see how it could be applied to more complex situations.

Another difficulty associated with the step $\overline{B\Upsilon_{ij}^*} = \overline{B}\,\overline{\Upsilon_{ij}^*}$ can be seen from the following consideration. For an isotropic medium $\overline{\Upsilon} = 1/3$; also if the medium contains a distribution of channel diameters, provided that the orientations of the various different channels are random. The value of $\overline{B\Upsilon}$ (i.e., the permeability k) will be different if the channels are connected according to different patterns, while always maintaining their random orientations. For the special case of only two different channel diameters, two such different patterns are (1) when channels of the same diameter are in series and channels of different diameters are in parallel, and (2) when the channels of different diameters are in series. As in the treatment of Bear and Bachmat there is no indication that $\bar{B}$ may be different when the same set of channels are connected according to different patterns (random distribution of the variables and volume-averaging are assumed); Eq. (3.8.3), when combined with Eq. (3.8.2), is seen to lack generality.

It is noted that Whitaker (1967) has defined a tortuosity vector $\vec{T}$ that includes the sinuousness of the channels as well as their expansions and contractions.

REFERENCES

Ahmed, N., and Sunada, D. K. (1969). *J. Hyd. Div. Proc. ASCE* **95** (HY6), 1947.

Allawi, Z. M., and Gun, D. J. (1987). *AIChE J.* **33**, no. 5, 766

Archie, G. E. (1942). *Trans. AIME* **146**, 54.

Ash, R., Barrer, R. M., and Pope, C. G. (1963). *Proc. Roy. Soc. London* Ser. A **271**, 1.

Azzam, M. I. S. (1975). Ph.D. Dissertation, Univ. of Waterloo, Canada.

Azzam, M. I. S., and Dullien, F. A. L. (1976). *IEC Fundamentals* **15**, 281.

Azzam, M. I. S., and Dullien, F. A. L. (1977). *Chem. Eng. Sci.* **32**, 1445.

Babbitt, J. D. (1950). *Can. J. Res.* **A28**, 449.

Barrer, R. N. (1967). *In* "The Solid-Gas Interface" (E. A. Flood, ed.), Vol. II, pp. 557–609. Dekker, New York.

Barrer, R. M., and Gabor, T. (1960). *Proc. R. Soc. London* Ser. A **256**, 267.

Barthels, H. (1972). *Brennst. Wärme Kraft* **24**, 233.

Batel, W. (1959). *Chemie-Ing-Techn.* Vol. 31, p. 388.

Bear, J. (1972). "Dynamics of Fluids in Porous Media." Elsevier, New York.

Bear, J., and Bachmat, Y. (1966). "Hydrodynamic Dispersion in Non-uniform Flow Through Porous Media, Taking into Account Density and Viscosity Differences" (in Hebrew with English summary). Hydraulic Lab., Technion, Haifa, Israel, IASH, P.N. 4/66.

Bear, J., and Bachmat, Y. (1967). *A generalized theory on hydrodynamic dispersion in porous media*. I.A.S.H. Symp. Artificial Recharge Management Aquifers, Haifa, Israel, IASH, P.N. 72, 7–16.

Beavers, G. S., and Sparrow, E. M. (1971). *Int. J. Heat Mass Transfer* **14**, 1855.

Berg, R. R. (1970). *Annu. Gulfcoast Assoc. Geol. Soc. Meeting*, 20th, Streveport, Louisiana, October **20**, 303.

Bird, R. B., Stewart, W. E., and Lightfoot, E. N. (1960). "Transport Phenomena." Wiley, New York.

Brauer, H. (1971). "Grundlagen der Einphasen- und Mehrphasenströmungen." Verlag Sauerländer, Aarau/Frankfurt.

Brenner, H. (1979). Book in preparation, personal communication.

Brownell, L., Dombrowski, H. S., and Dickey, C. A. (1950). *Chem. Eng. Progr.* **46**, 415.

Brutsaert, W. (1967). *Trans. ASAE* **10**, 400.

Brutsaert, W. (1968). *Water Resource Res.* **4**, 425.

Carman, P. C. (1937). *Trans. Inst. Chem. Eng. London* **15**, 150.

Carman, P. C. (1938). *J. Soc. Chem. Ind.* **57**, 225.

Carman, P. C. (1956). "Flow of Gases Through Porous Media." Butterworths, London.

Case, C. M., and Cochran, G. F. (1972). *Water Resour. Res.* **8**, 728.

Chatzis, I. (1980). Ph.D. Thesis, Univ. of Waterloo, Waterloo, Ontario.

Chen, C. Y. (1955). *Chem. Rev.* **55**, 595.

Cheng, P. Y., and Schachman, H. K. (1955). *J. Polym. Sci.* **16**, 19.

Childs, E. C., and Collis-George, N. (1950). *Proc. R. Soc. London* Ser. A **210**, 392.

Christiansen, E. B., Kelsey, S. J., and Carter, T. R. (1972). *AIChE J.* **18**, 372.

Cornell, D., and Katz, D. L. (1953). *Ind. Eng. Chem.* **45**, 2145.

Coulson, J. M. (1949). *Inst. Chem. Eng. Trans.* **27**, 237.

Cunningham, R. S., and Geankoplis, C. J. (1968). *I & Fundamentals* **7**, 535.

Dacey, J. R. (1965). *Ind. Eng. Chem.* **57**, 27.

Dauben, D. L., and Menzie, D. E. (1967). *Trans. Soc. Pet. Eng. AIME* **240** (1), 1065.

Davies, C. N. (1952). *Proc. Inst. Mech. Eng.* **1B**, 185.

Defay, R., Prigogine, Il., Bellemans, A. (1966). "Surface Tension and Adsorption." Everett, D. H. (trans.) Longmans Green, London.

deGroot, S. R., and Mazur, P. (1969). "Non-Equilibrium Thermodynamics." North-Holland Publ., Amsterdam.

De La Rue, R. E., and Tobias, C. W. (1959). *J. Electrochem. Soc.* **106**, 827.

Derjagin, B. B., and Bakanov, S. P. (1957). *Sov. Phys. -Dokl.* **2**, 326.

Doering, E. (1955). *Alg. Wärmetech.* **4**, 82.

Dominguez, J. G., and Willhite, G. P. (1977). *Soc. Pet. Eng. J.* **17**, 111.

Dudgeon, C. R. (1966). *Houille Blanche* **7**, 785.

Dullien, F. A. L. (1975). *The Chem. Eng. J.* **10**, 1.

Dullien, F. A. L. (1975a). *AIChE J.* **21**, 299.

Dullien, F. A. L. (1975b). *AIChE J.* **21**, 820.

Dullien, F. A. L. (1990). *In proceedings* "Physics of Granular Media," Feb. 20–Mar. 1, 1990, Les Houches, France.

Dullien, F. A. L. (1990a). *In proceedings* "SEG Research Workshop on Permeability, Fluid Pressure and Pressure Seals in Crust," August 5–8, 1990, Denver, Colorado.

Dullien, F. A. L., and Azzam, M. I. S. (1972). *AIChE J.* **19**, 222.

El-Sayed, M. S. (1978). Ph.D. Dissertation, Univ. of Waterloo, Canada.

Emanuel, G., and Jones, J. P. (1968). *Int. J. Heat Mass Transfer* **11**, 827.

Ergun, S. (1952). *Chem. Eng. Prog.* **48**, 98.

Ershaghi, I. (1972). Ph.D. Thesis, Univ. of Southern California, Los Angeles, California.

Evans, R. M., Watson, G. M., and Mason, E. A. (1961). *J. Chem. Phys.* **35**, 2076.

Fancher, G. H., and Lewis, J. A. (1933). *Ind. Eng. Chem.* **25**, 1140.

Ferrandon, J. (1948). *Génie Civil* **125**, 24.

Fischer, R. (1967). *VDI-Forschungsh* 524.

Flood, E. A. (1958). Special Rept. No. 40, Highway Research Board, Washington, D.C., p. 65.

Flood, E. A., Tomlinson, R. H., and Leger, A. E. (1952). *Can. J. Chem.* **30**, 348.

Fontugne, D. J. (1969). M. S. Thesis, Syracuse Univ., Syracuse, New York.

Forchheimer, P. H. (1901). *Z. Ver. Deutsch. Ing.* **45**, 1781.

Geankoplis, C. H. (1972). "Mass Transport Phenomena." Holt, New York.

Geertsma, J. (1974). *Soc. Pet. Eng. J.* **14**, 445.

Gilliland, E. R., Baddour, R. F., and Russell, J. L. (1958). *Am. Inst. Chem. Eng. J.* **4**, 90.

Gogarty, W. B. (1967). *Soc. Pet. Eng. J.* **7**, 161.

Goodknight, R. C., and Fatt, I. (1961). *J. Am. Chem. Soc.* **65**, 1709.

Green, L., and Duwez, P. (1951). *J. Appl. Mech.* **18**, 39.

Greenberg, D. B., and Weger, E. (1960). *Chem. Eng. Sci.* **12**, 8.

Greenkorn, R. A., Johnson, C. R., and Shallenberger, L. K. (1964). *Soc. Pet. Eng. J.* **4**, 124.

Grove, D. M., and Ford, M. G. (1958). *Nature* (London) **182**, 999.

Guin, J. A., Kessler, D. P., and Greenkorn, R. A. (1971a). *Fluids* **14**, 4.

Guin, J. A., Kessler, D. P., and Greenkorn, R. A. (1971b). *Chem. Eng. Sci.* **26**, 1475.

Gupte, A. R. (1970). *Dr. -Ing. Dissertation, Univ. of Karlsruhe*, Germany.

Happel, J., and Brenner, H. (1965). "Low Reynolds Number Hydrodynamics, with Special Application to Particulate Media." Prentice-Hall, Englewood Cliffs, New Jersey.

Happel, J., and Epstein, N. (1954). *Ind. Eng. Chem.* **46**, 1187.

Haring, R. E., and Greenkorn, R. A. (1970). *AIChE J.* **16**, 477.

Harleman, D. R. F., Mehlhorn, P., and Rumer, R. R. (1963). *Proc. ASCE, J. Hydraulics Div.* **89**, (No. HY-2) 67.

Haynes, H. W. Jr., and Brown, L. F. (1971). *AIChE J.* **17**, 491.

Heller, J. P. (1972). *Proc. Symp. Fundamentals Transport Phenomenon Porous Media*, 2nd, IAHR-ISSS Vol. I, p. 1. Univ. of Guelph, Ontario, Canada.

Henderson, J. H. (1949). *Producers Mon.* **14**, 32.

Hewitt, G. F. (1967). *In* "Porous Carbon Solids" (R. L. Bond, ed.), p. 203. Academic Press, New York.

Hewitt, G. F., and Sharratt, E. W. (1963). *Nature* (London) **198**, 952.

Hoogschagen, J. (1955). *Ind. Eng. Chem.* **47**, 906.

Horiguchi, Y. Hudgins, R. R., and Silveston, P. L. (1971). *Can. J. Chem. Eng.* **49**, 76.

Irmay, S. (1959). *Trans. Am. Geophys. Un.* No. 4 **39**, 702.

Johnson, W. E., and Hughes, R. V. (1948). *Producers Mon.* **13**(1), 17.

Johnson, W. E., and Breston, J. N. (1951). *Producers Mon.* **15**(4), 10.

Johnson, M. F. L., and Stewart, W. E. (1965). *J. Catal.* **4**, 248.

Katz, A. J., and Thompson, A. H. (1986). *Phys. Rev.* **B34**, 8179.

Kaufman, P. M., Dullien, F. A. L., Macdonald, I. F., and Simpson, C. S. (1983) *Acta. Stereol.* **2** (Suppl. 1), 145.

Keulegan, G. H. (1954). Rep. No. 3411, Natl. Bur. Stand., Washington, D.C.

Klock, G. O., Boersma, L., and DeBacker, L. W. (1969). *Soil Sci. Soc. Am. Proc.* **33**, 12.

Knudsen, M. (1909). *Ann. Physik* **28**, 75.

Koplik, J. (1982). *J. Fluid Mech.* **119**, 219.

Koplik, J. (1983). *J. Fluid Mech.* **130**, 468.

Koplik, J., Lin, C., and Vermette, M. (1984). *J. Appl. Phys.* **56**, 3127.

Kozeny, J. (1927). *Royal Academy of Science, Vienna, Proc. Class I* **136**, 271.

Kwiecien, M. J. (1987). M. A. Sc. Thesis, Univ. of Waterloo, Waterloo, Ontario.

Kwiecien, M. J., Macdonald, I. F., and Dullien, F. A. L. (1990). *J. Microsc.* **159**, 343.

Kyan, C. P., Wasan, D. T., and Kintner, R. C. (1970). *Ind. Eng. Chem. Fundamentals* **9**, 596.

Kyle, C. R., and Perrine, R. L. (1971). *Can. J. Chem. Eng.* **40**, 19.

Latini, R. J. (1967). M. S. Thesis, Syracuse Univ., Syracuse, New York.

Lin, C. (1983). *J. Math. Geology* **15**, 3.

Lin, C., and Cohen, M. H. (1982). *J. Appl. Phys.* **53**, 4152.

Lin, C., and Hamasaki, J. (1982). *J. Sed. Petrol.* **53**, 670.

Lin, C., and Perry, M. J. (1982). *IEEE Workshop on Computer Vision*, 38.

Lundgren, T. S. (1972). *J. Fluid Mech.* **51**, 273.

Luther, H., Abel, O., Rittner, K., Gieseler, M., and Schultz, H. -J. (1971). *Chem. Ing. Tech.* **6**, 376.

Macdonald, I. F., El-Sayed, M. S., Mow, K., and Dullien, F. A. L. (1979). *Ind. Eng. Chem. Fundam.* **18**, no. 3.

Macdonald, I. F., Kaufman, P. M., and Dullien, F. A. L. (1986). *J. Micros.* **144**, 277.

Macmullin, R. B., and Muccini, G. A. (1956). *AIChE J.* **2**, 393.

Marshall, T. J. (1958). *J. soil. Sci.* **9**, 1.

Marshall, R. J., and Metzner, A. B. (1967). *Ind. Eng. Chem. Fundamentals* **6**, 393.

Matthies, H. J. (1956). *VDI-Forschungsh.* 454.

Maxwell, J. C. (1881). "A Treatise on Electricity and Magnetism," 2nd ed. Oxford Univ. Press (Clarendon), London and New York.

McCarty, K. P., and Mason, E. A. (1960). *Phys. Fluids* **3**, 908.

Michaels, A. S. (1959). *AIChE J.* **5**, 270.

Miller, L., and Carman, P. C. (1960). *Nature* **186**, 549.

Miller, L., and Carmen, P. C. (1961). *Nature* **191**, 375.

Millington, R. J., and Quirk, J. P. (1961). *Trans. Faraday Soc.* **57**, 1200.

Morrow, N. R., Huppler, J. D., and Simmons, A. B. III (1969). *J. Sediment. Pet.* **39**, 312.

Ng, K. M., and Payatakes, A. C. (1985). *AIChE J.* **31**, no. 9.

Nocilla, S. (1961). "Rarefied Gas Dynamics" (Proc. Symp., 2nd), p. 169. Academic Press, New York.

Otani, S., Wakao, N., and Smith, J. M. (1965). *AIChE J.* **11**, 439.

Pahl, M. H. (1975), Dr. -Ing. Dissertation, Univ. of Karlsruhe, Germany.

Pakula, R. J., and Greenkorn, R. A. (1971). *AIChE J.* **17**, 1265.

Payatakes, A. C., Tien Chi, and Turian, R. (1973). *AIChE J.* **19**, 58, 67.

Petersen, E. E. (1958). *AIChE J.* **4**, 343.

Petersen, E. E. (1965). "Chemical Reaction Analysis." Prentice-Hall, Englewood Cliffs, New Jersey.

Philip, J. R. (1972). *J. Appl. Math. Phys.* **23**, 353.

Polubarinova-Kochina, P. YA. (1952). "Teoriya dvizheniya gruntovykh vod. Gosudarstv." Izdat. Tekh. -Teoret. Lit., Moscow [English transl.: R. deWiest, "Theory of Groundwater Movement." Princeton Univ. Press. Princeton, New Jersey, 1962].

Present, R. D. (1958). "The Kinetic Theory of Gases." McGraw-Hill, New York.

Pryor, W. A. (1971). Preprint for 46th Ann. Fall Meeting Soc. Pet. Eng. AIME, 46th, New Orleans, Louisiana, October 3–6. "Reservoir Inhomogeneities of Some Recent Sand Bodies."

Rice, P. A., Fontugne, D. J., Latini, R. J., and Barduhn, A. J. (1970). *In* "Flow Through Porous Media," (R. Nunge, ed.), p. 47. American Chemical Society, Washington, D.C.

Rothfeld, L. B. (1963). *AIChE J.* **9**, 19.

Roybal, L. A., and Sandler, S. I. (1972). *AIChE J.* **18**, 39.

Rumer, R. R. (1962). *Proc. ASCE, J. Hydraulics Div.* **88**, (HY-4) 147.

Rumpf, H., and Gupte, A. R. (1971). *Chem. Ing. Tech.* **43**, 367.

Runchal, A. K., Spalding, D. B. and Wolfshtein, M. (1969). *Phys. Fluids Suppl.* **12**, 21.

Sadowski, T. J. (1963). Ph.D. Thesis, Univ. of Wisconsin, Madison, Wisconsin.

Satterfield, C. N. (1970). "Mass Transfer in Heterogeneous Catalysis." MIT Press, Cambridge, Massachusetts.

Savins, J. G. (1970). *In* "Flow Through Porous Media," (R. Nunge, ed.), p. 11. American Chemical Society Washington, D.C.

Scheidegger, A. E. (1954). *Geofis. Pura Appl.* **28**, 75.

Scheidegger, A. E. (1956). *Geofis. Pura Appl.* **33**, 111.

Scheidegger, A. E. (1960), (1974). "The Physics of Flow Through Porous Media, 3rd ed." Univ. of Toronto Press, Toronto.

Schopper, J. R. (1966). *Geophys. Prospect.* **14**, 301.

Scott, D. S., and Dullien, F. A. L. (1962a). *AIChE J.* **8**, 293.

Scott, D. S., and Dullien, F. A. L. (1962b). *AIChE J.* **8**, 113.

Schwartz, J., and Probstein, R. F. (1969). *Desalination* **6**, 239.

Simons, S. (1967). *Proc. R. Soc. London* Ser. A. **301**, 387, 401.

Slattery, J. C. (1972)."Momentum, Energy and Mass Transfer in Continua." McGraw-Hill, New York.

Slichter, C. S. (1899). *U. S. Geol. Surv.*, 19*th Annu. Rep.*, Part II, p. 295.

Spiegel, M. R. (1959). "Theory and Problems of Vector Analysis." Schaum, New York.

Spiegler, K. S. (1966). *Ind. Eng. Chem. Fundamentals* **5**, 529.

Stanek, V., and Szekely, J. (1974). *AIChE J.* **20**, 974.

Suetin, P. E., and Volubuev, P. V. (1964). *Sov. Phys. -Tech. Phys.* **9**, 859.

Suetin, P. E., and Volubuev, P. V. (1965). *Sov. Phys. -Tech. Phys.* **10**, No. 2, 269.

Szabo, B. A. (1968). *Water Resour. Res.* **4**, 801.

Szabo, M. T. (1975a). *Soc. Pet. Eng. J.* **15**, 323.

Szabo, M. T. (1975b). *Soc. Pet. Eng. J.* **15**, 338.

Turner, G. A. (1958). *Chem. Eng. Sci.* **7**, 156.

Turner, G. A. (1959). *Chem. Eng. Sci.* **10**, 14.

Wakao, N., Otani, S., and Smith, J. M. (1965). *AIChE J.* **11**, 435.

Wakao, N., and Smith, J. M. (1962). *Chem. Eng. Sci.* **17**, 825.

Ward, J. C. (1964). *J. Hyd. Div. Proc. Am. Soc. Civil Eng.* **90**, HY5, 1.

Weber, S. (1954). *Kgl. Danske Videnskab. Selskab. Mat. Fys. Medd.* **28**, 2.

Whitaker, S. (1967). *AIChE J.* **13**, 420.

Whitaker, S. (1969). *Ind. Eng. Chem.* **61**, 14 (See also criticism by Chase, C. A. Jr. (1970). *Ind. Eng. Chem.* **62**, 183).

Wicke, E., and Vollmer, W. (1952). *Chem. Eng. Sci.* **1**, 282.

Wiggs, P. K. C. (1958). *In* "The Structure of Properties of Porous Materials" (D. H. Everett and F. S. Stone, eds.), p. 183. Academic Press, New York.

Wyllie, M. R. J. (1957). "Fundamentals of Electric Log Interpretation." Academic Press, New York.

Wyllie, M. R. J., and Spangler, M. B. (1952). *Bull. Am. Assoc. Pet. Geol.* **36**, 359.

Wyllie, M. R. J., and Gregory, A. R. (1955). *Ind. Eng. Chem.* **47**, 1379.

Youngquist, G. R. (1970). *In* "Flow Through Porous Media," p. 57. American Chemical Society, Washington, D.C.

4 | Selected Operations Involving Transport of a Single Fluid Phase through a Porous Medium

The scope of the brief reviews presented in this chapter is limited to some aspects of the physical mechanisms of the processes considered.

4.1. DEEP FILTRATION

It is customary to distinguish between two kinds of filtration mechanisms: (1) solid accumulation in front of the filter medium and (2) retention inside a deep porous bed. In the first case, a filter cake is built up. The mechanism of filtration by filter cakes has resisted efforts to analyze it quantitatively. In the second case, the suspension flows through the porous medium in which the particles are retained. Medium filtration is called, variously, "blocking," "surface," "micronic," and "deep" filtration.

4.1.1. Liquids

Herzig *et al.* (1970) reviewed the mechanism of liquid–solid separation by deep filtration and gave an extensive list of literature references. The use of deep filtration to treat polluted water is an old and well-established process.

Herzig *et al.* distinguished two types of deep filtration: a mechanical filtration for particles ≥ 30 μm and a physicochemical filtration for small particles $\sim 1 \, \mu$m. For particle sizes lying in between these two limits, both mechanical and physicochemical mechanisms are active.

The particles, when flowing through a porous medium, are brought into contact with the possible retention sites; they either are retained there or are carried away by the stream. Previously retained particles occasionally break loose and are carried away too. Herzig *et al.* gave a detailed account of the various capture and release mechanisms and the kinetic theories of particle capture and release.

Apart from a basic empirical formula, two main types of semiempirical models have been used to predict the variation of pressure drop in the course of deep filtration of liquids. In the first it is assumed that the internal surfaces are uniformly coated by small particles. The second model is based on the assumption that the filter is gradually clogged by large particles. Both models can be fitted to experimental data; they cannot be used, however, to predict the pressure drop in new situations.

The empirical equations for constant rate filtration can usually be written in the following form (Heertjes and Lerk, 1967; Hudson, 1948; Ives, 1962):

$$\Delta \mathcal{P}/(\Delta \mathcal{P})_0 = 1/(1 - j\sigma)^m, \tag{4.1.1}$$

where σ is the "retention" (volume of deposited particles per unit volume of filter) and j and m the constants. For $j\sigma \ll 1$,

$$\Delta \mathcal{P}/(\Delta \mathcal{P})_0 = 1 + mj\sigma. \tag{4.1.2}$$

This can be transformed into the following form:

$$\Delta \mathcal{P} = (\Delta \mathcal{P})_0 [1 + (mjv\phi_i t/L)], \tag{4.1.3}$$

where ϕ_i is the initial concentration of the suspension expressed as volume fraction, t is the time, and L is the filter bed depth. The validity of Eq. (4.1.3) has been verified on numerous occasions (e.g., Cleasby and Baumann, 1962; Eliassen, 1941; Ling, 1955), at least at the beginning of filtration (Heertjes and Lerk, 1967; Ives, 1962). The reported values of mj range from about 30 into the hundreds, but most of the values lie between 30 and 80.

The approach in which uniform coating of the internal pore surfaces is assumed utilizes the Carman–Kozeny equation for permeability [Eq. (3.3.7)]. It is an easy task to calculate the changed porosity and specific surface area (for the latter a specific pore shape must be assumed—Herzig *et al.* used the uniform cylindrical pore model). As a result, the following expression was obtained:

$$\Delta \mathcal{P}/(\Delta \mathcal{P})_0 = [1 - (\beta\sigma/\phi_0)]^{-2}, \tag{4.1.4}$$

which has the same form as Eq. (4.1.1) with ϕ_0 the porosity of the clean bed and β the inverse compaction factor of the retained particles (e.g., for a cubic packing of spheres $\beta = 6/\pi$).

Note that in the initial phases of filtration there are not enough particles deposited to cover the entire surface with a monolayer; hence introduction of β amounts to taking the "uniform coating" model too literally. In any event, it turns out that $mj = 2\beta/\phi_0 \approx 20$ (with $\beta = 6/\pi$ and $\phi_0 = 0.4$), which is too small as compared with the empirical values. A number of unsuccessful attempts have been made to improve the agreement; for references, see Herzig *et al.* (1970).

It should be noted, however, that as more particles are retained in the first layer of the bed than in the second one, and so forth (i.e., the retention σ is a decreasing function of distance measured from the inlet face of the bed), the total pressure drop must be calculated. This is because the integral (or sum) of the pressure drop has its minimum value in the case of uniform coating and, for the nonuniform case, it may dramatically exceed this minimum as σ increases.

Heertjes (1957), studying the blocking of cloth with particles, verified the relation

$$\Delta \mathscr{P}/(\Delta \mathscr{P})_0 = 1\Big/\big[1 - m'(\sigma/N_p)\big]^{m''}, \qquad (4.1.5)$$

where N_p is the number of pores per unit section area and m' and m'' are constants. Several other references to contributions using the pore blocking model can be found in the work of Herzig *et al.* (1970).

4.1.2. Gases

Gas filters using the filtering action of the solids accumulated in front of the filter medium are referred to as "bag filters" (see, for example, Iinoya and Orr, 1977). There are two main categories of deep filters: (1) "fibrous mat filters," commonly referred to simply as "air filters" (Davies, 1973); and (2) "granular bed separators."

Air filters are employed to eliminate atmospheric dust present in much lower concentration than industrial dust. Some air filters may be regenerated, but often they are thrown away after loading up with dust. They include "viscous" filters in which the filter medium is coated with a viscous material to retain the dust, "dry" filters that are usually of greater depth than the viscous filters, and automatic filters that are cleaned continuously and automatically.

Granular bed separators utilize packed beds of gravel or other granular material or fluidized beds (Perry–Chilton, 1973).

The pressure loss in clean fibrous mats can be calculated with the help of a number of empirical or semiempirical relationships [see Eqs. (3.2.13), (3.2.41), (3.2.32), (3.2.33), and (3.2.34)]. Additional empirical equations are as follows

(Kimura and Iinoya, 1959);

$$\Delta P = \left(0.6 + \frac{4.7}{(\mathrm{Re_c})^{1/2}} + \frac{11}{\mathrm{Re_c}}\right)\frac{2\rho v^2 L(1-\phi)}{\pi D_f \phi^2}, \qquad (4.1.6)$$

where

$$\mathrm{Re_c} = v_p D_f \rho/\mu \qquad (4.1.7)$$

and (Davies, 1952)

$$\Delta P D_f^2/L\mu v = 64(1-\phi)^{1.5}\left[1 + 56(1-\phi)^3\right]. \qquad (4.1.8)$$

The following differential equations, giving the relative number n_{out}/n_{in} of dust particles of a certain size removed by a layer of thickness L of the mat, was derived (see Davies, 1973; Strauss, 1975) for the purpose of calculating the collection efficiency of a fiber mat:

$$\ln\frac{n_{out}}{n_{in}} = -\frac{\eta L(1-\phi)}{D_f \phi} \qquad (4.1.9)$$

Here η is the collection efficiency of a single (individual) fiber in the mat for a particle of a given size and density, under the existing flow conditions.[1]

"Paper filters" usually contain zones that are almost impermeable, interspersed with areas of high porosity (Davies and Aylward, 1951). As the filter loads up with dust, more of the flow is transferred to the less permeable areas.

Modern "membrane filters" (see Davies, 1973) include nucleopore filters, aluminum oxide sheet filters, metallic membrane filters, etc. These have a homogeneous structure that can be approximated mathematically with a system of straight parallel uniform capillary tubes of length equal to the thickness of the filter.

4.2.   MASS TRANSFER THROUGH MEMBRANES

Initially the word "membrane" was used to denote thin pliable sheets of animal or vegetable origin (biological membranes), but later its meaning was extended to cover a variety of materials mostly made of plastics and other artificial materials. Most membranes contain only molecular interstices and are not porous in the customary sense of the word, because they are not permeable by viscous flow mechanism. Some membranes, however, may be classified as porous media containing micro- or ultramicropores. For a

---

[1]Eq. (4.1.9) has been written for random orientation of the fibres.

detailed discussion of mass transfer through biological membranes, including electrical effects, the reader is referred to the treatise by Lightfoot (1974).

Membranes find applications in various industrial separations, such as reverse osmosis, dialysis, electrodialysis, and ultrafiltration. Large-scale processes have been developed in the gaseous state for the preparation of ultrapure hydrogen and the separation of uranium isotopes.

In the phenomenological treatment of mass transfer, the membrane has often been treated as one of the chemical species participating in the transport process rather than a rigid structure through which the fluid species are being transported. Fick's law of diffusion has been used mostly for the description of the permeation of a porous septum by a single fluid, even though this treatment may be criticized on the grounds that this is the rigorous form of the "diffusion law" (de Groot and Mazur, 1969) for the case of binary counterdiffusion with one component stationary, only if the permeant–membrane system is very dilute in the permeant. There is, of course, physical justification for thinking of the membrane as a component because a certain analogy exists with diffusion of a soluble gas, vapor, or liquid through a nonporous solid or a stagnant (i.e., not counterdiffusing) liquid.

Barrer (e.g., 1967) has shown that by conveniently defining the diffusion coefficient and the permeability coefficient, Fick's law and Darcy's law can be made formally similar and either can be used for the description of transport of gases and vapors through porous media in general. This correspondence should not obscure the fact that in Fick's law molecular diffusion is the mechanism of permeation whereas in Darcy's law viscous flow is the mechanism.

It is customary to distinguish between "ideal" and "nonideal" membranes (Peterlin, 1974) in a thermodynamic sense of the word. By a purely formal application of "Fick's law" of diffusion, the volume flux N_v of a single penetrant relative to the membrane can be expressed as follows:

$$N_v \equiv v = -\mathscr{D}\,\partial\phi/\partial x, \tag{4.2.1}$$

where ϕ is the volume fraction of the fluid in the membrane and $\mathscr{D}$ is the diffusion coefficient.

Since in the usual experiment the membrane is fixed on a rigid support, it does not move after reaching steady state. Hence one does not need to consider the flow of the membrane material caused by swelling, if one is only interested in steady state. Therefore, in a plane membrane in steady state for an ideal permeant–membrane system,

$$v = \mathscr{D}(\phi_1 - \phi_2)/L \tag{4.2.2}$$

if $\mathscr{D}$ is independent of ϕ.

In the case of an ideal "gas–membrane pair" Henry's law applies, that is,

$$\phi = K_H P, \tag{4.2.3}$$

where K_H is Henry's law constant. Thus, from Eqs. (4.2.2) and (4.2.3), one obtains formally Darcy's law:

$$v = \mathscr{D}K_H(P_1 - P_2)/L = k_M(P_1 - P_2)/L, \qquad (4.2.4)$$

where k_M is the permeability coefficient used in conjunction with membranes:

$$k_M = \mathscr{D}K_H. \qquad (4.2.5)$$

Note that in Eq. (4.2.4) it is not specified whether v is measured at the pressure P_1 or P_2.

For the case of transport of a liquid through a membrane, one usually writes Darcy's law directly without going through the motions of first invoking Fick's law; for example, for liquids,

$$v = k_M(P_1 - P_2)/L. \qquad (4.2.6)$$

Regardless of the fact that in both the gas and the liquid transport formally Darcy's law was obtained, the mechanism of permeation is usually not viscous flow but molecular diffusion. The coefficient k_M is not a property of the membrane and, unless the permeation is by viscous flow, there is nothing to be gained from trying to introduce the fluid viscosity by letting $k_M = k/\mu$.

In the third case considered by Peterlin (1974), when a labeled component of the liquid is transported through the membrane under the concentration gradient of that component, the volume flux of the tracer N_v^*, is written as follows:

$$N_v^* = k_D'(\phi_1^* - \phi_2^*)/L, \qquad (4.2.7)$$

where ϕ^* is the volume fraction in the membrane of labeled molecules and k_D' is the "diffusive permeability coefficient," which is the same as the effective diffusion coefficient $\mathscr{D}_{\text{eff}}$ (cf. Section 3.4.2). In the case of Eq. (4.2.7) the "diffusion law," as defined by de Groot and Mazur (1969), was applied in a straightforward and rigorous manner to mass transport through a membrane.

Nonideality means that K_H and $\mathscr{D}$ are not constant but depend on the concentration. Peterlin (1974) derived, on thermodynamic grounds, the following general flux relation for permeation by a gas:

$$N_v \equiv v = -k_M \, dP/dx, \qquad (4.2.8)$$

where k_M is a function of x and P. Equation (4.2.8) is not unlike the usual form of Darcy's law for gases [cf. Eq. (1.1.6)] and is also valid for flow of a pure liquid across the membrane.

For the case of a tagged liquid Peterlin obtained for the molar flux J_m^*:

$$J_m^* = -\phi\mathscr{D} \, dC_m^*/dx, \qquad (4.2.9)$$

where C_m^* is the molar concentration of the tagged component. In Eq. (4.2.9) the volume fraction ϕ occupied by the liquid plays the same role as the porosity in the usual equations of diffusion in porous media [cf. Eq. (6.3.9)]. Note that in Eq. (4.2.9) the tortuosity factor X was not included.

For more information on mass transport through membranes see, for example, Hopfenberg (1974), Lakshminaraianaiah (1965), and Lightfoot (1974).

4.2.1. Dialysis and Electrodialysis

Dialysis (e.g., Tuwiner, 1962) is the transfer of a solute across a membrane by diffusion due to a concentration gradient. Separation of solutes in dialysis is due to differences in their diffusion rates. The membranes used can be either porous or nonporous. Membrane porosity and pore size govern membrane selectivity and transfer rate. The major mass-transfer resistance is the membrane; therefore, future improvements of dialyzer performance will depend on finding better ways of controlling porosity and pore size.

Electrodialysis is dialysis accelerated by an electromotive force applied to electrodes adjacent to opposite sides of the membrane. It is useful especially in removing electrolytes from proteins (e.g., Shaffer and Mintz, 1966).

4.2.2. Reverse Osmosis and Ultrafiltration

Reverse osmosis is the name applied to processes in which a true (as distinct from colloidal) solution is passed through a membrane (usually called a "semipermeable" membrane) under an operating pressure P, greater than the osmotic pressure, with the aim of obtaining an effluent containing the solute at a much lower concentration (ideally, a pure solvent is the objective). (See, e.g., Sourirajan and Agrawal, 1970; Lonsdale, 1970; Merten, 1966). Ultrafiltration is the same process used to remove macromolecules such as proteins from solutions.

4.2.2.1. *Maximum Possible Separation in Reverse Osmosis*

The maximum theoretically possible solute separation efficiency (f_{max}) corresponding to zero effective driving pressure for the system sodium chloride–water at different operating pressures and feed concentrations is shown in Fig. 4.1, with f defined as follows:

$$f = \frac{\text{molality of feed } (m_1) - \text{molality of product } (m_3)}{\text{molality of feed } (m_1)}. \quad (4.2.10)$$

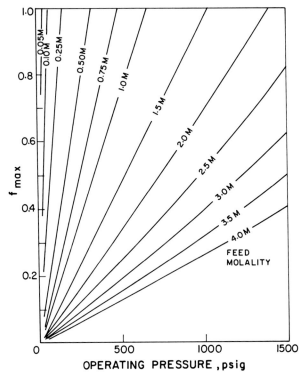

Figure 4.1.   Maximum possible solute separations for the sodium chloride–water system in the reverse osmosis process at different operating pressures and feed concentrations (Sourirajan and Agrawal, 1970).

The effective driving pressure ΔP for fluid flow through the membrane is

$$\Delta P = P - \Delta\pi, \tag{4.2.11}$$

where

$$\Delta\pi = \pi(x_{s2}) - \pi(x_{s3}) \tag{4.2.12}$$

with $\pi(x_{s2})$ the osmotic pressure of the concentrated boundary solution on the intake side of the membrane (see Fig. 4.2) and $\pi(x_{s3})$ to osmotic pressure of the product solution.

The osmotic pressure of a solution is given by the following relationships (Robinson and Stokes, 1959):

$$\pi = -\left(\mathscr{R}T/\bar{V}_{w}\right)\ln a_{w}, \tag{4.2.13a}$$

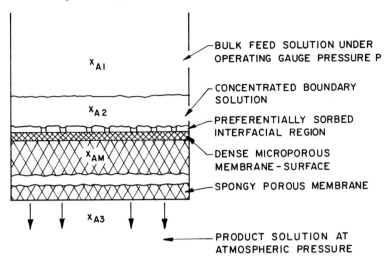

Figure 4.2. Reverse osmosis process under steady-state conditions (Sourirajan and Agrawal, 1970).

where a_w is the activity and $\bar{V}_w$ is the partial molar volume of water or:

$$\pi = \left(\mathscr{R}TM_w \sum_i \middle/ 1000\bar{V}_w\right)m\phi, \qquad (4.2.13b)$$

where M_w is the molecular weight of water, $\sum_i$ are the total number of moles of ions given by one mole of electrolyte, m is the solute molality, and ϕ is the osmotic coefficient.

Loeb and Sourirajan (six publications, for references see Sourirajan and Agrawal, 1970) developed a porous cellulose acetate membrane technology. A wide variety of other membranes are also in use, for example, graphitic oxide membranes (Flowers *et al.*, 1966); polysalt complex membranes (Michaels *et al.*, 1965a, b); and dynamically formed membranes (Kraus *et al.*, 1967; Kuppers *et al.*, 1966, 1967).

4.2.2.2. The Rate Equations in Reverse Osmosis

The rate of transport of water N_w (mol/m² s) through a membrane can be expressed as

$$N_w = \tilde{k}\,\Delta P = \tilde{k}\{P - [\pi(x_{s2}) - \pi(x_{s3})]\}, \qquad (4.2.14)$$

where $\tilde{k}$ is the "pure water permeability constant" (PWP), which varies with the operating pressure P and the viscosity μ_w as follows:

$$\tilde{k} = \tilde{k}_0 e^{-\alpha P} \qquad (4.2.15)$$

and

$$\tilde{k}\mu_w = \text{const.} \qquad (4.2.16)$$

It is noted that the membrane thickness L has been included into $\tilde{k}$. The justification for doing this is that the effective membrane thickness is not known.

When pure water is passed through the membrane, we have, from Eq. (4.2.14),

$$N_w^0 = \tilde{k}P. \qquad (4.2.17)$$

The rate of transport of solute N_s (mol/cm^2 s) through a membrane can be written as

$$N_s = (\mathscr{D}_{\text{eff}}/KL_e)(C_2 x_{s2} - C_3 x_{s3}), \qquad (4.2.18)$$

where L_e is the effective thickness of membrane where C_2 and C_3 are the molar concentration (mol per cubic centimeter) of the concentrated boundary solution and product solution, respectively (see Fig. 4.2), and K is defined as

$$K \equiv C x_s / C_M x_{sM}, \qquad (4.2.19)$$

where the subscript M refers to the membrane phase. The quantity $(\mathscr{D}_{\text{eff}}/KL_e)$, called the "solute transport parameter," is obtained empirically as a group. It has been found that at the given temperature and pressure $(\mathscr{D}_{\text{eff}}/KL_e)$ is independent of feed concentration, feed flow rate, and membrane compaction. Thus, as

$$N_s = [x_{s3}/(1 - x_{s3})]N_w, \qquad (4.2.20)$$

Eq. (4.2.18) becomes

$$N_w = (\mathscr{D}_{\text{eff}}/KL_e)[(1 - x_{s3})/x_{s3}](C_2 x_{s2} - C_3 x_{s3}). \qquad (4.2.21)$$

The solute transfer from the concentrated boundary solution on the high-pressure side of the membrane can be expressed by the general relation (Bird *et al.*, 1960)

$$N_s = x_s(N_s + N_w) - \mathscr{D}C_1 \, dx_s/dx. \qquad (4.2.22)$$

Eliminating N_s between Eqs. (4.2.20) and (4.2.22) and, subsequently, integrating over the thickness of the concentrated boundary solution ℓ, the following expression for N_w can be obtained:

$$N_w = \underline{k}C_1(1 - x_{s3}) \ln[(x_{s2} - x_{s3})/(x_{s1} - x_{s3})], \qquad (4.2.23)$$

where the mass transfer coefficient $\underline{k}$ was defined as

$$\underline{k} = \mathscr{D}/\ell. \tag{4.2.24}$$

At a given temperature $\underline{k}$ was found to be a function of feed concentration and feed flow rate.

Sourirajan and Agrawal (1970) pointed out that a single set of experimental PWP, production rate, and f data enables the prediction of both solute separation and product rate at that temperature and pressure, for all feed concentrations and feed flow rates.

4.2.2.3. *Ultrafiltration*

Ultrafiltration occurs at the arterial ends of vascular beds, and most particularly in the kidneys. Ultrafiltration is also an important part of the treatment of uremia by hemodialysis, and it is used commercially for the concentration of proteinaceous solutions.

The fundamental practical problem in ultrafiltration is that as the filtration pressure is increased the apparent permeability of the system decreases and the filtration rate approaches an asymptotic limit. This problem was analyzed theoretically by Kozinski and Lightfoot (1971, 1972) (see also Lightfoot, 1974). The main reason for the asymptotic behavior is concentration polarization of the protein at the boundary of the membrane. The asymptotic rate is different from the resistance effect of an ordinary filter cake because of the diffusional nature of mass transfer in a membrane. It probably corresponds to a standoff between any additional forward diffusion and a compensating backflow of water, which is caused by the backdiffusion of the protein from the membrane.

4.3. GEL CHROMATOGRAPHY

Gel chromatography has become, in a short period of time, widely accepted and popular as a new separation technique. In gel chromatography the separation is based on differences in molecular size. The theoretical principles of gel chromatography are not yet firmly established.

Gels consist of two components (e.g., Hermans, 1949): the dispersed substance or gel-forming material and the dispersing agent or solvent. Small molecules diffuse in the gel with practically the same velocity as in the solvent. The gel structure is usually formed by a network of macromolecules.

A practical introduction into gel chromatography has been given by Determann (1969).

In gel chromatography a packed bed is prepared of a granulated gel mixed with the solvent. A slug of the solution containing the substances to be separated is applied at the top of the column and is subsequently eluted with

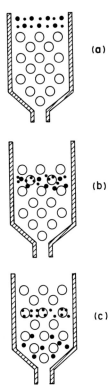

Figure 4.3. Schematic representation of gel chromatography. (Reprinted with permission from Determann, 1969.)

the solvent. It is understood that convective flow is restricted to the interparticle channels. The process is shown schematically in Fig. 4.3. The small molecules, indicated by small dots, can freely diffuse into the gel particles (open circles), whereas the large molecules (large dots) cannot ("exclusion" principle). Therefore, the large molecules move faster than the small ones. Molecules of intermediate size can enter only certain portions of the gel particles. The components of the mixture of solutes will leave the column in the order of decreasing molecular weight because larger molecules spend less time in the gel than the smaller ones.

According to the theory (loc. cit.) of gel chromatography the separation takes place because in the packed bed different volumes are accessible to molecules of different sizes. The largest molecules are restricted to travel in the void volume, i.e., in the interparticle channels. Intermediate size molecules will travel both in the void volume and in part of the volume occupied by the gel particles. At the other end of the size spectrum, the smallest molecules will occupy all the accessible solvent volume between, as well as inside, the gel particles (i.e., the void volume plus the "inner volume"). This theory, based on different accessible volumes, is very graphic; it explains the observations provided that the flow in the interparticle channels is slow enough for diffusion equilibrium to exist throughout the elution process. As the "elution volume" for the smallest molecules was found to be equal to the sum of the void volume and the inner volume, this must indeed be the case. (The elution volume is measured from the elution peak.)

It is of interest to note that in petroleum production research portions of the pore volume in sandstone plugs, sand packs, and artificial Teflon plugs have been found inaccessible to high molecular weight polymer molecules dissolved in water (e.g., Dawson and Lantz, 1972; Szabo, 1975a, b; Dominguez and Willhite, 1977).

When "flooding" porous plugs consisting of solid particles, such as sand, with polymer solutions, it was observed that the polymer "broke through" at an elution volume that was less than the total "pore volume" of the sample. Adding a "tracer" such as alcohol to the solution, the elution volume corresponding to the tracer was found to be equal to the "pore volume" of the sample. These observations are in agreement with other experience, which indicates that the presence of a gel is not absolutely essential for the "exclusion" principle to become operative. Porous glass beads have also been used, instead of gel, in "gel" chromatography.

There has been considerable speculation as to which portions of the pore space in a sand pack are inaccessible to large polymer molecules. It is almost certain that the volumes near the areas of contact between two touching grains as inaccessible to a large molecule, the same way as they are inaccessible to any large-sized object. This inaccessible volume is related to the volume occupied by the pendular (toroidal) rings of the wetting phase at irreducible wetting phase saturation.

REFERENCES

Barrer, R. N. (1967). *In* "The Solid–Gas Interface" (E. A. Flood, ed.), Vol. II, pp. 557–609. Dekker, New York.

Bird, R. D., Stewart, W. E., and Lightfoot, W. N. (1960). "Transport Phenomena," Wiley, New York.

Cleasby, J. L., and Baumann, E. R. (1962). Selection of sand filtration rates. *J. Am. Water Works Assoc.* 579–602.

Davies, C. N. (1973). "Air Filtration." Academic Press, New York.

Davies, C. N. (1952). *Proc. Inst. Mech. Engrs.* **1B**, 185.

Davies, C. N., and Aylward, M. (1951). *Brit. J. Appl. Phys.* **2**, 352.

Dawson, R., and Lantz, P. B. (1972). *Soc. Pet. Eng. J.* **12**, 448.

de Groot, S. R., and Mazur, P. (1969). "Non-Equilibrium Thermodynamics." North-Holland Publ., Amsterdam.

Determann, H. (1969). "Gel Chromatography." Springer-Verlag, Berlin and New York.

Dominguez, J. G., and Willhite, G. P. (1977). *Soc. Pet. Eng. J.* **17**, 111.

Eliassen, R. (1941). Clogging of rapid sand filters. *J. Am. Water Works Assoc.* **33**(5), 926–941.

Flowers, L. C., Sestrich, D. E., and Berg, D. (1966). "Reverse Osmosis Membranes Containing Graphitic Oxide." U.S. Dept. Interior, Office of Saline Water, Research and Development Progress Rep. No. 224.

Heertjes, P. M. (1957). Studies in filtration; blocking filtration. *Chem. Eng. Sci.* **6**, 190–203.

Heertjes, P. M., and Lerk, C. F. (1967). The functioning of deep bed filters. *Trans. Inst. Chem. Eng.* **45**, T129–T145.

Hermans, P. H. (1949). *In* "Colloid Science" (H. R. Kruyt, ed.), Vol. **2** p. 483. Elsevier, Amsterdam.

Herzig, J. P., Leclerc, D. M., and LeGoff, P. (1970). *In* "Flow Through Porous Media," pp. 129–157. American Chemicai Society, Publ., Washington, D.C.

Hopfenberg, H. D. (ed.) (1974). Permeability of plastic films and coatings to gases, vapors and liquids. *In* "Polymer Science and Technology," Vol. **6**. Plenum Press, New York.

Hudson, H. E., Jr. (1948). A theory of the functioning of filters. *J. Am. Water Works Assoc.* 868–872.

Iinoya, K., and Orr, C. Jr. (1977). *Air Pollut.* **4**, 149.

Ives, K. J. (1962) A theory of the functioning of deep filters. *Proc. Symp. Interaction between Fluids Particles, Inst. Chem. Eng.* 260–268.

Kimura, N., and Iinoya, K. (1959). *Kayaku Kagaku* **23**, 792.

Kozinski, A. A., and Lightfoot, E. N. (1971). *AIChE. J.* **17**, 81.

Kozinski, A. A., and Lightfoot, E. N. (1971). *AIChE. J.* **18**, 1030.

Kraus, K. A., Shor, A. J., and Johnson, J. S. (1967). *Desalination* **2**, 243.

Kuppers, J. R., Harrison, N., and Johnson, J. S. Jr. (1966). *J. Appl. Polym. Sci.* **10**, 969.

Kuppers, J. R., Marcinkowsky, O. O., Kraus, A. E., and Johnson, J. S. (1967). *Sep. Sci.* **2**(5), 617.

Lakshminaraianaiah, N. (1965). *Chem. Rev.* **65**, 491.

Lightfoot, E. N. Jr. (1974). "Transport Phenomena and Living Systems." Wiley, New York.

Ling, J. T. (1955). A study of filtration through uniform sand filters. *Proc. Am. Soc. Civil Eng.* **81**, Paper 751, 1–35.

Lonsdale, H. K. (1970). *Prog. Sep. Purif.* **3**, 191.

Merten, U. (ed.). (1966). "Desalination by Reverse Osmosis." MIT Press, Cambridge, Massachusetts.

Michaels, A. S., Bixler, H. J., Hausslein, R. W., and Fleming, S. M. (1965a). "Polyelectrolyte Complexes as Reverse Osmosis and Ion-Selective Membranes." U.S. Dept. Interior, Office of Saline Water, Research and Development Progress Rep. No. 149.

Michaels, A. S., Bixler, H. J., and Hodges, R. M. Jr. (1965b). *J. Colloid Sci.* **20**, 1034.

Perry, R. H., and Chilton, C. H. (1973). "Chemical Engineers Handbook," 5th ed. McGraw-Hill, New York.

Peterlin, A. (1974). Permeability of plastic films and coatings to gases, vapors, and liquids. *In* "Polymer Science and Technology" (H. B. Hopfenberg, ed.), Vol. **6** pp. 9–34. Plenum Press, New York.

Robinson, R. A., and Stokes, R. H. (1959). "Electrolyte Solutions," 2nd ed., pp. 476–490. Butterworths, London.

Shaffer, L. A. and Mintz, M. S. (1966). *In* "Principles of Desalination" (K. S. Spiegler, ed.), Chapter 5. Academic Press, New York.

Sourirajan, S., and Agrawal, J. P. (1970). *In* "Flow through Porous Media" (R. Nunge, ed.) pp. 219–241. Am. Chem. Soc., Washington, D.C.

Strauss, W. (1975). "Industrial Gas Cleaning," Vol. **8**. Pergamon, Oxford.

Szabo, M. T. (1975a). *Soc. Pet. Eng. J.* **15**, 323.

Szabo, M. T. (1975b). *Soc. Pet. Eng. J.* **15**, 338.

Tuwiner, S. B. (1962). "Diffusion and Membrane Technology." Van Nostrand-Reinhold, Princeton, New Jersey.

5 | *Multiphase Flow of Immiscible Fluids in Porous Media*

5.1. INTRODUCTION

This chapter deals with problems of flow through porous media involving two or more fluid phases present simultaneously in the pore spaces and separated from one another by interfaces. It is taken for granted that solution phase equilibrium has been reached (i.e., all phases are saturated with respect to all the various components present).

The subject of flow of immiscible fluids has been reviewed in a number of monographs (Muskat, 1937; Scheidegger, 1974; Bear, 1972; de Wiest, 1969; Collins, 1961; Craig, 1971; Childs, 1969) and papers (Marle, 1965; Oroveanu, 1966; Morel-Seytoux, 1969; Wooding and Morel-Seytoux, 1976; Philip, 1970, 1973; Greenkorn, 1983).

Flow of immiscible fluids in porous media may be conveniently subdivided into two categories: (1) steady-state (i.e., all macroscopic properties of the system are time invariant at all points); and (2) unsteady state (i.e., properties change with time).

In steady flow of immiscible fluids the saturation of the medium with respect to all the fluids contained in the system is constant at all (macroscopic) points. Hence, in steady flow there is no displacement of any fluid by any of the other fluids in the pores. On the other hand, in unsteady state the saturation at a given point in the system will, in general, change. Therefore, displacement phenomena fall in this category.

One can distinguish also co-current flow, when both phases flow in the same direction and the countercurrent flow, when different phases flow in opposite directions. The latter takes place, for example, when a system saturated with a nonwetting phase is brought in contact with a wetting fluid. The latter will imbibe and drive out some of the nonwetting fluid in countercurrent flow (Graham and Richardson, 1959). Another example is when capillary forces are small as compared with gravitational and viscous forces, which is the case when heavy oil drains from a formation in counter-current flow with steam.

5.1.1. Steady Flow

There seem to be two major conceptions of the microscopic picture of simultaneous steady flow of two immiscible fluids in porous media. According to one of these both the wetting and the nonwetting phase flow simultane-ously in all capillaries, the wetting phase on the outside surrounding the nonwetting phase, which is on the inside and occupies the central portion of each capillary (see, e.g., Scheidegger, 1974, pp. 56–57 and 248). This so-called funicular flow regime is the distribution of two continuous phases at high nonwetting phase saturations when the flow channels in the medium have irregular cross sections. Increasing saturation of the nonwetting phase im-plies an increase in the average diameter of the funicular entities of this phase.

The other conception of the microscopic picture is also based on direct visual observation of two-phase flow (see, e.g., Craig, 1971, pp. 15–16) in synthetic porous matrices. The photomicrographs showed that each phase moved through its own separate network of interconnecting channels. As the nonwetting phase saturation increased there was an increase in the number of channels carrying nonwetting phase and a corresponding decrease in the number of channels carrying only the wetting phase. This picture of two-phase flow is often called the "channel flow concept" of fluid flow. It was confirmed in these studies that there is also some wetting fluid in the pores carrying nonwetting fluid, corresponding to the funicular flow concept. It appears, however, that—at least when one phase has a strong wetting preference for the solid, the viscosity of neither fluid is very great, the interfacial tension between the two fluids is not very low, the saturation of neither phase is very low, and the pores are not all identical—then most of the flow takes place by the channel flow mechanism.[1]

---

[1]There is evidence (e.g., Geffen *et al.*, 1951) that the nonwetting phase is capable of contributing to flow at much lower saturations (i.e., less than 1%) than the "breakthrough" (see Section 2.5.2) value. It is not conceivable that there may be a continuous network of a fluid at such low saturations and, therefore, the flow of the nonwetting phase probably takes place by a "blob" or "slug"-flow mechanism, under the influence of the viscous forces exerted on it by the wetting phase.

Under these conditions the nonwetting fluid tends to occupy the larger pores and the wetting fluid the finer ones. The two separate networks, each containing its own fluid phase, which are separated from each other by a multitude of menisci between the two fluids, are very stable.

A few possibilities as far as the distribution of the two fluids in the pore space is concerned are shown in Fig. 5.1. Each fluid may flow in separate

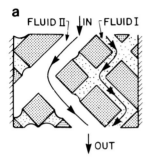

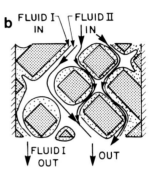

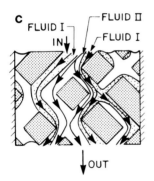

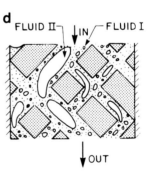

Figure 5.1. Two-dimensional representation of three-dimensional co-current steady two-phase flow in porous media.

   (a)   Both fluids flow in separate channels. Fluid I wets the uniformly wet solid surface preferentially.

   (b)   Both fluids in the same channels. Fluid I wets the uniformly wet solid surface preferentially.

   (c)   Both fluids flow in the same channels. Each fluid wets preferentially different portions of the mixed-wet solid surface.

   (d)   Both fluids flow in the same channel, but one fluid is continuous and the other dispersed. Fluid I wets the uniformly wet solid surface preferentially. (Dullien, 1988).

pore channels, which form two continuous subsets of the pore network of the porous medium; or both fluids may flow in the same pores, in which case either both flowing fluids are continuous or only one fluid is continuous, whereas the other is discontinuous (i.e., it is dispersed).

In three-phase flow, there is a hierarchy of wetting; for example, fluid I wets the solid surface most, fluid II wets it less, and fluid III the least. (In a "water-wet" medium, fluid I is water, fluid II is oil, and fluid III is gas.) Evidently many more possibilities exist for the distribution of three phases than of two phases. One particularly interesting distribution has been noted recently (Chatzis *et al.*, 1988) where the intermediate wetting fluid, for example oil, spreads between the other two fluids.

In steady co-current flow, the fluid pressures are constant at any point in the system, and their difference is equal to the capillary pressure as given by the Laplace equation at that particular point. Hence, the stable position of a meniscus anywhere is determined by the difference between the steady state pressures in the two phases and the pore geometry.

It is obvious that there cannot be any flow of the nonwetting phase in a capillary that contains a static meniscus. It has been shown experimentally that at relatively low saturations of the nonwetting phase, corresponding to conditions just after breakthrough in a drainage process, about 50% by volume of the nonwetting phase is thus prevented from flow (El-Sayed and Dullien, 1977). Geffen *et al.* (1951) and Fatt (1961, 1966) has pointed out this phenomenon previously. Such immobile, but continuous, nonwetting phase is sometimes called "dendritic," because of its treelike branching appearance.

Under the conditions listed above, it has been observed that the networks occupied by each phase are independent of the total flow rate across the sample within a wide range of values.

5.1.2. Displacement

If, say, the nonwetting fluid is pumped into the system at an increased rate, while the flow rate of the wetting phase is kept unchanged, a temporary unsteady state will result, while the menisci separating the two phases move to new equilibrium positions (i.e., the nonwetting phase penetrates some smaller pores from which it displaces the wetting phase).

During the displacement process there is no capillary equilibrium in the system, but the pressure difference between the two sides of a meniscus at any microscopic point in the system has been assumed to be equal to the capillary pressure as predicted by Laplace's equation for the conditions existing at that point. [Dynamic capillary pressures, differing by a wide margin from the static values have been reported recently by Calvo *et al.* (1990). This issue will be discussed in Section 5.3.4.] There are variations in the pressure along the capillaries, resulting in flow and displacement of the interfaces.

Owing to various different effects of pore morphology, the displacing phase tends to surround and cut off portions of the phase originally present in the pore space. Once cut off and isolated, these so-called blobs or ganglia of the displaced phase normally become stationary because of the presence of interfacial forces that are too great to be overcome by the viscous or gravitational forces present.

As pores in real porous media usually have irregular cross sections and a rough surface, some of the wetting phase is always trapped along the surface in grooves, edges, and wedges. The thickness of this film is much greater than the thin film of oil or water left on a smooth glass surface.

When the flow rate of the displacing phase is increased and/or the value of the interfacial tension between the two phases is decreased sufficiently, the blobs or ganglia may start flowing along with the displacing phase. Under usual conditions, however, all the residual displaced phase in the pores is immobile. At this stage, the two-phase flow degenerates into one-phase flow in a pore network, which contains obstacles in the form of the blobs or ganglia of the disconnected displaced phase.

5.1.3. The Fundamental Differential Equations on the Pore Level

Theoretically speaking, it is possible to treat the problems of two-phase flow in porous media by starting at the pore level and integrating the constant ρ and constant μ Navier–Stokes equations

$$\rho \, \frac{D\mathbf{u}}{dt} = -\nabla P + \mu \nabla^2 \mathbf{u} + \rho \mathbf{g} \tag{5.2.1}$$

$(D/Dt) =$ "substantial time derivative," or "derivative following the motion") over the medium with the following boundary conditions:

(1) On the solid surface, the velocity is zero.
(2) At the fluid/fluid interface, the velocities are equal.
(3) The difference between forces exerted by the fluids on each side of the fluid/fluid interface is balanced by capillary forces.

Contrary to the claims made in some texts on two-phase flow in porous media (that the above information, together with the equations of state of the two fluids and the equation of continuity, give a complete statement of the problem), a great deal of additional information would be required in order to attempt a theoretical solution of the problem of two-phase flow.

Pore morphology aside, as shown in Fig. 5.1 in steady co-current flow, the two fluids

(1)   may flow in separate channels and a portion of each fluid may be trapped, (i.e., it is not flowing); in this case information is needed on the distribution of the two fluids over the different flow channels of the pore network;

(2)   may flow side-by-side in the same pores; in this case information is needed on the distribution of the two fluids in each and every pore channel (i.e., the fraction of the pore cross-section occupied by each fluid must be known at every point);

(3)   may flow together in the same pores, but only one fluid is continuous, the other is dispersed; in this case, the properties of the dispersion (emulsion) are required and the problem reduces to single-phase flow.

Combinations of the above types of flow also exist. The existing flow pattern must be known or predicted on the basis of the properties of the system: solid matrix/fluid 1/fluid 2, before any *a priori* calculation of two-phase flow can be attempted. Evidently, the situation is even much more complicated in nonuniformly wetted or mixed-wet media, where the distribution of the contact angle over the surface should also be known.

5.2.   MACROSCOPIC (PHENOMENOLOGICAL) DESCRIPTION OF FLOW

5.2.1.   Macroscopic Equations: Relative Permeabilities

In practice, macroscopic equations are used to describe two-phase flow in porous media, which are generalizations of Darcy's law of single-phase flow in porous media. Darcy's law is limited to viscous (or creeping flow), that is, to pore or particle Reynolds numbers of less than 1 (or, strictly, less than 0.2), Newtonian fluids, absence of physical or chemical changes due to the fluid, no slip, and isotropic media.

Analogously to Darcy's law the following equations have been written for two-phase flow in porous media, under steady-state conditions:

$$Q_i = \left(\frac{k_i A}{\mu_i}\right)\left(\frac{\Delta P_i}{L}\right), \qquad (i = 1, 2), \qquad (5.2.2)$$

where Q_i is the volumetric flow rate, ΔP_i is the pressure drop, and μ_i is the viscosity of fluid i. A and L have the same meaning as in Eq. (1.1.5); k_i is referred to as "effective permeability" or "phase permeability" of the porous medium to fluid i.

It is customary to introduce "relative permeability" $k_{ri} = k_i/k$ and rewrite Eq. (5.2.2) as follows:[2]

$$Q_i = \left(\frac{k_{ri}kA}{\mu_i} \right) \left(\frac{\Delta P_i}{L} \right), \qquad (i = 1, 2). \qquad (5.2.3)$$

Extension of Darcy's law appears to have been first suggested by Muskat and coworkers (Muskat and Meres, 1936; Muskat *et al.*, 1937). The relative permeability k_{ri} is not determined by the pore structure of the porous medium alone but depends also on other parameters characterizing the system: solid matrix/fluid 1/fluid 2.

Darcy's law has been further generalized (see, e.g., Marle, 1981) by assuming that it applies at all points in the porous medium and also in nonsteady flow:

$$\mathbf{v}_i = \left(\frac{kk_{ri}}{\mu_i} \right) (\nabla P_i - \rho_i \mathbf{g}), \qquad (5.2.4)$$

where $\mathbf{v}_i = (\delta Q_i/\delta A)\mathbf{n}$, with $\mathbf{n}$ the unit normal vector of the surface δA, $\mathbf{g}$ is the gravitational acceleration vector, and ρ_i is the density of fluid i. The rationale behind generalization of Darcy's law to a point in the medium is that the medium is imagined to be subdivided into blocks small enough to permit the use of approximately constant values of $\mathbf{v}_i$, P_i, kk_{ri}, ρ_i, and μ_i for each fluid within each block, but large enough for Darcy's law to apply in its macroscopic form in each block. The introduction of the effect of gravity is standard procedure in hydraulics.

The pressures P_i in any two phases at any point in the porous medium are assumed to be related to each other via the capillary pressure P_c. Hence, two pressure gradients are related to each other by the gradient of the capillary pressure (Leverett, 1941):

$$\nabla P_2 - \nabla P_1 = \nabla P_c. \qquad (5.2.5)$$

Although this system of equations has been used universally to describe two-phase flow in porous media, it does not follow that use of these equations is always justified, as discussed in the next section.

5.2.2. Problems of Relative Permeabilities

Marle (1981) has listed the following dimensionless groups, involving all the parameters that enter into a steady-state relative permeability experiment: μ_1/μ_2, $(\rho_1 - \rho_2)g\ell^2/\sigma$, $\mu_i v_i/\sigma$, $\rho_i \ell v_i/\mu_i$, θ, and S_w. ℓ is a characteristic pore size. The relative permeabilities may be functions of any or all of

---

[2] Often, relative permeabilities are obtained by dividing the effective permeabilities by the effective permeability to oil at the lowest water saturation reached in the experiment.

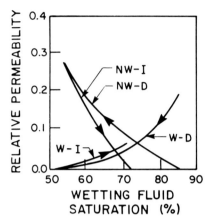

Figure 5.2. Relative permeability curves (Osoba *et al.*, 1951 © SPE-AIME). The values shown are for drainage ●, $(k_r)_0$ and ○, $(k_r)_g$; and for imbibition ▲, $(k_r)_0$; and △, $(k_r)_g$.

these dimensionless groups. Under certain special conditions, discussed below, they depend only on a few of these parameters.

There are so-called strongly water-wet systems, such as clean glass (or quartz) porous matrix/water/mineral oil, for example, where the contact angle θ on the water side is very small. Hence, in this case water is the wetting fluid everywhere in the porous medium. If, in addition, the co-current flow is in the viscous (or creeping) flow regime, the viscosity of neither fluid is very high; the interfacial tension is not extremely low, (i.e., the "capillary number" $\mu_i v_i / \sigma$ is very small) and the water saturation is not very low either then capillary forces control two-phase flow. In this eventuality the flow pattern shown schematically in Fig. 5.1a is predominant,[3] where

(1) the wetting fluid occupies preferentially the relatively small pores and the nonwetting fluid the relatively large ones; and

(2) the two fluids are separated from each other by stable interfaces which, in steady state, are stationary and behave like rigid partitions.

Where this flow pattern dominates, the two fluids flow in separate 3-dimensional networks of pore channels where the flow of each fluid can be regarded as hydrodynamically independent of the other fluid. As a result, Darcy's law of flow of a single fluid is applicable to each fluid. Equations (5.2.2) and (5.2.3) are valid and the relative permeabilities are functions only of the saturation S_w, as shown in Fig. 5.2. Due to capillary hysteresis,

---

[3]A film of varying thickness of the wetting fluid (fluid I) is usually present on the solid surface also in those pores occupied by the nonwetting fluid (fluid II).

however, for a fixed S_w, there is a range of values of P_c and also of relative permeabilities. Usually, only two curves are determined experimentally for each fluid, the primary drainage and the (secondary) imbibition relative permeabilities. Therefore, one distinguishes four relative permeability curves, that is, the drainage relative permeability curves of the wetting (displaced) fluid and of nonwetting (displacing) fluid, respectively, and the imbibition relative permeability curves of the nonwetting (displaced) and the wetting (displacing) fluid, respectively. In summary, when capillary forces are controlling, in the steady co-current flow the relative permeability curves are independent of the fluid pair and the flow rates, but they do depend on the pore structure and the history of the system (e.g., Wyckoff and Botset, 1936; Hassler *et al.*, 1936; Reid and Huntington, 1938; Botset, 1940; Terwilliger and Yuster, 1946, 1947; Calhoun, 1951a, b).

In the systems discussed above, there is a "thick film" of wetting fluid between the pore walls and the nonwetting fluid in those pores in whose center nonwetting fluid is flowing, as shown in Fig. 5.1b. Note that the nonwetting fluid is thus not in direct physical contact with the pore walls (i.e., the solid phase). The thickness of the film of the wetting fluid is much higher values in pores of a rough surface or irregular cross section than in the case of water in a smooth-walled capillary tube. It has been demonstrated by experiments by Dullien *et al.* (1986, 1989) and Melrose (1988) that the wetting fluid present as "thick films" on rough or angular pore surface is mobile, because it often forms a 3-dimensional network, extending over the surface of the entire pore network. As the capillary pressure is increased (i.e., the nonwetting fluid is subjected to a higher pressure made possible by a semipermeable membrane at the exit), there corresponds to the new capillary equilibrium a smaller radius of curvature of the interface nonwetting fluid/wetting fluid. Therefore, a certain amount of wetting fluid is squeezed out from the pore space via the continuous network of "surface capillaries." At saturations where the wetting fluid flows through the porous medium in capillaries that it fills completely, the flow of wetting fluid in surface channels can usually be neglected. However, in the range of saturations where the continuity of the wetting fluid filling complete pores has been interrupted everywhere, corresponding to the steeply ascending portion of the capillary pressure curve, the wetting fluid can only flow via the network of surface capillaries. Given enough time and sufficiently high capillary pressures, the wetting fluid saturation can be reduced to very low values.

5.2.2.1. *The Effect of Contact Angle on Relative Permeabilities*

The effect of wettability on relative permeabilities in uniformly wetted systems at very small capillary numbers has been investigated by several authors. Detailed studies by Owens and Archer (1971) and McCaffery and Bennion (1974) are noted. The first authors used outcrop sandstone, whereas the latter used synthetic polytetrafluoroethylene and varied the wettability by using different pairs of fluids.

Owens and Archer (1971) used surfactants, added either to the oil or the water, in order to change the wettability. They fired the cores in an electric furnace at 1600°F in order to stabilize any clay minerals present in the rock space and to provide an internal rock surface of as near constant properties as possible. Both gas–oil (in the direction of gas displacing oil) and water–oil (in the direction of water displacing oil) relative permeabilities were obtained for Torpedo sandstone. The fluids used in the water–oil tests were a mild sodium chloride brine solution and a 1.7-cP refined oil to which were added varying amounts of barium dinonyl napthalene sulphonate (BDNS). The wetting conditions in the core during some of these tests were altered and controlled by changing the concentration of the BDNS in the oil phase.

The contact angles were measured on a flat quartz surface. The Penn State type steady-state method was used in the water–oil relative permeability tests. The relative permeability curves obtained in the Torpedo sandstone samples with the water–oil system as shown in Fig. 5.3 for five different wetting conditions (i.e., contact angles of 0, 47, 90, 138, and 180°).

The manner in which relative permeability curves at different contact angles θ have been presented by Owens and Archer (1971) and others invites false conclusions to be drawn on the effects of θ, for the following reasons. The customary way of presenting the data has been to plot the curves obtained for fluids 1 and 2 at different values of θ as functions of the saturation of fluid 1, which was always the displacing phase in all the experiments. [In one instance, all the results were plotted versus the saturation of the displaced fluid (McCaffery, 1973).] Let us suppose that θ is measured through fluid 1. Then, for fluid 1, the curve $\theta = 0°$ is an *imbibition wetting phase* relative permeability curve, whereas the curve $\theta = 180°$ is a *drainage nonwetting phase* relative permeability curve, with the important difference that the first curve is plotted versus the wetting phase saturation, whereas the second is plotted versus the nonwetting phase saturation. Similarly, for fluid 2, the curve with $\theta = 0°$ is an *imbibition nonwetting phase* curve plotted versus the wetting phase saturation, whereas the curve with $\theta = 180°$ is a *drainage wetting phase* curve plotted versus the nonwetting phase saturation.

It is well known from studies of the distribution of the wetting and nonwetting phases in the pore space of strongly wetted systems (Yadav *et al.*, 1987; Chatzis and Dullien, 1983; Lenormand *et al.*, 1983) that, in general, the wetting phase preferentially occupies the smaller pores and it is also present near the pore walls in those pores that have been penetrated by the nonwetting phase. By contrast, in general, the nonwetting phase preferentially occupies the central portion of the larger pores.

In imbibition, at the outset, the nonwetting phase is usually continuous, and it becomes gradually disconnected by the wetting phases. Part of the

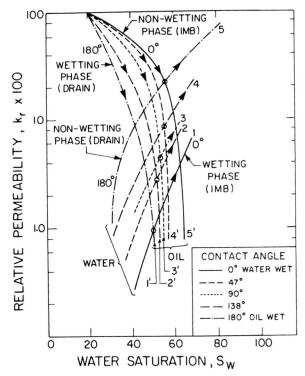

Figure 5.3. Relative permeabilities for the spectrum of wetting conductions for Torpedo sandstone. The contact angles measured through water were 0 and 180° (—); 47° (– – –); 90° (···); 138° (— —) (Owens and Archer, 1971 © SPE-AIME).

nonwetting phase is trapped, and it blocks some pore junctions and some larger pores. At the same time, the wetting phase becomes increasingly interconnected and occupies ever larger pores, but it is always hindered by the trapped blobs of the nonwetting phase.

In primary drainage, at the outset, there is only wetting phase present and flow of the nonwetting phase starts only after reaching the breakthrough capillary pressure (bubbling pressure) which, in sandstones, corresponds to a nonwetting phase saturation of about 10 pore volume percent (Chatzis and Dullien, 1985). From that point on, both phases are continuous, with the nonwetting phase occupying the central portion of relatively large pores. Increasing portions of the wetting phase become disconnected until finally all wetting fluid is discontinuous, except for the thick films of the wetting phase present in the surface capillaries. In drainage, the interfaces penetrate the pore throats from adjacent pore bodies by piston-like motion, and the displacement process moves from pore throat to the next narrower throat with occasional trapping of the wetting fluid. In imbibition, however, the wetting fluid is also transported to the pore throats in the thick films between the nonwetting fluid and the solid walls; and it fills the throats by choking off the nonwetting fluid. It then moves by piston-type motion into adjacent pores where break-off and snap-off events disconnect the nonwetting phase and provide new interfaces in throats that undergo further piston-type motions (Li and Wardlaw, 1986).

Secondary drainage is started in the presence of disconnected, trapped blobs of the nonwetting phase, which becomes gradually reconnected as the drainage process progresses (Chatzis and Dullien, 1981).

In the light of the above outlined differences, it is to be expected that *imbibition wetting phase* relative permeability curves should be different from *drainage nonwetting phase* relative permeability curves and also *imbibition nonwetting phase* relative permeability curves should be different from *drainage wetting phase* permeability curves, as indeed it has been found in experiments where the contact angle θ was varied from 0° to 180°.

There is no question that ultimately the different contact angle (i.e., 0° or 180°) is the most important factor behind the differences existing between these curves. However, the mechanisms responsible for obtaining different curves (i.e., "imbibition of wetting phase" versus, "drainage of nonwetting phase" and "displacement of nonwetting phase in imbibition" versus "displacement of wetting phase in drainage") are so different that comparing these curves is like comparing apples with oranges. What is of real interest is how each of these four curves behaves when the contact angle measured through either phase is varied from 0° to 90° (while that measured through the other phase changes from 180° to 90°). Whereas the range of possible values of θ extends from 0° to 180°, as θ_1, measured through fluid 1, is related to θ_2, measured through fluid 2, by the relationship $\theta_2 = 180° - \theta_1$, it is sufficient to vary θ from 0° to 90° in order to cover the entire range of possible values of θ. Using the whole range from 0° to 180° results in duplication of the conditions as, for example, the water relative permeability curve with $\theta_{H_2O} = 138°$ in Fig. 5.3 was obtained at analogous wettability conditions as the oil relative permeability curve with $\theta_{H_2O} = 42° = 180° - \theta_{oil}$ (i.e., $\theta_{oil} = 138°$), etc. Now, one reason why these "corresponding" curves in Fig. 5.3, assigned numbers from 1 to 5 and 1' to 5', do not appear symmetrical is that in these experiments all relative permeability tests were started at an initial water saturation of about 20% and in all experiments water was displacing oil. An additional reason is that 20% oil saturation was seldom reached. Nevertheless, the "corresponding curves" intersect near 50% (between 50% and 60%) water saturation.

The initial 20% water saturation was achieved by saturating the dry core with brine, then flooding it with viscous mineral oil to about 20% water saturation. Finally, the viscous mineral oil was replaced with 1.7 mPa refined mineral oil containing the desired amounts of detergent (BDNS). As at the end of the viscous oil flood the fluids distribution never corresponded to capillary equilibrium, every time the viscous oil was replaced with the refined oil, there must have been a major rearrangement of the oil and water present in the pores. This is reflected in the values obtained for the effective oil permeabilities at the initial water saturation, shown in the table. The 16% decrease in effective permeability for $\theta = 47°$, compared with the case $\theta = 0°$, is quite significant and it is probably due to some breaks occurring in the oil network, whereas for $\theta = 0°$, at 80% oil saturation the oil network can be expected to be complete, without any breaks in it. As apparent from Fig. 5.3, up to a water saturation of about 40%, the oil relative permeability curves for $\theta = 47°$ and $\theta = 0°$ practically coincide and then they part because at $\theta = 47°$ more oil is trapped than at $\theta = 0°$, something which is consistent with initial breaks in the oil network for $\theta = 47°$. The water relative permeabilities for $\theta = 47°$ are greater than for $\theta = 0°$ because breaks in the oil network open up additional flow paths for the water.

Effective oil permeability	Contact angle [°]
561	0
472	47
459	90
380	138
357	180
571 (air permeability)	

Results obtained by several authors (e.g., Schneider and Owens, 1970; Leverett and Lewis, 1941) have indicated that for a system having a strong preference for either water or oil, the wetting phase relative permeability is only a function of its own saturation; that is, it shows little or no hysteresis. It is interesting, in this context, to compare in Fig. 5.4 the gas–oil drainage and the oil–water imbibition relative permeability results obtained by Owens and Archer (1971) in the fired Torpedo core. Oil was the nonwetting phase in the imbibition runs, whereas gas was the nonwetting phase in the drainage runs. The wetting phase curves appear to be continuations of each other, indicating that the wetting phase imbibes into the same pore network where it also retreats when driven out of the larger pores by a nonwetting phase.

It can be expected that if experiments had been done starting with 100% nonwetting phase saturation and if the wetting phase relative permeabilities could have been measured down to at least 20% wetting phase saturations or below, one would have found a difference between the imbibition branch and the drainage branch because of different spatial distributions of the wetting phase. The wetting phase relative permeabilities in this range, however, are very small.

For the nonwetting phase, the separation of the drainage and the imbibition branches is evident from Fig. 5.4. This effect is measurable because for

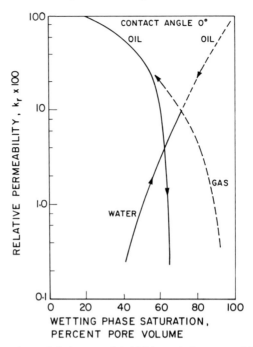

Figure 5.4. Comparison of drainage and imbibition relative permeability relationships for Torpedo sandstone (Owens and Archer, 1971 © SPE-AIME).

the condition $(S_{nw}$ drainage$) = (S_w$ imbibition$)$ the relative permeability of the nonwetting phase along the drainage branch is much greater than that of the wetting phase along the imbibition branch. The reason for the existence of the two separate branches is that along the imbibition branch more and more of the nonwetting phase becomes disconnected, whereas along the drainage branch (after breakthrough) all of the nonwetting phase is continuous.

It appears that there were no breaks in the nonwetting phase network at the end of primary drainage, at least up to $\theta = 49°$, according to the results of McCaffery (1973) in artificial Teflon plugs, because his relative permeability curves were unchanged in the range of $\theta = 0°$ to $\theta = 49°$. Over the range of $\theta = 49°$ to $\theta = 90°$, a shift in the relative permeability curves was reported also by McCaffery, as can be seen in Figs. 5.5 and 5.6.

The results of McCaffery are consistent with observations that the end of free imbibition of air into a glass capillary network saturated with mercury $(\theta_{air} \approx 42°)$ there was the same distribution of the two phases as at the end of free imbibition of oil against water $(\theta_{air} \approx 0°)$ into the same network, made oil-wet (Chatzis and Dullien, 1981).

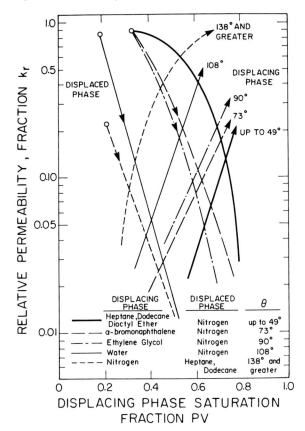

Figure 5.5. Effects of wettability on relative permeability in artificial Teflon plug with nitrogen and pure liquids. Relative permeabilities are based on absolute permeability. The contact angle was measured through the displaced phase on a smooth flat plate (McCaffery, 1973).

At sufficiently low capillary number $\mu_i v_i / \sigma$, in steady state, when the two fluids flow in two separate networks at a fixed saturation the stationary equilibrium liquid/liquid interfaces separating the two fluids from each other in a multitude of pores, in principle should have the same positions as long as the effective contact angle $(\theta + \phi)$ (see Section 2.3.1) is in the range $0° \leq (\theta + \phi) < 90°$, because the fluid through which θ is in this range remains the preferentially wetting fluid and the other fluid, the nonwetting fluid. The capillary pressure P_c, of course, varies, but this has no effect on the relative permeabilities. The situation is different for $(\theta + \phi) = 90°$, because at this point the interface is flat, and therefore, pores of different sizes have an equal chance to be penetrated by either fluid. The value of ϕ changes along

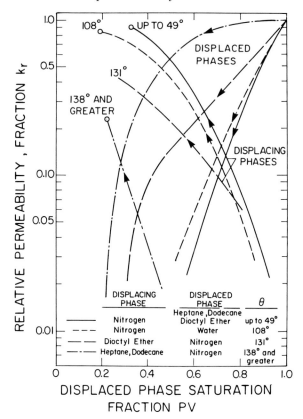

The legend contained within the figure:

DISPLACING PHASE	DISPLACED PHASE	θ
Nitrogen	Heptane, Dodecane Dioctyl Ether	up to 49°
Nitrogen	Water	108°
Dioctyl Ether	Nitrogen	131°
Heptane, Dodecane	Nitrogen	138° and greater

Figure 5.6. Effects of wettability on relative permeability in artificial Teflon plug with nitrogen and pure liquids. The saturation of the displacing phase was initially at the irreducible value. The contact angle was measured through the displacing phase (McCaffery, 1973).

every pore between a maximum value ϕ_{max} and 0°. Therefore, it can be expected that the relative permeability curves will remain unchanged over the range of contact angles $0° \leq \theta < (90° - \phi_{max})$ and change in the range $(90° - \phi_{max}) \leq \theta \leq 90°$.

As pointed out in Chapter 2, drainage capillary pressure curves were found not to vary noticeably in the range of contact angles $0° \leq \theta \leq 42°$. By this criterion one can estimate $\phi_{max} = 90° - 42° = 48°$. Imbibition curves, however, varied over the entire range of contact angles $0° \leq \theta \leq 90°$. There was no spontaneous imbibition beyond $\theta = 60°$, indicating another critical value of $\phi < \phi_{max}$. These values are not consistent with the range $0° \leq \theta \leq 49°$ over which no change was found in the relative permeability curves.

The same criticism of the manner in which the data were plotted, elaborated on with regard to Fig. 5.3, applies also to Figs. 5.5 and 5.6. The data of

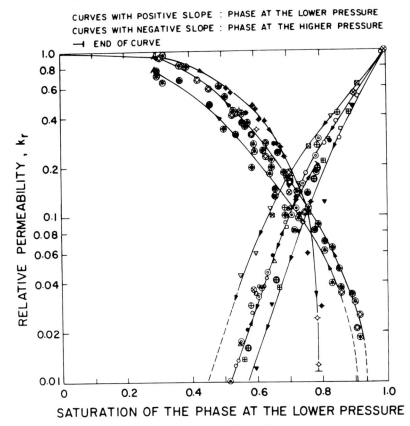

CURVES WITH POSITIVE SLOPE : PHASE AT THE LOWER PRESSURE
CURVES WITH NEGATIVE SLOPE : PHASE AT THE HIGHER PRESSURE
⊣ END OF CURVE

Figure 5.7. (a) Replot of data plotted in Figs. 5.5 and 5.6.
Symbols designate the fluid being displaced when it is at the lower pressure:

System	θ (measured through I)
• = heptane (I)/nitrogen (2)	20°
⊙ = dodecane (I)/nitrogen (2)	42°
⊕ = dioctyl ether (I)/nitrogen (2)	49°
⊞ = dodecane (I)/water (2)	14° (20°)
⊗ = α-bromonaphthalene (I)/ ▢ = nitrogen (2)	73°
⊠ = ethylene glycol (I)/ △ = nitrogen (2)	90°
▼ = nitrogen (I)/ ▽ = water (2)	72°

The displaced fluid when it is at the higher pressure is denoted by the same symbol with "arms." The symbol used for a displaced fluid is encircled to designate the same fluid when that is displacing. For example, in the system heptane/nitrogen • denotes heptane (which is at the lower pressure) when it is being displaced, ⊙ denotes heptane when it is displacing nitrogen, ✦ denotes nitrogen (which is at the higher pressure) when it is being displaced, and ⊕ denotes nitrogen when it is displacing heptane.

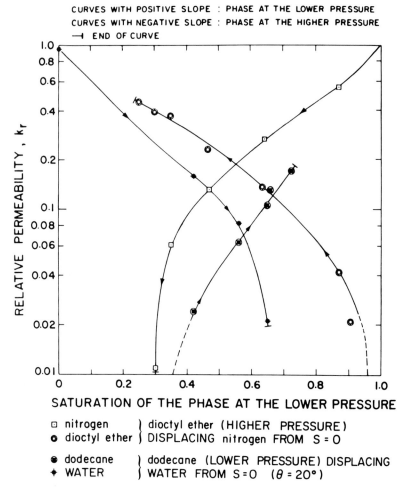

CURVES WITH POSITIVE SLOPE : PHASE AT THE LOWER PRESSURE
CURVES WITH NEGATIVE SLOPE : PHASE AT THE HIGHER PRESSURE
⊢ END OF CURVE

SATURATION OF THE PHASE AT THE LOWER PRESSURE

□ nitrogen   } dioctyl ether (HIGHER PRESSURE)
⊙ dioctyl ether } DISPLACING nitrogen FROM S = 0

◉ dodecane   } dodecane (LOWER PRESSURE) DISPLACING
✦ WATER   } WATER FROM S = 0   ($\theta = 20°$)

Figure 5.7. (b) Effects of 100% initial nonwetting phase saturation on relative permeability curves.

McCaffery (1973) shown in Figs. 5.5 and 5.6 have been replotted by the author in Fig. 5.7a in a manner that is suitable when the contact angle is varied over a wide range of values, including a region of "neutral or intermediate wettability" in the vicinity of 90° where neither of the two fluids can be considered the wetting fluid.

In the neutral wettability region, the terms "wetting phase" and "nonwetting phase" have no meaning. In every displacement or relative permeability experiment, however, one of the two phases is under a higher pressure than the other. Therefore, it is useful to distinguish between the two phases quite generally by referring to "the phase at the higher pressure" and "the phase at the lower pressure."

When specifying a relative permeability curve in the most general terms, two attributes have to be stated: whether the fluid is at the higher or at the lower pressure and whether the fluid is the displacing or the displaced phase. Accordingly, the following four types of relative permeability curves may be distinguished (in parentheses the conventional terminology is indicated):

(1) The fluid at the lower pressure, being displaced (wetting fluid, in drainage).

(2) The fluid at the higher pressure, displacing (nonwetting fluid, in drainage).

(3) The fluid at the lower pressure, displacing (wetting fluid, in imbibition).

(4) The fluid at the higher pressure, being displaced (nonwetting fluid, in imbibition).

For the neutral wettability region, only the curves (1) and (2) exist, because the displacing fluid is always at the higher pressure (and the displaced fluid is always at the lower pressure) in this case.

Generalization of the conventional way of plotting the relative permeability curves versus the wetting phase saturation (*or,* the nonwetting phase saturation) consists of plotting the four curves versus the saturation of the phase at the lower pressure (*or,* the saturation of the phase at the higher pressure).

In Fig. 5.7a the data representing the fluid at the lower pressure, being displaced from 100% pore volume saturation, can be fitted by three curves. The middle curve fits heptane and dodecane, being displaced by nitrogen, and also nitrogen, being displaced by α-bromonaphthalene or ethylene glycol. The upper curve fits ethylene glycol and water, being displaced by nitrogen. The lower curve fits nitrogen and dodecane, being displaced by water. Finally, dioctyl ether and α-bromonaphthalene, being displaced by nitrogen, lie between the upper and the middle curves. The two sets of dodecane data, as well as the heptane and the dioctyl ether data correspond to a wetting phase, whereas the two sets of nitrogen data, as well as the ethylene glycol, the water and the σ-bromonaphthalene data represent neutral wetting conditions; but no corresponding trend of the data with the contact angle is discernible.

The data, representing the displacing fluid at the higher pressure, lie even closer together and can be fitted by two curves. The upper curve fits nitrogen, displacing heptane, dodecane, dioctyl ether, α-bromonaphthalene, or ethylene glycol, as well as water, displacing nitrogen water, as well as α-bromonaphthalene and ethylene glycol, displacing nitrogen. There is no trend with the contact angle discernible in this case, either.

The data of McCaffery such that the displacing phase is at the lower pressure are limited to the four systems heptane/nitrogen, dodecane/nitrogen, dioctyl ether/nitrogen, and dodecane/water. The experiments were started at the "irreducible" saturation of the fluid at the lower pressure. In the other three systems, which were of neutral wettability, displacement was possible only with the displacing phase at the higher pressure, no matter which of the two fluids was the displacing phase. In other words, no spontaneous imbibition could be observed in these systems. The fluids at the higher pressure (i.e., nitrogen) displaced by heptane, dodecane, or dioctyl ether, and water, displaced by dodecane, can be fitted by a single curve that first lies above, then intersects with, and then lies below the two curves corresponding to the displacing fluid at the higher pressure. The fluids at the lower pressure (i.e., dodecane) displacing nitrogen, or water, as well as heptane, and dioctyl ether, displacing nitrogen practically follow the same path traced in those experiments when the same fluids were displaced by their opposite number. In other words, as usual, there is very little if any hysteresis for the wetting phase. No trend with the contact angle is apparent in this case, either. The data obtained in a special run, in which displacement of nitrogen by heptane was started at zero percent heptane saturation, coincided with those obtained in the other test where displacement was started at 30% heptane saturation.

Interesting, and anomalous, behavior is exhibited by two systems [i.e., dioctyl either (1)/nitrogen (2) and dodecane (1)/water (2)] when fluid (1) displaced fluid (2) starting at zero saturation of fluid (1). (see Fig. 5.7b). In the case of the first of these two systems, dioctyl ether would not initially penetrate the core plug unless forced by positive pressure differential (McCaffery and Bennion, 1974), whereas it was imbibed spontaneously in the other test, which was started at "irreducible" ether saturation. Only initially does the nitrogen curve follow the uppermost of the three curves corresponding to the displaced phase at the lower pressure in Fig. 5.7a, and then it rapidly assumes much higher values, which is characteristic of the phase at the higher pressure. Similarly, the dioctyl ether curve follows initially the same trend as displayed by a displacing phase at the higher pressure and then it gradually assumes lower values, typical of the phase at the lower pressure.

In the case of the other system, spontaneous imbibition of dodecane into the core plug took place, much like when the run was started at "irreducible" dodecane saturation. However, the water relative permeability curve now lies very much lower and the dodecane relative permeability curve is very much higher than the corresponding curves obtained when starting at "irreducible" dodecane saturation. In this case, the cycle of relative permeability tests was started with the core plug 100% saturated with dodecane. It is likely that, at the "irreducible" dodecane saturation, the pore walls were covered with a film of dodecane and hence, imbibition of dodecane took place (Dullien *et al.*, 1986) via the surface channels containing dodecane. When the experiment was started at zero dodecane saturation, however, there was no surface film of dodecane present at the beginning of the run and apparently at 42° contact angle the formation of a surface film was less favored energetically than the direct piston-like imbibition into the pores. As, in this case, the

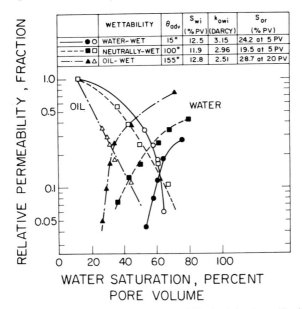

Figure 5.8.   Effects of wettability on relative permeability in dolomite pack with water and oil treated with octanoic acid. Relative permeabilities are based on the effective oil permeability at the initial water saturation (Morrow *et al.*, 1973).

dodecane always occupies the entire pore cross section, rather than only the space next to the pore walls, it is logical that, at a given saturation, the relative permeability to it will be greater. As for water, the nonwetting fluid in this case, at a given saturation more of its channels can be expected to be cut off by the dodecane than if part of the dodecane were present in the form of a film on the pore walls.

The existence of distinct "neutral wettability" relative permeability curves has been clearly demonstrated by Morrow *et al.* (1973) in dolomite packs with water and oil treated with octanoic acid (see Fig. 5.8).

5.2.2.2.   *The Effect of Viscosity Ratio on the Relative Permeabilities*

Normally, whenever the capillary number is low (e.g., less than about 10^{-6}) one of the fluids is strongly wetting (i.e., capillary forces are in control), and the saturation is in the intermediate range, in co-current steady flow each fluid flows in a separate network of pores and the two streams do not influence each other. The flow of the wetting fluid in the film present between the pore walls and the nonwetting fluid, shown in Fig. 5.1b, is then negligible. Under these conditions, a viscosity ratio of nonwetting fluid/

wetting fluid greater than 1 has no effect on the relative permeabilities (Osoba *et al.*, 1951; Richardson *et al.*, 1955; Schneider and Owens, 1980).

The conditions are quite different when the wetting phase flows *only* in the form of thick films between the pore walls and the nonwetting fluid. This situation exists in just about any porous medium in a relative permeability test at wetting phase saturations near the lower end of the saturation range, where all pores have been penetrated by the nonwetting phase, and even at intermediate wetting phase saturations in porous media characterized by uniform pore throat sizes. The wetting phase is flowing in thick films and is in strong hydraulic coupling with the nonwetting phase flowing in the central portion of the pores. This coupling increases with decreasing pore throat size (Odeh, 1959). As a result, the nonwetting fluid may experience an apparent hydraulic slip. In such an instance, Eqs. (5.2.2) to (5.2.4) do not apply, because they assume independence of the two flows. If these equations are applied regardless, using the actual viscosity of each fluid instead of an effective viscosity of the nonwetting fluid then an apparent relative permeability of the nonwetting fluid is found that increases with viscosity ratios nonwetting fluid/wetting fluid > 1 (Odeh, 1959). This is because the effect of an effective viscosity different than the value used in Eqs. (5.2.2) to (5.2.4) is simply transferred to the calculated value of the relative permeability. At the

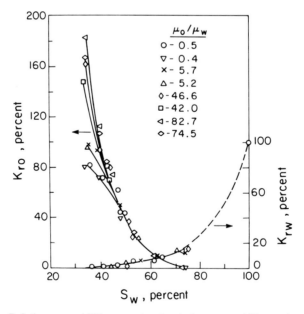

Figure 5.9.   Relative permeabilities vs. saturation in low permeability sandstone core (Odeh, 1959).

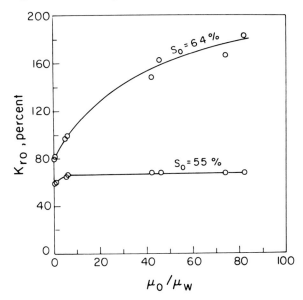

Figure 5.10. Relative permeability to oil vs. viscosity ratio at two low water saturations in the core illustrated in Fig. 5.9 (Odeh, 1959).

same time, the flow of the less viscous wetting fluid is unaffected by the viscosity of the nonwetting fluid and, therefore, the relative permeability of the wetting fluid does not vary with the viscosity ratio (Odeh, 1959). Some of Odeh's results are shown in Figs. 5.9 and 5.10. Odeh's results have been confirmed also by the data of Danis and Jacquin (1983) as illustrated in Fig. 5.11.

Templeton (1954) measured viscosities of oil in capillaries of 3 to 40 μm inner diameters, after the oil displaced water from the tubes. Assuming no capillary pressure hysteresis effects and perfect displacement of water, lower values were obtained than the known viscosity of the oil (which was higher than that of water). This discrepancy could be reconciled by assuming the existence of an annular water film of 20–260 Å thickness left on the capillary wall, after the bulk of the water from the core of the capillary tube was displaced by the oil. The remaining film of water apparently had a "lubricating" effect on the flow of oil.

If either fluid is very viscous, capillary forces do not control the process any more and the less viscous fluid will channel through the centers of the pores, as shown in Fig. 5.1b. Regardless of whether the viscous fluid is preferentially wetting or not, if introduced into the sample first, it will adhere to the pore walls and it is impossible to create an arbitrary uniform saturation in the sample. In such situations, the viscosity ratio is evidently a very important parameter.

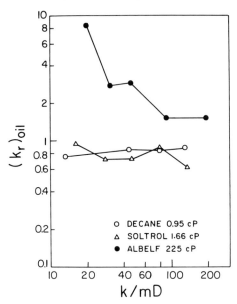

Figure 5.11. Relative permeabilities to oil of Rouffach limestone at connate water saturation (Danis and Jacquin, 1983).

5.2.2.3. *The Effect of Interfacial Tension and Total Flow Rate on the Relative Permeabilities*

As long as the capillary number is sufficiently low, and both viscosities are low, at the usual slow flow rates used, the relative permeabilities do not vary with the interfacial tension (Amaefule and Handy, 1982), because the fluid/fluid interfaces are stable and the independent networks, in which the two fluids flow, remain the same. As the interfacial tension is sufficiently lowered, however, the networks of the two fluids begin to break up, first in the relatively large pores and then, at even lower interfacial tension, also in the finer pores. As a result, slugs of both fluids start traveling in the same pores with the wetting fluid continuous and the nonwetting fluid discontinuous. Under these conditions, the flows of the two fluids are coupled and Eqs. (5.2.2) to (5.2.4) do not apply, because they assume that the two flows are independent. The relative permeabilities calculated by these equations are apparent values. Under such conditions, the apparent relative permeability of a given nonwetting fluid can be expected to decrease with increasing viscosity of the wetting fluid at a fixed intermediate-to-high wetting fluid saturation because, with everything else unchanged, the flow rate of the slugs of the nonwetting fluid embedded in the wetting fluid decreases with increasing viscosity of the continuous wetting fluid.

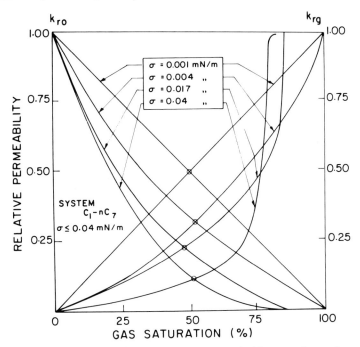

Figure 5.12. Gas–oil "relative permeability" of Fontainebleau sandstone for very low interfacial tension values with a C_1–nC_7 system for $\sigma \leq 0.04$ mN/m (Bardon and Longeron, 1978 © SPE-AIME).

At low interfacial tensions and intermediate-to-high wetting fluid saturations, increased total flow rates result in increasing apparent relative permeabilities of each fluid, in much the same way as this has been observed to happen at a fixed total flow rate and decreasing interfacial tensions (Amaefule and Handy, 1981; Fulcher *et al.*, 1985). As the interfacial tension approaches zero, at a viscosity ratio of one, the two apparent relative permeability curves become straight diagonal lines as each fluid flows in every pore in proportion to its saturation in the sample (Bardon and Longeron, 1978). (See Fig. 5.12.) While at high interfacial tensions, variation of the fluid velocity in the limited range of 4.9 to 24 m/d did not result in a noticeable change in relative permeabilities (Fulcher *et al.*, 1985), it has been shown in water-wet sandstone, with permeabilities that varied over two orders of magnitude, that the trapped residual oil (i.e., zero relative permeability to oil) could be made to flow by increasing the flow rate of water through the core by one to three order of magnitude (thereby changing the apparent relative permeability to oil from zero to finite values) (Chatzis andMorrow, 1984). Following mobilization of oil, water relative permeabilities increased markedly with reduction in oil saturation (Morrow *et al.*, 1985). The equivalence of increasing total flow and decreasing interfacial tension has been amply demonstrated, at least at relatively high wetting phase saturations (Taber, 1969; Taber *et al.*, 1973), where the main concern was to mobilize and recover trapped residual oil. The criterion of mobilization and complete recovery has been expressed in terms of the ratio of viscous-to-capillary forces (i.e., the "capillary number," which has been written in many different forms). In Chatzis and Morrow (1984), it was found that the criterion of mobilization and complete recovery of residual oil corresponded to values of the capillary number $k_i \Delta P_i / L\sigma$ of about 2×10^{-5} and 1.5×10^{-3}, respectively (ΔP_i is the pressure drop in the displacing fluid, i.e., water).

5.2.2.4. *Apparent Relative Permeabilities*

In summary, it may be stated that at high capillary numbers and when viscosities are great, capillary forces do not control the flow and the nonwetting fluid is broken up and dispersed in the wetting fluid, as shown in Fig. 5.1d. In these situations, the capillary number $\mu_i v_i / \sigma$ ($i = 1, 2$), the Bond number $(\rho_1 - \rho_2)g\ell^2/\sigma$, and also ρ_1/ρ_2 and μ_1/μ_2 become important parameters.

Darcy's law, generalized for two-phase flow, is a linear relationship between the fluid velocities v_i, on the one hand, and the reciprocal viscosities $(1/\mu_i)$, as well as the pressure drop over the sample, ΔP_i, on the other. Whenever the relative permeabilities change with the capillary number, they also change with μ_i and ΔP_i (or, equivalently, with μ_i and v_i), for these are contained in the capillary number [i.e., $k_{ri} = k_{ri}(\Delta P_i, \mu_i)$]. Hence, v_i is no longer a linear function of $(1/\mu_i)$ and ΔP_i. The term "apparent relative permeability" has been introduced here to indicate the fact that $k_{ri}(\Delta P_i, \mu_i)$ in Eqs. (5.2.2) to (5.2.4) depends on ΔP_i and μ_i.

5.2.2.5. *Extension of Darcy's Law for Two-Phase Flow with Coupling*

The coupling effects pointed out in the previous sections have led to the formulation of coupled Darcy's laws by Rose (1972, 1974, 1988), de Gennes (1983), de la Cruz and Spanos (1983), Auriault and Sanchez-Palencia (1986), Whitaker (1986), Auriault (1987), Spanos *et al.* (1986), Kalaydjian (1987), Kalaydjian and Legait (1987), Spanos *et al.* (1988), and Kalaydjian (1990). Equivalent formulations of these relationships are as follows:

$$\mathbf{v}_1 = -\frac{k_{11}}{\mu_1} \nabla P_1 - \frac{k_{12}}{\mu_2} \nabla P_2, \qquad (5.2.6a)$$

$$\mathbf{v}_2 = -\frac{k_{21}}{\mu_1} \nabla P_1 - \frac{k_{22}}{\mu_2} \nabla P_2, \qquad (5.2.6b)$$

and

$$-\nabla P_1 = \mu_1 \left(\frac{1}{k'_{11}} \mathbf{v}_1 - \frac{1}{k'_{12}} \mathbf{v}_2 \right), \qquad (5.2.7\text{a})$$

$$-\nabla P_2 = \mu_2 \left(\frac{1}{k'_{22}} \mathbf{v}_2 - \frac{1}{k'_{21}} \mathbf{v}_1 \right), \qquad (5.2.7\text{b})$$

where

$$k'_{11} = k_{11} - \frac{k_{12} k_{21}}{k_{22}}, \qquad (5.2.8\text{a})$$

$$k'_{12} = \frac{k_{11} k_{22}}{k_{12}} - k_{21}, \qquad (5.2.8\text{b})$$

$$k'_{22} = k_{22} - \frac{k_{12} k_{21}}{k_{11}}, \qquad (5.2.8\text{c})$$

$$k'_{21} = \frac{k_{11} k_{22}}{k_{21}} - k_{12}. \qquad (5.2.8\text{d})$$

Note that for $k_{12} = k_{21} = 0$ we have $k'_{11} = k_{11}, k'_{22} = k_{22}, 1/k'_{12} = 1/k'_{21} = 0$, and thus Eqs. (5.2.6) and (5.2.7) reduce to the conventional, uncoupled forms of Darcy's law for two-phase flow. The physical meaning of the equations with co-upling is simple. The leading terms, containing the effective phase permeabilities k_{ii} and k'_{ii}, establish the relationship between the flow rate and the pressure gradient in phase i that would exist if the other phase j were a solid. This assumption is a good one for steady co-current two-phase flow at intermediate saturations under conditions of sufficiently low capillary number and low viscosities of both fluids, as has been demonstrated by Yadav *et al.* (1987) in experiments where either one or the other of the two fluid phases was replaced by a solid using "phase immobilization" techniques and the permeabilities of samples containing different saturations of the immobilized phase were measured. As shown in Fig. 5.13, these permeabilities, when expressed as relative permeabilities, compared quite well with the conventional relative permeabilties of similar Berea sandstone.

Equations (5.2.6) show that the flow rate of fluid i *may be* altered, if the other fluid j is also flowing, by an amount that is proportional to the pressure gradient of fluid j, ∇P_j, and inversely proportional to μ_j, the viscosity of fluid j. The only requirement of the proportionality coefficient k_{ij} is that is must be independent of ΔP_j and μ_j. It is logical that the value of k_{ij}, for flow of two immiscible fluids through a porous medium at a given saturation will depend on the type of flow, the way the two fluids are distributed in the

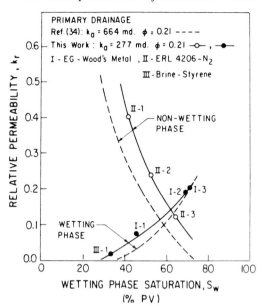

Figure 5.13. Relative-permeability vs. saturation curves for Berea sandstone sample using three different fluid pairs. The wetting phases, ethylene glycol/Wood's metal (System I), ERL 4206 resin/N_2 (System II), and brine/styrene (System III) were all strongly wetting the rock surface. The capillary pressure (in kilopascals) for the points on the plot are I-1 = 100.05; I-2 = 89.7, I-3 = 62.1, II-1 = 241.0, II-2 = 282.9, II-3 = 414, and III-1 = 34.5 kPa (Yadav *et al.*, 1987). (The dashed lines are from Shankar and Dullien, 1979.)

pores of the pore network, and on the values of the capillary number and the viscosity ratio.

The simplest case is steady, co-current flow established in the course of a drainage type displacement. Over the saturation range extending from the breakthrough capillary pressure to low wetting phase saturations, the wetting phase (fluid 2) flows predominantly in pore channels it does not share with the nonwetting fluid (fluid 1), whereas the nonwetting phase flows in the central part of pore channels where it is surrounded by the wetting fluid. Naturally, some of the wetting fluid in these pores is also flowing (because there is no hydraulic slip at the wetting fluid/nonwetting fluid interface) but this flow contributes practically nothing to the total flow rate of the wetting fluid as long as there are continuous pore channels that are completely filled with wetting fluid, because the flow in the latter short-circuits the flow of the wetting fluid in the pores containing also nonwetting fluid. Therefore coupling is very weak; k_{12} is very small. As the saturation level is reached, however, where there are no more pore channels left in which only the wetting phase flows, there is a pressure gradient ∇P_2 set up in the "film" of wetting fluid contained *throughout the sample*, caused by momentum transfer by the nonwetting fluid. As a result, the wetting phase will flow within the sample; and even if there is very little or no wetting fluid leaving the sample, coupling will become noticeable at viscosity ratios $\mu_2/\mu_1 \gg 1$, because of an apparent hydraulic slip (lubricating) effect. Under these conditions k_{12} may become significant, as illustrated by the example taken from the data of Danis and Jacquin (1983). Naturally, the physical picture of hydraulic slip presented here implies the existence of a slowly changing saturation gradient in the direction of flow.

As it is apparent from Fig. 5.11, at $k = 20$ md, for $\mu_1/\mu_2 = 225$, $k_{r1} = 9$ whereas for $\mu_1/\mu_2 \approx 1$, $k_{r1} = 0.8$. Assuming that $\nabla P_1 = \nabla P_2 = \nabla P$, one can write (d = decane, A = Albelf):

$$\frac{k_d}{\mu_d} = \frac{k_{11}}{\mu_d} + \frac{k_{12}}{\mu_w}, \quad \text{or,} \quad 0.8 \times 20 = k_{11} + k_{12}$$

and

$$\frac{k_A}{\mu_A} = \frac{k_{11}}{\mu_A} + \frac{k_{12}}{\mu_w}, \quad \text{or} \quad \frac{9 \times 20}{225} = \frac{k_{11}}{225} + k_{12},$$

Simultaneous solution of these equations gives $k_{11} = 15.268$ mD and $k_{12} = 0.732$ mD.

Danis and Jacquin (1983) used numerical model calculations of two-phase flow in bundles of periodically constricted capillary tubes of various irregular cross sections to assess the effect of viscosity ratio on the conventional relative permeabilities. Their calculations showed, in agreement with experiment, no effect of the viscosity ratio on the wetting phase relative permeability and an increasing trend of the nonwetting phase relative permeability with increasing oil/water viscosity ratio over the entire saturation range, extending from 0 to 100%. Experiments, however, show absolutely no lubrication effect, except at very low (connate) water saturations. The probable reasons for this discrepancy are that Danis and Jacquin (1) used a bundle of tubes model that displayed hydraulic slip at all saturations, whereas in a pore network slip becomes significant only at saturation such that water does not flow any more by itself; (2) assumed absolutely no macroscopic flow of water in those tubes containing oil. As a result of this assumption they calculated relative permeabilities to oil at low water saturations that were much less than the measured values.

The original explanation of oil relative permeabilities greater than one at connate water saturations, due to Odeh (1959), assumes a constant film thickness of water independent of the capillary pressure; and it hinges also on the argument that the effect of apparent hydrodynamic slip in relatively large pore channels is negligibly small and only the finest pore channels contribute the extra flow that results in the anomalously high oil relative permeabilities. In view of the fact that the finest pores contribute only very little to the overall flow, Odeh's explanation is implausible. Odeh's theory, however, qualitatively explains the observation that the anomalous oil relative permeability effect is the greater the lower the permeability of the media, because it is plausible to assume that the slip effect should increase with decreasing pore size.

It is interesting to contemplate application of Eqs. (5.2.6) to the case when one of the two fluids is emulsified in the other, as shown in Fig. 5.1d. Let us assume that fluid 2 is emulsified in fluid 1. In this case, Eqs. (5.2.5) become, as $\nabla P_2 = 0$:

$$\mathbf{v}_1 = -\frac{k_{11}}{\mu_1} \nabla P_1 \quad \text{and} \quad \mathbf{v}_2 = -\frac{k_{21}}{\mu_1} \nabla P_1.$$

As long as fluid 1 remains the continuous phase, $k_{11} = S_1 k$ and $k_{21} = S_2 k$, where S_1 and S_2 are the saturations of the two fluids, provided that $\mu_1 = \mu_2$. If $\mu_1 \neq \mu_2$ the viscosity of the emulsion will not be μ_1.

5.2.2.6.   *Fractionally Wet and Mixed-Wet Systems*

The above discussions were limited to the case when the porous medium is uniformly wetted (i.e., θ is the same everywhere). There have been instances reported in the literature (e.g., Salathiel, 1973; Dullien *et al.*, 1990) where natural porous media were "mixed-wet"; that is, certain parts of the pore surface were preferentially wetted by fluid 1 and the rest by fluid 2. The reason for this phenomenon is probably the deposition of certain components of crude oil on portions of the surface, thus rendering them oil-wet. The distribution of the preferentially water-wet and oil-wet portions of the pore surface varies from case to case. If both types of pore surface form a continuum, it may happen that both fluids will flow side by side in the same pore, as sketched in Fig. 5.1c. Braun and Blackwell (1981) have measured the steady-state relative permeability curves shown in Fig. 5.14 in Berea sandstone core artificially aged with crude oil. Note the unusually high permeabilities to water at low water saturations and the low permeabilities to oil that become 0 past 48% water saturation that parallel the trends found in Pembina Cardium cores (Dullien et al., 1990).

5.2.2.7.   *Three-Phase Relative Permeabilities*

In reservoir engineering there is often simultaneous flow of oil, water and gas. The mechanism of three-phase flow has been studied on the pore level only in special cases (Chatzis *et al.*, 1988). It was found that oil often spreads at the water/gas interface with the creation of an oil/water and an oil/gas interface (see Fig. 5.15). In this case strong coupling between the three flows can be expected, depending on the conditions of flow, and measured conventional (apparent) three-phase relative permeabilities can be expected to be

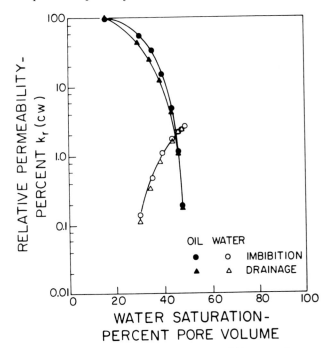

Figure 5.14. Steady-state relative permeability curves of Berea sandstone core aged with crude oil (Braun and Blackwell, 1981).

valid only for the particular oil/water/gas system for which they were determined and only for the conditions, including history of the system, under which they were measured.

Measured three-phase relative permeability curves can be occasionally very complicated owing to the dependence on two saturations rather than just one. A good review of this difficult field, including description of experimental techniques and data presentation, has been given by Oak *et al.* (1988).

5.2.2.8. *Effect of Pore Structure on Relative Permeabilities*

As pointed out by Melrose (1965), among others, the most likely reason for the lack of spontaneous imbibition of a "wetting phase," as observed, for example, by McCaffery and Bennion (1974), is the fact that the size of the cross-sectional area varies along the axis of the capillary and the surface is uneven, rough. When the contact angle in the "wetting" phase is greater than a certain critical value, the curvature of the meniscus may become zero (or negative) in a divergent capillary segment (see Fig. 5.16). The value of the critical angle is determined by pore geometry. At such a point where the

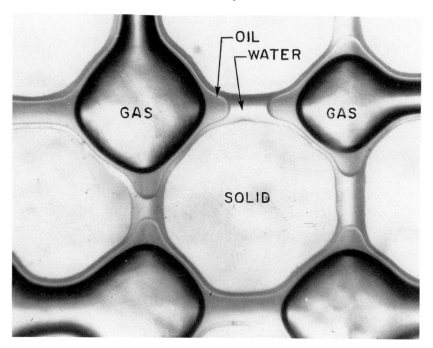

Figure 5.15. Spontaneous spreading of oil blobs on the surface of connate water in a capillary micromodel after being contacted by air (Chatzis *et al.*, 1988).

meniscus has zero curvature, the capillary suction becomes zero and the imbibition comes to a halt.

The influence of pore structure on water–oil relative permeabilities has been studied qualitatively by Morgan and Gordon (1970). They investigated photomicrographs of thin sections of a large number of different natural porous media, such as sandstones and limestones, and compared their microscopic observations with the imbibition relative permeability characteristics found with these samples. Some of the different curves obtained by them have been plotted in Fig. 5.17 to show the range of the variations. It is obvious from this figure that pore geometry can have a most profound effect on relative permeabilities.

Morgan and Gordon (1970) have concluded that rocks with large pores and correspondingly small specific surface areas have low "irreducible" water saturations (~ 0.18) that leave a relatively large amount of pore space available for the flow of fluids. Hence, in such samples the end points of the relative permeability curves are high and a large saturation change (0.18–0.70) may occur during the two-phase flow. By contrast, rocks with small pores have larger surface areas and "irreducible" water saturations that leave less

WETTING | NONWETTING
PHASE | PHASE

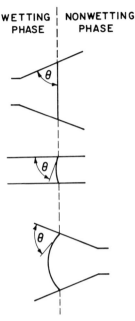

Figure 5.16. Effect of position of meniscus in a convergent–divergent capillary tube on the value of capillary pressure.

room for the flow of fluids. Under these conditions the end points of the relative permeability curves are lower and the saturation change during two-phase flow is smaller (0.33–0.73). Finally, rocks containing some relatively large pores connected by small pores have large surface area, resulting in high connate water saturation (0.38) and a relative permeability behavior similar to the case of small pores only. The complete range of S_{wi} was found to extend from 0.15 to 0.65.

A few additional rules of thumb may be given on the effects of pore structure on relative permeabilities. In drainage, the nonwetting phase displaces the wetting phase from the pores in a piston-like manner. The hierarchy of displacement is in the direction going from large to small pores and the invading nonwetting phase by-passes and cuts off the path of the wetting phase. As a result, the relative permeability to the nonwetting phase at a given saturation, say, $S_{nw} = 0.4$ is much higher than the relative permeability to the wetting phase at the corresponding saturation $S_w = 0.4$. In imbibition the displacement is not piston-like. Wetting phase "chokes off" the nonwetting phase in pore throats and in this way reduces its relative permeability as compared with the drainage case, while the relative permeability to the wetting phase usually is not very different because the bulk of

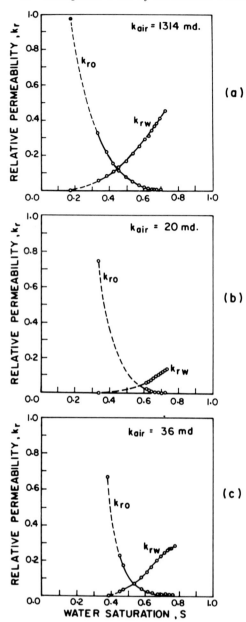

Figure 5.17.   (a) Sandstone containing large, well-connected pores with k_{air} = 1314 md. (b) Sandstone containing small, well-connected pores with k_{air} = 20 md. (c) Sandstone containing few large pores that are connected by small pores with k_{air} = 36 md (Morgan and Gordon, 1970 © SPE-AIME).

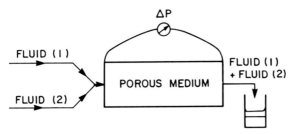

Figure 5.18. Schematic diagram of steady-state relative permeability apparatus (Dullien, 1988).

the wetting phase imbibes into the same pores from which it was displaced in drainage. Evidently, a great deal more work remains to be done in order to relate the relative permeability behavior quantitatively to the pore geometry.

5.2.3. Experimental Methods of Measuring Relative Permeabilities

The methods available for the measurement of relative permeabilities can be divided into two categories: steady and unsteady flow tests.

5.2.3.1. *Steady Flow Tests*

In a typical steady-state method the two liquids are injected simultaneously at a fixed ratio and known, metered flow rates (see Fig. 5.18). The criterion of steady state is determined by the condition that the inflows equal the outflows and/or that constant pressure drop has been reached across the sample. The attainment of steady state may take anywhere from 2 to 40 hours or even longer, depending on the sample permeability and the method used.

If one of the fluids is preferentially wetting, it will flow into the sample under zero pressure differential; but for the nonwetting fluid to be able to penetrate the porous medium, first a pressure difference $P_c = P'' - P'$, at least as great as the breakthrough pressure P_{cb}, must develop between the two fluids at the inlet face of the sample. This pressure difference increases with increasing pumping rate ratio: nonwetting fluid/wetting fluid. At a constant ratio of pumping rates, eventually constant flow rates of both fluids are established along the entire length of the sample that correspond to steady two-phase flow.

It is customary to assume that, in steady two-phase flow through a homogeneous sample, the saturation S_w is uniform in the entire sample, except near the outflow end, where the boundary condition of zero capillary pressure on the outside results in imbibition of the wetting phase (i.e., increased S_w). There are no physical grounds, however, to generally assume uniform S_w far from the outlet face either. Uniform S_w implies uniform value of the capillary pressure P_c which, in turn, implies identical pressure gradients in both the wetting and the nonwetting fluid [see Eq. (5.2.5)]. Assuming that for some particular ratio of flow rates the pressure gradients are equal, one would expect that increasing the flow rate of the nonwetting fluid while keeping the flow rate of the wetting fluid unchanged might result in a greater pressure gradient in the nonwetting fluid than in the wetting fluid, because the concomitant shift toward decreased S_w may not be sufficient to offset the effect of increased ratio of flow rates: nonwetting fluid/wetting fluid. The fact that there is no automatic adjustment for the pressure gradients to remain equal was demonstrated by Osoba *et al.* (1951) where the sample was placed between two porous discs (semipermeable membranes) that were permeable only to the oil over the range of capillary pressures used in the experiment (Hassler method). The nonwetting phase was gas which flowed into and out of the sample through radial grooves in the face of the porous discs lying against the sample. In this arrangement, the pressure drops in the oil and in the gas, respectively, could be measured and regulated independently. For every oil flow rate, the pressure drop in the gas was regulated by a valve in the gas line downstream from the sample to set it equal to the pressure drop in the oil. Without the use of semipermeable membranes, there is no way to regulate the two pressure drops independently and they self-adjust to values that may generally be different from each other. There are some instances in the literature where the saturation profile in the sample has been determined either by electrical conductivity measurement (Chatzis and Morrow, 1984) or by x-ray computed tomography (Withjack, 1987). In both instances, small gradients in S_w have been found that kept changing with the flow rate ratio. For high flow ratios nonwetting fluid/wetting fluid, S_w tends to be lower at the inlet end than at the outlet end, whereas for low flow ratios the converse is the case, as expected.

If the porous medium is initially saturated with the wetting fluid and, subsequently, nonwetting fluid is pumped into the sample either at a fixed flow rate or at a fixed pressure, there is nonsteady flow as long as there is displacement, and the saturation keeps changing. At the end of the displacement process, a certain amount of residual wetting fluid remains in the sample much like at the end of a drainage capillary pressure test. When there is no displacement, there is a steady single-phase flow of the nonwetting fluid. The special effects due hydraulic coupling have been discussed in Sections 5.2.2.2, 5.2.2.4, and 5.2.2.5.

Displacement of the nonwetting fluid by the preferentially wetting fluid, pumped into the sample at the residual wetting phase saturation, runs a similar course and the process ends with stationary residual nonwetting fluid remaining in the sample. The residual nonwetting phase saturation in a given sample depends on the capillary number, which can be expressed in a number of different ways—for example, $\mu_i v_i / \sigma$, where μ_i and v_i, are the viscosity and the superficial velocity, respectively, of the wetting fluid (Chatzis and Morrow, 1984). In addition, the residual nonwetting phase saturation depends also on the pore structure.

The pressure drop is usually measured in the fluid under the higher pressure (i.e., the nonwetting fluid) and it is assumed that the pressure drop in the wetting fluid is the same, on the grounds that the saturation is the same everywhere under steady flow conditions. Richardson *et al.* (1952) measured the pressure both in the oil, the wetting phase, using porous porcelain semipermeable membranes and in the helium gas, the nonwetting phase, by means of piezometric rings. In the case of oil/water systems there is no known record of measuring the pressures in both phases. From the measured pressure drop across the sample, the flow rates, and the dimensions of the sample, the relative permeability is calculated by using Eq. (5.2.3).

The methods available for the determination of saturations include direct weighing, x-ray absorption, electrical resistivity measurement, and use of a volumetric balance on the fluids passing through the core. In addition, there are several more recent methods, such as the gamma-ray absorption technique introduced by Reid (1958) and the neutron bombardment method introduced by Snell (1962). According to Fatt and Saraf (1967), these two methods have the disadvantage that the total liquid saturation is all that can be measured and, therefore, another method is required to determine the saturation of at least one of the phases. Caudle *et al.* (1951) used vacuum distillation, and Fatt and Saraf (1967) applied the technique of nuclear magnetic resonance (NMR) for measuring fluid saturations. The NMR technique is based on the fact that when a sample containing atoms whose nuclei have a nonzero spin quantum number is placed in a static magnetic field and is simultaneously irradiated with a small radio-frequency field rotating in a place perpendicular to the static magnetic field, the nuclei absorb energy from the radio-frequency field at the resonant frequency. The energy absorbed is proportional to the number of nuclei present. The most recent methods of saturation measurement include x-ray computed tomography (CAT scanning) (e.g., Withjack, 1987; Wellington and Vinegar, 1985, 1987).

A large number of different techniques have been developed for the measurement of relative permeability saturation relations in the laboratory. These essentially differ from one another in the manner in which boundary or end effects are minimized or eliminated and in the manner in which the two fluids are introduced into the core. The relative merits of the various methods have been discussed and presented in tabular form by Scheidegger (1974).

Richardson *et al.* (1952) compared drainage relative permeabilities obtained by the following six methods: Penn State, single-core dynamic, dispersed feed, Hafford technique, Hassler technique, and gas-drive technique. The conclusion was reached that determinations of relative permeability saturations by the Penn State, Hassler, Hafford, and dispersed feed systems all yield essentially the same drainage relative permeability-saturation relations for all sample lengths.

The Hassler method (1944) used the most direct approach to eliminate end effects by using membranes permeable only to the wetting phase at both ends of the core in order to control and equalize the capillary pressure (Osoba and coworkers, 1951). The other steady-state methods rely either on large pressure gradients, as in the dispersed feed and the single core dynamic techniques (Osoba *et al.*, 1951; Loomis and Cromwell, 1962), or on the three-section core assembly technique used in the Penn State method (Morse *et al.*, 1947; Geffen *et al.*, 1951) to eliminate or render insignificant the end effects.

It has been observed that if there is no saturation gradient present in the core due to end effects and if the flow rate is not high enough to cause inertial effects, laboratory relative permeabilities are independent of flow rates over a reasonable range of values.

Recently, the three-core Penn State technique has acquired popularity. The method is flexible in the sense that low flow rates may be employed because the end effects are outside the central test section. This technique can be applied to either liquid–liquid or liquid–gas systems for either drainage or imbibition process. It also provides for easily controlled stepwise changes in saturation. It has been effectively employed in permeability studies of systems covering all possible wettabilities (McCaffery and Bennion, 1974). In addition, the three-section core assembly permits easy dismantling of the test section for saturation determination by weighing. All the same, the method is slow and takes anywhere from 8 hours upward to complete a relative permeability curve, depending on the permeability of the sample. A diagrammatic view of the experimental equipment is shown in Fig. 5.19.

The boundary or end effects were investigated by Richardson *et al.* (1952). For horizontal, unidirectional flow, assuming that P_c is a function only of the saturation, there follows:

$$\frac{dS_w}{dx} = \left(\frac{Q_w \mu_w}{k_w} - \frac{Q_{nw}\mu_{nw}}{k_{nw}} \right) \frac{1}{A} \frac{1}{dP_c/dS_w}. \qquad (5.2.9)$$

Richardson *et al.* (1952) arbitrarily selected rates of flow of the two fluids. This fixed the saturations at an infinite distance from the outflow end. Next, about 10 saturations in incremental steps were chosen between the saturation at infinite distance from the outflow end and the saturation at which the permeability to the nonwetting fluid becomes greater than zero. Experimental evidence has shown that this saturation exists at the outflow face. At each saturation the quantity dS_w/dx was determined from Eq. (5.2.9). Then the reciprocal dx/dS_w was plotted as a function of saturation and by graphical integration of this relation, the plot of the saturation as a function of the distance from the outflow end was obtained. By repeating this calculation for a number of fluid flow rates, a set of curves relating saturation and distance from the outflow end was obtained. These curves, representing the saturation

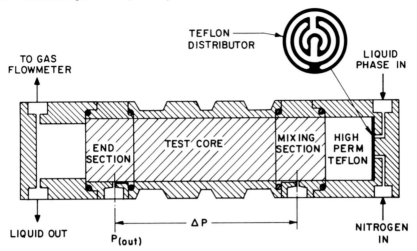

Figure 5.19. Core holder for steady-state relative permeability tests by Penn State method (McCaffery and Bennion, 1974).

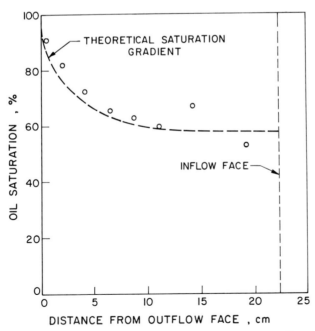

Figure 5.20. Comparison of experimental and theoretical saturation gradients due to boundary effect with $Q_g = 0.15$ cc/sec and $Q_0 = 0.000336$ cc/sec (Richardson *et al.*, 1952 © SPE-AIME).

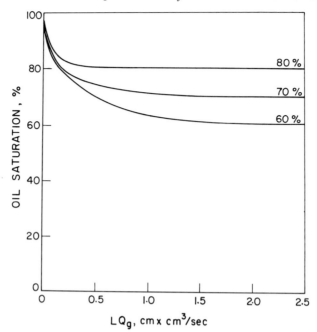

Figure 5.21. Calculated saturation gradients in Berea sandstone due to boundary effects with $A = 35.8 \text{ cm}^2$, $\mu_{\text{oil}} = 1.74$ cp, $\mu_{\text{gas}} = 0.0185$ cp, and L is the distance from the outflow face (Richardson *et al.*, 1952 © SPE-AIME).

gradient due to the end effect are shown in Fig. 5.20. Note that in these calculations oil was the wetting fluid and gas the nonwetting fluid.

Richardson *et al.* (1952) also studied experimentally the saturation distribution in a sample of porous material due to the boundary effect when two fluids were flowing simultaneously through the sample. An example of the resulting saturations compared with the saturation distribution that was predicted by Eq. (5.2.9) is shown in Fig. 5.21 for a Berea sandstone sample.

The same authors also analyzed the effect of gas expansion to produce a saturation gradient. They obtained good agreement between the theoretically calculated and experimentally measured gradients due to gas expansion in the sample. They concluded that the gas expansion effects have no important bearing on relative permeabilities under conditions where the gas expands severalfold along the flow path.

5.2.3.2.   *Unsteady Flow Methods*

In the unsteady-state (external drive) method, one phase is displaced from a core by pumping in the other phase and the relative permeability ratio is calculated from the produced fluid ratios. The method is due originally to

Welge (1952), who based it on an extension of the Buckley–Leverett frontal advance equation (see Section 5.3.3.3 where this method is discussed in detail).

Advantages of the unsteady-state method include speed and the fact that the relative permeabilities measured in this fashion will exactly reproduce the experimental oil recovery data (from which they were derived).

Johnson *et al.* (1959) presented a method whereby individual relative permeabilities could be obtained from the results of an external drive experiment.

A great deal of work has been done to check whether the unsteady-state relative permeabilities agree with those determined by steady-state techniques. The outcomes of these tests tend to confirm that the unsteady-state tests often give the same results as the steady-state techniques; however, there have also been numerous exceptions to this rule (e.g., Schneider and Owens, 1970; Owens *et al.*, 1965; Loomis and Crowell, 1962; Archer and Wong, 1971).

According to Craig (1971), the unsteady-state method has serious limitations for water–oil tests arising mainly from two conditions: (1) The pressure differentials used are in excess of 50 psi in order to eliminate outlet end effects; and (2) viscous oils are generally used in order to obtain relative permeabilities over the maximum possible saturation range. Otherwise, in imbibition-type tests the displacements at high flow rates would be pistonlike and the unsteady-state relative permeability curves would be limited to points at the connate (irreducible) water and the residual oil saturations (Archer and Wong, 1971). According to Craig, as a consequence of these two conditions, preferentially water-wet porous media frequently behave as if they were oil wet during an external water drive.

With the advent of modern techniques, such as sonic, microwave attenuation, CAT-scanning, permitting *in situ* saturation measurements, unsteady-state phase permeabilities can now be measured inside the sample as a function of position and time (e.g., Islam and Bentsen, 1986).

5.2.4. Mathematical Models of Relative Permeability

The various mathematical models of relative permeability may be conveniently classified under the following categories:

(1) capillaric models,
(2) statistical models,
(3) empirical models, and
(4) network models.

Capillaric models ignore the interconnected nature of the capillaries and, therefore, cannot give a realistic description of two-phase flow phenomena. The statistical models, unless they take into account and make use of the interconnectedness of the pores, cannot be more successful than the capillaric models. In order to bring the results of capillaric and statistical models in line with reality, empiricism had to be resorted to. At the present, the practically useful mathematical models of two-phase flow in porous media are basically empirical. Network models, however, have the best potential for understanding and quantitative description of two-phase flow phenomena in porous media, both because they take into account the interconnectedness of pore structure and because microscopic flow mechanisms are considered in them. Review of network models is delayed to Section 5.3.7, until after the discussion of microscopic displacement mechanisms.

5.2.4.1.   *Capillaric Models*

The model due to Fatt and Dykstra (1951) is based on the assumption of a bundle of capillary tubes in which the fluid path length L_e is greater than the length of the sample L in the macroscopic flow direction. In contrast with the usual way of modeling single-phase flow, in the case of two-phase flow it has been assumed that the effective fluid path length may be different in capillaries of different diameters. If Q is the total flow rate through all the tubes and N the number of the tubes, then one can write for the average flow rate per tube in the interval dN:

$$q_{av} = dQ/dN. \tag{5.2.10}$$

Straightforward derivation, involving differentiation of Darcy's law to obtain dQ/dk_i, gave the following result:

$$\frac{dk_i}{dS_i} = \frac{D_e^2 \phi L^2}{32 L_e^2} = \frac{D_e^2 \phi}{32 \Upsilon}. \tag{5.2.11}$$

This equation can be integrated only if Υ is known as a function of S_i. Various empirical relations have been used for this purpose, the physical meaning of which remains obscure.

Other models, using somewhat different manipulations of the capillary tubes, are discussed next.

5.2.4.2.   *Statistical Models*

The model of Wyllie and Gardner (1958) is discussed along with the "cut-and-random-rejoin" model due to Childs and Colis-George (1950).

Consider a bundle of capillary tubes with diameters ranging from a minimum value of $(D_e)_1$ to a maximum value of $(D_e)_2$ the frequencies of which are distributed according to a density function $\gamma(D_e)$.

Using the symbol S_{wi} to denote the "irreducible" wetting phase saturation, let us define the "drainable" porosity ϕ_{eff} as

$$\phi_{eff} = \phi(1 - S_{wi}) \tag{5.2.12}$$

and the saturation S_{eff} based on it as

$$S_{eff} = (S - S_{wi})/(1 - S_{wi}), \tag{5.2.13}$$

sometimes called also "reduced" saturation. Evidently, it is based on the pore space available at "irreducible" wetting phase saturation.

The bundle of capillary tubes is imagined to be cut into a large number of thin slices by planes perpendicular to the axis of the bundle, the short pieces of tubes in each slice are imagined to be rearranged randomly, and finally the slices are reassembled.

The reduced saturation S_{eff} of the wetting phase of the model has been expressed as follows:

$$S_{eff} = \int_{(D_e)_1}^{D_e} D_e^2 \gamma(D_e)\, dD_e \bigg/ \int_{(D_e)_1}^{(D_e)_2} D_e^2 \gamma(D_e)\, dD_e. \tag{5.2.14}$$

In any slice of area A of the model, the area $\phi_{eff} S_{eff} A$ is occupied by water in pores of diameter between $(D_e)_1$ and D_e. An equal area is occupied by the wetting fluid in neighboring slices; however, parts of these areas are not connected because of the random distribution of the pores in the model in each slice. Considering a point on the interface between the two neighboring slices, the probability that it lies in the wetting fluid in each slice is $\phi_{eff} S_{eff}$ and, hence, the probability that it lies in both slices simultaneously, thus making an interconnecting pore, is $(\phi_{eff} S_{eff})^2$. By similar reasoning, it can be shown that, since the probability of a pore filled with wetting fluid in a slice is $\phi_{eff} S_{eff}$, the area common to a single pore of cross-sectional area $\pi D_e^2/4$ in one slice and all the pores filled with wetting fluid in a neighboring slice is $\pi D_e^2 \phi_{eff} S_{eff}/4$.

It has been assumed that the passage of water from a pore of cross-sectional area $\pi D_e^2/4$ always takes place to a constricted area $\pi D_e^2 \phi_{eff} S_{eff}/4$. The constricted area is visualized as a pore of diameter D_e'' such that

$$D_e'' = (\phi_{eff} S_{eff})^{1/2} D_e. \tag{5.2.15}$$

The velocity v_w is obtained by integrating the flows expressed with the aid of the Hagen–Poiseuille equation over unit area of the section:

$$\begin{aligned}
v_w &= \frac{\phi_{eff}^3 S_{eff}^2}{32\mu} \frac{\partial P}{\partial x} \int_0^{S_{eff}} D_e^2\, dS_{eff} \\
&= -\frac{\phi_{eff}^3 S_{eff}^2 \sigma^2 \cos^2\theta}{2\mu} \frac{\partial P}{\partial x} \int_0^{S_{eff}} \frac{dS_{eff}}{P_c^2(S_{eff})}.
\end{aligned} \tag{5.2.16}$$

Combining this result with Darcy's law, the following expression is obtained for the effective wetting phase permeability:

$$k_w = \frac{\phi_{eff}^3 S_{eff}^2 \sigma^2 \cos^2 \theta}{2} \int_0^{S_{eff}} \frac{dS_{eff}}{P_c^2(S_{eff})}. \tag{5.2.17}$$

Assuming that the permeability of the sample is unaffected by the irreducible water saturation and that there is no residual nonwetting phase saturation (i.e., S_{eff} can be increased to unity) the following expression is obtained for the permeability of the sample at $S_{eff} = 1$:

$$k = \frac{\phi_{eff}^3 \sigma^2 \cos^2 \theta}{2} \int_0^1 \frac{dS_{eff}}{P_c^2(S_{eff})}. \tag{5.2.18}$$

The expressions of relative permeabilities follow immediately:

$$k_{rw} = S_{eff}^2 \frac{\displaystyle\int_0^{S_{eff}} \frac{dS_{eff}}{P_c^2(S_{eff})}}{\displaystyle\int_0^1 \frac{dS_{eff}}{P_c^2(S_{eff})}}, \tag{5.2.19a}$$

$$k_{rnw} = (1 - S_{eff})^2 \frac{\displaystyle\int_{S_{eff}}^1 \frac{dS_{eff}}{P_c^2(S_{eff})}}{\displaystyle\int_0^1 \frac{dS_{eff}}{P_c^2(S_{eff})}}. \tag{5.2.19b}$$

It is apparent from these equation, which were obtained by Burdine (1953) using hydraulic radius theory, that the relative permeability varies as S_{eff}^2. This behavior has been interpreted as the "tortuosity" in immiscible flow, whereas in reality, it is a result of the incomplete connectivity of the network formed by each fluid and the fluids being trapped in the pores and, therefore, not contributing to the measured flows.

The approach of Wyllie and Gardner previously described is quite similar to the so-called cut-and-random-rejoin model introduced originally by Childs and Collis-George (1950), then modified by Marshall (1958), and later by Millington and Quirk (1961). A good account of this method has been given more recently by Brutsaert (1967). Apparently the published literature on the cut-and-random-rejoin model escaped the attention of Wyllie and Gardner.

The cut-and-random-rejoin model is reviewed in some detail in Section 3.3.2.1 for the special case of saturated, single-phase flow. In this treatment the pore velocity is assumed to be given by the average velocity in the smaller of the two intersecting pores rather than in a pore of diameter D_e'' [Eq. (5.2.15)]. As a result, there is a difference between the two models, amounting to the factor $\phi_{eff} S_{eff}$.

Both models have been used extensively, with varying success. It is noted that the permeability predictions by these models are based on a slice having an effective areal porosity of either ϕ_{eff}^3 or ϕ_{eff}^2. This departure from the known porosity ϕ_{eff} is usually interpreted as a result of the "tortuosity" which, in the light of the present day knowledge of flow patterns on the pore level, is without any physical meaning.

5.2.4.3.   *Empirical Relations*

In order to describe successfully the experimental relative permeabilities, one had to resort to empiricism. For example, Corey (1954) proposed the following equations for the wetting phase and the nonwetting phase relative permeabilities, respectively:

$$k_{\text{rw}} \propto S_{\text{eff}}^4; \quad k_{\text{rnw}} \propto (1 - S_{\text{eff}})^2 (1 - S_{\text{eff}}^2). \tag{5.2.20}$$

The value of the exponent in these equations has been found to vary for different porous media.

Brooks and Corey (1964) have proposed the following more general empirical relationships for the relative permeabilities:

$$k_{\text{rw}} = (S_{\text{eff}})^{(2+3\lambda)/\lambda}, \tag{5.2.21a}$$

$$k_{\text{rnw}} = (1 - S_{\text{eff}})^2 (1 - S_{\text{eff}})^{(2+\lambda)/\lambda}, \tag{5.2.21b}$$

where

$$S_{\text{eff}} = (P_{\text{cb}}/P_{\text{c}})^{\lambda} \quad (P_{\text{c}} \geq P_{\text{cb}}), \tag{5.2.22}$$

with P_{cb} the "bubbling pressure" and λ the "pore size distribution index." Only the parameters λ and P_{cb} need to be determined experimentally. These are obtained by plotting S_{eff} versus the capillary pressure P_{c} on the log-log scale and fitting a straight line to the points. The slope of this line is λ and the intercept at $S_{\text{eff}} = 1$ is P_{cb}. Brooks and Corey (1964) demonstrated the applicability of this technique for various porous media.

Naar and Henderson (1961) proposed an imbition relative permeability model based on the idea that, during imbition, portions of the nonwetting phase become trapped or blocked.

Raimondi and Torcaso (1964) determined the amount of trapped oil at various saturations along the imbition curve by using miscible displacement technique. A saturation correction was applied to the imbition relative permeability curve of the oil, which brought it closer to the drainage curve.

Land (1968) also developed equations for the prediction of imbition relative permeability for two- and three-phase flow. His theoretical development involves the use of the following empirical relationship between initial and residual gas saturations:

$$(1/S_{gr}^*) - (1/S_{gi}^*) = C, \qquad (5.2.23)$$

where the constant C is a characteristic of the porous medium and S_{gr}^* and S_{gi}^* are the residual and initial saturations, respectively, expressed as fractions of the pore volume, excluding the pore volume occupied by the "irreducible" wetting phase. In these experiments the initial gas saturation was established by withdrawal of wetting fluid (i.e., drainage). According to Eq. (5.2.23), when withdrawing a smaller amount of wetting phase (leaving a larger initial gas saturation), the residual gas saturation after imbibition will also be bigger. Land assumed that Eq. (5.2.23) applies at any point in an imbibition process, using the untrapped (i.e., still flowing) gas saturation in place of S_{gi}^* and the saturation of gas yet to be trapped in the imbibition process in place of S_{gr}^*. The value obtained in this way for the flowing gas saturation was used for $(S_g)_{eff}$, the effective saturation of the gas in "Burdine–Corey" type equations [Eqs. (5.2.19) and (5.2.20)]. Land (1968) thought that S_{gi}^* was 100% flowing, which is in contradiction with the experimental results of El-Sayed (1979) and Mohanty (1981). He also assumed that the entire wetting phase remains mobile, which is also unlikely to be true in the light of the network analysis of dendritic pore fractions by Chatzis (1980), Mohanty and Salter (1982), Chatzis and Dullien (1982), (1985).

5.2.5. Resistivity Index and Archie's Law

The effect of partially desaturating a porous medium initially saturated with a conducting wetting fluid is to increase the resistivity of the system, provided that the conducting fluid displaced from the medium is replaced by a nonconducting fluid. The ratio F_e/F, where F_e is the effective formation factor of a porous medium at partial saturation, is called the "resistivity index" I at the saturation S_w, that is,

$$I = F_e/F. \qquad (5.2.24)$$

Providing that surface conduction is negligible, it has been shown experimentally by several authors (see, e.g., Wyllie and Spangler, 1952) that the relationship between I and S_w is of very simple form, and it would appear that in the majority of cases involving reasonably homogeneous porous media it is as follows:

$$I = S_w^{-n}, \qquad (5.2.25)$$

where the exponent n is independent of the value of the saturation. This relationship is often referred to as "Archie's law." The n values for a number of natural and synthetic, consolidated as well as unconsolidated porous media are shown in Fig. 5.22. It has been shown by data presented by Dunlap

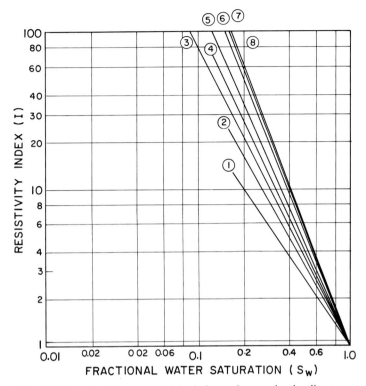

Figure 5.22. Relationship between resistivity index and saturation by direct measurement. (Reprinted with permission from Wyllie and Spangler, 1952.)

Type medium	No	k md	F	ϕ	n	Reference
Pennsylvanian sandstone	1	59.0	20.3	0.17	1.42	*Loc. cit.*
Berea sandstone	2	383.0	16.0	0.19	1.72	*Loc. cit.*
Unconsolidated sand	3	—	—	0.45	1.96	Manegold *et al.* (1931)
Unconsolidated sand	4	—	—	0.20	2.04 (Mean)	Morcom (1946)
Tuscaloosa sandstone	5	870	7.5	0.28	2.20	*Loc. cit.*
Corning Pyrex	6	380	5.2	0.38	2.42	*Loc. cit.*
Alundum	7	1935	8.3	0.30	2.49	*Loc. cit.*
Corning Pyrex	8	212	7.4	0.29	2.55	*Loc. cit.*

and co-authors (1949) that in a particular porous medium the value of the exponent n depends on the manner in which the wetting phase saturation is obtained (i.e., it depends on the saturation history of the sample that is the usual behavior involving phenomena of capillary pressure hysteresis). In capillary pressure desaturation, however, it has been demonstrated that the nature of the nonconducting, nonwetting phase and the value of the interfacial tension between the two phases did not affect the values of the exponent n.

5.3.   IMMISCIBLE DISPLACEMENT

5.3.1.   Introduction

In the preceding discussion of multiphase flow of immiscible fluids, it was mostly taken for granted that the flow was at steady state and, in particular, that neither phase was in the process of displacing other phases. In this section, unsteady multi- (two-) phase flow is discussed, where at a given point in the system the saturation is a function of time.

The terminology used in this area tends to be confusing sometimes. The meaning of some of the terms is probably best explained by giving examples.

When a porous medium containing a nonwetting phase is brought in contact with a wetting phase and the wetting phase spontaneously displaces some of the nonwetting phase, the process is called "free" or "spontaneous" imbibition, or "imbibition" for short. Spontaneous imbibition may be "controlled" by means of regulating externally the capillary pressure or by a throttling valve or a constant rate pump. The latter case is sometimes referred to as "forced" imbibition in general, whereas this term should be reserved for the special case when the pressure in the wetting phase exceeds the value it has under conditions of spontaneous imbibition. The process of penetration of a wetting phase is also called "wetting," "saturation," or "absorption."

When a wetting phase is drained out of the pores, either under the influence of gravity or by forcing in a nonwetting phase under pressure, it is called "drainage," "gravity drainage," "desaturation," "dewetting," or "desorption."

Immiscible displacement, in general, no matter whether a wetting phase is displacing a nonwetting phase, or vice versa, and regardless of the type of forces that are responsible for it, can be referred to as "infiltration."

"Unsaturated flow" is a general term that may mean merely flow of a phase not occupying the entire pore space. More specifically, however, the term "unsaturated flow" has often been applied to the phenomenon of penetration of the medium by a phase, while displacing another phase. Often the other phase was air and no attention was paid to the (small) resistance it offered to the penetration.

When a nonwetting phase (e.g., oil in sand) is being displaced by a wetting phase (e.g., water) pumped into the system at a constant rate, under a pressure in excess of the capillary pressure, characteristic of the system, one speaks of "water flooding." Conversely, when water is being displaced by a nonwetting phase (e.g., oil) at a constant rate under a pressure much in excess of the capillary pressure necessary for penetration into the system, it is customary to speak of "oil flooding."

5.3.2. Similitude and Scaling in Immiscible Displacement

The problems of scaling and similitude in capillary flow have been discussed by a number of authors. Probably the most lucid approach to this problem is due to Miller and Miller (1956) who realized that if one is to calculate the flow patterns within the liquid, the detailed microscopic geometry of the liquid-filled space must be known; the microscopic liquid geometry is determined by the shapes and positions of the gas–liquid interfaces within the pores. The analysis of Miller and Miller (1956) is outlined below. As these authors were interested in flow of groundwater, capillary number and viscosity ratio were left outside their consideration.

The interface geometry within a given porous medium usually depends on three constant factors—the pore geometry, the surface tension, and the contact angle—and a variable factor—the capillary pressure. The microscopic liquid geometry in a porous medium depends on capillary pressure history rather than on the momentary value of the capillary pressure alone and, therefore, the microscopic liquid geometry is a "functional," rather than a function of capillary pressure. Miller and Miller (1956) introduced the term "hysteresis function" for such quantities and they pointed out that since the microscopic liquid geometry determines the macroscopic properties, such as saturation and phase permeability, these also must reflect the characteristic capillary pressure history dependence of the former. In other words, they must be also "hysteresis functions" (subscript h) of capillary pressure.

The variables—capillary pressure, filter velocity, position, and time—must be distinguished from the system parameter viscosity, surface tension, contact angle, one microscopic and one macroscopic length, and body forces.

The various quantities entering Darcy's law and the continuity equation have been replaced by corresponding reduced quantities that contain combinations of system parameters with no more than one variable. The combinations are so arranged that any two systems for which the reduced medium properties and the reduced macroscopic boundary conditions are identical must necessarily exhibit identical reduced flow behavior.

Because of the linearity of the two microscopic equations (i.e., Laplace's equation and the Stokes equations) it is possible to obtain detailed similitude

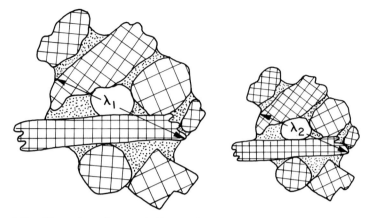

Figure 5.23. Illustration of two "similar media" in "similar states." Note that the two characteristic lengths, λ_1 and λ_2, connect corresponding points in the two media. (Reprinted with permission from Miller and Miller, 1956).

of interface shapes and of microscopic flow patterns between two media whose solid geometries differ only by a constant magnifying factor. Two such media are called similar media by analogy to the familiar term "similar triangles." When the interface geometries are also similar, the two media are said to be in *similar states*. A pair of similar media in similar states is illustrated in Fig. 5.23. When each of these geometries is reduced (i.e., expressed in terms of a characteristic length Λ), the resulting reduced geometries are identical.

In practice, the occurrence of detailed similarity throughout the microscopic geometries of two media has zero probability. Nevertheless, a macroscopic result can be derived from the assumption of exact microscopic similarity and then noted that these results are necessarily the same as those obtained from a statistical equivalence criterion. This is so because experimentally it is possible to synthesize media with different characteristic lengths that closely satisfy the statistical criterion for similarity.

A necessary condition for two similar media to be in similar states is that the reduced interface curvatures must be the same. Since curvature is reciprocal length, it is reduced by multiplication by Λ. Applying this reduction to Laplace's equation, we have the following equation:

$$2(\Lambda/r_{\mathrm{m}}) = \Lambda P_{\mathrm{c}}/\sigma, \qquad (5.3.1)$$

where r_{m} is the mean radius of curvature of the interface.

It is clear from this equation that two media in similar states must have equal values of the group $\Lambda P_{\mathrm{c}}/\sigma$ and that this group is the desired reduced pressure parameter.

A second condition for similarity of state is that the contact angle at each point in one medium must match the contact angle at the corresponding point of the other medium. Hence, comparisons can be made only between systems having the same contact angle.

It is obvious that the reduced microscopic liquid geometries for two similar media must be identical hysteresis functions of reduced pressure.

The foregoing statement leads directly to reduced expressions for the two functionals that characterize the porous media.

The saturation of a phase S being the ratio of two volumes can be calculated just as well from reduced liquid geometry as from actual geometry. Two media having the same reduced liquid geometry must therefore have identical values of S. Combining this with the foregoing argument on reduced pressure dependence of reduced microscopic geometry, it follows that two similar media must be represented by identical functionals $S_h(\Lambda P_c/\sigma)$.

The phase permeability k_h can also be calculated from the microscopic liquid geometry; but whereas S is determined solely by the reduced geometry, k_h is sensitive also to the scale factor Λ. Simple dimensional considerations lead to the result that k_h must vary proportionally to Λ^2. Thus, two similar media in similar states must have identical values of k_h/Λ^2.

Adding now the reduced pressure dependence, used previously for the saturation S, it is seen that two similar media must be represented by identical functionals of the reduced phase permeability $(1/\Lambda^2)k_h(\Lambda P_c/\sigma)$.

Next, the macroscopic flow equation that is Darcy's law is reduced by multiplying through with appropriate constants, giving

$$\frac{\mu}{\Lambda\sigma}\mathbf{v} = \frac{1}{\Lambda^2}k_h\left(\frac{\Lambda P_c}{\sigma}\right)\left[\frac{\Lambda}{\sigma}\rho\mathbf{g} - \nabla\left(\frac{\Lambda P_c}{\sigma}\right)\right]. \qquad (5.3.2)$$

A characteristic macroscopic length L will have to be introduced to obtain the gradient operator and the macroscopic velocity in reduced form. As a result, the final reduced form of Darcy's law is as follows:

$$\frac{\mu L}{\Lambda\sigma}\mathbf{v} = \frac{1}{\Lambda^2}k_h\left(\frac{\Lambda P_c}{\sigma}\right)\left[\frac{\Lambda L}{\sigma}\rho\mathbf{g} - (L\nabla)\left(\frac{\Lambda P_c}{\theta}\right)\right]. \qquad (5.3.3)$$

The continuity equation, that is,

$$\nabla\cdot\mathbf{v} = \partial S/\partial t \qquad (5.3.4)$$

becomes

$$(L\nabla)\cdot\left(\frac{\mu L}{\Lambda\sigma}\mathbf{v}\right) = \frac{\partial[S(\Lambda P_c/\sigma)]}{\partial[(\Lambda\sigma/\mu L^2)t]}. \qquad (5.3.5)$$

It is clear that the entire reduced flow behavior of two systems must be identical if the two systems are composed of similar media and if they are

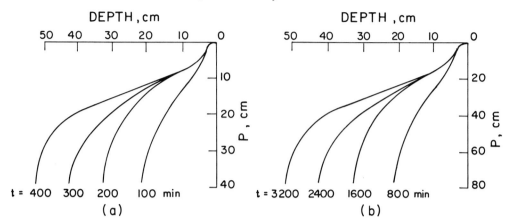

Figure 5.24. Principle of flow-system similitude. The characteristic pore size Λ for the graph in (a) is twice as great as that for the graph in (a), the only difference in the graphs themselves being in the labeling of curves and axis scales. (Reprinted with permission from Miller and Miller, 1956).

subjected to macroscopic boundary conditions and initial conditions that are identical when expressed in terms of reduced variables introduced here. Therefore, the two systems are designated as similar flow systems. The behavior of two such systems, representing infiltration of water into soil, is shown in Fig. 5.24. The only difference between the two sets of curves in this figure is the labelling of curves and axes. If these had been labeled in terms of reduced coordinates, there would have been no difference between them whatsoever. Summing up, the following reduced macroscopic variables have been defined:

$$\mathbf{r} \equiv \mathbf{r}/L \qquad \text{(position vector)}, \qquad (5.3.6)$$

$$\hat{\nabla} \equiv L\nabla, \qquad (5.3.7)$$

$$\hat{\nabla}^2 \equiv L^2\nabla^2, \qquad (5.3.8)$$

$$\widehat{\rho\mathbf{g}} \equiv (L\Lambda/\sigma)\rho\mathbf{g} \qquad \text{(gravity)}, \qquad (5.3.9)$$

$$\widehat{\rho\omega^2 r} \equiv (L\Lambda/\sigma)\rho\omega^2 r \qquad \text{(centrifugal force)}, \qquad (5.3.10)$$

$$\hat{v} \equiv (\mu L/\Lambda\sigma)v, \qquad (5.3.11)$$

$$\hat{t} \equiv (\Lambda/\mu L^2)t. \qquad (5.3.12)$$

Klute and Wilkinson (1958) tested the similar media concept of capillary flow by using five size fractions of sand. The size distribution of each fraction

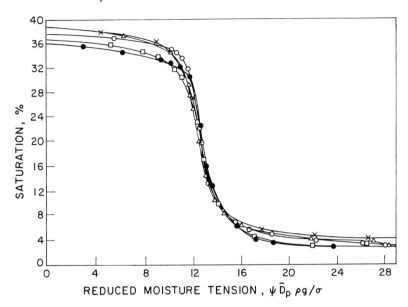

Figure 5.25. Reduced dewetting data for each sand size fraction (measured in microns): 104–175 (△); 125–149 (○); 149–177 (×); 177–210 (□); 210–250 (●). (Adapted with permission from Klute and Wilkinson, 1958).

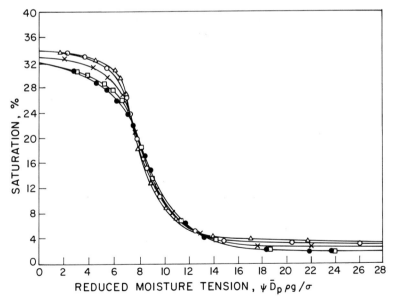

Figure 5.26. Reduced wetting data for each sand size fraction as noted in Fig. 5.25. (Adapted with permission from Klute and Wilkinson, 1958).

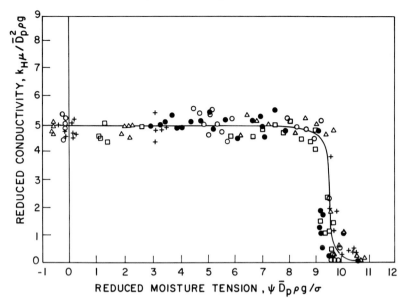

Figure 5.27.   Reduced hydraulic conductivity–moisture tension curves for each sand fraction as noted in Fig. 5.25. (Adapted with permission from Klute and Wilkinson, 1958).

was determined microscopically. The size distribution curves were matched at their peak values and the relative size values for each fraction were obtained by dividing the actual particle diameters by the diameter corresponding to the peak of the distribution curve. The relative quantities for each fraction were obtained by dividing the actual number of particles by the number corresponding to the peak of the distribution curve. The normalized size distribution curves obtained in this manner were very nearly identical and, therefore, the fractions could be considered similar. The bulk densities and the particle densities of the sand packs were practically identical.

Saturation and desaturation curves were obtained and the capillary pressures were reduced in the manner proposed by Miller and Miller (1956). In this way, the reduced saturation and desaturation curves coincided for all fractions.

The unsaturated phase permeabilities were also determined experimentally and the values obtained were reduced according to the formula given by Miller and Miller (1956). Reduced unsaturated permeability versus reduced capillary pressure curves also coincided. The reduced curves are shown in Figs. 5.25–5.27. The analysis of Miller and Miller applies in the case of low capillary numbers.

5.3.3. Unidirectional Capillary Penetration, Infiltration, Immiscible Displacement

In the infiltration process, usually liquid enters the porous material under the influence of capillary and gravity forces. In the soil, infiltration mainly amounts to displacement of air by water. Experiment has shown that the infiltration rate is very large initially and decreases to a limiting value at large times. The high initial rates result from capillary forces; as infiltration proceeds, the relative role of capillary pressure is being reduced and ultimately the gravitational force becomes dominant.

Infiltration is mostly unsteady state, but steady-state infiltration has also been used, meaning simply "unsaturated flow."

5.3.3.1. *The Washburn Equation*

The simplest equation to model the rate of capillary penetration into a porous medium is attributed to Washburn (1921) (see also West, 1911). It is usually considered to be rigorous for the case of capillary penetration into a uniform capillary tube or a bundle of uniform capillary tubes.

In the case when the axis of a capillary tube encloses the angle α with the vertical direction, the pressure drop $\Delta \mathscr{P}$ between the meniscus and the bulk liquid displacing air can be expressed as follows:

$$\Delta \mathscr{P} = P_c - \rho g h = [(2\sigma \cos \theta)/R] - \rho g \ell \cos \alpha, \qquad (5.3.13)$$

where h is the elevation of the meniscus over the bulk liquid surface and P_c is the pressure difference between the two sides of the moving interface (normally assumed to be given by the Laplace equation).

Substituting Eq. (5.3.13) for $\Delta \mathscr{P}$ in the Hagen–Poiseuille equation, with $\alpha = 0°$, the Washburn equation

$$v \equiv \frac{dh}{dt} = \frac{1}{8\mu h} \left[\frac{2\sigma \cos \theta}{R} - \rho g h \right] R^2 \qquad (5.3.14)$$

is obtained. It relates the rate of advance of the meniscus at a given vertical height to the other properties of the system. It is noted that the presence of the meniscus imposes a condition of plug flow at the front of the liquid column, whereas use of the Hagen–Poiseuille equation implies a parabolic velocity distribution. This difference has been assumed to be reconciled by a fountain-type motion of the fluids on both sides of the interface (e.g., Dussan, 1977; Huh and Scriven, 1971) without any additional resistance. (For a discussion see Section 5.3.4.)

Laughlin and Davies (1961) studied experimentally the rate of capillary rise of oil in fibrous textile wicking. They found that the Washburn equation without the gravity term applied in some but not all cases. The modified equation

$$h = Ct^k, \qquad (5.3.15)$$

however, applied for all the samples, except at very large times. The value of the exponent varied from 0.50 down to 0.41 for various samples. The mass of fluid absorbed m as a function of time was given as

$$m = Bt^\ell \qquad (5.3.16)$$

(ℓ was always smaller than k).

The applicability of Washburn's equation to the description of capillary rise in porous media has been tested by van Brakel and Heertjes (1977, 1977a), who observed capillary rise in cylindrical beds with a diameter of 23 mm and heights ranging from 10 to 25 cm. Most experiments were carried out with glass beads or sand as a solid phase, comprising sieve fractions in the size range from 50 to 1000 μm. Mixtures of sand and glass beads, crushed glass, fused silica particles, copper beads, and polystyrene beads were also used.

The liquid phases used were water, toluene, h-hexane, n-heptane, paraffin oil, ethanol, glycol, dichloroacetic acid, etc. The beds were checked for homogeneity by x-ray technique.

Each experiment was conducted by submerging the bottom of the medium 2 mm below the surface. The liquid front was observed visually, but sometimes x-ray absorption was used to measure the saturation of the porous medium as a function of time in 1-mm sections of the medium.

Various patterns were observed, which were subdivided into four different categories depending on whether the rate of capillary rise was slow or fast and whether a saturation gradient was developing or not.

Whenever pronounced saturation gradients started developing, Eq. (5.3.14) also started to break down. However, in some cases, Eq. (5.3.14) was disobeyed also in the absence of a saturation gradient.

Van Brakel and Heertjes (1977, 1977a) found that in some experiments the following form of the hydraulic radius theory

$$\phi \frac{dh}{dt} = \frac{k_{CK}}{\mu h} \left(\frac{\sigma \cos \theta}{r_H} - \rho g h \right) \qquad (5.3.17)$$

was in order-of-magnitude agreement with the measured rates of rise.

Van Brakel and Heertjes (1977, 1977a) concluded that the main factor in determining the observed behavior was the contact angle. A wide saturation gradient at capillary equilibrium corresponded to a contact angle approxi-

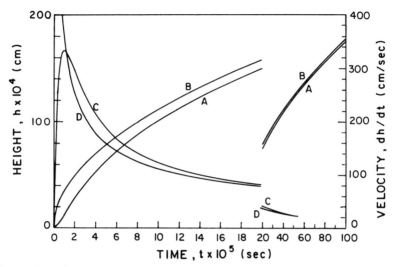

Figure 5.28. Plot of the position of the meniscus and the rate of rise against time for $R = 0.001$ cm, $P_c = 50/R$, $\rho = 1$ gm/cm^3, and $\mu = 10^{-2}$ gm/cm sec. Curve A, plot of h vs. t according to the full set of equations; curve B, plot of h vs. t according to the Washburn equation; curve C, plot of dh/dt according to the full set of equations; curve D, plot of dh/dt according to the Washburn equation (Szekely *et al.*, 1971).

mately equal to zero, a sharp front at equilibrium, however, to a contact angle greater than some critical value. Relatively slow capillary rise was due to a slow decrease in the contact angle.

Szekely *et al.* (1971) pointed out that in Eq. (5.3.14), at very short times, the penetrated length goes to zero and the rate of rise to infinity, whereas as the time goes to zero, the velocity should do likewise. At very small times the liquid accelerates from zero velocity to a quasi-steady-state value. The Washburn equation is not applicable to this unsteady-state process. Szekely *et al.* (1971) developed the following differential equation, applicable also to the initial unsteady-state capillary rise:

$$\left(h + \frac{7R}{6}\right)\frac{d^2h}{dt^2} + 1.225\left(\frac{dh}{dt}\right)^2 + \frac{8\mu h}{\rho R^2}\frac{dh}{dt} = \frac{1}{\rho}(P_c - \rho g h) \quad (5.3.18)$$

and solved this numerically, using the boundary conditions $h = 0$ at $t = 0$ and $dh/dt = 0$ at $t = 0$. Typical calculated results are shown in Fig. 5.28 from which it is evident that after a few seconds the Washburn equation becomes applicable in its usual form.

Apparently unaware of the previous work of Szekely *et al.* (1971), Levine *et al.* (1976) analyzed the rates of penetration predicted at small times by the

Washburn equation, Darcy's law, and the Forchheimer equation. At larger times all these equations gave the same rates of penetration.

Levine *et al.* (1976) also paid attention to the fact that in most pores wider sections alternate with narrower ones in a quasi-periodic manner.

Considering a periodic tube consisting of alternating conical sections, Levine *et al.* pointed out that the capillary pressure undergoes periodic variations because the curvature of the interface varies periodically, depending on the position of the meniscus in the alternating conical sections. They concluded that the volume flow rate in capillary rise in a tube consisting of alternating conical sections undergoes periodic fluctuations. It is noted that this is true only for a short liquid column. In realistic situations the liquid column behind the meniscus is too long, as compared with the length of the period of the alternating sections, to accelerate and decelerate periodically in the time interval in which the meniscus traverses a period of the alternating conical sections. The frontal rate of advance of the liquid meniscus in a periodically constricted tube will alternate even at a constant volume flow rate, however, because an incompressible fluid passes through a cross section of varying size.

Dullien *et al.* (1977) calculated an equivalent mean square capillary radius, with the help of the Washburn equation, from measured rates of capillary rise in various sandstones and found it to be smaller by about two orders of magnitude than the neck sizes determined by mercury intrusion porosimetry in the same sample. (In most sandstones the height of rise of brine could be followed visually.) These results were explained in terms of capillaries with periodic step changes in the diameter, as follows. (It may be necessary to point out that the mathematical singularities introduced by a step change do not have a physical meaning because a real capillary tube with step changes in diameter would not have such singularities, owing to rounded corners at the step change. This model, therefore, should be thought of as having such rounded corners.)

The average rate of advance of the meniscus over a period of the tube (see Fig. 5.29) is

$$\frac{\overline{d\ell}}{dt} \equiv \frac{\text{length of a period}}{\begin{array}{c}\text{time it takes meniscus to}\\ \text{traverse a period}\end{array}} = \frac{\sum_{k=1}^{n}\ell_k}{\sum_{k=1}^{n}t_k}. \tag{5.3.19}$$

A period is assumed to be very short as compared with the entire tube and, therefore, $\overline{d\ell}/dt$ is treated as a point function along the tube.

Application of the Hagen–Poiseuille equation to each capillary segment, neglecting contraction and expansion losses, results in the following expression:

$$Q_k = \frac{\pi}{128\mu}\frac{(4\sigma/D_k) - \rho g h}{\ell}\frac{\sum_{j=1}^{n}\ell_j}{\sum_{j=1}^{n}\ell_j/D_j^4}, \tag{5.3.20}$$

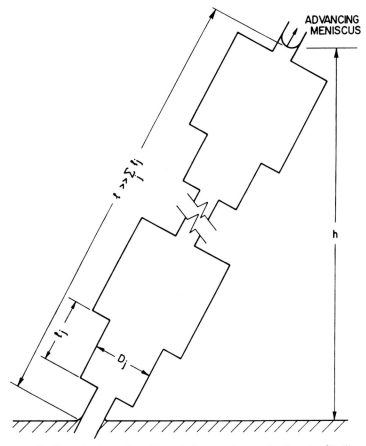

Figure 5.29. Capillary rise in a tube with periodic step changes in diameter (Dullien *et al.*, 1977).

where ℓ_j and D_j are the length and diameter, respectively, of segment j ($j = 1, 2, \ldots, n$) of the basic unit constituting the periodically repeating element of the tube, $\sum_j \ell_j$ is the length of a period, and ℓ is the length of the tube, consisting of a large number of basic units that have been penetrated by the liquid and over which the pressure drop is $(4\sigma/D_k) - \rho g h$ [see Eq. (5.3.13)]. It was assumed that $\theta = 0$.

The time t_k spent by the meniscus in segment k is found by combining Eq. (5.3.20) with

$$Q_k = (D_k^2 \pi / 4)(\ell_k / t_k), \tag{5.3.21}$$

resulting in

$$t_k = 32\mu \ell \ell_k \sum_j \left[\frac{(\ell_j/D_j)(D_k/D_j)^3}{(4\sigma - \rho g h D_k)\Sigma_j \ell_j} \right]. \tag{5.3.22}$$

As long as $\rho g h D_k \ll 4\sigma$, t_k increases as D_k^3, and, therefore, the meniscus spends most of the time in the largest segments.

On combining Eq. (5.3.21) with Eq. (5.2.19) one obtains

$$\frac{\overline{d\ell}}{dt} = \left(\sum_k \ell_k \right)^2 32\mu \ell \sum_k \left[\frac{\ell_j \Sigma_j (\ell_j/D_j)(D_k/D_j)^3}{4\sigma - \rho g h D_k} \right]^{-1}. \tag{5.3.23}$$

The value of $\overline{d\ell/dt}$ is very close to the rate of advance of the meniscus in the largest capillary segment.

The rate of capillary rise $\overline{dh/dt}$ in a tube, the axis of which encloses the angle α with the vertical direction is

$$\left(\frac{\overline{dh}}{dt} \right)_{\text{tube}} = \frac{\overline{d\ell}}{dt} \cos \alpha. \tag{5.3.24}$$

With $h = \ell \cos \alpha$, there follows from Eqs. (5.3.23) and (5.3.24) for such conditions that $\rho g h D_k \ll 4\sigma$:

$$\left(\frac{\overline{dh}}{dt} \right)_{\text{tube}} = \frac{\sigma \cos^2 \alpha}{8\mu h} \left(\sum_k \ell_k \right)^2 \left[\sum_k \ell_k \sum_j \left(\frac{\ell_j}{D_j} \right) \left(\frac{D_k}{D_j} \right)^3 \right]^{-1}. \tag{5.3.25}$$

The rate of capillary rise in the model dh/dt is the average $\langle (\overline{dh/dt})_{\text{tube}} \rangle$ taken over random orientations of the repeating periods. As

$$\overline{\cos^2 \alpha} = (4\pi)^{-1} \int_0^{2\pi} \int_0^{\pi/2} \cos^2 \theta \sin \theta \, d\sigma \, d\phi, \tag{5.3.26}$$

one obtains

$$\frac{dh}{dt} = \frac{\sigma}{(8 \times 3)h} \left(\sum_k \ell_k \right)^2 \left[\sum_k \ell_k \sum_j \left(\frac{\ell_j}{D_j} \right) \left(\frac{D_k}{D_j} \right)^3 \right]^{-1}. \tag{5.3.27}$$

Comparing Eq. (5.3.27) with the equivalent form of the Washburn equation, that is,

$$dh/dt = (\sigma/8\mu h)D_{\text{app}}, \tag{5.3.28}$$

TABLE 5.1   Results of Rate of Capillary Rise Measurements in Sandstones[a,b]

Sandstone	h^2/t (cm²/sec)	$(D_e)_{inf.pt.}$ (μm)	D_{max} (μm)	D_{app} (μm)	$(D_{app})_{model}$ (μm)	ϕ	k (cm² × 10¹¹)	k/D_{app} (μm)
Bandera	0.0010	5.0	120	0.0055	0.0069	0.191	8.6	1.564
Clear Creek	0.0055	8.0	110	0.031	0.029	0.191	95	3.065
Noxie 129	0.0060	8.5	120	0.033	0.035	0.283	148	4.485
Cottage Grove	0.010	11.0	120	0.056	0.073	0.216	116	2.071
Big Clifty	0.010	9.0	100	0.056	0.061	0.192	164	2.929
Torpedo	0.020	12.0	110	0.11	0.1	0.246	184	1.673
St. Meinrad	0.025	14.1	120	0.14	0.14	0.247	210	1.500
Berea 108	0.025	26.0	180	0.14	0.47	0.178	613	4.378
Boise	0.6	47.0	180	3.3	3.1	0.232	5370	1.628

[a]From Dullien *et al.*, 1977.
[b]$(D_{app})_{model}$ was calculated by Eq. (5.3.32), using the neck diameter corresponding to the inflection point of the mercury porosimetry curve, $(D_e)_{inf.pt.}$, the biggest bulge diameter measured photomicrographically, D_{max}, and the mean $[(D_e)_{inf.pt.} + D_{max}]/2$; $a_j = 1$ was used throughout.

the following expression for $(D_{app})_{model}$ is obtained:

$$(D_{app})_{model} = \frac{1}{3} \sum_k \ell_k \left[\sum_k \ell_k \sum_j \left(\frac{\ell_j}{D_j} \right) \left(\frac{D_k}{D_j} \right)^3 \right]^{-1}. \qquad (5.3.29)$$

The value of D_{app}, determined from the rate of capillary rise data using the integrated form of the Washburn equation,

$$h^2/t = \sigma D_{app}/4\mu = \text{const.}, \qquad (5.3.30)$$

and the corresponding values of $(D_{app})_{model}$, obtained purely from pore structure information via Eq. (5.3.29), are compared for a number of sandstones in Table 5.1.

In evaluating $(D_{app})_{model}$ numerically, the length-to-diameter ratio a_j of a capillary segment

$$a_j \equiv \ell_j/D_j \qquad (5.3.31)$$

was introduced into Eq. (5.3.29), giving

$$(D_{app})_{model} = \left(\sum_k a_k D_k \right)^2 \left[\sum_k a_k D_k \sum_j \left(\frac{D_k}{D_j} \right)^3 \right]^{-1}. \qquad (5.3.32)$$

It was found that $(D_{app})_{model}$ was insensitive to the value of a_j, as shown in Table 5.2, and $a_j = 1$ has been used throughout.

TABLE 5.2   Sample Calculation of $(D_{app})_{model}$ from Pore Structure Data[a, b]

D_j (μm)	11	60	120	$(D_{app})_{model}$ (μm)
	54[c]	20[c]	1[c]	0.064
	20	6	1	0.038
a_j	10	3	1	0.030
	5	2	1	0.033
	3	1.5	1	0.038
	2	1	1	0.041
	1	1	1	0.073

[a]From Dullien *et al.*, 1977.
[b]Sandstone sample: Cottage Grove.
[c]Calculated from bivariate pore size distribution (see Section 3.3.1.2). The rest of the entries of a_j were assumed arbitrarily.

Qualitative explanation of the fact that $(D_{app})_{model}$ is very much smaller than the neck diameter $(D_e)_{inf.pt.}$ is as follows. The resistance to capillary rise is determined by necks in the neighborhood of $(D_e)_{inf.pt.}$, the average capillary pressure driving the process, however, corresponds to a radius of curvature of the meniscus in the largest segments in series with the necks.

The data in Table 5.1 show an approximate proportionality between $(D_{app})_{model}$ and the permeability of the sample. The mean value of the ratio $k/(D_{app})_{model}$ is 2.6 μm $\pm$ 40%.

Using thermodynamical considerations and neglecting gravity, Deryagin (1946) derived a simple relationship between the rate of capillary penetration, the permeability, and the specific surface. This equation, which can be obtained also by applying Darcy's law to capillary penetration and substituting for the pressure drop the capillary pressure corresponding to the mean curvature in a tube of a hydraulic radius equal to that of the medium, is as follows [cf. Eq. (5.3.30)]:

$$\ell^2/t = 2(k/\mu)(S_v/\phi^2)\sigma\cos\theta, \qquad (5.3.33)$$

where S_v is the ratio of the internal surface to the total volume of the porous body. Deryaguin *et al.* (1952) calculated the specific surface of porous bodies by Eq. (5.3.33) from the measured rate of capillary penetration and permeability. As independent checks, the specific surface was also determined by the gas adsorption method and the Carman–Kozeny formula. In the case of quartz sand, Eq. (5.3.33) consistently gave significantly smaller specific surface than the other two methods, which agreed well (see Table 5.3).

Deryaguin *et al.* (1952) explained this discrepancy by assuming that the contact angle in the penetration process was not zero as it was in equilibrium. An alternate explanation, however, is that in periodically convergent–diver-

TABLE 5.3   Specific Surface of Quartz Sand[a,b]

Fraction	Gas Method	Carman–Kozeny formula	Capillary penetration by water
Fraction 7–10 μm	0.90	0.73	0.34
Fraction, after washing with chromic mixture	1.08	1.04	0.53
Fraction 10–14 μm	0.61	0.63	0.23
Fraction, after washing with chromic mixture	—	0.57	0.22
Fraction 14–20 μm	0.40	0.40	0.12
Fraction, after washing with chromic mixture	0.39	0.38	0.12

[a] From Deryagin *et al.*, 1952.
[b] In m^2/g

gent channels the rate of penetration (i.e., ℓ^2/t) is smaller than the value calculated from the hydraulic radius, or from S_v, by Eq. (5.3.33).

5.3.3.2.   Richards' Equation

Much as in the case of the Washburn equation, the simplifying assumption is used that the resistance of the displaced phase (air) is negligible, but Richards' equation does not imply a sharp front.

The basic equations are as follows:

$$v_w = -[k_w/\mu_w]\nabla(P_w + g\rho_w z) \qquad \text{(Darcy's law)}, \qquad (5.3.34)$$

$$\phi\, \partial S_w/\partial t = -\nabla \cdot v_w \qquad \text{(continuity equation)}, \quad (5.3.35)$$

$$P_w = -P_c(S_w) \qquad \text{(capillary pressure function)}, \qquad (5.3.36)$$

The pressure in the air has been assumed constant and equal to zero, hence P_w is gauge pressure. Equations (5.3.34)–(5.3.36) can be combined in the following form:

$$\phi\, \partial S_w/\partial t = \nabla \cdot (k_H \nabla \phi_c) = -\nabla \cdot (k_H \nabla \psi) + (\partial k_H/\partial z), \quad (5.3.37)$$

where Eqs. (2.3.3), (2.3.4), and (3.1.4) have been used. For simplicity, the same symbol (i.e., k_H) is used for the hydraulic conductivity in unsaturated as well as in saturated flow, and k_H depends on z through S_w. It is noted that often the "volumetric moisture content" $\theta \equiv \phi S_w$ is used. The signs on the right-hand side of Eq. (5.3.37) may vary, depending on the orientation of the positive z axis (upward or downward) and the type of potential function used.

Equation (5.3.37) was proposed in 1931 by Richards and it has been used in several alternate forms. For example,

$$\phi \, \partial S_w / \partial t = \nabla \cdot (\mathscr{D}_m \nabla S_w) + (dk_H / dS_w)(\partial S_w / \partial z), \qquad (5.3.38)$$

where $\mathscr{D}_m$ is the "moisture diffusivity" or "hydraulic diffusivity":

$$\mathscr{D}_m \equiv -k_H \, d\psi / dS_s, \qquad (5.3.39)$$

which was introduced for the sake of analogy with the diffusion equation.

It has become a widespread custom to call any process that can be described by the "diffusion equation" a diffusion process, regardless whether or not random molecular motion is involved in it. This custom has resulted in misrepresenting deterministic processes such as infiltration as a stochastic process. For example, the "hydraulic diffusivity" was analyzed in terms of Markov's stochastic process theory by Laroussi and DeBacker (1975). According to this interpretation, Darcy's law becomes a special case of Fick's law with the saturation gradient as the driving force and the hydraulic diffusivity as the transport coefficient.

Experimental study for packs of glass beads (Laroussi *et al.*, 1975) has indicated four ranges of variation of the hydraulic diffusion coefficient as a function of water saturation. An example for this dependence of the hydraulic diffusion coefficient is shown in Fig. 5.30. An attempt at the theoretical interpretation of this relationship should be preceded by an analysis of the statistical significance of the boundaries of the four zones, as mixing of arithmetic logarithmic scales may be responsible for their apparent sharpness.

In a series of papers Philip (1957a, b, c, d) developed a theory of infiltration and obtained an approximate solution to Richards' equation under the boundary conditions of constant water content at the upper surface. The following equation for infiltration rate v_0 measured at the upper surface was obtained:

$$v_0 = \tfrac{1}{2} s t^{-1/2} + (k_H + b) + \tfrac{3}{2} dt^{1/2} + 2 e t^{3/2} + \cdots, \qquad (5.3.40)$$

where s, b, d, and e are constant coefficients. The constant k_H is the hydraulic conductivity associated with the initial water content. As this equation diverges for large times, Philip proposed also a large time approximation.

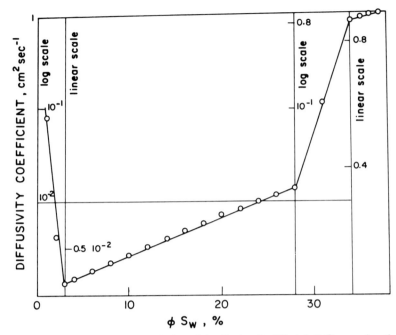

Figure 5.30. The four ranges of variation of the "hydraulic diffusivity" $\mathscr{D}_m$ as a function of water content θ. The ordinate scale is logarithmic for $0 < \theta < 3\%$ and $28 < \theta < 34\%$ (glass bead fraction, 125–180 μm in diameter) (Laroussi *et al.*, 1975 © The Williams & Wilkins Co., Baltimore, Maryland).

More recently, Parlange (1971a, b) proposed a different approximate solution to Richards' equation. Numerical agreement between his solution and Philip's results has been very good. Stimulated by Parlange's contribution, Philip (1973b; Knight and Philip, 1974; Philip and Knight, 1974) worked out a new quasi-analytical technique for solving the infiltration equation.

As an example of infiltration, saturation profiles obtained in drainage as a function of time are shown in Fig. 5.31.

5.3.3.3. *Immiscible Displacement while Considering the Flow of Both Phases*

It has long been recognized that immiscible displacement is a process involving two-phase flow, even though much of the work on the displacement of air has been based on a one-phase equation. Neglecting the resistance of air to flow has been justified in many cases because of the small viscosity of air as compared to that of water.

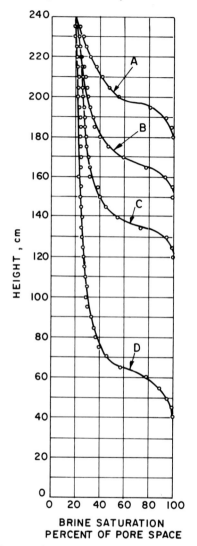

Figure 5.31. Change in saturation distribution in gravity drainage with time. Curve A, 0 hr; curve B, 24 hr; curve C, 52 hr; curve D, 120 hr (Terwilliger *et al.*, 1951 © SPE-AIME).

When water is displacing oil or vice versa, the resistance of both phases must always be considered. An ingenious and simple approach to this problem is due to Buckley and Leverett (1942).

The Buckley–Leverett Method. The Buckley and Leverett (1942) frontal displacement theory gained popularity in the petroleum industry, partly because of its great simplicity and partly because often it is surprisingly accurate.

The Buckley–Leverett Saturation Profile. Material balance over an infinitesimal element of the system in the direction of unidirectional macroscopic flow gives

$$(\partial S_w/\partial t)_x = -(Q_T/\phi A)(\partial F_w/\partial x)_t, \qquad (5.3.41)$$

where Q_T is the constant total flow rate through the cross section A and $F_w = |Q_w/Q_T|$ the fraction of flowing stream comprising displacing fluid (i.e., usually water). The main and very fruitful idea of Buckley and Leverett (1942) was to transform Eq. (5.3.41) into the following form

$$(\partial x/\partial t)_{Sw} = (Q_T/\phi A)(\partial F_w/\partial S_w)_t, \qquad (5.3.42)$$

stating that the rate of advance of a plane of fixed saturation S_w is proportional to the rate of change in composition of the flowing stream with saturation at that moment.

Were F_w known as a function of S_w and t, either from experiment or from theory, Eq. (5.3.42) could be used to calculate the accurate time history of the saturation profile. In fact, by neglecting the effects due to capillary pressure and gravity, Buckley and Leverett (1942) estimated F_w. They defined the "fractional flow function" f_w as follows:

$$f_w \equiv \left[1 + \frac{k_{ro}\mu_w}{k_{rw}\mu_o}\right]^{-1} \qquad (5.3.43)$$

and assumed that F_w was sufficiently well approximated by f_w.

As f_w is not an explicit function of t, Eq. (5.3.42) can be integrated to give the position of a particular saturation as a function of time:

$$x_{Sw} = (Q_T t/\phi A)(df_w/dS_w) + x_0, \qquad (5.3.44)$$

where x_0 is the position of the water saturation at time $t = 0$.

It is assumed that at $t = 0$ the water saturation in the system is uniform, equal to the "irreducible" water saturation, and all water saturations are represented at $x = x_0$, from S_{wi} to $(1 - S_{or})$. These assumptions are approximately correct only when starting to inject water vertically upward into a system after gravity drainage (see Fig. 5.31) as was done originally by Buckley and Leverett (1942). In a horizontal system the water saturation at $x = x_0$ and $t = 0$ is S_{wi} and no other saturation is present. According to Eq. (5.3.44), each saturation will advance into the system at a rate in direct proportion to $f_w' = df_w/dS_w$ (see Fig. 5.32). As shown in Fig. 5.33, the shape of the saturation profile will become similar to that of f_w' (i.e., the saturation profile

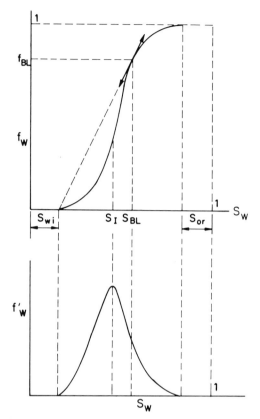

Figure 5.32. Fractional flow function and derivative (Morel-Seytoux, 1969).

is not unique, but triple valued. This being physically meaningless, the law of conservation of mass has been used to construct a sharp leading edge for the saturation front (Buckley and Leverett, 1942; Brinkmann, 1948).

Morel-Seytoux (1969) expressed Eqs. (5.3.42) and (5.3.44) in a form that is applicable to the case of a saturation discontinuity at the front. He considered the elemental volume, shown in Fig. 5.34, bounded by two surfaces $+$ and $-$ on either side of the discontinuity or front. Application of the conservation of mass in finite difference form over the front yielded

$$\frac{dx}{dt} = \frac{1}{A\phi} \frac{Q_w^+ - Q_w^-}{S_w^+ - S_w^-} = \frac{Q_T}{A\phi} \frac{f_w^+ - f_w^-}{S_w^+ - S_w^-}. \tag{5.3.45}$$

In the case of discontinuity one can still proceed to the limit for x and t because the limit of $(Q_w^+ - Q_w^-)/(S_w^+ - S_w^-)$ exists.

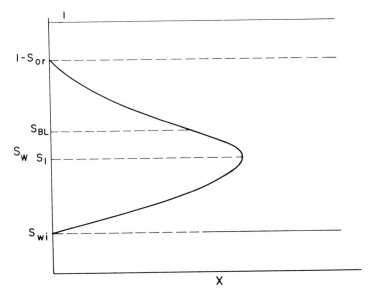

Figure 5.33.   Saturation profile (Morel-Seytoux, 1969).

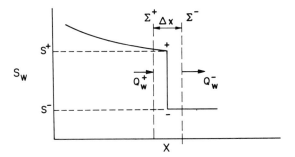

Figure 5.34.   Material balance at a front (Morel-Seytoux, 1969).

The integral of Eq. (5.3.45) gives the position x_f of the front as a function of time:

$$x_f = \frac{Q_T}{A\phi} \frac{f_w^+ - f_w^-}{S_w^+ - S_w^-} t + x_{f0}. \qquad (5.3.46)$$

It is noted that introducing the discontinuity is equivalent to replacing the fractional flow curve by a straight line between the saturation values S^- and

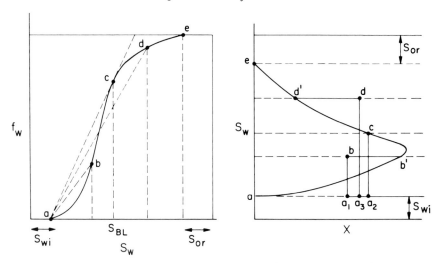

Figure 5.35.   Tentative saturation profiles (Morel-Seytoux, 1969).

S^+ because all saturations within that range must move with the same velocity.

In Fig. 5.35 four tentative solutions for the saturation profile at a given time t are illustrated. Morel-Seytoux (1969) explained that the profile $a_2cd'e$ is the proper solution because it uses the continuous solution to the maximum extent compatible with the physical constraint that the solution must be single valued. This profile is obtained by the graphical construction due to Welge (1952), which consists of drawing a tangent to the fractional flow curve from point a corresponding to the initial saturation in the core. The saturation at the tangent point S_{BL} is the saturation established in the core immediately behind the front. The velocity at which the Buckley–Leverett front travels is proportional to the slope of tangent ac. The velocities at which saturations greater than S_{BL} travel are proportional to the slopes of the tangents to the fractional flow curve at these saturations.

The shape of the resulting profile, the Buckley–Leverett profile, does not change with time, except for a uniform stretching in the x direction. The Buckley–Leverett front is created instantaneously. Its size $(S_w^+ - S_w^-)$ and its rate of advance do not change with time.

Levine (1954) calculated fractional flow versus saturation curves in which the capillary pressure and gravity effects were taken into account. Nevertheless, his curves had an inflection point that might have been caused by sample inhomogeneity.

McWhorter (1971) published a calculational method of infiltration that takes capillarity into account but otherwise is quite similar to the

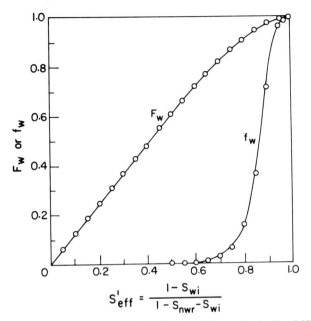

Figure 5.36. F_w and f_w curves used for calculation of the profile in Fig. 5.37 (McWhorter, 1971).

Buckley–Leverett approach. An integral equation for F_w was developed and used to calculate F_w from relative permeability and capillary pressure data. The results are shown in Fig. 5.36 (f_w is also shown for comparison) for horizontal infiltration of water into a column of sand initially containing air. The dimensionless saturation profile calculated with the help of F_w is shown in Fig. 5.37.

Rapoport and Leas (1953) combined Eqs. (5.2.4), (5.2.5), (5.3.41), and (5.3.43) and obtained, for a strictly one-directional process in the absence of gravity, the following equation:

$$\frac{\partial S_w}{\partial \nu_p} + \frac{df_w}{dS_w}\frac{\partial S_w}{\partial \hat{x}} - \frac{k}{\kappa}\frac{1}{L\nu\mu_w}\frac{\partial}{\partial \hat{x}}\left[k_{ro}f_w\frac{dP_c}{dS_w}\frac{\partial S_w}{\partial \hat{x}}\right] = 0, \quad (5.3.47)$$

where the following boundary condition was used:

$$k_{ro}\left[1 + \frac{k}{L\nu\mu_w}k_{rw}\frac{dP_c}{dS_w}\frac{\partial S_w}{\partial \hat{x}}\right] = 0 \qquad \text{at } \hat{x} = 0 \quad \text{for any } \nu_p, \quad (5.3.48)$$

where $\nu_p = Q_T/\phi AL$ is the volume of water injected in units of pore volume, $\kappa = \mu_o/\mu_w$, and $\hat{x} = x/L$.

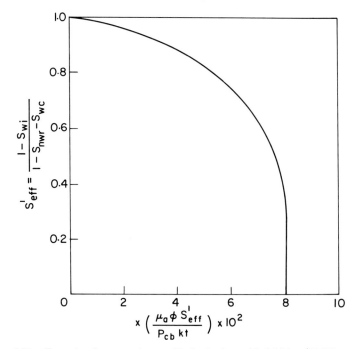

Figure 5.37. Example of a saturation profile for horizontal imbibition (McWhorter, 1971).

Equations (5.3.47) and (5.3.48) are scaled for the special case of identical porous media, fixed value of the oil–water viscosity ratio, and the same value of the "scaling coefficient" $Lv\mu_w$, provided that the capillary pressure and the relative permeabilities are independent of the rate of injection and the viscosity of water. According to these equations the flood history in two identical porous media, at identical oil–water viscosity ratio, identical boundary conditions, and system geometry is the same, provided the group $Lv\mu_w$ also has the same value in both systems.

Rapoport and Leas (1953) pointed out that if the value of the scaling coefficient is sufficiently large, the second-order term in Eq. (5.3.47) becomes negligible as compared with the first-order term; that is, the role played by capillary pressure will become negligible and the flow will "stabilize." Under these conditions Eqs. (5.3.47) and (5.3.48) reduce as follows:

$$\partial S_w / \partial \nu_p = -(df_w / dS_w)(\partial S_w / \partial \hat{x}) \qquad (5.3.49)$$

and

$$k_{ro} = 0 \quad \text{at} \quad \hat{x} = 0 \quad \text{for any} \quad \nu_p, \qquad (5.3.50)$$

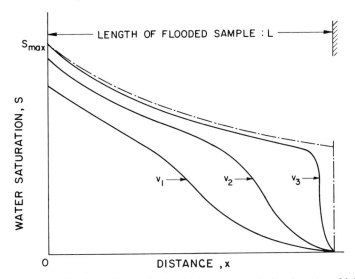

Figure 5.38. Stabilization of linear floods with increasing velocity, the rates of injection are $v_1 < v_2 < v_3$ (Rapoport and Leas, 1953 © SPE-AIME).

which correspond to the conditions assumed in the Buckley–Leverett approach. Some of the results obtained by Rapoport and Leas (1953) are shown in Figs. 5.38 and 5.39. It is apparent that at low velocities there is a gradual saturation change instead of a Buckley–Leverett front.

An interesting aspect of this work is that the experiments were performed either with dry-filmed or oil-treated cores in an attempt to obtain oil-wet media. The runs were performed with water as the nonwetting phase displacing oil, which was the wetting phase. No attempt seems to have been made to use the wetting phase as the displacing phase.

As in these experiments the water was the displacing nonwetting phase, it is natural that it formed a continuum everywhere where it was present. Hence the effect of any saturation change at the inlet end of the system is expected to be transmitted almost instantaneously throughout the entire system.

Recent experiments by Bacri *et al.* (1985, 1990) where the wetting phase (i.e., water) was pumped into about 10 cm long sample at different, constant rates have shown similar behavior as the one illustrated in Fig. 5.38 (see Figs. 5.40a, b). Apparently, in controlled imbibition (i.e., a waterflood when the rate of injection is slow) the displacing wetting phase establishes a continuum not only via surface grooves and edges of the pores but through a network of pores entirely filled with the wetting fluid, resulting in very gradual saturation profiles, and the effects of saturation changes are transmitted almost instan-

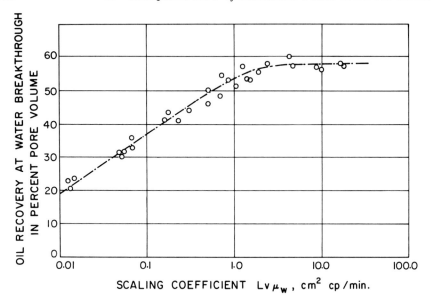

Figure 5.39. Relation between oil recovery at breakthrough and scaling coefficients, dri-filmed alundum cores, no connate water (Rapoport and Leas, 1953 © SPE-AIME).

taneously throughout the entire length of the system. Some aspects of these results are discussed in Section 2.3.5.

It may be concluded that the Buckley–Leverett model approximates the real one-dimensional waterflood front both in water-wet and in oil-wet systems only at relatively high flow rates and favorable viscosity ratios. Otherwise, due to capillary forces, the displacement front will tend to be gradual, resulting in "early" water breakthrough and concomitant poor oil recovery.

Calculation of Petroleum Production. The amount of oil produced is calculated as follows. For a sample of uniform cross section A and length L, Eq. (5.3.42) can be written in the following form:

$$dx/L = f'_w Q_T \, dt/\phi AL = f'_w Q_T \, dt/V_p = f'_w \, d\nu_p, \qquad (5.3.51)$$

where V_p is the pore volume and $d\nu_p$ is the volume of water injected in units of pore volume. Since the saturation of Eq. (5.3.51) is constant, the equation can be integrated:

$$x = L f'_w \nu_p \qquad (5.3.52)$$

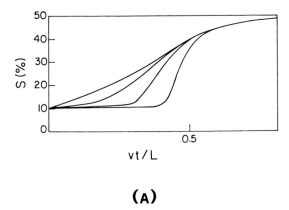

(A)

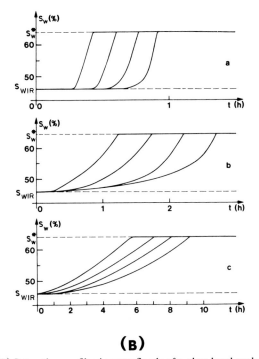

(B)

Figure 5.40. (A) Saturation profiles in waterfloods of a glass bead pack, from left to right, at flow rates Q = 2.4, 10, 40, and 160 cm^3/h at a distance x = 18 cm from the inlet (Bacri *et al.*, 1990). (B) Saturation profiles in the time domain, measured at four different positions (from left to right: x = 3.8, 6.8, 9.8, and 12.8 cm from the inlet end) in waterfloods of a sandstone core of 110 mD permeability at three different constant injection rates (a : 12, b : 3, and c : 1.2 cm^3/h). Viscosity ratio $\kappa = \mu_0/\mu_w$ = 1.5 (Bacri *et al.*, 1985).

giving the distance swept by saturation S_w for ν_p pore volumes of water injected.

Prior to breakthrough of the Buckley–Leverett front, only oil is produced at the outlet face. After breakthrough, the incremental pore volumes of oil produced in $d\nu_p$ pore volumes of total fluids is $[1 - f_w(S_w)]d\nu_p$, where S_w is the prevailing saturation at the outlet face.

The cumulative oil recovery, in units of pore volume R_p, is then

$$R_p = \int_0^{\nu_p} (1 - f_w)\, d\nu_p. \tag{5.3.53}$$

Integrating by parts gives

$$R_p = (1 - f_w)\nu_p + \int_0^{\nu_p} \nu_p\, df_w. \tag{5.3.54}$$

At the outlet face, where $x = L$, there follows, from Eq. (5.3.52)

$$1 = f_w'\nu_p \tag{5.3.55}$$

or

$$dS_w = \nu_p\, df_w. \tag{5.3.56}$$

Combining Eq. (5.3.56) with Eq. (5.3.51) gives

$$R_p = (1 - f_w)\nu_p + \int_{S_{wi}}^{S_w} dS_w = (1 - f_w)\nu_p + S_w - S_{wi}. \tag{5.3.57}$$

This equation makes it possible to calculate the oil recovery after breakthrough in units of pore volume, using the irreducible water saturation, the water saturation at the outlet face at the time of interest, the number of pore volumes of fluid produced (or injected) in units of pore volume, and the value of the simplified fractional flow function at the saturation of interest.

Relationship between $\bar{S}_w$ and S_w. A simple relationship can be derived between the average saturation of water $\bar{S}_w$ at time t in the flooded system and the saturation S_w at the outlet face at the same time, as follows (Morel-Seytoux, 1969).

Because the volume of oil recovered is equal to the increase of volume of water in the system, one can write

$$R_p = \bar{S}_w - S_{wi}. \tag{5.3.58}$$

Comparison of Eqs. (5.3.57) and (5.3.58) yields

$$\bar{S}_w = S_w + (1 - f_w)\nu_p. \tag{5.3.59}$$

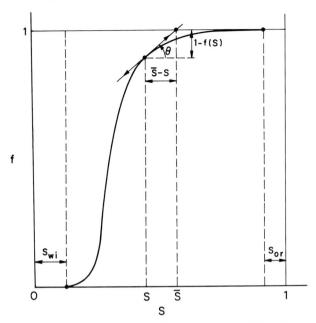

Figure 5.41. Graphical determination of flood performance (Morel-Seytoux, 1969).

Combining Eq. (5.3.55) with Eq. (5.3.59) gives the result

$$\bar{S}_{\mathrm{w}} = S_{\mathrm{w}} + \left[(1 - f_{\mathrm{w}})/f'_{\mathrm{w}} \right]. \tag{5.3.60}$$

A graphical solution of this equation, given by Jones-Parra and Calhoun (1953) is illustrated in Fig. 5.41.

Unsteady-State or Dynamic Relative Permeabilities. The technique discussed above is often reversed and used to obtain curves of relative permeability ratios. One proceeds as follows: The sample is initially saturated with water and then flushed with oil until water is no longer produced. By difference, the initial irreducible water saturation S_{wi} is obtained. Then the sample is flooded with water, and the relationship among the number of pore volumes of water injected ν_{p}, the instantaneous water cut F_{w}, and the oil recovery R_{p} in units of pore volume is determined. From Eq. (5.3.57) the value of the water saturation S_{w} at the outlet face of the sample, corresponding to the values of the quantities listed above, is as follows:

$$S_{\mathrm{w}} = R_{\mathrm{p}} + S_{\mathrm{wi}} - (1 - F_{\mathrm{w}})\nu_{\mathrm{p}}. \tag{5.3.61}$$

Equation (5.3.43) is solved for the relative permeability ratio, that is,

$$k_{rw}/k_{ro} = (\mu_w/\mu_o)[f_w/(1 - f_w)], \qquad (5.3.62)$$

where f_w is equated to the measured water cut F_w to calculate the permeability ratio. The value of the corresponding saturation is given by S_w as obtained from Eq. (5.3.61). Hence, from the flood data, the k_{rw}/k_{ro} versus S_w curve can be constructed point by point. This procedure is due to Welge (1952) and is the so-called unsteady-state, dynamic, or Welge method of determining relative permeabilities. It is obvious that if the permeability ratio calculated with the substitution $f_w = F_w$ is used in Eq. (5.3.43), it will yield the "correct" value (i.e., F_w) for the fractional water flow. The reader who is interested in a detailed discussion of waterflooding is referred to Craig's monograph on this subject (1971).

Effects of Wettability and Viscosity on Water Flood Histories.   Water flood curves tend to have very different appearances in strongly water-wet versus strongly oil-wet media (Anderson, 1987). At the same time the viscosity ratio also has a definite effect. As it is apparent from Fig. 5.42 of Dullien *et al.* (1990) (note that in this reference Figs. 5 and 4 have been inadvertently mixed up) there has been no more oil production after water-breakthrough in water-wet Berea sandstone cores, even at unfavorable viscosity ratios. It is noted also that in core No. 2, which was first saturated with oil, in the first waterflood the displacement pattern was very different, but in the second waterflood there was the same pattern also in this core as in all the others that were initially saturated with brine, with the important difference that the ultimate oil recovery was much less.

There is obviously a different displacement mechanism as a result of the core being initially saturated with oil. Dullien *et al.* (1990) could explain all the results obtained in core No. 2 quantitatively by assuming that in this core, in the first waterflood, (1) the displacement mechanism in individual pores was piston-like (there was no "pinch-off" of oil); and (2) regions of oil-saturated pores were bypassed from which the oil could not be recovered in the second water flood, either, because these regions were completely surrounded by water.

The waterfloods obtained in the same reference (Dullien *et al.*, 1990) in Berea sandstone cores made oil-wet either by aging for three months while saturated with Pembina crude oil (core No. 6), or by treating with a 2% solution of dichloromethylsilane (drifilm) in heptane (core No. 4), are shown in Fig. 5.43. The general trend of these curves is markedly different than in the water-wet samples, where there was practically no more oil production after water breakthrough. Water is not produced from the water-wet cores until all the continuous oil has been produced. This does not mean that there was a sharp, step-like saturation front in the sample throughout the flood

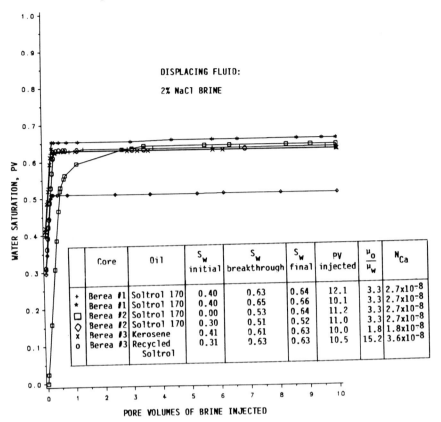

	Core	Oil	S_w initial	S_w breakthrough	S_w final	PV injected	$\frac{\mu_o}{\mu_w}$	N_{Ca}
+	Berea #1	Soltrol 170	0.40	0.63	0.64	12.1	3.3	2.7×10^{-8}
*	Berea #1	Soltrol 170	0.40	0.65	0.66	10.1	3.3	2.7×10^{-8}
□	Berea #2	Soltrol 170	0.00	0.53	0.64	11.2	3.3	2.7×10^{-8}
◇	Berea #2	Soltrol 170	0.30	0.51	0.52	11.0	3.3	2.7×10^{-8}
x	Berea #3	Kerosene	0.41	0.61	0.63	10.0	1.8	1.8×10^{-8}
o	Berea #3	Recycled Soltrol	0.31	0.63	0.63	10.5	15.2	3.6×10^{-8}

DISPLACING FLUID:

2% NaCl BRINE

WATER SATURATION, PV

PORE VOLUMES OF BRINE INJECTED

Figure 5.42. Waterfloods in water-wet Berea sandstone (Dullien *et al.*, 1990).

(e.g., Bacri *et al.*, 1985, 1990), because the saturation could have been building up gradually to the final value of about 0.64. It appears natural that there is no water production under conditions such that water would imbibe into the core. In the oil-wet samples there was significant oil production after water breakthrough in every run. In core No. 6 the effect of viscosity ratio is as expected, but in core No. 4 the flood behavior is altogether anomalous. Drifilming also reduced the permeability to heptane from 200 md to 127 md.

It is interesting to compare in Fig. 5.44 the waterflood behavior of mixed-wet Pembina Cardium core samples (Pembina No. 1: $k_{N2} = 3.9$ md; Pembina No. 2: $k_{N2} = 6.5$ md) with that of water-wet Berea samples (of much higher permeabilities). The mixed wet cores are characterized by very early water breakthrough and oil production continuing (practically) indefinitely. Extraction of the Pembina cores, using toulene and methanol alternately, however, drastically changed (improved) their flooding pattern, as

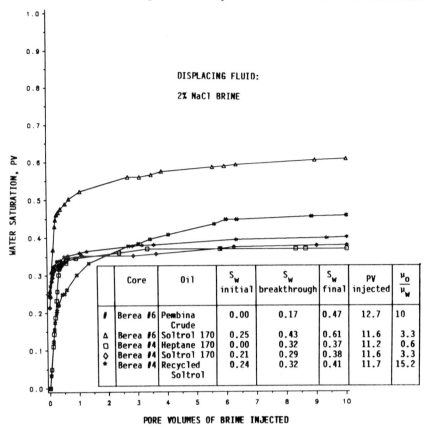

Figure 5.43.   Waterfloods in oil-wet Berea sandstone (Dullien *et al.*, 1990).

shown in Fig. 5.45. The fractional flow curve in Pembina core No. 2 exhibited similar behavior as the curve of Braun and Blackwell (1981), shown in Fig. 5.14.

Countercurrent versus Co-current Imbibition.   Laboratory countercurrent imbibition of water, displacing oil starting at connate water saturation from a sandstone block was first reported by Graham and Richardson (1959). This process is of considerable practical significance because of oil production by this mechanism in fractured reservoirs. It has been demonstrated in a capillary micromodel network by Lenormand (1981) that countercurrent imbibition does not take place if all the pores have the same size. Therefore it follows that in countercurrent imbibition, in the case of porous media containing distributed pore sizes, the water imbibes into relatively fine pores

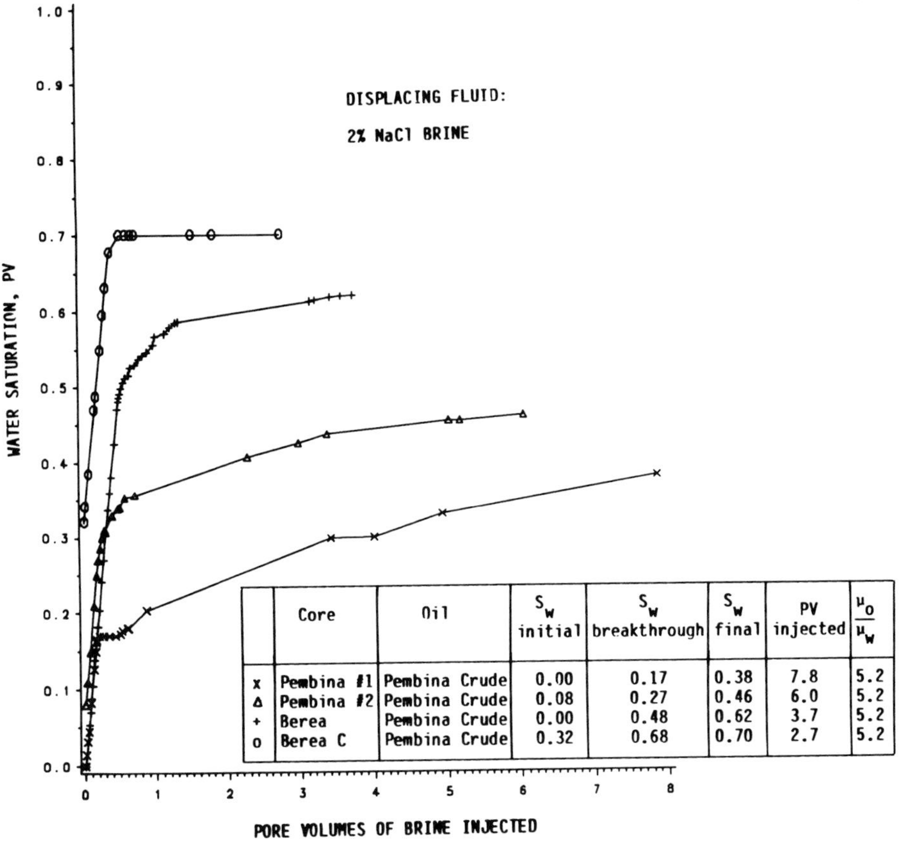

	Core	Oil	S_w initial	S_w breakthrough	S_w final	PV injected	μ_o/μ_w
x	Pembina #1	Pembina Crude	0.00	0.17	0.38	7.8	5.2
Δ	Pembina #2	Pembina Crude	0.08	0.27	0.46	6.0	5.2
+	Berea	Pembina Crude	0.00	0.48	0.62	3.7	5.2
o	Berea C	Pembina Crude	0.32	0.68	0.70	2.7	5.2

Figure 5.44. Waterfloods in Pembina Cardium cores No. 1 and No. 2 and two water-wet Berea sandstone cores (Dullien *et al.*, 1990).

and the oil is produced (at the same face) through relatively large pores. It would appear from this that the displacement mechanisms are different in co-current and countercurrent imbibition, respectively. Another question that arises is whether there is any change in the extent of recovery by countercurrent imbibition with increasing depth of imbibition into the material. These questions have been investigated in the author's laboratory (unpublished notes) by performing imbibition tests into cube-shaped Berea sandstone samples of different sizes, wholly submerged in water, and by comparing the residual oil saturations reached in all these tests and also in tests where the imbibition was strictly co-current (Dullien *et al.*, 1990). The co-current imbibition apparatus is shown in Fig. 5.46. All these tests yielded about the same residual oil saturations, the value obtained also in a typical water flood.

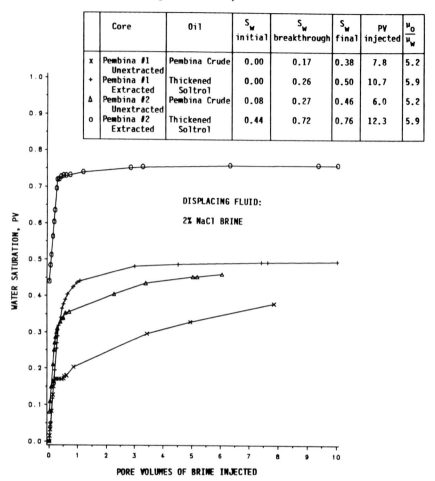

	Core	Oil	S_w initial	S_w breakthrough	S_w final	PV injected	$\frac{\mu_o}{\mu_w}$
x	Pembina #1 Unextracted	Pembina Crude	0.00	0.17	0.38	7.8	5.2
+	Pembina #1 Extracted	Thickened Soltrol	0.00	0.26	0.50	10.7	5.9
Δ	Pembina #2 Unextracted	Pembina Crude	0.08	0.27	0.46	6.0	5.2
o	Pembina #2 Extracted	Thickened Soltrol	0.44	0.72	0.76	12.3	5.9

Figure 5.45. Waterfloods in extracted Pembina Cardium cores No. 1 and No. 2 (Dullien *et al.*, 1990).

In a recent paper, Bourbiaux and Kalaydjian (1988) ran co-current and countercurrent imbibition tests into a long (290 mm) sandstone core of rectangular cross section (61 mm by 21 mm). The core was standing on one end. The saturations were determined *in situ* by x-ray absorption. Predominantly co-current imbibition was ensured by producing oil (lower density than water) at the top end while permitting water to imbibe at the bottom end; whereas in the countercurrent imbibition test the lower was sealed off while both water imbibition and oil production took place at the top end. The evolution of water saturation profiles in the two types of tests is shown in Fig.

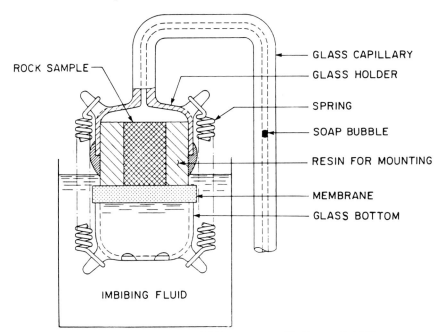

ROCK SAMPLE

GLASS CAPILLARY
GLASS HOLDER
SPRING
SOAP BUBBLE
RESIN FOR MOUNTING
MEMBRANE
GLASS BOTTOM

IMBIBING FLUID

Figure 5.46. Co-current imbibition apparatus (Dullien *et al.*, 1990).

5.47 and Fig. 5.48. It is very interesting to see in Fig. 5.48 that oil was produced along the entire length of the core, all the way to the bottom end.

Production of oil far from the producing face, of course, involved a greater resistance to flow to be overcome by capillary and gravity forces, driving the process, resulting also a slower production rate. Final recovery was 8.5% less than in the co-current case. Numerical simulation was used to fit the data to relative permeability curves, which turned out to be lower for the countercurrent case. This was attributed by the authors to an increased degree of hydraulic coupling between the two liquids in the countercurrent case. Hydraulic coupling, however, is not consistent with the physical picture of the oil and the water flowing in different pores. It is likely that the distribution of the two liquids over the pores of the network was different in the two types of tests, and this might have resulted in a greater resistance to the flow of each liquid in the countercurrent case.

5.3.4. Dynamic Capillary Pressure

Throughout this chapter, it has been assumed, as has been the custom, that the static capillary pressures apply also under dynamic or flow conditions.

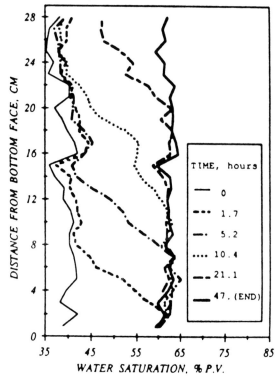

Figure 5.47. Saturation profiles measured in (predominantly) co-current imbibition run (Bourbiaux and Kalaydjian, 1988).

Important departures from equilibrium (static) capillary pressures P_c^s, under dynamic displacement conditions, have been reported by Calvo *et al.* (1990). The dynamic capillary pressure P_c^d was calculated from measurements as a function of time t of the position x of the interface between two immiscible fluids, 1 and 2, in the process of fluid 1 displacing fluid 2 in a uniform capillary tube, with a constant pressure differential ΔP maintained between the two extremities of the tube, from an integrated form of the Washburn equation applied to horizontal displacement where both fluids have nonnegligible viscosities in imbibition

$$\int_0^t P_c^d \, dt = \frac{32}{D^2} \left[\frac{\mu_1 - \mu_2}{2} x^2 + \mu_2 Lx \right] - \Delta Pt \qquad (5.3.63a)$$

and in drainage

$$\int_0^t P_c^d \, dt = -\frac{32}{D^2} \left[\frac{\mu_1 - \mu_2}{2} x^2 + \mu_2 Lx \right] + \Delta Pt \qquad (5.3.63b)$$

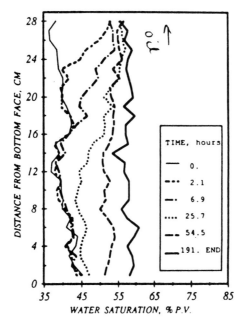

Figure 5.48.   Saturation profiles measured in countercurrent imbibition test (Bourbiaux and Kalaydjian, 1988).

where $P_c^d \equiv P_{nw} - P_w$, μ_1 and μ_2 are the viscosities, L and D are the length and the diameter of the capillary, respectively and $\Delta P \equiv P_1 - P_2$. In this formulation the effects of non-Poiseuille flow are lumped into P_c^d.

Both imbibition and drainage type tests have been carried out at different velocities.

In the first test, where kerosene was displacing air, $P_c^d \approx P_c^s \approx 1300$ dyne/cm^2 was found.

In the test with water displacing kerosene (imbibition) $P_c^s = 550$ dyne/cm^2, $P_c^d = 212$ dyne/cm^2, whereas with kerosene displacing water (drainage) $P_c^d = 920$ dyne/cm^2, respectively, were found. As in these tests the velocity kept changing, owing to the different viscosities of the two liquids, the system water–cyclohexane of equal viscosities was selected by Calvo *et al.* (1990) for further studies, where in each run the velocity was kept constant. Several runs were carried out, each at a different constant velocity.

First, imbibition of water into "dry" tubes containing cyclohexane was studied while varying the velocity over a range covering three orders of magnitude. With $P_c^s = 370$ dyne/cm^2 the dynamic values decreased with increasing velocity from $P_c^d = 250$ dyne/cm^2 to $P_c^d = -600$ dyne/cm^2 (!), as shown in Fig. 5.49 (solid squares). $P_c^s = 370$ dyne/cm^2 was obtained by

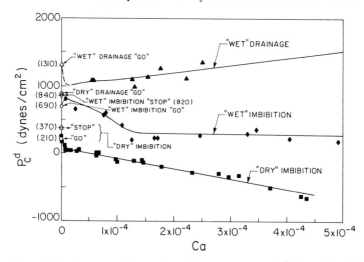

Figure 5.49.   Summary of dynamic capillary pressure results (Calvo *et al.*, 1990).

gradually increasing the opposing pressure differential and thus stopping the imbibition. After equilibrium, the opposing pressure differential was gradually decreased until imbibition started once again. This happened at a pressure differential $P_c^a = 210$ dyne/cm^2, which is somewhat less than $P_c^d = 250$ dyne/cm^2 measured at the lowest steady velocity. If, after stopping imbibition, the opposing pressure differential was increased until the interface started moving in the opposite direction (drainage), the least value of this pressure differential was found to be $P_c^r = 840$ dyne/cm^2. It is noted that at this point the tube walls on the nonwetting side of the interface had not been contacted by water (i.e., they were "dry"). Immediately after the interface started moving in the direction of the wetting fluid, however, the situation changed because the walls on the nonwetting phase side of the interface now had been contacted by water. Therefore, the value $P_c^r = 840$ dyne/cm^2 is singular and is not representative of the dynamic capillary pressures measured in subsequent drainage in which P_c^d increased with velocity from about 1100 dyne/cm^2 to about 1300 dyne/cm^2, as shown in Fig. 5.49 (solid triangles). These values should be compared with the static capillary pressure measured in a tube whose walls had been contacted by water ("prewet") on both sides of the interface. This value was found to be $P_c^s = 820$ dyne/cm^2, which is much higher than 370 dyne/cm^2, found in the "dry" tube. In the "prewet" tube the opposing pressure differential had to be increased gradually to $P_c^r = 1310$ dyne/cm^2 for drainage to start. This value is greater than the least value of $P_c^d = 1100$ dyne/cm^2, found at the lowest velocity. The static contact angle measured through water was calculated from P_c^s, D and σ to 75°.

Finally, imbibition in the prewetted tube started when the opposing pressure differential was gradually reduced to $P_c^a = 690$ dyne/cm². The values of the dynamic capillary pressure found under steady conditions decreased from about 810 dyne/cm² to about 200 dyne/cm² as the velocity was increased, as shown in Fig. 5.49 (solid diamonds). Again, $P_c^a = 690$ dyne/cm² is less than 810 dyne/cm², similarly as in the dry tube $P_c^a = 210$ dyne/cm² was less than $P_c^d = 250$ dyne/cm², measured at the lowest velocity. It is remarkable that the three ratios of this kind, corresponding to the three different types of tests are the same value (i.e., 210/250, 690/810, and 1100/1310 $\approx$ 0.84).

Calvo *et al.* (1990) have correctly pointed out that the differences between the capillary pressures measured in the "dry" and the "prewet" tubes, respectively, are due to different static contact angles existing under the two different conditions. The static contact angle in the "prewet" case works out to 55°. The rest of these results, however, could not be explained in terms of existing theories of contact angle changes. Consistent explanation of the results can be given by assuming that all the changes observed, excluding the differences between the "dry" and the "prewet" tests, are due to effects associated with an excess force required to make the contact line move and keep it moving. The physical reason for this excess force requirement is that under displacement conditions the entire interface including the contact line moves with the same velocity, whereas the velocity distribution in the fluids on either side of the fluid/fluid interface is parabolic, with no slip at the tube wall. Whereas it is logical to assume that an excess force is required to make the contact line move along the tube wall, to the author's knowledge no satisfactory theory has been produced as yet to account for this force.

A slip boundary condition of the contact line has been introduced *ad hoc* by Dussan (1976, 1979), Ngan and Dussan (1982), and also by Cox (1986). These contributions, as also practically all of the rest of the pertinent literature, have been mainly concerned with the change of the contact angle *per se* under dynamic conditions. Unless the displaced fluid is moved at the contact line, a film of the displaced fluid will always remain between the displacing fluid and the solid, something that is contrary to experience when a preferentially wetting fluid is displacing a nonwetting fluid (i.e., in imbibition). In drainage, "slip" may take place at a small distance from the wall, corresponding to the thickness of the film of the wetting fluid remaining of the solid surface after drainage. Assuming for the sake of argument that the thickness of this film on the surface of smooth glass is 100 Å in Poiseuille flow, the velocity at this distance from the wall is only 0.004% of the average velocity in the capillary. Therefore, the fluid velocity must be increased 25,000 times at this distance to bring it in line with the velocity of the interface. Even though the fluid is evidently moved away from the wall rather than parallel to it, such a relatively large velocity near the wall nevertheless results in very high shear stresses that can be expected to cause non-Newtonian behavior.

Rose and Heins (1962) were among the first to study the rate dependency of the contact angle. Precision-bore Pyrex tubing was used with a Nujol-air system. The motion of the Nujol slug was observed, including its rate of advance, and the pressure drop and the contact angles were also measured. The data could be represented by the following equation:

$$\bar{u} = (R^2/8L\mu)(\Delta P + (2\sigma/R)\{\cos\theta_A - \cos\theta_R\}) \qquad (5.3.64)$$

where $\bar{u}$ is the velocity of the interface and L is the slug length. The receding contact angle was found to be zero, whereas the advancing contact angle was a function of the velocity of the interface. Advancing angles as high as $\theta_A = 65°$ were measured.

An historic review of dynamic contact angles was published by Huh and Scriven (1971). Several theoretical and experimental studies on dynamic contact angles were reviewed by Schwartz and Tejada (1972).

In the very careful experiments of Hoffman (1975) with an advancing liquid–air interface in a capillary of about 2 mm inside diameter, it was demonstrated that the interface would "bow out" into the air phase at high enough velocities, thus covering contact angles on both sides of 90° in the same experiment. All the dynamic contact angles determined with five different liquids at various velocities could be correlated by the function $\mu\bar{u}/\sigma + F(\theta_s)$, where θ_s is the static contact angle in the system concerned. The values of $F(\theta_s)$ for the different systems ranged from zero up to 2.35×10^{-2}. In all of Hoffman's experiments the interface remained spherical, and he expressed his opinion that departures from spherical interface are due to inertial effects.

The results of Calvo *et al.* (1990) with kerosene displacing air were found to be compatible with those of Hoffman (1975); however, the deviations of the dynamic capillary pressures from the static value found in the experiments with the water–kerosene and water–cyclohexane systems were not compatible with those of Hoffman and were one order of magnitude greater than those predicted by Cox (1986).

In the author's laboratory experiments with oil–water systems in grooves between glass plates have shown that when the viscosities of the two fluids are comparable the liquid/liquid interface may not remain spherical. Instead it may deform indefinitely. Under the conditions of these experiments the contact line often did not move at all, and the contact angle did not change perceptibly during the displacement either (see Fig. 5.50). One can speculate that strong departures from spherical interface shape took place because of much greater viscous resistances in the displaced phase near the contact line than far from it. Such resistances do not exist if the displaced fluid is air.

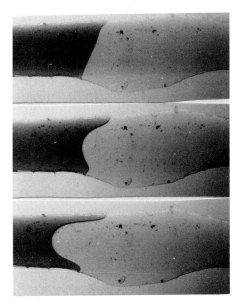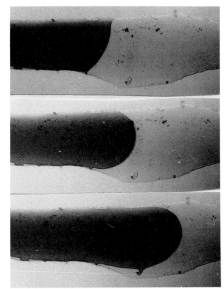

Figure 5.50. Immiscible displacement in an etched glass capillary under neutral wettability conditions: The contact line is stationary and the meniscus changes its shape (even "turns inside out" if the phase with lighter shading is the displacing fluid) as the displacing fluid advances (Dullien, unpublished work).

The values of the capillary pressure in the four types of experiment carried out by Calvo *et al.* (1990) at 0 velocity (I, stop), at 0 velocity (II, "go"), and the lowest measured steady velocity (III) are interesting to compare:

	$P_c/$dyne cm^{-2}		
	I	II	III
	(static	("go")	(dynamic
	or stop)		$dx/dt \to 0$)
(1) Imbibition into "dry" tube	370	210	250
(2) Displacement of interface in "dry" tube in the direction of wetting fluid	—	840	—
(3) Imbibition into "prewet" tube	820	690	810
(4) Drainage from "prewet" tube	—	1310	1100

Qualitatively speaking, the extreme position occupied by the starting pressure indicates the necessity of overcoming an activation energy barrier when starting the interface to move. The fact that the three ratios 210/250, 690/810, and 1100/1310 are the same value (~ 0.84) seems to indicate that the factors controlling the process after overcoming the barrier are of purely hydrodynamical nature. The "excess displacement pressures" in imbibition are $370 - 210 = 160$ dyne/cm^2 and $820 - 690 = 130$ dyne/cm^2 in the "dry" and "prewet" tube, respectively. The corresponding values in drainage are $840 - 370 = 470$ dyne/cm^2 and $1310 - 820 = 490$ dyne/cm^2. These values correlate with the static contact angle measured in the displaced liquid and make an interpretation of the data in terms of chemical contamination of the glass surface unlikely. Assuming excess hydrodynamic resistance to the displacement of the contact line permits consistent interpretation of all the results and it also offers an explanation of dynamic contact angles (Dullien, 1991).

5.3.5. Techniques Used to Calculate Immiscible Displacement in Two and Three Dimensions

The models of immiscible displacement discussed in Section 5.3.3 were limited to one-dimensional macroscopic flow. Methods are available to solve problems of immiscible displacement in two and sometimes three dimensions. These can be conveniently classified under two categories. In the first category, it is assumed that there is an abrupt interface separating the two phases and under these so-called elementary displacement conditions, certain mathematical techniques are available to calculate the displacement. In the second category, the equations of change (i.e., Darcy's law) and the equation of continuity combined with Laplace's equation are solved on a digital computer by numerical stepwise calculations. The methods in the latter category are often referred to as the "advanced methods" of calculating immiscible displacement.

The methods in the first of these categories have been discussed by Scheidegger (1974) and in even greater detail by Bear (1972) and deWiest (1969). They are considered to lie outside the scope of the present monograph.

The "advanced methods" category utilizes specialized, digital computer-oriented techniques, some of which are appropriately discussed in texts on numerical methods using the digital computer (e.g., Nobles, 1974; Peaceman, 1977). These methods also lie outside the scope of the present work.

5.3.6. Microscopic Displacement Mechanisms

A great deal of work has been done in the past decade on the study of microscopic displacement mechanisms, mostly in two-dimensional capillary micromodels. The scope of the studies may be conveniently divided into two parts: (1) displacement of a continuous phase and (2) displacement of discontinuous, that is, trapped oil. In both instances the emphasis lay on trying to establish the laws that govern immiscible displacement on the pore level.

5.3.6.1. *Displacement of a Continuous Phase*

There is relatively little work available on correlating displacement efficiency with pore structure. A classical work on water-flood tests with long cores was performed by Morse and Yuster (1946). They performed 42 water-flooding tests on three different long Second Venango sand cores of widely different permeabilities. Two series of tests were conducted one in which the sand was saturated completely with Bradford crude oil before water flooding and the other series consisting of flooding the same cores in which interstitial water was present with the oil. They concluded that no measurable effect of pressure gradient or velocities upon the residual oil saturations attained by water flooding could be detected.

Bethel and Calhoun (1953) studied the displacement efficiency in glass bead packs as a function of calculated apparent contact angle. The glass beads were surface treated with dry film to render them oil wet by n-octane to varying degrees. They found that the n-octane residual saturation decreased with decreasing apparent contact angle, as shown in Fig. 5.51.

Warren and Calhoun (1955) studied the water flood efficiency in oil-wet systems consisting of short pieces of consolidated Pyrex glass cores rendered oil-wet by chemical treatment. The contact angle, interfacial tension, permeability, and porosity were varied. Oil and water viscosities, core length, and velocity of flooding were held constant. They found that the breakthrough recovery correlated with the group $\sigma \cos \theta \sqrt{k}$, and ultimate recovery at 20 pore volumes of throughput correlated with the group $(\sigma / \cos \theta) \sqrt{\phi / k}$. In each case the recovery decreased with increasing value of the correlating group.

Kennedy *et al.* (1955) also investigated the effect of wettability on oil recovery by water flooding. In the first part of the work in which Woodbine outcrop sandstone was used in conjunction with East Texas crude oil, the scatter of the data was too great to establish any clear trend of wettability.

In the second part of the work in which the interfacial tensions were held constant, an artificial core made from pure silica sand was used. In this case the oil recovery had a slight maximum under conditions where the oil and the brine had equal tendency to wet the rock.

Mungan (1966) studied the effects of wettability, interfacial tension, and viscosity ratio on the immiscible displacement of a liquid by another liquid. Porous plugs made of compressed polytetrafluorethylene (TFE) powder were used. Displacement of a wetting by a nonwetting liquid was always found less efficient than the other way around, all other things being equal. In the former case the recovery efficiencies could be increased substantially by

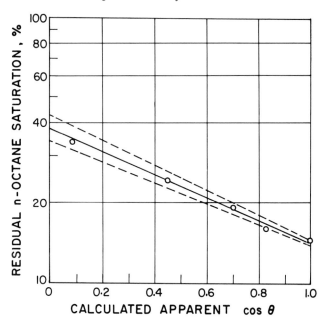

Figure 5.51.   Residual *n*-octane saturation as a function of the calculated apparent contact angle (based upon the assumption that complete *n*-octane wetting was obtained with beads treated with 1% drifilm). Apparent contact angles were calculated at 60% *n*-octane saturation (Bethel and Calhoun, 1953 © SPE-AIME).

either reducing the interfacial tension or increasing the viscosity of the displacing fluid.

Salathiel (1973) found that very low residual nonwetting phase saturations can be reached under special conditions of so-called mixed wettability (i.e., when the large pores are oil-wet and the small ones are water wet). The reason for mixed wettability to occur in nature is supposed to be the fact that surface active materials deposit from the oil in the larger pores where the oil is present, rendering the surface of these pores oil wet. According to Salathiel, it could be demonstrated that the oil keeps draining in tiny rivulets along the surface of the mixed wettability pore, resulting in very low residual oil saturations. This mechanism of two-phase flow is very different from the customary one observed usually in porous media.

The "irreducible saturation" was investigated theoretically by Iczkowski (1968) in the case of drainage from random sphere packings. His results are compared with experiment for the case of a thin bed consisting of 10 layers in Fig. 5.52.

Lefebvre du Prey (1973) studied extensively the effect of the dimensionless group $\pi = \sigma/\mu v$ on both irreducible wetting fluid saturation and residual

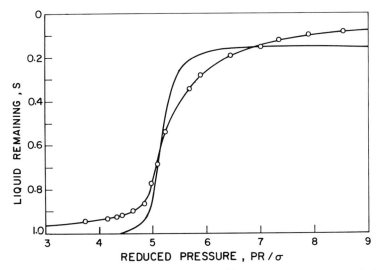

Figure 5.52. Calculated (solid curve) and observed (○, Harris and Morrow, 1964) displacement curves. (Reprinted with permission from Iczkowski, 1968 © by American Chemical Society).

nonwetting fluid saturations. In his experiments, using Teflon cores with a viscosity ratio of one, he found a significant decrease in the irreducible wetting fluid saturation and a corresponding increase in the residual nonwetting fluid saturation with decreasing π. While π varied from about 200 to about 6,000,000, the residual wetting fluid saturation increased from about 30 to 60%, whereas the residual nonwetting fluid saturation increased from under 10 to about 20%.

It looks from Fig. 5.53 that further increase of π will not result in any further displacement of either phase. Similar effects were obtained with a number of other materials, including alumina and stainless steel, using a number of different fluid pairs.

Interpretation of these results is not clear. Possibly as the capillary forces increase relative to the viscous forces, in a drainage or dewetting process, increasingly large capillaries are swept only, resulting in an increasing residual wetting fluid saturation. In the imbibition or wetting process, however, as π increases; the penetration increasingly favors narrow capillaries in the sample, resulting in an increasing residual nonwetting phase saturation.

Ehrlich and Crane (1969) proposed a version of their own of the pore doublet model for the prediction of displacement of one immiscible phase by another. They analyzed both steady-state (i.e., quasistatic) drainage and imbibition, and unsteady-state drainage and imbibition processes. Their

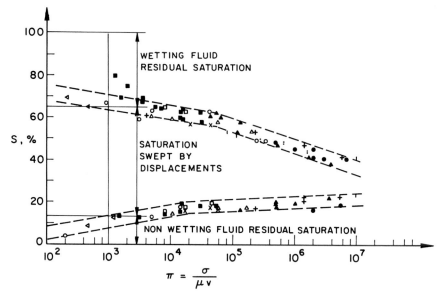

Figure 5.53. Effect of $\sigma/\mu v$ on residual saturations. Viscosity ratio = 1; porous medium is Teflon (Lefebvre du Prey, 1973 © SPE-AIME).

considerations do not include the important role played by pore topology and, therefore, the conclusions reached by them are probably often violated in real porous media.

The Pore Doublet Model. A simple model of immiscible displacement, the so-called pore doublet model, was introduced a long time ago by Rose and Witherspoon (1956), Rose and Cleary (1958), and, independently by Moore and Slobod (1956). A conventional pore doublet is shown schematically in Fig. 5.54. Both of these groups of authors were interested in oil displacement by water in strongly water-wet media and both predicted trapping of oil in

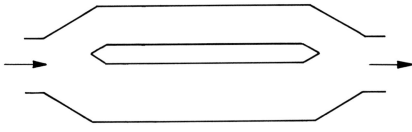

Figure 5.54. Pore doublet.

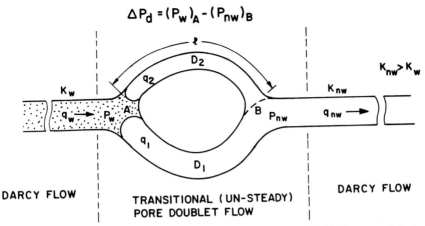

Figure 5.55. Conventional pore doublet model representation of imbibition-type of displacement (Chatzis and Dullien, 1983).

either the larger or the smaller capillary of the doublet, depending on the conditions of the displacement. More recently, Stegemeir (1974) and Doscher (1981) also analyzed this problem without, however, shedding light on the true mechanisms and the laws governing them.

A thorough analysis of immiscible displacement from a conventional pore doublet has been presented by Chatzis and Dullien (1983), who substantiated the results of their analysis visually by experiments performed in transparent capillary micromodels. Conventional pore doublet model representation of imbibition type displacement[4] is shown schematically in Fig. 5.55. By elementary considerations, using the Hagen–Poiseuille equation and Laplace's equation of capillarity, the following relationship was obtained for the ratio of the interface velocities λ, in the two branches of the doublet

$$\lambda \equiv \frac{\bar{u}_2}{\bar{u}_1} = \frac{N_{c1} + \beta^4 N_{c2}}{\beta^2 (N_{c1} - N_{c2})}, \qquad (5.3.65)$$

where $\beta = D_1/D_2$, $N_{c1} = 32\bar{u}_2^0 \mu/\sigma$, and $N_{c2} = D_2^2 \Delta P_c/\ell\sigma$ are dimensionless groups and $\bar{u}_2^0 = 4q_w/\pi D_2^2$ is a pore velocity of the water, based on the capillary diameter D_2, $\Delta P_c = P_{c2} - P_{c1}$.

---

[4]*Spontaneous imbibition*: The driving force of the process is the capillary pressure; i.e., the pressure drop over the system is equal to the capillary pressure. *Forced imbibition*: The pressure drop over the system is greater than the capillary pressure. *Free spontaneous imbibition*: The rate of spontaneous imbibition is determined by the resistance of the porous medium to flow. *Controlled spontaneous imbibition*: The rate is *less* than in free imbibition because an extra resistance is offered by a throttling valve or because the rate is determined by a pump.

428   5. Multiphase Flow of Immiscible Fluids in Porous Media

An analysis of Eq. (5.3.65), treated as a function $\lambda = \lambda(\bar{u}_2^0)$ shows that below a certain critical value of $\bar{u}_2^0$, $\lambda < 0$, at that critical value, $\lambda = \infty$ and, above that critical value, $\lambda > 0$. Using data in Eq. (5.3.65), which are typical of water-flooding operations of petroleum production, Chatzis and Dullien (1983) have shown that, under these conditions, $\lambda < 0$, with a wide margin of safety. $\lambda < 0$ means countercurrent displacement of the two interfaces, with the interface in the large capillary retreating rather than advancing. The physical reason for this is that for low values of $\bar{u}_2^0$, the magnitude of the viscous pressure drop from point A to point B in capillary 2, ΔP_2, is less than $(P_{c2} - P_{c1})$. The condition $\lambda < 0$, when applied to the physical situation depicted in Fig. 5.55, evidently means that the interface in capillary 1 will remain at point A, whereas the interface in capillary 2 will move toward point B, where it is shown by the dashed line in Fig. 5.55. Further advance of this interface is not possible unless the thread of oil ruptures at point B. The conditions of "snap-off," however, do not exist in the case of the configuration shown here, because the pore cross section gradually expands and has a local maximum at point B. As soon as the interface in capillary 2 has come rest, as shown by the dashed line in Fig. 5.55, imbibition into capillary starts and it continues all the way from point A to point B. As a result, contrary to the arguments that have been put forward in the literature, there is no trapping of the nonwetting phase in conventional pore doublets under conditions characterizing typical water-flood operations. An example of experimental results vindicating the results of theoretical analysis is shown in Fig. 5.56.

Nevertheless, there is trapping of the nonwetting phase in imbibition, but by a completely different mechanism that does not necessarily involve the existence of pore doublets. It is due to the strongly wetting phase advancing in grooves and edges in the solid surface *between* the solid and the nonwetting fluid and then giving rise to so-called "snap-off" of threads of the nonwetting phase at points of pore constriction. Once a portion of the nonwetting fluid is separated from the continuum communicating with the nonwetting fluid outside the medium as a result of one or several "snap-offs," it is trapped. The conditions of "snap-off" have been studied extensively by several workers and this subject will be discussed in the next section, after having briefly reviewed the findings of Chatzis and Dullien (1983) on displacement of the wetting fluid by nonwetting fluid (i.e., drainage) from a conventional pore doublet.

The situation in drainage is shown schematically in Fig. 5.57. The ratio λ^* of the interface velocities is given in this case by the following equation:

$$\lambda^* \equiv \bar{u}_1/\bar{u}_2 = \frac{\beta^2(N_{c1} + N_{c2})}{N_{c1} - \beta^4 N_{c2}}. \tag{5.3.66}$$

For conditions likely to be encountered in most real situations, $N_{c1} \ll \beta^4 N_{c2}$

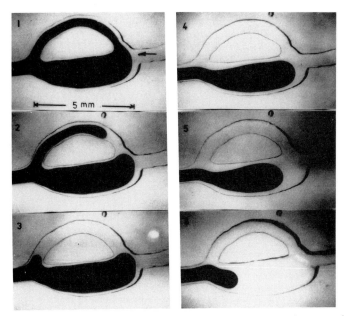

Figure 5.56. Experimental results illustrating the mechanism of displacement during free imbibition of water against *n*-decane in pore doublets (Chatzis and Dullien, 1983).

and, hence $\lambda^* < 0$ and $|\lambda^*| > 1$ (i.e., only the large capillary will be invaded). As the pore cross section at point B is greater than that of capillary 1, the nonwetting phase penetrates the node at B and cuts off and traps the wetting phase in pore 2, as shown both in the sketch in Fig. 5.57 and in the photographs of the experiment in Fig. 5.58.

Drainage and Imbibition Mechanisms in Pore Networks. Trapping of the nonwetting phase by the displacing wetting fluid by snap-off or choke-off mechanism was pointed out by Mohanty *et al.* (1980) when referring to the event taking place at the retreating end of an oil bank. Penetration of the wetting phase deep inside the oil bank by "film spreading," and choking off the oil there in pore constrictions was, to the author's knowledge, first observed in capillary micromodels and reported by Chatzis (1980).

Lenormand (1981) and Lenormand *et al.* (1983) did fundamental work on mechanisms of immiscible displacement on the pore level. These studies were carried out in capillary micromodel networks made of polyester resin, using a photographically etched mould (Bonnet and Lenormand, 1977). All the capillaries consisted of rectangular ducts that met at points of intersection in a pattern as of intersecting streets in a city map. There were no bulges

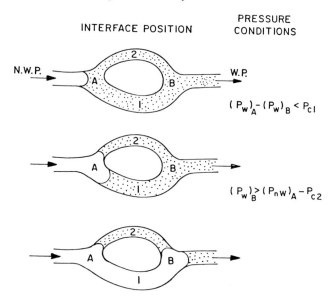

INTERFACE POSITION

PRESSURE
CONDITIONS

N.W.P.

W.P.

$(P_w)_A - (P_w)_B < P_{c1}$

$(P_w)_B > (P_{nw})_A - P_{c2}$

Figure 5.57. Conventional pore doublet model representation of drainage-type of displacement (Chatzis and Dullien, 1983).

at the intersections. The duct measurements were according to design. Fluids were chosen to assure perfect wettability ($\theta = 0°$).

Two kinds of displacement of the meniscus was observed in a duct:

(a) Piston-type displacement of the wetting fluid by the nonwetting fluid (i.e., drainage) under the influence of a displacement capillary pressure $P_p = P_{nw} - P_w$. The situation is sketched in Fig. 5.59. The value of the displacement pressure has been calculated by Lenormand (1981) and Legait and Jacquin (1982) for a duct of infinite length:

$$P_p = F(\varepsilon)2\sigma\left(\frac{1}{x} + \frac{1}{y}\right), \qquad \text{with } \varepsilon = \frac{x}{y} \qquad (5.3.67)$$

where $F(\varepsilon)$ is close to one.

(b) "Snap-off" of the nonwetting fluid already present in the duct occurs when a thread of the nonwetting fluid is created that is unstable. Lenormand *et al.* (1983) assumed that "snap-off" occurs when the nonwetting fluid first loses contact with the wall, as shown in Fig. 5.60 and, therefore the "snap-off" capillary pressure P_s is as follows:

$$P_s = \frac{2\sigma}{x}, \qquad (x < y) \qquad (5.3.68a)$$

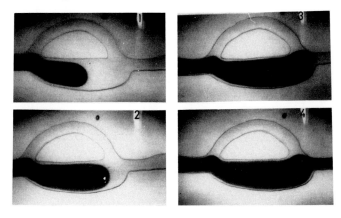

a.

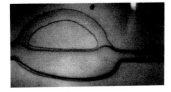

b.

Figure 5.58. Experimental results illustrating the mechanism of displacement under dynamic drainage conditions in a pore doublet. (a) *n*-Decane (black) displacing water (b) Air displacing water. (Chatzis and Dullien, 1985).

or

$$P_s = \frac{2\sigma}{y}, \qquad (x > y) \tag{5.3.68b}$$

Evidently, we have $P_p > P_s$. Hence, "snap-off" occurs in drainage only if there is a local reduction of pressure for topological reasons. In imbibition, however, "snap-off" is one of the main displacement mechanisms.

The following displacement behavior was observed at the intersection of four ducts, that is, at a node of the network:

(1) In drainage, the fluid when arriving at a node penetrates this rapidly and without a breakup, followed by penetration into other adjoining duct(s) of a displacement pressure less than or equal to the prevailing applied capillary pressure.

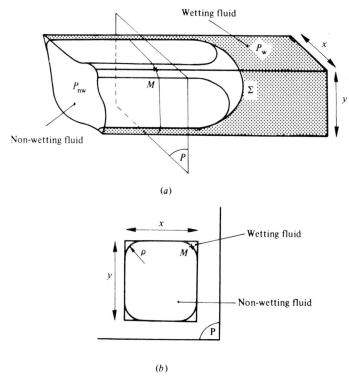

Figure 5.59.   Situation of the fluids in the section of a duct when the capillary pressure P_c is equal to the threshold pressure $P_P = 2\sigma(1/x + 1/y)$: (a) perspective view, (b) sectional view (P) (Lenormand, 1981).

(2)   In imbibition, two different displacement mechanisms at a node have been distinguished by Lenormand *et al.* (1983), depending on the topology of the menisci.

(a)   Imbibition I1 is the case, shown in Fig. 5.61, when the wetting phase has reached the node in three adjacent ducts. As the capillary pressure P_c is decreased by the experimenter [in the case of Lenormand *et al.* (1983) this was achieved by increasing the pressure on the wetting phase while maintaining the outside pressure on the nonwetting phase constant], position 2 in Fig. 5.61 is reached where the meniscus loses contact with the walls, resulting in instability. The capillary pressure corresponding to this point, P_{I1}, represents a local minimum in the evolution of capillary pressure. By the time the meniscus has reached position 4 in Fig. 5.61 the original capillary pressure P_p in the ducts before the jump of the interface across the node is reestablished. As the externally applied capillary pressure has been reduced in the process

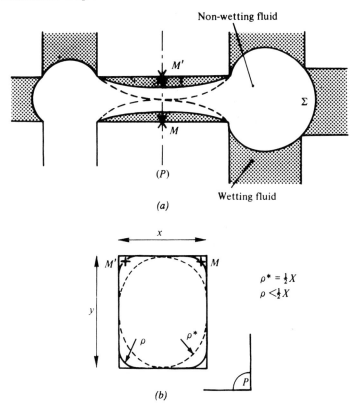

Figure 5.60. "Snap-off" in a duct: (a) in the plane of the network; (b) sectional view. The dashed curve shows the critical position (Lenormand *et al.*, 1983).

to P_{I1}, the meniscus will not be stationary at position 4, but will keep moving to the left in Fig. 5.61.

In the special case of four ducts with the same width x, we have

$$P_{I1} = \sigma\left(\frac{\sqrt{2}}{x} + \frac{2}{y}\right). \tag{5.3.69}$$

(b)   Imbibition I2 is the case, shown in Fig. 5.62a when the wetting fluid has reached the node in two adjacent ducts. In this case, the capillary pressure must be lowered even more by the experimenter for the wetting fluid to be able to cross the node (i.e., to P_{I2}) corresponding to the radius of curvature R at point A in Fig. 5.62a. For the special case of four ducts with the same width, we have

$$P_{I2} = 2\sigma\left(\frac{0.15}{x} + \frac{1}{y}\right). \tag{5.3.70a}$$

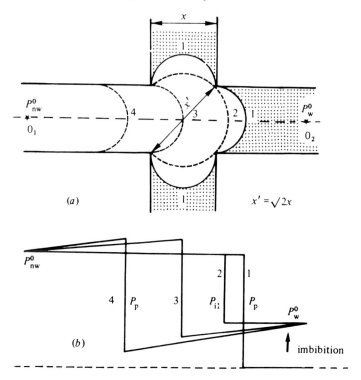

Figure 5.61. (a) Behavior of the meniscus during a type-I1 imbibition in a node; (b) variation of the pressure along the axis O_1O_2 (Lenormand *et al.*, 1983).

A "degenerate" case of I2, with the same meniscus topology but involving only three ducts, two of which intersect at an angle 2α and the third contains the wetting phase shown in Fig. 5.62b, gives rise to the following displacement pressure P_{12}:

$$P_{12} = 2\sigma\left[\frac{1 - \sin\alpha}{2x} + \frac{1}{y}\right]. \qquad (5.3.70b)$$

In effect, the topologies shown in Fig. 5.62c and d represent special cases of Eq. (5.3.70b) (i.e., when $\alpha = 0$). In this case Eq. (5.3.70b) reduces to Eq. (5.3.68b) and snap-off will occur in the duct containing the wetting phase.

In the rectangular network represented by the capillary micromodel of Lenormand *et al.* (1983) $y = 1$ mm and the various dimensionless displacement pressures, corresponding to different values of x and calculated by Eqs. (5.3.67)–(5.3.70), are listed in Table 5.4.

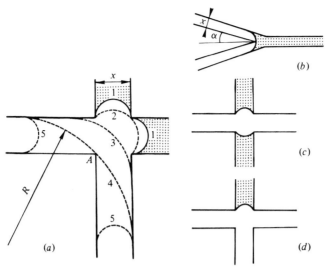

Figure 5.62. (a) Type I2 imbibition, the collapse takes place when the meniscus reaches the wall to the duct at the point A $[R = (2 + \sqrt{2})x]$. (b) More general case when the two ducts cross at an angle 2α. (c), (d) These other configurations are very stable, and the imbibition only occurs by "snap-off" inside the ducts (Lenormand *et al.*, 1983).

The magnitude and the structure of residual oil saturation was studied in detail experimentally by Chatzis *et al.* (1982, 1984). Experiments were performed in random packs of equal spheres, heterogeneous packs with heterogeneities of different scales, two-dimensional capillary networks and Berea sandstone. In the bead packs and in Berea sandstone brine and styrene monomer, containing percent-by-weight benzoyl peroxide, were used. The

TABLE 5.4 Values of the Various Displacement Pressures P/σ (mm^{-1}) in the Experimental Network[a]

Class	Width x (mm)	P_P/σ	P_s/σ	P_{I1}/σ	P_{I2}/σ
1	0.2	12	10	9	3.4
2	0.4	7	5	5.5	2.7
3	0.6	5.3	3.3	4.3	2.5
4	0.8	4.5	2.5	3.8	2.4
5	1	4	2	3.4	2.3
6	1.2	3.7	2	3.2	2.2
7	1.4	3.4	2	3	2.2

[a]From Lenormand *et al.* (1983)

trapped styrene was polymerized at 80°C for 48 hours with the core submerged in water and pressurized at 70 psi. The solidified styrene blobs were recovered by leaching the sample with a series of mineral acids. Figure 5.63 shows typical large oil blobs. Blob size distributions were determined. In the Berea sandstone, on a number basis, 50% of the blobs were in the range from 30 μm to 120 μm, which is consistent with the range of pore sizes

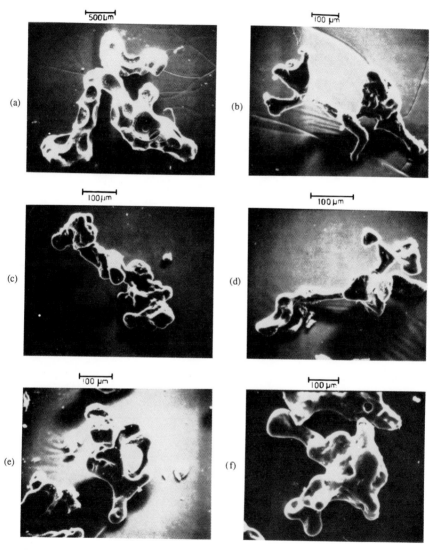

Figure 5.63.   Typical large oil blobs in (a) bead packs and (b–f) Berea sandstone (Chatzis, *et al.*, 1982).

measured. Larger blobs occupied more than one pore body. In the packs of equal beads, the residual oil saturation was independent of bead size.

A large number of different network patterns were created and trapping was studied as a function of the morphology of the networks. The following main conclusions were reached:

(1) The two main mechanisms of trapping are piston-like advance resulting in by-passing and trapping islands consisting of many pores and surface film flow, resulting in snap-off and trapping of smaller units of the nonwetting phase.

(2) The amount of trapping and the size distribution of trapped blobs depends to a very large extent on the aspect ratio and on the manner different size pores are distributed relative to each other. Large aspect ratios cause very high residual oil saturations. Large pores and clusters of large pores surrounded by smaller pores retain all their oil.

Wardlaw (1982) studied the effects of geometry, wettability, viscosity, and interfacial tension on trapping. He made the important observation that trapping is at a minimum in the case of 90° contact angle. Li and Wardlaw (1986) studied the minimum pore-to-throat effective diameter ratio ("aspect ratio") necessary for snap-off in throats as a function of advancing contact angle and found no snap-off at contact angles in excess of $\sim 70°$.

Wardlaw and Li (1988) performed imbibition experiments in a carefully constructed micromodel of etched capillary network (McKellar and Wardlaw, 1982) where the pores and the throats were "tailor-made." Experiments with mercury and air, lowering the pressure on mercury with the model 100% filled with mercury, resulted in snap-off in a few throats close to the air source boundary, followed by frontal drive of air without practically any trapping of mercury. In a similar experiment, started at 92% initial mercury saturation, there was no frontal drive and emptying proceeded by cluster growth all over the micromodel, resulting in a great deal of trapping and a residual mercury saturation of 55%. In the case of initial saturations between 100% and 93% there was a combination of frontal drive and cluster growth.

Other imbibition experiments were carried out in the same micromodel, using a glycerol–water mixture as the wetting phase and a refined oil as the nonwetting phase ($\theta_A \approx 30°$). The viscosities of both fluids were in the neighborhood of 50 cP (glycerol/water: 68 cP and oil 46 cP). Starting the imbibition at 80–90%, at a capillary number of about 10^{-4}, "water" always advanced frontally with no cluster growth observed at all. Typical residual oil saturation was about 40%. Wardlaw and Li (1988) do not explain the striking observation of frontal displacement of mercury by air without (practically) any trapping versus frontal displacement of oil by "water," but with about 50% trapping. The mercury/air case appears to represent an example of imbibition without trapping, explained by Chatzis and Dullien (1983), and illustrated in Fig. 5.64. The reason why all the pores acted in

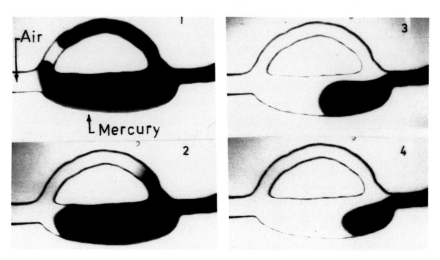

Figure 5.64.   Experimental results illustrating the mechanism of displacement during free imbibition of air against mercury (Chatzis and Dullien, 1983).

concert as a single capillary system might be the very low viscosity of the wetting phase (i.e., air), coupled with the high surface tension of mercury. (The mechanism of imbibition has not been investigated systematically as a function of capillary number and viscosity ratio.)

The situation was different in the case of the viscous water/oil system where local pressure changes were not equalized instantaneously between pores at a certain distance from each other. Some of the effects of interaction or no interaction between parallel capillaries are illustrated in Fig. 5.65 (Chatzis and Dullien, 1983). In the case of no interaction, the water advanced faster along certain, more conductive pathways present in the micromodel than along others, resulting in cutting off threads of the nonwetting fluid and trapping some of the same. (Unfortunately, Wardlaw and Li did not publish any photographs of the front.) This mechanism operates on a larger scale than imbibition with complete displacement on the level of pore doublets. As far as one can judge from the results reported by Wardlaw and Li (1988), there was no "surface film flow" in any of their experiments. Cluster growth in the mercury/air tests does not require any "surface film flow," because mercury can vacate pores without any influx of air from the outside.

The detailed statistics of fluid topology of pores emptied or occupied at various stages of the experiments of Wardlaw and Li (1988) is interesting, but it does not offer any insight into or any explanation of the observed course of the experiments. These authors also refer to many experiments of Morrow *et al.* (1981) with sandpacks or disordered packs of (smooth) glass beads where in imbibition consistently about 15% residual nonwetting phase satura-

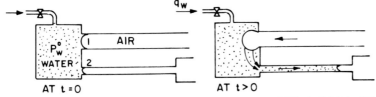

a) FREE IMBIBITION AGAINST AIR, WITH LIMITED SUPPLY OF THE WETTING PHASE, q_w = CONSTANT.

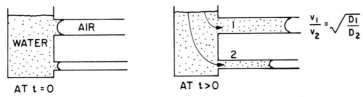

b) FREE IMBIBITION AGAINST AIR, WITH UNLIMITED SUPPLY OF THE WETTING PHASE

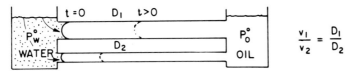

c) FREE IMBIBITION OF WATER AGAINST OIL, $P_o^\circ = P_w^\circ$, $\mu_w = \mu_o$

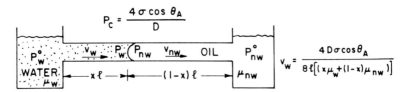

d) FREE IMBIBITION OF A WETTING PHASE IN A CAPILLARY TUBE, $P_w^\circ = P_{nw}^\circ$, $\mu_w \neq \mu_{nw}$

Figure 5.65. Free imbibition into parallel, noninteracting capillary tubes (Chatzis and Dullien, 1983).

tion was found, under displacement conditions completely controlled by capillary forces. Wardlaw and Li (1988) state: "This large displacement efficiency is incompatible with withdrawal from pores in a sequence related to size." They refer to their result (Li *et al.*, 1986) of 69% residual nonwetting phase saturation in imbibition, obtained by computer simulation of a three-dimensional model with $z = 6$ and random arrangement of pores.

Nonwetting phase was withdrawn from the pores in a sequence strictly related to pore size. Wardlaw and Li (1988) suggest, without offering any specifics, that a possible explanation of the high withdrawal efficiency obtained for sandpacks is that fluid topology exerts a major control on withdrawal sequence. While not trying to deny the importance of fluid topology, it should be pointed out that trapping predicted by percolation theory (Diaz *et al.*, 1987) is about 32%, on a number basis. As disordered packs of fairly uniform size beads have also a fairly narrow pore size distribution, trapping of the nonwetting phase is predicted by this theory to be about 30% of the void volume. The experimental results showing about 15% trapping are much lower than these predictions, which may be due to the following two causes. First, it is recalled that in percolation theory an infinite percolation cluster is formed at the percolation threshold that corresponds in finite size samples to the displacing phase penetrating the entire length of the sample (i.e., to "breakthrough"). On further displacement, the percolation cluster branches out in every direction and traps about 32% of the other fluid. Immiscible displacement tests in bead packs have been carried out traditionally in a gravity stable manner. In the case of imbibition the water was imbibing vertically upward in the bead pack. Under these conditions of displacement there does not exist any infinite percolating cluster. Instead, there is only an irregular front of the imbibing water. Such a front can be expected to trap less nonwetting fluid in its upward progress than a cluster extending over the entire length of the sample, particularly in the case of a bead pack consisting of nearly uniform beads. In this case the advancing front of water is not even very irregular, because all the paths in the pack have nearly the same hydraulic diameter. As the hydraulic diameter distribution becomes narrower, trapping can be expected to decrease and go towards zero in the limit as the width of the hydraulic diameter distribution becomes vanishingly small. It would be interesting to check on these effects by running appropriate experiments, including immiscible displacements in bead packs using liquids of equal density so as to eliminate gravity effects. Trapping of the nonwetting fluid can be expected under these conditions to be higher.

The other cause that may contribute to less trapping, on a volumetric basis, than predicted on a number basis, is that the trapped nonwetting phase in bead packs is present in the form of round globules that do not fill the entire void space where they are located.

Vizika and Payatakes (1989) published a systematic study of what they called "forced" imbibition into a network type glass capillary micromodel initially *completely* saturated with the nonwetting phase ("dry" model). Their terminology is different than the one used in this text. It appears arbitrary to draw the line at any particular value of the capillary number at which viscous forces become dominant. For example, in the case of free, spontaneous imbibition into a horizontal cylindrical capillary with matched viscosities, for $\theta = 0°$, the capillary number Ca $= \mu \bar{u}_w / \sigma$ reduces to Ca $= R/4L$, where R is the radius and L the length of the capillary. Evidently, Ca can have practically any value in this case, depending *only* on the values of R and L, and it does not seem to make any physical sense to ask the question whether viscous or capillary forces control this process. If the rate of the imbibition process is indicated, however, then the distinction between viscosity and capillary control becomes physically meaningful, because if the velocity of the interface in the capillary is greater than $\bar{u}_w = R\sigma/4L\mu_w$ then evidently viscous forces dominate and, conversely, if it is less than this value, then capillary forces dominate. The former is called forced imbibition in this text and the latter, controlled spontaneous imbibition. Absolute dominance, however, exists only as $\bar{u}_w$ approaches either 0 or infinity. Accordingly, the watershed is the value of the velocity in free spontaneous imbibition, which in a porous media is $v_w = (k_w/\mu_w)(P_c/L)$. (See Section 5.3.6.3.)

Vizika and Payatakes (1989) varied systematically the capillary number in the range from 10^{-4} to 10^{-7} in each run, and they ran constant-rate experiments with systems of equilibrium contact angles in the range from 5° to 68° and of viscosity ratios μ_{nw}/μ_w in the range from 0.01 to 1.52. At the end of every run the residual nonwetting phase saturation was determined. Two different networks, one with aspect ratio 5:1 and another with aspect ratio 3:1, were used. The trend shown by the results generally confirms previous findings; that is, the residual nw phase saturations increased with decreasing capillary number, they decreased with increasing contact angle, and they were greater in the case of 5:1 than in the case of 3:1 aspect ratio. The apparently surprising result was that the residual nw saturations were strongly influenced by the viscosity ratio. They were much higher over the entire range of capillary numbers at $\mu_{nw}/\mu_w = 0.5$. This is the expected effect of unfavorable versus favorable viscosity ratio under conditions when viscous forces are important, but at a capillary number of 10^{-7} this was not expected to be the case.

Vizika and Payatakes (1989) have observed two distinct displacement fronts: a primary front, formed by "microfingers" of the displacing phase that fills the entire pore in this case, and a secondary front that advances ahead of the primary one along the pore edges and surface roughness. The secondary front caused severe disconnection of the nonwetting fluid by pinch-off. The average distance between the two fronts was found to increase with decreasing capillary number, and for $\theta = 5°$ and $Ca = 10^{-6}$ the length of the "wetting film" was 20–25 "grain" diameters long. The dynamics of transport of the wetting phase in surface edges and grooves is not understood well enough at the present to make unequivocal statements on the relationship between the pumping rate of the wetting fluid and the rate of advance of the "surface film." The first thing to investigate would be what happens to the "surface film" if the pumping is stopped at some intermediate stage of imbibition? Will the "surface film" advance farther at the expense of the wetting phase present in the micromodel, filling entire pores, or will it stop advancing? The little that has been disclosed by Vizika and Payatakes (1989) makes it likely that the length of the "wetting film" would continue increasing if the capillary number (i.e., the pumping rate) were reduced even more. It is also likely that the "wetting film" uses the wetting phase present in the primary front as a source, and its velocity relative to this front may be, if not constant, then only a weak function of the velocity of the primary front. According to this picture, the advance of the "wetting film" is driven by capillary forces and is resisted by viscous forces, regardless how low the capillary number is.

It must be pointed out at this stage that the territory between the primary front and the secondary front (i.e., the "wetting film" region) contains many pores that have been completely filled with the wetting film as a result of snap-off of the nonwetting phase in these pores. In other words, the flow of the wetting fluid in "surface capillaries" is frequently interrupted by flow through entire pores. Conditions for snap-off evidently could not be established if the "film" thickness were constant. Variable film thickness necessitates introduction of a more realistic model of flow in surface grooves than the simplistic picture of capillary tubes. The surface "film" does not only advance along the surface but it also expands toward the center of the capillary or pore.

In summary, during the past 10 years a great deal of knowledge has been acquired on the pore scale mechanism of imbibition. The emphasis has been on the advance of the wetting phase in surface grooves and edges in what is often called a "wetting film" form and "snap-off," "pinch-off," or "choke-off" of threads of the nonwetting phase caused by the presence of the film of the wetting phase on the pore walls. Under conditions of strong preferential wetting, in quasi-static displacement and in displacement at very low capillary numbers on the order of 10^{-7}, movement of the "wetting film" plays a major role in the displacement mechanism and the trapping of the nonwetting phase. At higher capillary numbers and larger contact angles the role played by the "wetting film" progressively diminishes.

The experimental observations have been made, so far, in two-dimensional micromodels the geometry and topology of which are oversimplified as compared with the case of real porous media. The capillary surface in these models is much smoother than the pore surface in real porous media. Therefore, it is necessary to study the effects of more realistic pore geometries and topologies, and of surface roughness, on imbibition mechanisms and trapping. Some preliminary work has been done in this direction, under quasistatic conditions, by Dullien *et al.* (1990).

An interesting contribution by Pathak *et al.* (1982) related the residual nonwetting phase saturation to the genus, representing the degree of interconnectedness of the pore space. Copper powder was sintered to produce cores of different pore topology. The genus was calculated by analyzing two-dimensional sections. Displacement experiments in the cores were carried out using water and Soltrol 50. Fair-to-good correlations between $(S_{nw})_r \phi$ and the genus per unit volume of core and the genus per initial particle (before sintering) were obtained. These results are important because they suggest that different porous media may be divided into classes according to their genus (topology), and within each class certain similarity of behavior of capillary and transport properties can be expected. The effect of topology must, however, be complemented with the effects of aspect ratios of adjacent pores and throats and of spatial correlations between pores.

Distribution of Fluid Phases in the Pores of Real Porous Media, In Situ Phase Immobilization. Impregnation of the pore space of porous media by resins has been used for quite some time in the process of preparation of thin sections (samples of a small thickness, with a depth-effect) and polished sections (where only the polished surface of the sample is viewed and there is no depth-effect). Impregnated samples have been used in photomicrographic pore morphology studies. Recently this technique has been extended to the study of the distribution of immiscible fluids in the pores by Yadav *et al.* (1985, 1987). A suitable pair of immiscible fluids has been used as wetting and nonwetting phases for conducting the usual displacement experiments in core samples so that at least one of the fluids could be conveniently solidified *in situ* at the end of the experiment without a significant alteration of the equilibrium pore-level shape, size, or position acquired in the liquid state. After solidifying one of the phases, the other one was displaced with conventional methods and macroscopic properties of this particular network could be determined (phase permeability, dispersion coefficient, electrical resistivity, etc.). The empty phase was filled by another liquid, which was subsequently solidified *in situ*. Finally, the rock matrix was also replaced with a resin by etching. In all the experiments macroscopically uniform saturation of the phases was established by using a semipermeable membrane at the exit face of the core.

The following fluid pairs, representing the wetting and the nonwetting phases, respectively, have been investigated and found to be satisfactory. System I: ethylene glycol/Wood's metal (alloy 158); System II: epoxy resin ERL 4026/N_2 gas; System III: brine/styrene monomer (containing benzoyl peroxide as catalyst). Details of the various procedures used are described in *loc. cit.*

A technique was developed using color photography to obtain photomicrographs in which the three phases were clearly distinguishable. The samples were serially ground and polished and serial sections were photographed. Sample photomicrographs are shown in Figs. 5.66 and 5.67.

These photomicrographs can be used to reconstruct the three-dimensional geometry of the two fluid phases and this, in turn, can be very useful in realistic network modeling of two-phase flow in porous media.

This technique may also be adapted to the case of immiscible displacement in progress, with a saturation gradient present if one of the phases can be immobilized in a very short time, practically instantaneously.

5.3.6.2. *Displacement of Trapped Nonwetting Fluid*

Displacement of the nonwetting phase may occur according to a primary mechanism, involving flow of a continuous nonwetting phase, or by a sec-

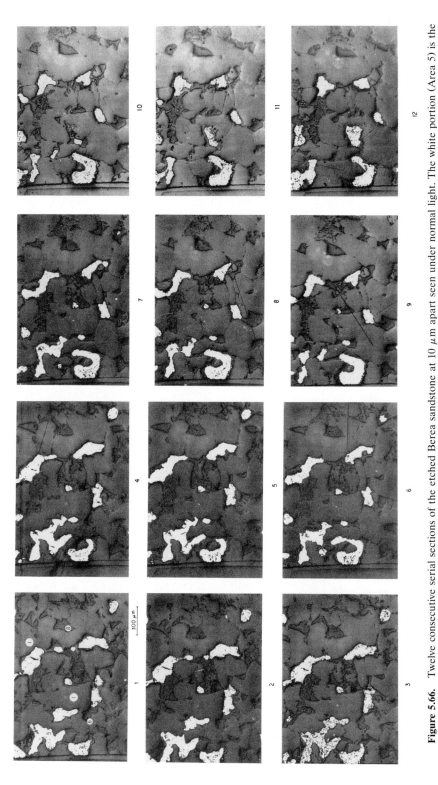

Figure 5.66. Twelve consecutive serial sections of the etched Berea sandstone at 10 μm apart seen under normal light. The white portion (Area 5) is the Woods' metal representing the nonwetting phase. Ethylene glycol, the wetting phase that is replaced by the resin ERL 4206 in the photograph, is the dark gray portion (Area 6). It is seen as strongly wetting the rock (Area 1), which is replaced by Buehler resin (Yadav *et al.*, 1987).

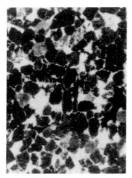 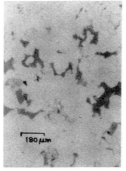 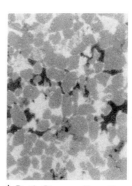

a) Wetting Phase Only   b) Non-Wetting Phase Only   c) Both Phases Together

Figure 5.67. Visualization of microscopic distribution of fluids in a typical serial thin section of Berea sandstone at a wetting-phase saturation of 53%, showing (a) the wetting phase only (white portion of Area 1); (b) the nonwetting phase only (dark portion of Area 2); and (c) the wetting phase, nonwetting phase, and rock (Areas 1, 2, and 3, respectively) (Yadav *et al.*, 1987).

ondary mechanism involving a disconnected, so-called trapped, nonwetting phase. The two mechanisms can be separated if the primary displacement is conducted under such conditions that the displacing phase is strongly wetting, the capillary number is low, and the viscosities are not very high.

If these conditions are met, the trapped phase is present in the form of disconnected "blobs" or "ganglia" of various sizes and shapes. These entities are held in the intergranular spaces by capillary forces. A displacement of the trapped phase is possible by a number of different techniques: The value of the interfacial tension can be decreased, the viscous forces increased, or the two phases can be made miscible so that the trapped phase will be dissolved by the flooding phase. This latter technique, however, is outside the scope of this chapter (see Chapter 6).

Early Contributions. The earliest explanations of the nature of the forces holding a disconnected "blob" trapped were based on Jamin's (1860) observation that in a capillary tube containing a number of drops of liquid interspersed with gas, a large difference in pressure between the ends of the tube may be required to make the drops move. This effect has been ascribed to a difference between the advancing and the receding contact angles. It was studied in some detail by Schwartz *et al.* (1964), who found that the minimum force per unit length of boundary line ("critical line force") increased linearly from zero as cos θ_{equil} decreased from unity.

The capillaries in most porous media have a cross section of varying size, and under these conditions a considerable force may be required to displace a trapped drop because Laplace's equation predicts a higher curvature for the meniscus at the front end of the drop, as it is being squeezed through a capillary neck, than at the rear. This effect was pointed out by Gardescu (1930).

Early contributions in this area are due to Ojeda *et al.* (1953) and Paez *et al.* (1954), who water-flooded cores containing iso-octane, starting at irreducible water saturation. Each flood was stopped after the rate of water flow leaving the core became constant. The iso-octane saturation after the water flood was determined, the water was replaced by a surfactant solution, and the flood was contained. The surfactant flood was stopped when the surface tension of the effluent approximated that of the solution being injected. A number of different Berea cores were used, but all recovery tests were made at a single value $\sigma/\Delta P = 0.163$ cm.

In another set of tests, no surfactants were used. Toluene was the oil-phase. The pores were not saturated with water first but instead were filled with carbon dioxide gas. Then, by displacement of the gas, they were completely filled with toluene that had been shaken thoroughly with water and ethyl alcohol. The alcohol added was sufficient to lower the interfacial tension to the desired value. The cores were flooded with water from the same container in which the toluene, alcohol, and water had been mixed. At all times the flooding water was kept in contact with the toluene–alcohol mixture. Saturation of toluene after a water flood was determined, subtracting measured produced toluene from original toluene.

The residual saturations of iso-octane in the surfactant tests were correlated, using $\sqrt{k/\phi}$ as a correlating parameter. Approximately linear correlation was found in a semilog plot.

For those tests using toluene–water with the interfacial tension lowered by alcohol, the residual saturations were plotted versus $\sigma/\Delta P$ for each case.

An interesting set of experiments was done by Taber (1969), who found that the displacement by water of discontinuous residual oil is a function of the ratio $\Delta P/L\sigma$. There was a minimum critical value of this ratio at which significant oil production started in the range of 2–5 (psi/ft)/(dyn/cm) for a variety of fluids and surfactants. He found that it was possible to displace nearly 100% of the residual oil if the value of this ratio was raised sufficiently. In order to obtain a significant removal of the trapped oil phase, the value of $\Delta P/L\sigma$ had to exceed the critical value of this quantity at least by one order of magnitude.

In a later paper, Taber *et al.* (1973) confirmed Taber's earlier findings. They found large differences in the critical value of $\Delta P/L\sigma$ if rocks of different permeabilities were used, but the viscosity of the fluids had relatively little effect. For example, in a permeable sandstone core, the value of the critical ratio ranged from 1.7 to 2.9 (psi.ft)/(dyn/cm) when brine displaced residual oils with viscosities ranging from 0.63 to 13.33 cP. On the other hand, the critical value for the displacement of oil dropped from 23.8

(psi/ft)/(dyn/cm) for a core with a permeability 95 md to 0.31 for a highly permeable sandstone core of 2190 md.

Dullien *et al.* (1969, 1972, 1973, 1976) have concentrated on the effect pore structure has on the recoverability of oil trapped after a water flood. They kept the value of Taber's group $\Delta P/L\sigma$ constant in their experiments, which were conducted in a large number of different sandstones with widely different permeabilities, and determined the so-called tertiary oil displacement efficiency (i.e., the percent of the water-flood residual oil saturation recovered in a surfactant flood).

Based on elementary model considerations similar to those used by Gardescu (1930) a group called the "structural difficulty index" of residual oil recovery *DI* was defined as

$$DI = \left(1/\overline{D}_\mathrm{e} - 1/\overline{D}\right), \qquad (5.3.71)$$

where $\overline{D}_\mathrm{e}$ and $\overline{D}$ are the mean pore entry diameter and median bulge diameter, respectively. For the former, the pore entry diameter corresponding to the inflection point in the mercury intrusion porosimetry curve of the sample was used; and for the latter, the mean bulge size determined photomicrographically in sections of the sample was used. The "structural difficulty index" correlated all the tertiary oil recoveries well, as shown in Fig. 5.68; however, the group introduced by Paez *et al.* (1954), $\sqrt{k}/\phi$, also correlated all but two of the data points.

From a combination of the Taber number and the difficulty index, Macdonald and Dullien (1976) constructed the dimensionless "tertiary oil recovery number" N_SAA defined as

$$N_\mathrm{SAA} = \frac{(\Delta P_\mathrm{w}/L)\bar{\ell}}{\sigma DI}, \qquad (5.3.72)$$

where $\bar{\ell}$ is the mean length of oil ganglia.

Melrose and Brandner (1974) used a capillary number N_Ca, defined as

$$N_\mathrm{Ca} = \mu_\mathrm{w} v_\mathrm{w}/\phi\sigma, \qquad (5.3.73)$$

in conjunction with Darcy's law to obtain a criterion for the recoverability of trapped oil saturation. As a result, they obtained the following expression:

$$N_\mathrm{Ca} = [k_\mathrm{w}/\phi\sigma](\Delta P_\mathrm{w}/L). \qquad (5.3.74)$$

A long list of different formulations of the "capillary number" by different authors has been compiled by Taber (1981).

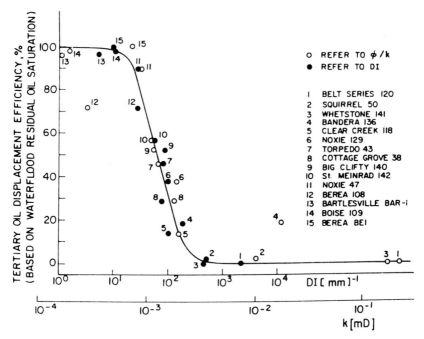

Figure 5.68.   Correlation of tertiary oil displacement efficiency versus ϕ/k ($\bigcirc$) and DI ($\bullet$). The following points are plotted: 1, Belt series 120; 2, Squirrel 50; 3, Whetstone 141; 4, Bandera 136; 5, Clear Creek 118; 6, Noxie 129; 7, Torpedo 43; 8, Cottage Grove 38; 9, Big Clifty 140; 10, St Meinrad 142; 11, Noxie 47; 12, Berea 108; 13, Bartlesville BAR-1; 14, Boise 109; 15, Berea BE1 (Macdonald and Dullien, 1976 © SPE-AIME).

Blob Mobilization.   Experimental studies of various aspects of the mobilization and motion of trapped oil blobs have been reported, among others, by Batycky and Singhal (1978), Ng *et al.* (1978), Rapir (1980), Egbogah and Dawe (1980), Hinkley (1982), Chatzis *et al.* (1983), Dawe and Wright (1983), Lenormand *et al.* (1983), Mohanty and Salter (1983), Mahers and Dawe (1985), Wardlaw and McKellar (1985), Legait (1981, 1983), Legait and Jacquin (1982), and Olbricht and Leal (1982). Excellent reviews on blob (ganglia) dynamics have been published by Payatakes (1982) and Payatakes and Dias (1984).

Ng *et al.* (1978) have presented a theory for the mobilization of a blob in the pore space of a random packing of equal spheres. According to this theory a blob will move when the capillary pressure difference between the ends of the blob is less than the hydrodynamic pressure produced by the viscous fluid flowing in the pore space alongside the blob. The theory was tested in visual experiments in a column of Perspex spheres by matching its refractive index and that of the wetting phase. The blobs were observed to move in small, quantum-like jumps, because in a sphere packing the menisci at the advancing end of the blob are forced through constrictions of various different sizes, while the retreating end of the blob moved through larger cavities. There is a jump each time when a meniscus is forced through a constriction. The theory of Ng *et al.* (1978) was confirmed in similar experiments by Yadav and Mason (1983).

The same physical principles have been applied by Chatzis and Morrow (1981) to arrive at the condition of mobilization of a blob of length ℓ:

$$\frac{k \, \Delta P}{L\sigma} = \frac{\phi D_e}{24 \, \ell} \left(\cos \theta_R - \frac{\cos \theta_A}{\beta} \right) \tag{5.3.75}$$

where $\beta = D/D_e$, an aspect ratio. Equation (5.3.75) was used to correlate residual oil saturations obtained in sandstone cores by increasing ΔP step-by-step. There is a very close relationship between Eq. (5.3.75) and Eq. (5.3.72) because in the derivation of the former the relationship $k = \phi D_e^2/96$ has been used. Substituting this relationship back into Eq. (5.3.75), on rearrangement Eq. (5.3.72) is obtained with $N_{SAA} = 4$ (rather than 1, as it should). Equation (5.3.72) does not express the condition of mobilization of blobs of length $\bar{\ell}$, and N_{SAA} is merely the *ratio* of viscous to capillary forces acting on the blob. Hence, in Eq. (5.3.72) N_{SAA} is a variable that could be used to compare mobilization in different porous media characterized by different values of the ratio ℓ/DI. It is noted that Chatzis and Morrow (1981) did not determine either the contact angles or β. It was not explained why it follows from Eq. (5.3.75) that the left-hand side should result in a unique correlation of reduced residual oil saturation for geometrically similar systems. A plausible explanation would be that ℓ is the same function of D_e and the blob populations differ only by a magnification factor. The data show good correlations between the group $k \, \Delta P/L\sigma$ and the residual oil saturation for individual sandstone samples; and the curves of different sandstones come closer if the residual saturations are reduced by dividing each residual saturation obtained in a run by its limiting value measured at low ΔP.

In a subsequent paper by the same group (Morrow *et al.*, 1985) it was demonstrated that in packs of beads of equal size, plotting the reduced residual saturations versus $v\mu/\sigma$ has resulted in the same curve for four different size beads and four different interfacial tensions. In this case the practically perfect geometrical similarity of pore morphology of the bead packs was responsible for the excellent correlation (see Fig. 5.69). On the other hand, plotting the reduced residual saturations of different media versus $v\mu/\sigma$ has resulted in widely different relationships, as illustrated in Fig. 5.70. This is rather disappointing, because if the capillary number truly represented the typical value of the ratio of viscous-to-capillary forces in the pores of the medium under study, then one would expect correlations with different spreads but approaching the same mean value. It is possible,

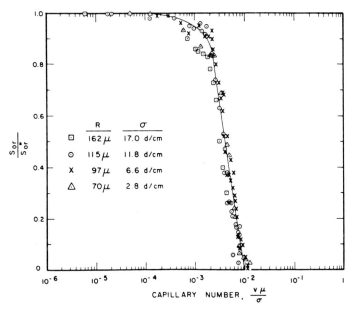

Figure 5.69.   Reduced residual saturation vs. capillary number in bead packs (Morrow *et al.*, 1985).

however, that in consolidated media the resistance to the motion of blobs is not due to capillary forces alone, but there is a contribution by "steric hindrance"; that is, the blob may be pushed by the viscous force into a corner without a window. Another observation is that, for sandstones, the left-hand side of Eq. (5.3.75) varies by a factor of a thousand, whereas the blob size, on the right-hand side varies only by a factor of 30.

Wardlaw and McKellar (1985) confirmed the findings of Morrow and Chatzis (1981) and Chatzis *et al.* (1983, 1984) in oil displacements from bead packs. They derived their version of the right-hand side of Eq. (5.3.75) (with $\theta_R = \theta_A = 0°$) and evaluated it numerically for the case $D = \ell$. The value of this "critical capillary number" was found to be 2.2×10^{-3}, which is consistent with experiment in bead packs. It is certainly peculiar that when the viscous and capillary forces acting on a blob are equal the capillary number that is supposed to be equal to the ratio of viscous-to-capillary forces is equal to 2.2×10^{-3}.

Lenormand and Zarcone (1988) carried out water floods, followed by blob mobilization experiments in two-dimensional etched networks containing more than 40,000 interconnected capillaries with various known sizes distributed randomly over the network. The displaced phase was air and the wetting phase was low-viscosity oil. Many runs were done. Each run consisted

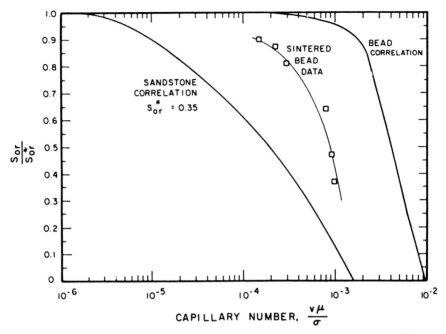

Figure 5.70. Reduced residual saturation vs. capillary number in various media (Morrow *et al.*, 1985).

of a water flood, dubbed as "imbition," at a known, constant capillary number, defined as

$$Ca = Q\mu/A\sigma, \tag{5.3.76}$$

where Q is the volumetric flow rate and A, the cross-sectional area, is equal to the product of the depth of the etched channels (i.e., 1 mm) by the width of the network (i.e., 135 mm). The corresponding "porosity" has not been disclosed. Each water flood was followed by more water flood at increasing flow rates and correspondingly higher capillary numbers, until at the end of each flood all the air was displaced. The results of these experiments are plotted in Fig. 5.71. It is apparent by comparing this plot with the "sandstone correlation" curve in Fig. 5.70 that the range of capillary numbers is about the same in both plots, although the ranges of residual air saturations are quite different. The authors analyzed their results in terms of the same critical capillary number of blob mobilization used previously by Chatzis and Morrow (1981) and Wardlaw and McKellar (1985), using the measured blob lengths and the capillary pressure differences calculated between the drainage (downstreams) end and the imbition (upstream) end of the blob from the known capillary measurements and displacement mechanisms. The drainage

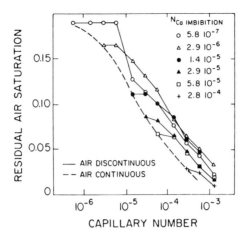

Figure 5.71.   Residual saturation vs. capillary number in capillary micromodel (Lenormand and Zarcone, 1988).

pressure was assumed to correspond to piston-type displacement whereas the imbibition, to rupture of "loops" in the network by "snap-off" mechanism. Their general observation was that the dominant mechanism of blob displacement was by "snap-off" at all but the highest capillary numbers. On the other hand in the "imbibition" runs "snap-off" at random locations over the entire network was dominant only at the lowest flow rate (Ca = 5.8 × 10^{-7}). Under these conditions "quasi-static imbibition" was a percolation process consisting of randomly removing bonds.

De la Cruz and Spanos (1983) presented a refreshingly different approach to blob mobilization in an in-depth study. Their criterion of mobilization is the point when Darcy's law breaks down for the continuous phase, because the permeability increases more than linearly with the pressure gradient, as a result of increasing flow of the discontinuous phase.

Weinhardt and Heinemann (1985) have studied water flood residual oil saturation obtained in sandstone cores of different diameters and different pressure gradients (i.e., flow rates). The size distribution of the trapped oil ganglia was also determined by the technique using styrene monomer as oil and subsequently polymerizing the trapped styrene. There appears to have been a decreasing trend of S_{or} with increasing pressure gradient.

The size distribution of trapped blobs generally shifted in the direction of smaller sizes with increasing pressure gradient. It was pointed out that small core diameters tend to result in increased trapping.

Hinkley *et al.* (1987) studied the velocities of oil ganglia (blobs) of different sizes in monolayers of equal size glass beads arranged in a regular square array and held between two glass plates. Runs have been carried out both in axial and in diagonal directions. The authors distinguished between quasi-static displacement of ganglia, occurring at capillary numbers slightly greater than the critical value necessary for mobilization and dynamic displacement at capillary numbers, only greater than the critical value by a factor of 3 and up, when the ganglia breaks up into many pieces. In the quasi-static process in displacements in the axial direction the ganglia line up with and flow parallel to the channels between rows of beads. In the case of diagonal flow the ganglia line up with and flow parallel to the macroscopic flow direction which, in this case, necessitates snake-like motion of the ganglia.

The results of this study have been expressed in terms of the ganglion volume expressed in units of the number of pores occupied by the ganglion v^* and the average velocity of the ganglion relative to the interstitial velocity of the aqueous phase v_p. The viscosity ratio $\kappa = \mu_0/\mu_w$ played an important role. The following are some of the conclusions reached:

(1)   For $\kappa > 1$ (unfavorable viscosity ratio) $v_p < 1$ for all v^*.
(2)   For $\kappa < 1$ (favorable viscosity ratio) $v_p > 1$ at the higher capillary numbers and also at the low capillary numbers for relatively large values of v^*.
(3)   For a fixed v^*, v_p increases with the capillary number.
(4)   For $\kappa > 1$ only, blobs with $v^* \ll 1$ move much faster than ganglia with $v^* \geq 1$.

Legait (1983) did some interesting mathematical modeling and experimental work on the passage of an oil slug through a constriction in a capillary of square cross section. The wetting fluid was forced to flow past the slug in the four corners of the tube. As a result, circulation ensues in the slug that has the interesting effect that the slug is forced through the constriction at a lower capillary number at a more unfavorable viscosity ratio, because the viscous drag acting on the drop increases with the viscosity ratio $\kappa = \mu_0/\mu_w$.

Oil Film Spreading.   New and highly efficient method of blob mobilization has been discovered recently (Kantzas *et al.*, 1988a, 1988b; Chatzis *et al.*, 1988; Dullien *et al.*, 1989).

In a water-wet medium, the trapped blobs of oil when contacted with a gas will spontaneously spread at the water/gas interface, as shown in Fig. 5.15, whenever the spreading coefficient S_{ow} is positive, that is (see Section 2.2.1),

$$S_{ow} = \sigma_{wg} - \sigma_{og} - \sigma_{ow} > 0 \qquad (5.3.77)$$

which is the case with most oils. This mechanism of blob mobilization can be conveniently combined with gravity drainage to produce a large percentage of the residual oil. As the water drains out, its place is occupied by the inert gas (air or N_2) which, on contact, makes the oil blobs spread; then the oil film drains downward on the water surface and eventually builds up a continuous

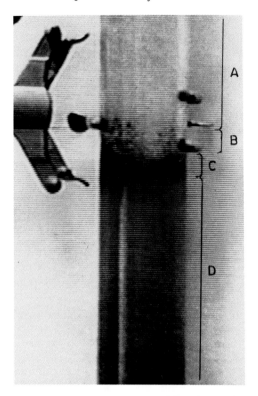

Figure 5.72.   Oil bank formed by spontaneous spreading of waterflood residual oil blobs in the course of gravity drainage from a bead pack (A: Air and connate water; B: Transition zone; C: Oil bank; D: Water and waterflood residual oil) (Dullien *et al.*, 1989).

oil bank that can be produced at the bottom of the core by the application of sufficient pressure (usually a few psi) to overcome the capillary end effect. In this stage of the process it is useful to apply at the producing end a mixed wet semipermeable membrane that will pass water and oil, but prevents the production of gas up to a certain "bubbling pressure." A paste consisting of about equal parts of finely ground calcium carbonate powder and active carbon powder has been developed by Dullien and Catalan (1991) that prevents gas breakthrough up to about 6 psi pressure. The evolution of this type of blob mobilization is illustrated in Fig. 5.72 in the case of a bead pack.

5.3.6.3.   *Capillary Number and Critical Capillary Number of Displacement*

In this chapter the capillary number has played a very prominent role. It is important, therefore, to analyze and discuss this important parameter.

Capillary number is usually introduced as a measure of the ratio of viscous-to-capillary forces. In order to see what kind of a measure the capillary number is of the relative importance of viscous and capillary forces it is useful to derive the ratio of viscous-to-capillary forces in the case when a preferentially wetting fluid is displacing a nonwetting fluid of the same viscosity in a uniform capillary tube of radius R and length L. The viscous force F_v is equal to the wall shear stress τ_w multiplied by the surface area of the tube, that is,

$$F_v = \tau_w (2\pi RL) \qquad (5.3.78)$$

and, assuming viscous (Poiseuille) flow

$$\tau_w = 4\mu\bar{u}/R, \qquad (5.3.79)$$

where $\bar{u}$ is the average velocity in the tube. Combining Eqs. (5.3.78) and (5.3.79) we obtain

$$F_v = 8\pi\mu\bar{u}L. \qquad (5.3.80)$$

The capillary force F_c is equal to the capillary pressure P_c times the cross-sectional area of the tube (i.e., πR^2). As

$$P_c = \frac{2\sigma \cos\theta_A}{R} \qquad (5.3.81)$$

we obtain

$$F_c = 2\sigma \cos\theta_A (\pi R). \qquad (5.3.82)$$

Finally, the ratio of the viscous-to-capillary forces in the tube, a capillary number that is going to be denoted by CA_{imb} is

$$CA_{imb} = \frac{4\mu\bar{u}}{\sigma \cos\theta_A} \frac{L}{R} = \frac{4\,Ca}{\cos\theta_A} \frac{L}{R}. \qquad (5.3.83)$$

Equation (5.3.83) shows that the true ratio of viscous-to-capillary forces in a tube is four times the conventional capillary number Ca times the length of the tube L, divided by $\cos\theta_A$ and the radius R of the tube. It is clear from this that stating only the value of Ca in an imbibition process is not sufficient to specify the dynamic conditions of the system and, therefore, two imbibition processes taking place in two capillaries characterized by the same value of Ca, but arbitrary values of $\cos\theta_A$, L, and R are not in the same dynamic state.

It can be readily shown that for the case of free spontaneous imbibition in a tube

$$\bar{u}_{free\,imb} = \frac{R\sigma \cos\theta_A}{4L\mu}, \qquad (5.3.84)$$

and, combining Eq. (5.3.84) with Eq. (5.3.83) yields $CA_{imb} = 1$; that is, in free spontaneous imbibition the viscous and the capillary forces acting in the tube are equal. This is also self-evident from force balance or energy balance considerations.

If the spontaneous imbibition is controlled by means of an external device such as a throttling valve or a constant rate pump or an opposing pressure head and $\bar{u}$ is reduced below the value given by Eq. (5.3.84), then $CA_{imb} < 1$, because the viscous forces have been reduced but not so the capillary forces. In forced imbibition, however, where $\bar{u}$ is caused to be greater than the value given by Eq. (5.3.84) by means of a constant rate pump or a pressure head assisting the flow, $CA_{imb} > 1$ because the viscous forces have increased, while the capillary forces did not change.

It is also noted that for neutral wettability (i.e., $\theta_A = 90°$) Eq. (5.3.83) predicts $CA_{imb} = \infty$, because with a flat meniscus present the capillary forces are equal to zero—they don't exist.

Equation (5.3.83) can be generalized for imbibition into a pore network, or a core of length L, as follows:

$$CA_{imb} = \frac{4\mu v_p}{\sigma \cos \theta_A} \frac{L}{R_{eq}}, \qquad (5.3.85)$$

where v_p is the average interstitial velocity and R_{eq} is an equivalent pore radius, the radius of a uniform capillary tube of length L in which free spontaneous imbibition, with the same values of μ, σ, and θ_A, would result in the same velocity v_p as in the core. R_{eq} can be calculated from the condition $CA_{imb} = 1$, that is,

$$R_{eq} = \frac{4\mu L (v_p)_{free\ imb}}{\sigma \cos \theta_A}. \qquad (5.3.86)$$

It is important to note that R_{eq} is independent of L, the core length because, everything else being equal, we have $L(v_p)_{free\ imb} = $ constant.

Having determined R_{eq}, Eq. (5.3.85) can be used to calculate CA_{imb} in the same porous medium also for different values of L, μ, v_p, σ, and also θ_A, as long as the strong wetting preference is there. The importance of R_{eq} lies in making it possible to compare the conditions of imbibition and water flood in different porous media: Systems with the same value of CA_{imb} are in the same dynamic state. It is expected that R_{eq} will be related to D_{app} calculated from rate of capillary rise measurements, the results of which are listed in Table 5.1. In sandstones these are typically smaller by about two orders of magnitude than the "threshold" or "breakthrough" diameters calculated from bubbling pressures. This means that $Ca = 10^{-6} - 10^{-7}$ may correspond to $CA_{imb} = 1$, that is, to the condition of equality of viscous and capillary forces in imbibition into sandstone cores, whereas in bead packs $CA_{imb} = 1$ can be expected to correspond to values of Ca that are greater by several orders of magnitude.

In Eq. (5.3.85) the viscous forces act over the length L which, in imbibition, while displacing a continuous nonwetting phase, is naturally taken to be the length of the core being flooded. It is also noted that in imbibition at least part of the driving force of displacement is due to the capillary pressure. The situation is *entirely* different when water flood is used to mobilize trapped blobs of the nonwetting phase. Fixing our attention to any particular blob of length ℓ in the direction of average macroscopic flow, we are interested in this case in the viscous forces acting on this blob (i.e., the viscous forces acting over a distance ℓ, rather than L). The role played by the capillary pressure is, in this case, actually opposite to the case of imbibition, because now the capillary forces oppose the viscous forces, rather than the other way around (i.e., viscous forces oppose capillary forces). Therefore Eq. (5.3.85) does not apply for this different physical situation. In the case of mobilization one can still maintain Eq. (5.3.80) by simply replacing L with ℓ, but Eqs. (5.3.81) and (5.3.82) must be replaced with

$$(F_c)_{dr} - (F_c)_{imb} = 2\pi\sigma\left(\frac{R_e^2}{r_{dr}} - \frac{R^2}{r_{imb}}\right), \tag{5.3.87}$$

where r_{dr} and r_{imb} are the equilibrium radii of curvature of the menisci of the blob at the threshold of mobilization, and R_e and R are the pore throat and pore body radii, respectively. By using Eqs. (5.3.80) and (5.3.86) we obtain for the ratio of viscous-to-capillary forces in blob mobilization:

$$\text{CA}_{mob} = \frac{4\mu v_p}{\sigma}\frac{\ell}{\Delta(R^2/r)} = \frac{4\,\text{Ca}}{\phi}\xi, \tag{5.3.88}$$

where the parenthetical expression in Eq. (5.3.87) has been abbreviated by $\Delta(R^2/r)$. In this case, unlike R_{eq}, $\Delta(R^2/r)$ is a variable in the same core much as ℓ is. For each blob (i.e., each different value of ξ) there applies the displacement condition $\text{CA}_{mob} = 1$ for which there correspond different values of Ca. The reason why different porous media yield different S_{or} versus Ca curves is mainly that the distribution function $f(\xi)$ is different for different media.

Let CA_{imb} be a certain value ≥ 1 for which an "imbibition" run in a given core was completed, leaving a certain trapped nonwetting phase saturation in the sample. In this system, at the same velocity at which the "imbibition" was completed, $\text{CA}_{mob} < 1$ for all the trapped blobs, because these were trapped at this very velocity. Hence, $\text{CA}_{mob} < \text{CA}_{imb}$, both being evaluated in the same system and at the same velocity. On increasing the velocity, mobilization of ganglia will start to take place. To the mobilization of any particular ganglia, characterized by a certain value of the ratio $\ell/\Delta(R^2/r) = \xi$, corresponding to the condition $\text{CA}_{mob}(\xi) = 1$, there corresponds also a value of CA_{imb}, which can be readily calculated because v_p and all the rest of the parameters in Eq. (5.3.85) are known. Thus S_{or} in the sample can be determined as a function of CA_{imb} and the residual oil saturations in different porous media can thus be compared with reference to CA_{imb} characteristic of each sample. As a result the curves in Fig 5.70 are expected to move closer together because the value of R_{eq} is likely to be the smallest in the sandstones and the largest in the packs.

5.3.6.4. *Relative Permeabilities at Residual Oil Saturations*

Handy and Datta (1966) have carried out experiments for a wetting phase displacing a nonwetting phase and found that the relative permeabilities depended on the displacement conditions. In all their tests the free imbibition relative permeabilities were found to be equal to or greater than those measured in the controlled (i.e., quasi-static) imbibition tests at the same saturation. Handy and Datta interpreted these results to be in agreement with the predictions of the pore doublet model that in the case of free imbibition the water must be somewhat more in the larger pores than it is after controlled imbibition.

There is, however, at least another possible interpretation of the data of Handy and Datta. Inspection of typical graphs obtained by these authors (e.g., Fig. 5.73) shows that in the free imbibition and the water flood tests the residual nonwetting phase conditions were reached at a saturation where, in the controlled imbibition tests, part of the nonwetting phase was still being displaced, and, therefore, it must have formed an interconnected network. Now, at a fixed saturation, the relative permeability to the wetting phase can be expected, for topological reasons, to be greater under such conditions that the nonwetting phase is present entirely in the form of isolated "blobs" than when part of this phase forms an interconnected network.

The explanation of the fact that Handy and Datta (1966) found the lowest residual saturations under conditions of controlled imbibition is that in these experiments the displacement is more efficient for two main reasons. First, in these experiments the capillary pressure and the saturation are uniform everywhere and, therefore, every time the capillary pressure is decreased there is more displacement taking place in every macroscopic portion of the medium, even back near the injection face. Second, in controlled imbibition enough time is allowed for all the possible displacement to take place at each fixed value of the capillary pressure.

By contrast, under conditions of free imbibition all displacement takes place in a relatively narrow zone behind the flood front and only limited time is available for the nonwetting phase contained in this zone to be trapped.

The following comments are in order on the mechanism of free imbibition. The lowest water saturation is found at the front of the invading–imbibing wetting phase. The lowest water saturation, however, corresponds to the

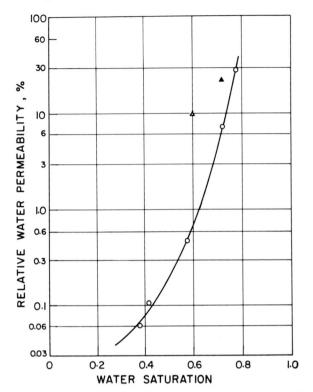

Figure 5.73. Imbibition relative permeabilities, in a water–air system for Berea sandstone core 202. The values shown are for free imbibition with $S_{wi} = 0$ (△) and $S_{wi} = 0.48$ (▲) and for controlled imbibition (○) (Handy and Datta, 1966 © SPE-AIME).

highest value of the capillary pressure on the capillary pressure–saturation curve. On the other hand, the largest capillary pressure corresponds to the penetration by the water of the narrowest capillaries. Hence, under conditions of free imbibition (and probably also in water flood), the wetting phase penetrates the narrowest pores first, exactly in the same way as in quasi-static imbibition tests, except the penetration does not extend past the leading edge of the invading front.

Morrow *et al.* (1985) measured relative permeabilities to water in sandstone cores at different residual oil saturations, established by water flooding the cores at various, increasing rates. The relative permeabilities were found to increase strongly with increasing water saturation. The authors also tested the effect of the pressure gradient applied in the relative permeability test, on the value of the measured relative permeability. They found an increase of "relative permeability" to water at pressure gradients past the point of mobilization of trapped and blobs, but the measured values were fairly constant below this threshold.

5.3.7. Network Models of Two-Phase Flow

Network models were discussed in Section 2.5.2 from the point of view of predicting capillary pressure curves. An inevitably incomplete list of the literature of network models includes Fatt (1956), Dodd and Kiel (1959), Wardlaw and Taylor (1976), Chatzis and Dullien (1977, 1982, 1985), Mohanty (1981), Lin and Slattery (1982), Heiba *et al.* (1982), Diaz *et al.* (1987), Kantzas and Chatzis (1988a, 1988b), Lapidus *et al.* (1985), Lenormand *et al.* (1988), Dias and Payatakes (1986a, 1986b), Koplik and Lasseter (1982), Larson *et al.* (1981), Chandler *et al.* (1982), Winterfield *et al.* (1981), Koplik *et al.* (1984), Dias and Wilkinson (1986), and Li *et al.* (1986).

5.3.7.1. *Early Network Modeling*

Fatt (1956) was the first to show that network models can be used to qualitatively describe the unsaturated permeability behavior of porous media (see Section 2.5.2). Fatt used networks of resistors as analog computers to model the relative permeability and the resistivity index curves. Both single-size tube networks and networks with tube radius distribution were used.

Because the penetration was carried out from all sides of the network, the establishment of continuous paths in the nonwetting phase did not necessitate cutting off all the conductive paths of the wetting phase. Had the networks been of very much larger size, more closely corresponding to a macroscopic network, Fatt's simulation experiments would have run a different course. At the beginning of penetration, one would have measured boundary or edge effects, and after "breakthrough" (see Section 2.5.2), there would not have been conducting paths simultaneously in both phases.

Probine (1958) used an electrical analog method for predicting the effective permeability to *the wetting phase* of unsaturated porous materials. The hydraulic resistance of a water channel in the porous material was represented by an electrical resistor and the three-dimensional network of channels by a corresponding three-dimensional network of resistances. *Desaturation or dewetting* of the medium was simulated by progressively removing resistors representing the largest of the remaining water filled channels.

Two different analog models were used in this work, one of them purported to model a three-dimensional network of interconnecting capillary tubes (see Fig. 5.74a) and the other one a packed bed of spheres (see Fig. 5.74b). In the capillary model, each resistor represented a single pore and the emptying of a pore is simulated by opening the circuit at points a and b,

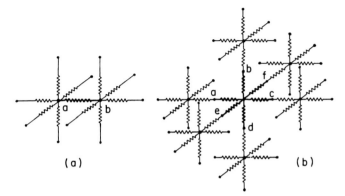

Figure 5.74. (a) Heavy line delineates a resistor representing a capillary tube in the capillary type model. Normal lines indicate the position of adjacent resistors (or capillary tubes). To simulate the emptying of a pore as material dries out, the circuit is broken at points *a* and *b*.

(b) Heavy line delineates a network representing a single pore in a packed-sphere-type of model. Normal lines indicate the position of adjacent networks (or pores). To simulate the emptying of a pore as material dries out, the circuit is broken at points *a–f* (Probine, 1958).

whereas in the case of the packed sphere model, the circuits were broken at a, b, c, d, e, and f.

Analogs based on both the capillary tube model and the packed sphere model have been tested against actual permeability measurements on two granular-type porous materials, A and B. Material A was graded sand and material B ungraded glass beads.

The results of the analog predictions of the effective permeabilities for materials A and B are shown in Figs. 5.75 and 5.76, respectively. In each case the predictions have been compared with both the measured permeability and the permeability computed by the method of Childs and Collis-George (1950). The packed sphere model was definitely superior to the capillary tube model in the case of material A, whereas for material B the difference was not as marked. The smaller difference is due to the fact that in randomly packed ungraded material the argument requiring six entry points, which is based on ideal packing, will no longer be completely valid and, therefore, the capillary tubes concept probably becomes more nearly correct.

Another type of network model of the effective permeability of porous media to a wetting phase, as a function of saturation, is provided by the so-called conductance theories developed first by Maxwell (1873), later by Bruggeman (1935), and more recently by Bottcher (1945). The last theory was adopted by Farrell and Larson (1972) to model "hydraulic conductivity" of unsaturated porous media.

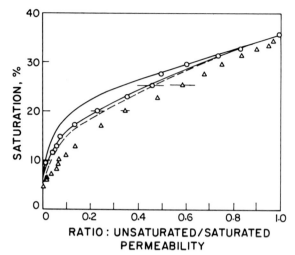

Figure 5.75. Permeability ratio for material A measured (solid line) and computed (dashed line) by Childs and Collis-George. Both the packed sphere model (○) with 64 pores and the capillary tube model (△) with 44 pores were used. Note that the permeability was made equal to unity at saturation (35.7%) (Probine, 1958).

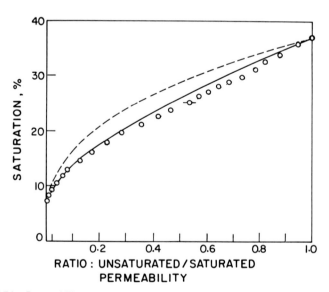

Figure 5.76. Permeability ratio for material B measured (solid line) by Probine and computed (dashed line) by the Childs and Collis-George method. The packed sphere (○) with 64 pores is also shown. Note that the permeability was made equal to unity at saturation (37%) (Probine, 1958).

It was assumed that the medium may be envisaged as a random assemblage of so-called pore domains, defined as an elemental volume of the pore space that has a specific desorption potential (see Section 2.3.2.2). Pore domains may include pore clusters or groups of pores characterized by the same desorption potential.

In Bottcher's theory, the contribution of the flux through any domain to the total conductance may be determined by assuming that the domain is surrounded by a medium having the average conductivity of the heterogeneous medium. The results obtained by Farrell and Larson (1972), using Bottcher's theory, are compared with those of Childs and Collis-George for various porous materials in Fig. 5.77. It is evident from this figure that the predictions differed, often substantially, from those of the Childs and Collis-George model. No attempt was made to compare the measured conductivities with the predicted ones.

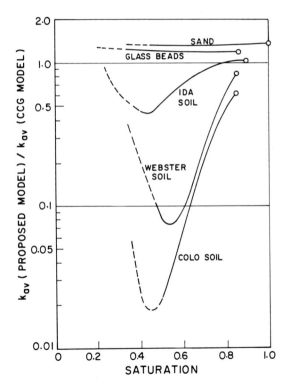

Figure 5.77. Ratio of conductivities predicted by the proposed model and by the Childs and Collis-George (CCG) model of pore structure for a range of porous media. The water desorption characteristics of sand were taken from Jackson *et al.* (1965) and of other materials were taken from Kunze *et al.* (1968). The dashed curve indicates that the isotherm required extrapolation (Farrell and Larson, 1972 © American Geophysical Union).

In a completely different type of two-phase flow, a gas is either diffusing or flowing through a porous medium, and at the same time there is an accompanying surface flow or surface diffusion of the same fluid (see Section 3.4.4). Nicholson and Petropoulos (1968, 1971) have developed three-dimensional network models for the purpose of modeling this type of two-phase flow phenomena.

5.3.7.2. *Network Models of Relative Permeability*

The current basis for predicting relative permeabilities with the help of network models is simulation of the distributions of the two fluid phases over the bonds (pore throats) and the sites (pore bodies) of a network model of the pore structure as a function of the saturation (this was discussed in Section 2.5.2). Each fluid phase consists, in general, of a conductive and a nonconductive portion. The latter has a "dendritic" and a "trapped" contribution. For the calculation of relative permeability to a phase, the conductivity of the conductive subnetwork constituted by this phase and the conductivity of the fully saturated network need to be calculated, in addition to the saturation. As the permeabilities are controlled entirely by the pore throats, which contribute to the saturation only to a small extent determined mostly by the pore bodies, for successful relative permeability prediction representative throat size and pore body size distributions are required. In addition, the network model must be representative of the porous medium, the behavior of which it purports to simulate, also as far as the spatial distribution, relative to each other, of different size pore throats and pore bodies is concerned. Last, but not least, the network must correctly represent the topology of the pore network of the sample. It is comparatively easy to "curve fit" experimental relative permeability curves by means of the percolation accessibility functions, discussed in Section 2.5.2, by choosing suitable pore size distribution functions and relationships between the size and the volume of a pore, without any reference to the actually existing pore topology in the particular sample. Exercises of this type, however, are purely of academic interest because they do not serve any practical purpose such as predicting transport properties of porous media from pore structure data. Unfortunately, the vast majority of the contributions falls in this category. Li *et al.* (1986) present a detailed study of the drastic effects of the various parameters and the pore-to-pore correlations chosen in the model, on the predictions. Notwithstanding the difficulties faced when the objective is to make true predictions of relative permeabilities, the results obtained by Chatzis and Dullien (1982, 1985) and by Kantzas and Chatzis (1988a, 1988b) are encouraging.

The problem of solving the equations of flow in a subnetwork occupied by a phase is equivalent to the problem of calculating the conductivity of a network of electrical resistors, for which techniques are readily available

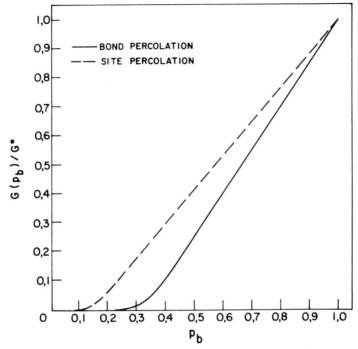

Figure 5.78. Relative conductivity curves of the cubic network (Kirkpatrick, 1973).

(Kantzas and Chatzis, 1988a). The difficulty lies in modeling consistently the subnetwork occupied by the phase.

In order to illustrate that very simple mathematical methods may also be used successfully to predict relative permeability curves of sandstones, the approach used by Chatzis and Dullien (1982, 1985) is sketched here. The basis of the calculation is formed by the relative conductivity curves of the cubic network calculated by Kirkpatrick (1973), shown in Fig. 5.78. These curves represent the variation of the electrical conductivity of the network consisting of equal conductors when removing, step by step, an increasing number of conductive bonds, according to the pattern of either the bond or the site percolation problem. For the latter, sites were picked at random and all the bonds meeting at the chosen site were removed from the network. The curves in Fig. 5.78 are to be compared with Fig. 2.58 to appreciate the very low values of $G(p_b)/G^0$ in the vicinity of the percolation threshold, as compared with the values of the pore fraction Y in the same region. This means that the threshold percolation cluster contributes very little to the conductivity. The reason for this is the very low degree of interconnectedness of the conducting network in the vicinity of the percolation threshold (see

Fig. 2.55) which outweighs the effect that in drainage this skeleton consists of the most conductive (i.e., the largest) throats in a porous medium. Therefore, the *fluid velocity* in this skeleton is greater than along the rest of conductive paths of the pore network, which has important consequences for hydrodynamic dispersion. The contribution of this skeleton to the *volume flow*, however, is very small indeed.

$d[G(p_b)/G^0]/dp_b$ is the rate of change of the relative conductivity of the subnetwork owing to penetration of more bonds by the nonwetting phase. $f_b(D_b) = -dp_b/dD_b$ is the number-based distribution density of bond diameters. Hence, $\{d[G(p_b)/G^0]/dp_b\}f_b(D_b)dD_b$ is the contribution of the penetrated bonds with diameters between D_b and $D_b + dD_b$ to the relative conductivity of the subnetwork in the case when all bonds have the same conductivity. The volume flow in cylindrical capillaries, however, is proportional to D_b^4, and for slit-shape pores in sandstone we let it be proportional to D_b^3. Hence, the drainage relative permeability to the nonwetting phase is calculated as

$$k_{rnw} = \frac{\int_{Db}^{Dbc} \dfrac{d[G(p_b)/G^0]}{dp_b} f_b(D_b) D_b^3 \, dD_b}{\int_{D\,bmin}^{Dbc} \dfrac{d[G(p_b)/G^0]}{dp_b} f_b(D_b) D_b^3 \, dD_b}. \tag{5.3.89}$$

The same probability density function $f_b(D_b)$ used to calculate the mercury porosimetry curve and the same saturations calculated for the mercury porosimetry curve (see Section 2.5.2.7) were used in predicting the relative permeability curve of mercury in Berea sandstone. The site percolation case,

$$\frac{G(p_b)}{G^0} = \frac{(zp_b - 1)}{z - 1}, \tag{5.3.90}$$

with $z = 6$ was used. For the nonlinear portion of the curve drawn in dashed lines in Fig. 5.78 the relationship

$$\frac{G(p_b)}{G^0} = 2(p_b - p_{bc})^{1.6} \tag{5.3.91}$$

was used. Intermediate results of these calculations are shown in Table 5.5 and the final results are plotted in Fig. 5.79 along with the experimental curves obtained by Batra (1973) for three different sandstones. If the pore size distribution functions $f(D_b)$ and $f(D_s)$ of different sandstones were identical when expressed in terms of the nondimensional variables $x = D_b/D_{bc}$ and $y = D_s/D_{bc}$, then the same relative permeability curves would be predicted by these calculations for the different sandstones. This seems to be the case for the three sandstones shown in Fig. 5.79.

TABLE 5.5 Calculation of the Relative Permeability Curve of Mercury in a Berea BE-1 Sample[a]

S^*_{nwk}	D_{bk}	p_{bk}	$\dfrac{d(G/G^0)}{dp}$	$f_b(D_{bk})$	$\dfrac{d(G/G^0)}{dp} D^3_{bk} f_b(D_{bk})$	k_{rnwk}
0.113	29.5	0.1090	0	0.1175	0	0
0.164	28.8	0.1225	0.1613	0.1265	478	0.005
0.286	27.0	0.1600	0.8050	0.1502	2,380	0.078
0.365	25.5	0.2025	1.1580	0.1701	3,266	0.197
0.443	24.0	0.2500	1.2	0.1900	3,152	0.333
0.496	22.4	0.3025	1.2	0.2105	2,939	0.468
0.522	20.7	0.3600	1.2	0.7312	2,461	0.595
0.605	19.2	0.4225	1.2	0.2479	2,206	0.691
0.659	27.6	0.4900	1.2	0.2634	1,723	0.778
0.708	76.0	0.5625	1.2	0.2756	1,355	0.847
0.764	14.3	0.6400	1.2	0.2840	996	0.903
0.818	77.5	0.7225	1.2	0.2860	670	0.945
0.869	10.6	0.8100	1.2	0.2767	395	0.974
0.918	8.5	0.9025	1.2	0.2463	131	0.991
0.972	5.0	1.0000	1.2	0	0	1.000

[a]From Chatzis and Dullien (1982, 1985).

In these calculations the effects of trapping the wetting phase were not included because in the mercury/rarified air/system there is no trapping of the wetting phase. Similar calculations as the one carried out above for the relative permeability of the nonwetting phase can be performed also for the relative permeabilities of the wetting phase, but the results would have no practical application without taking trapping into account. Qualitatively speaking, however, it can be pointed out that the first point on the relative permeability curve of the wetting phase at the percolation threshold (breakthrough) of the nonwetting phase would be much lower value than the final point on the relative permeability curve of the nonwetting phase, corresponding to complete trapping of the wetting phase, because the non-wetting phase occupies, already at the percolation threshold, a great proportion of the largest pore throats, leaving only smaller pore throats for the use of the wetting phase. At increasing penetration, the nonwetting phase becomes better interconnected and, consequently, it can take advantage more and more of the larger pore throats occupied by it, bringing its relative permeability to much higher values than the (highest) value of the relative permeability of the wetting phase at breakthrough. This is the explanation of the usual lack of symmetry, *as far as the magnitude of relative permeabilities is concerned*, of the relative permeability curves of the nonwetting phase and the wetting phase, respectively.

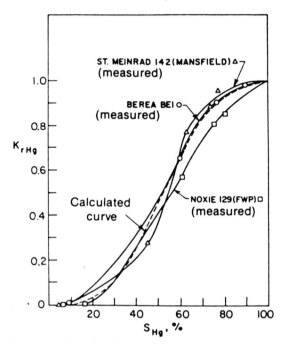

Figure 5.79. Comparison of predicted and measured mercury-relative permeability curves of sandstones (Chatzis and Dullien, 1985).

Another type of lack of symmetry often exists in sandstone samples,[5] in the *positions* occupied by the relative permeability curves of both the nonwetting phase and the wetting phase on the saturation axis: Often both curves lie in the upper half of the wetting phase saturation range. This behavior, which, of course, is due to trapping a large percentage of the wetting phase, is incompatible with random distribution of pores of different sizes over the pore network, usually assumed in percolation-type of network modeling. Excess trapping of the wetting phase may take place in secondary (micro) pores, present in cementing materials, which were not included in the network model; or it may be due to heterogeneities of the type where there are regions or clusters, consisting of relatively small pores, and surrounded by larger pores present in the sample, where all the wetting phase present is trapped (see Fig. 1.18).

Kantzas (1985) and Kantzas and Chatzis (1988a, b) used the accessibility data and pore size distributions of Diaz (1984) and Diaz *et al.* (1987) to calculate in sandstones the relative permeability curves of both phases, in

---

[5]These comments are limited to the case of strong water wetness. The position of the relative permeability curves may shift in one and the same sample if the sample wettability is altered (see Fig. 5.80).

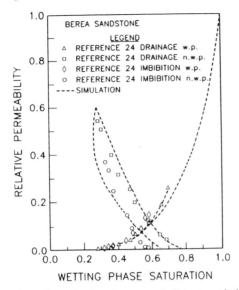

WETTING PHASE SATURATION

Figure 5.80. Comparison of predicted and measured oil/water relative permeability curves in Berea sandstone (Kantzas and Chatzis, 1988) (Shankar and Dullien, 1979).

both drainage and imbibition with trapping. The preconditioned conjugate gradient method implemented in the HARWELL library was used to solve the large sparse set of linear equations. The conductivity q of each node and throat has been expressed as

$$q = aD^3, \tag{5.3.92}$$

where a is a constant parameter and D is the diameter. Figure 5.80 shows the predicted relative permeability curves along with the corresponding measured data. The drainage-type curves agree satisfactorily with experiment, but the trends of the imbibition-type curves are different. The experimental nonwetting phase curve is much steeper along its entire length than the predictions. This can be explained by the difference between the effects of piston-type and "snap-off"-type displacement on the effective phase permeability of the nonwetting phase. In the modeling, piston-type displacement was assumed. In reality, the displacement is controlled by the "snap-off" mechanism, pinching off continuous threads of the nonwetting phase throughout the network. It is readily seen that "pinching-off"-type of disconnections cause a greater percentage reduction of the nonwetting phase permeability at a given saturation level than piston-type displacement, because it decreases the degree of interconnectedness of the subnetwork consisting of the nonwetting phase to a greater degree. The experimental wetting phase relative permeability curve agrees, at its beginning, with the predictions; but later on it shows a faster rate of increase. This behavior is probably due to the increasing degree of coalescence of isolated threads of the wetting phase, created by "snap-off," present all over the network, forming an interconnected network, as the wetting phase saturation is increased. It would be in order to repeat these simulations based on the "snap-off" mechanism of imbibition.

5.3.7.3. *Network Models of Immiscible Displacement*

Network models of immiscible displacement so far have been two-dimensional and small, modeling the displacement process in a thin two-dimensional layer, consisting of 10-to-100 grains, at best, in the direction of macroscopic flow. The displacing front enters this thin zone on one side, passes through the zone, and leaves through the opposite side. Displacing fluid is made to pass through this zone until there is no more displacement from it (i.e., until the saturation becomes constant). The number of thin-layer pore volumes required for this is usually not stated, but in some contributions a time scale is attached to the progress of simulation that is proportional to the number of pore volumes. The surprising information disclosed by this time scale is that the displacement is predicted to be complete after the passage of a couple of thin-layer pore volumes of displacing fluid. This means that the saturation is predicted to drop from the initial value (often assumed 100%) to the residual value over a layer of a thickness of a few scores of grains; that is, the saturation profile is practically a step function in macroscopic terms. The measurements of Bacri *et al.* (1985, 1990), shown in Figs. 5.40a and 5.40b, contradict this pattern.

Koplik and Lasseter (1982) used an electrical analog to calculate imbibition type immiscible displacement, for equal viscosities, in a two-dimensional 10×10 network model, consisting of ball-and-stick type of pores. The effect of capillary pressure on the pressure drop was handled by using the Washburn approximation. The constraint causing trapping of the nonwetting phase was modeled by a pair of differently biased diodes in parallel. The computation was carried out numerically. The capillary number $Ca = \mu v / \sigma$ was varied over the range extending from 10^{-8} to 10^{-4}. The assumptions used imply piston-type displacement and the absence of snap-off. Therefore, the results can be expected to be more representative of real displacement behavior at higher values of Ca. The displacement efficiency increased steadily with increasing value of Ca, as shown in Fig. 5.81. This trend is in agreement with the results of Lefebvre du Prey (1973) (see Fig. 5.5.53), but the rate of change with capillary number is much faster and the residuals are also much greater. This disagreement is of a fundamental nature and it indicates that the model of Koplik and Lasseter (1982) is not at all representative of water floods of long sandstone cores. It is important to find out the reasons for this. It would be instructive to run imbibition type displacement tests under the conditions used by Koplik and Lasseter (1982) in a capillary

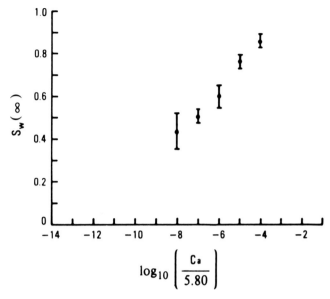

Figure 5.81. Variation of average predicted asymptotic (final) water saturation with capillary number (Koplik and Lasseter, 1982).

micromodel network of the same design used by these authors. By comparing experiment with prediction in the same system one would be able to tell whether the source of disagreement between predictions in two-dimensional ball-and-stick type of models and experiment in long sandstone cores lies in the difference between the network model and long sandstone cores or in the calculations of Koplik and Lasseter (1982). A very important factor is the small size of the network model, representing only a tiny portion of a sandstone core. Possibly, different boundary conditions would have resulted in a better agreement with reality.

A much simpler approach to the problem of immiscible displacement in a pore network has been proposed by Wilkinson and Willemsen (1983), who introduced "invasion percolation." The idea behind this mechanism is that in drainage-type of displacement always the largest available pore gets penetrated next, whereas in imbibition-type displacement the smallest available pore gets penetrated next, no matter where these may be situated in the network—as long as they are in contact with the invading phase. Trapping is included by the condition that pores from which there is no escape route cannot be invaded. The difference between "invasion percolation" and ordinary percolation is supposed to be that in the latter (e.g., in drainage) once a pore of a certain size has been invaded, all the pores in the network that are larger than or of the same size as this pore and are not blocked from this pore by any smaller pore, are also invaded *at the same time.* In invasion percolation, after the invasion of a pore there are two possibilities: Either there is a pore (or there are pores) of the same size or larger size and joining the pore that has been just invaded, in which case this pore (these pores) will be invaded *next,* or there aren't any, in which case the next largest pore(s) somewhere else, in contact with the invading fluid, will be penetrated *next.* If there is more than one pore that is eligible to be invaded through the pore that has just been invaded (e.g., one exactly the same size and another bigger) then the bigger one has priority and will be invaded first, whereas in ordinary percolation both pores are supposed to be penetrated at the same time. It is a practical impossibility, however, to have simultaneous penetration either in a simulation or in a real displacement process, because there is inevitably a hierarchy of sequence in any set of operations. The outcome of the simulation will depend on the arbitrary choice made in selecting a sequence from among two or more sequences of equal order of priority. Owing to this practical limitation in carrying out several steps simultaneously, the two types of percolation processes can be expected to yield equivalent results within the tolerances of these procedures.

It is entirely another question how closely the invasion percolation model simulates actual displacement processes. Lenormand and Zarcone (1985) carried out drainage-type displacement experiments of glycerol by air in a two-dimensional micromodel network and determined the fractal dimension of the trapped clusters, which they found in agreement with the theoretical value of ~ 1.82 obtained by Wilkinson and Willemsen (1983). It is well known, however, that the mechanism of displacement depends on the capillary number, the viscosity ratio, the rate supply of the displacing fluid, and the geometry and topology of the porous medium. The mechanisms are different also in imbibition and in drainage. There is no "general" displacement mechanism and, therefore, invasion percolation is not it, either.

Lenormand *et al.* (1988) did very extensive study, both by network simulation and by micromodel network experiments, of the different mechanisms of drainage-type displacement as a function of the capillary number Ca = $q\mu_2/A\sigma \cos \theta$ and the viscosity ratio $\kappa = \mu_2/\mu_1$ (2 = displacing fluid). Even though all of this work is limited to two-dimensional networks of ball-and-stick type pores and pore throats, their conclusions can be expected to be qualitatively representative of drainage, in general. The patterns of imbibition, however, are very different from those of drainage. In the numerical work, allowance was made for the different viscosities of the two fluids and the usual Washburn approximation was used. The role of capillarity was restricted to the cylindrical throats, and the spherical pore bodies were used only for fluid storage. A relaxation technique was used to solve the nonlinear flow problem of searching for the threshold pressure at which a particular throat is invaded, and then the time required to fill a pore completely was calculated in the knowledge of the total flow rate Q. This procedure was

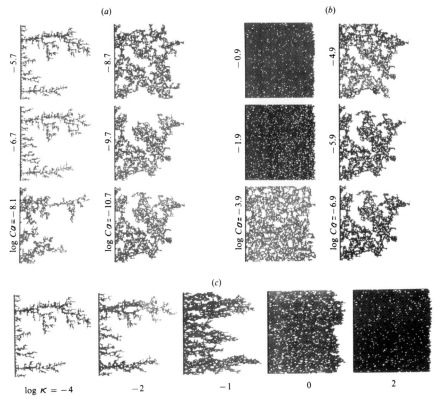

Figure 5.82. Network (100 × 100) simulations of drainage-type displacement at various viscosity ratios κ and capillary numbers Ca:
(a) log κ = − 4.7 (from viscous to capillary fingering)
(b) log κ = 1.9 (from stable displacement to capillary fingering)
(c) log Ca = 0 (from viscous fingering to stable displacement) (Lenormand *et al.*, 1988).

repeated every time the meniscus reached a throat. There was no trapping included in the model. The results of these simulations are shown in Fig. 5.82. Three patterns emerge: (1) dendritic (tree-like) viscous fingering, (2) capillary fingering characterized by loops, and (3) stable displacement. Quantitatively, the results of the simulations in a 25 × 25 mesh network could be represented as shown in Fig. 5.83. Within each of the three regions the saturation S of the invading fluid was found constant, with smooth transitions in the saturation existing between the regions. (These saturations are generally higher than the values found in real porous media.)

The displacement experiments, carried out with the fluid pairs consisting of air/very viscous oil, mercury/hexane, mercury/air, glucose solution/oil, yielded results that were in good *qualitative* agreement with the simulations.

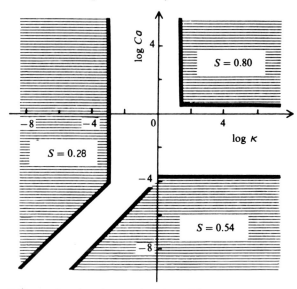

Figure 5.83. Different domains, characterized by a different drainage displacement mechanism and a corresponding different breakthrough saturation in drainage (simulation in 25×25 network) (Lenormand *et al.*, 1988).

The saturations at breakthrough appeared to be generally lower than the simulation values.

The physical interpretation of the various mechanisms, given by Lenormand *et al.* (1988), was as follows. In stable displacement the principal force is due to viscosity of the injected fluid ($\log \mathrm{Ca} = -0.9(!)$ in Fig. 5.82). In viscous fingering the principal force is due to the viscosity of the displaced fluid ($\log \mathrm{Ca} = -5.7(!)$ in Fig. 5.82). In capillary fingering viscous forces are negligible. (It is noted that the numerical values of the conventional capillary number make little physical sense. For example, at $\log \mathrm{Ca} = -0.9$ viscous forces dominate stable displacement *completely*. In addition, always using the viscosity of the displacing fluid in Ca introduces a bias. For example, at $\log \mathrm{Ca} = -5.7$ viscous forces dominate viscous fingering completely!)

It would be very interesting to extend this kind of simulation also to imbibition and to include the effects of gravity, that is, the role played by the Bond number, $\mathrm{Bo} = (\rho_w - \rho_0)gD^2/4\sigma$.

Dias and Payatakes (1986a) modeled imbibition-type displacement in two-dimensional networks of the type shown in Fig. 5.84 (CESV ≡ conceptual elemental void space; UC ≡ unit cell; GUC ≡ ganglion unit cell). The pore shapes were the same as those used in the work of Payatakes *et al.* (1973).

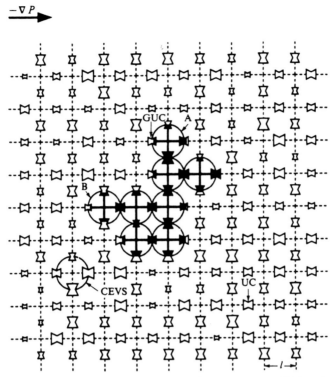

Figure 5.84. Schematic diagram of square network (see CEVS, UC, GUC and a 8-CEVS ganglion) (Dias and Payatakes, 1986).

Details of the numerical procedure can be found in the reference (Dias and Payatakes, 1986a).

Free imbibition was simulated by applying a very small macroscopic pressure gradient across a "short" (15 × 30 mesh) network. It was realized that, owing to the change in the viscous resistance in the course of the displacement of an oil of different viscosity than water, the capillary number keeps changing continually. In another set of simulations the capillary number was kept constant in each displacement by replacing the exit unit cells of the network with a "tail" of uniform unit cells of very small hydraulic conductance. All of the simulations were modeled after a sandpack studied by Leverett (1941).

The residual oil saturations obtained are shown as a function of the capillary number in Fig. 5.85. The trend of these curves is in agreement with the experimental results of Chatzis *et al.* (1983). The evolution of the

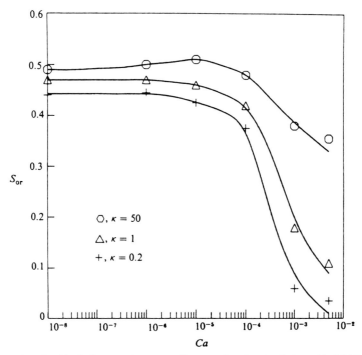

Figure 5.85. Residual oil saturation vs. capillary number for simulation of imbibition into a 15 × 30 network for various values of $\kappa = \mu_0/\mu_w$ (Dias and Payatakes, 1986).

displacement process is similar for all values of Ca and μ_0/μ_w studied; that is, oil ganglia of different sizes were disconnected and trapped by the water in a displacement process that might be called capillary fingering (microfingering). At the favorable viscosity ratio of $\mu_0/\mu_w = 0.2$ there is an indication of transition to stable displacement.

In a companion paper Dias and Payatakes (1986b) model the motion of ganglia in the same type of network. Good agreement was found between the predictions and the experimental results of Hinkley *et al.* (1987).

In a more recent paper Vizika and Payatakes (1988) incorporated into their network simulator the role played by the advancing wetting film in imbibition-type displacement. The wetting film advances in grooves of surface roughness of the pore walls. Two types of surface roughness were assumed: In the first kind of roughness the wetting film is driven by capillary forces, whereas in the second kind, it is driven by external pressure drop. This second mechanism seems difficult to justify on physical grounds. The advancing wetting film chokes off threads of continuous nonwetting phase ahead of the main front of the invading wetting phase, and results in much higher residual oil saturations.

5.3.8. Drying of Porous Solids

Drying of a porous solid involves evaporation of a liquid and mass transport from the interior to the surface of the system.

Drying of porous solids is of great importance both in chemical engineering and in soil science. The physical laws are the same in both cases, but the problem has been approached somewhat differently because of the different technologies involved.

Soil usually dries in the downward direction, starting from the surface. As long as there is a "water table" maintained at a constant depth, there can be steady-state evaporation at the surface because the moisture distribution in the soil remains unchanged. Under these conditions the filter velocity v is determined as follows:

$$v = -k_H \nabla \Phi_c = -k_H \nabla(z - \psi) = -k_H[1 - (d\psi/dz)], \quad (5.3.93)$$

where Eqs. (2.3.3) and (2.3.4) have been used.

Now, k_H is a function of the saturation S, or alternatively, of the "suction head" ψ. It can be shown (Schwartzendruber, 1969) that the maximum filter velocity v_m, corresponding to $\psi \to \infty$ at the evaporating soil surface, can be expressed as follows:

$$v_m(k_H^0 + v_m) = k_H^{02}\pi^2/4cH^2, \quad (5.3.94)$$

where k_H^0 is the hydraulic conductivity of the saturated soil, H is the depth of the water table, and c is an empirical constant. For more discussion of this problem, the reader is referred to the review of Schwartzendruber (1969).

When there is no continuous supply of water to the surface, Richards' equation [Eq. (5.3.37) or (5.3.38)] must be used, where $\partial S_w/\partial t$ is nonzero.

It is interesting to note (Wiegand and Taylor, 1961) that even when ψ and S_w change with time, an initial period exists during which the evaporation velocity at the soil surface remains constant, provided that the relative humidity of the atmosphere and the temperature remain unchanged. Schwartzendruber (1969) obtained an expression for the time t_c during which the constant evaporation velocity will be sustained. The criterion used in this calculation is that the saturation at the soil surface should be reduced to the "air-dry" or "equilibrium" value.

After the surface moisture is in equilibrium with the air, the rate of evaporation starts falling. The falling rate evaporation from soils has been reviewed and discussed by Schwartzendruber (1969).

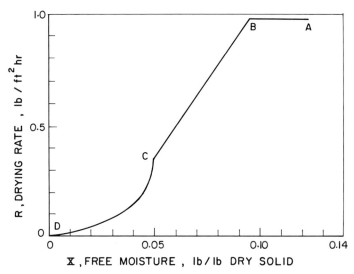

Figure 5.86.   Drying rate versus free moisture.

The same behavior as observed in the case of evaporation from soils plays a fundamental role in the drying of porous solids in the chemical industry. This field has been reviewed, for example, by McCabe and Smith (1976). The drying rate as plotted versus the "free moisture" (i.e., moisture content in excess of the equilibrium value) shows a constant rate period and a fall rate period (see Fig. 5.86) which can be further subdivided. In the first falling rate period, the moisture transport to the surface of the system is still mostly by capillary suction; that is, there still exists a continuous network of pores filled with the liquid. The rate keeps falling because the connectivity of the network becomes less and less. Finally, when the network ceases to be continuous throughout the sample, moisture suction will also cease to be able to pump the liquid to the surface. At this point diffusional transport in the vapor phase takes over.

REFERENCES

Amaefule, J. O., and Handy, L. L. (1981). SPE/DOE 9783 presented at SPE/DOE Symposium, Tulsa, Oklahoma.

Anderson, W. G. (1987). *JPT* **39**, No. 12.

Archer, J. R., and Wong, S. W. (1971). Interpretation of laboratory waterflood data by use of a reservoir simulator. *Preprint for the Ann. Soc. Pet. Eng. Fall Meeting*, 46th, New Orleans, Louisiana.

Auriault, J. L. (1987). *Transport in Porous Media* **2**, 45.

Auriault, J. L., and Sanchez-Palencia, E. (1986). *J. Theor. Appl. Mech.* Special Issue, 141.

Bacri, J. C., Leygnac, C., and Salin, D. (1985). *J. Phys. (Paris) Lett.* **46**, L467.

Bacri, J. C., Rosen, M., and Salin, D. (1990). *Europhys. Lett.* II (2), 127.

Bardon, C., and Longeron, D. (1978). Influence of very low interfacial tensions on relative permeability. *Preprint for Ann. Fall. Tech. Conf. Exhibit. SPE of AIME*, 53rd, Houston, Texas.

Batra, V. K. (1973). Ph.D. Dissertation, Univ. of Waterloo, Canada.

Batycky, J. P. and Singhal, A. K. (1978). *Res. Rep. RR*-35, Petroleum Recovery Institute, Calgary, Canada.

Bear, J. (1972). "Dynamics of Fluids in Porous Media." American Elsevier, New York.

Bethel, F. T., and Calhoun, J. C. (1953). *Pet. Trans. Am. Inst. Min. Eng.* **198**, 197.

Bonnet, J., and Lenormand, R. (1977). *Rev. Inst. Fr. Petr.* **42**, 477.

Botset, H. G. (1940). *Trans. Am. Inst. Min. Eng.* **136**, 91.

Bottcher, C. J. F. (1945). *Rev. Trav. Chim. Pays Bas* **64**, 47.

Bourbiaux, B., and Kalaydjian, F. (1988). SPE 18283, presented at 63rd SPE Conference, Houston, Texas.

Braun, E. M., and Blackwell, R. J. (1981). SPE 10155, SPE-AIME 56th Conference, Texas.

Brinkman, H. C. (1948). *Appl. Sci. Res.* **A1**, 333.

Brooks, R. H., and Corey, A. T. (1964). Hydraulic properties of porous media. Colorado State Univ., Hydrology Papers, Fort Collins, Colorado.

Bruggeman, D. A. G. (1935). *Ann. Phys. Liepzig* **24**, 636.

Brutsaert, W. (1967). *Trans. ASAE* **10**, 400.

Buckley, S. E., and Leverett, M. C. (1942). *Trans. Am. Inst. Min. Eng.* **146**, 107.

Burdine, N. T. (1953). *Trans. Am. Inst. Min. Eng.* **198**, 71.

Calhoun, J. C. (1951a). *Oil Gas J.* **50**, 117.

Calhoun, J. C. (1951b). *Oil Gas J.* **50**, 308.

Calvo, A., Paterson, I., Chertcoff, R., Rosen, M., and Hulin, J. P. (1990). *In proceedings* "Fundamentals of Fluid Transport in Porous Media," May 14–18, Arles, France.

Caudle, B. H., Slobod, R. L., and Brownscombe, E. R. (1951). *Trans. Am. Inst. Min. Eng.* **192**, 145.

Chandler, R., Koplik, J., Lerman, K., and Willemsen, J. (1982). *J. Fluid Mech.* **119**, 249.

Chatzis, I. (1980) Ph.D. Thesis, University of Waterloo, Ontario.

Chatzis, I., and Dullien, F. A. L. (1977). *J. Can. Pet. Tech.* **15**, 97.

Chatzis, I., and Dullien, F. A. L. (1981). *Powder Tech.* **29**, 117.

Chatzis, I., and Dullien, F. A. L. (1982). Revue de l'Institut Français du Pétrole, March/April.

Chatzis, I., and Dullien, F. A. L. (1983). *J. Colloid Interface Sci.* **91**, 199.

Chatzis, I., and Dullien, F. A. L. (1985). *ICE* **25**, 47.

Chatzis, I., Kantzas, A., and Dullien, F. A. L. (1988). SPE 18284, SPE 63rd Meeting, Houston, Texas.

Chatzis, I., Kuntamukkula, M. S., and Morrow, N. R. (1984) SPE 13213, presented at 59th Conference, Houston, Texas.

Chatzis, I., and Morrow, N. R. (1981). SPE 10114, presented at 56th SPE Conference, October, San Antonio, Texas.

Chatzis, I., and Morrow, N. R. (1984). *SPEJ* (Oct. 84), 555.

Chatzis, I., Morrow, N. R., and Lim, H. T. (1982). SPE/DOE 10681, presented at SPE/DOE 3rd Symposium, Tulsa, Oklahoma.

Chatzis, I., Morrow, N. R., and Lim, H. T. (1983). *SPEJ* **23**, 311.

Childs, E. C. (1969). "An Introduction to the Physical Bases of Soil Water Phenomena." Wiley (Interscience), New York.

Childs, E. C., and Collis-George, N. (1950). *Proc. R. Soc. London* Ser. A **201**, 392.

Collins, R. E. (1961). "Flow of Fluids through Porous Materials." Van Nostrand-Reinhold, Princeton, New Jersey.

Corey, A. T. (1954). *Producer's Monthly* **18**, 38.

Cox, R. G. (1986). *J. Fluid Mech.* **168**, 169.

Craig, F. F. Jr. (1971). "The Reservoir Engineering Aspects of Waterflooding." Society of Petroleum Engineers of AIME, New York.

Danis, M., and Jacquin, Ch. (1983). *Revue de l'IFP* **38**, no. 6.

Dawe, R. A., and Wright, R. I. (1983). Oil reservoir behavior. The Microscale. *R. Sch. Mines JL.* **33**, 25.

de Gennes, P. G. (1983). *Phys. Chem. Hydr.* **4**, 175.

de la Cruz, V., and Spanos, T. J. T. (1983). *AIChE J.* **29**, 854.

Deryagin, B. V., Melnikova, M. K., and Krylova, V. I. (1952). *Colloid J.* **14**.

Deryagin, B. V. (1946). *Colloid J.* **8**, 27.

deWiest, R. J. M. (1969). "Flow Through Porous Media." Academic Press, New York.

Dias, M., and Payatakes, A. (1986a). *J. Fluid Mech.* **164**, 305.

Dias, M., and Payatakes, A. (1986b). *J. Fluid Mech.* **164**, 337.

Dias, M. M., and Wilkinson, D. (1986). *J. Phys.* A **19**, 3131.

Diaz, C. E. (1984). M.A.Sc. Thesis, University of Waterloo, Ontario.

Diaz, C. E., Chatzis, I., and Dullien, F. A. L. (1987). *Transport in Porous Media* **2**, 215.

Dodd, C. G., and Kiel, O. G. (1959). *J. Am. Chem. Soc.* **63**, 1646.

Doscher, T. M. (1981). *Amer. Scientist* **69**, 183.

Dullien, F. A. L. (1969). *J. Pet. Tech.* January, p. 14.

Dullien, F. A. L. (1988). *Chem. Eng. and Tech.* II, 407.

Dullien, F. A. L., Allsop, H. A., Chatzis, I., and Macdonald, I. F. (1990). *Can. J.P.T.* **29**, no. 4, 63.

Dullien, F. A. L., and Batra, V. K. (1973). *Soc. Pet. Eng. J.* **13**, 256.

Dullien, F. A. L., and Catalan, L. (1991). Submitted to JCPT.

Dullien, F. A. L., Chatzis, I., and Kantzas, A. (1988). *In Proceedings of 3rd International Symp. on EOR*, February, Maracaibo, Venezuela.

Dullien, F. A. L., Dhawan, G. K., Gurak, N., and Babjak, L. (1972). *Soc. Pet. Eng. J.* **12**, 289.

Dullien, F. A. L., El-Sayed, M. S., and Batra, V. K. (1977). *J. Colloid Interface Sci.* **60**, 497.

Dullien, F. A. L., Lai, F., and Macdonald, I. F. (1986). *Colloid and Interface Sci.* **109**, 201.

Dullien, F. A. L., and Macdonald, I. F. (1976). *SPEJ* **16**, 48.

Dullien, F. A. L., Zarcone, C., Macdonald, I. F., Collins, A., and Bochard, R. D. E. (1989). *J. Colloid Interface Sci.* **127**, 362.

Dullien, F. A. L. (1991). *In* "Physical Chemistry of Colloids and Interfaces in Oil Production." Saint-Raphael, France. Sept. 4–6, 1991.

Dunlap, H. F., Bilhartz, H. L., Shuler, E., and Bailey, C. R. (1949). *Trans. Am. Inst. Min. Eng.* **186**, 259.

Dussan, V. E. B. (1976). *J. Fluid Mech.* **77**, 665.

Dussan, V. E. B. (1977). *AIChE J.* **23**, 131.

Dussan, V. E. B. (1979). *Ann. Rev. Fluid Mech.* **11**, 371.

Egbogah, E. O., and Dawe, R. A. (1980). *Bull. Can. Pet. Technol.* **28**, 200.

Ehrlich, R., and Crane, F. E. (1969). *Soc. Pet. Eng. J.* **9**, 221.

El-Sayed, M. S. (1979). Ph.D. Thesis, University of Waterloo, Ontario.

El-Sayed, M. S., and Dullien, F. A. L. (1977). Investigation of transport phenomena and pore structure of sandstone samples. *Preprint for the Ann. Tech. Meeting Petrol. Soc. CIM*, 28th, Edmonton, May 10–June 3.

Farrell, D. H., and Larson, W. E. (1972). *Water Resources Res.* **8**, 699.

Fatt, I. (1956). *Trans. Am. Inst. Min. Eng.* **297**, 144.

Fatt, I. (1961). *Science* **134**, 1750.

Fatt, I. (1966). *J. Inst. Pet.* **52**, 231.

Fatt, I., and Dykstra, H. (1951). *Trans. Am. Inst. Min. Eng.* **192**, 249.

Fatt, I., and Saraf, D. N. (1967). *Soc. Pet. Eng. AIME* **7**, 235.

Fulcher, R. A., Ertekin, T., and Stahl, C. D. (1985). *JPT* **37**, no. 2.

Gardescu, I. I. (1930). *Trans. Am. Inst. Min. Eng.* **86**, 351.

Geffen, T. M., Owens, W. W., Parrish, D. R., and Morse, R. A., (1951). *Trans. Am. Inst. Min. Eng.* **192**, 99.

Graham, J. W., and Richardson, J. G. (1959). *J. Pet. Tech.* **11**, 65.

Greenkorn, R. A. (1983). "Flow Phenomena in Porous Media." Marcel Dekker, New York and Basel.

Handy, L., and Datta, P. (1966). Fluid Distributions During Immiscible Displacements in Porous Media, Vol. 6. Chevron Research Co., La Habra, California.

Hassler, G. L., Rice, R. R., and Leeman, E. H. (1936). *Trans. Am. Inst. Min. Eng.* **188**, 116.

Hassler, G. L. (1944). U.S. Patent No. 2,345,935.

Heiba, A. A., Sahimi, M., Scriven, L. E., and Davis, H. T. (1982). SPE 11015, presented at 57th SPE Conference, September, New Orleans.

Hinkley, R. (1982). M. S. Thesis, University of Houston, Texas.

Hinkley, R. E., Dias, M. M., and Payatakes, A. C. (1987). *PhysicoChemical Hyrodynamics* **8**, no. 2, 185.

Hoffman, R. L. (1975). *J. Colloid Interface Sci.* **50**, No. 2, Parts I and II.

Huh, C., and Scriven, L. E. (1971). *J. Colloid Interface Sci.* **35**, 85.

Iczkowski, R. (1968). Displacement of Liquids from Random Sphere Packings, Vol. 8, No. 4. Research Division, Allis-Chalmers, Milwaukee, Wisconsin.

Islam, M. R., and Bentsen, R. G. (1986). *JCPT* **25**, 39.

Jackson, R. D., Reginato, R. J., and van Bavel, C. H. M. (1965). *Water Resour. Res.* **1** (3), 375.

Jamin, M. J. (1860). *C.R. Acad. Sci. Paris* **50**, 172.

Johnson, E. F., Bossler, D. P., and Naumann, V. O. (1959). *Trans. Am. Inst. Min. Eng.* **216**, 370.

Jones-Parra, J., and Calhoun, J. C. Jr. (1953). *Petrol. Trans., AIME* **198**, 335.

Kalaydjian, F. (1987). *Transport in Porous Media* **2**, No. 6, 537.

Kalaydjian, F. (1990). *Transport in Porous Media* **5**, No. 3.

Kalaydjian, F., and Legait, B. (1987). *C.R. Acad. Sc. Paris* Ser. II **304**, 869.

Kantzas, A. (1985). M.A.Sc. Thesis, University of Waterloo, Ontario.

Kantzas, A., and Chatzis, I. (1988a). *Chem. Eng. Comm.* **69**, 169.

Kantzas, A., and Chatzis, I. (1986b). *Chem. Eng. Comm.* **69**, 191.

Kennedy, H. T., Burja, E. O., and Boykin, R. S. (1955). *J. Phys. Chem.* **59**, 867.

Kirkpatrick, S. (1973). *Rev. M. Phys.* **45** (4), 574.

Klute, A., and Wilkinson, G. E. (1958). *Soil Sci. Soc. Proc.* 278.

Knight, J. H., and Philip, J. R. (1974). *Soil Sci.* **116**, 407.

Koplik, J., and Lasseter, T. J. (1982). SPE 11014, presented at 57th SPE Conference, September, Dallas, Texas.

Koplik, J., Lin, C., and Vermette, M. (1984). *J. Appl. Phys.* **56**, 3127.

Kunze, R. J., Vehara, G., and Graham, K. (1968). *Soil Sci. Soc. Amer. Proc.* **32**, 760.

Land, C. S. (1968). *Soc. Pet. Eng. J.* **8**, 149.

Lapidus, G. R., Lane, A. M., Ng, K. M., and Corner, W. C. (1985). *Chem. Eng. Comm.* **38**, 33.

Laroussi, C., Vandervoorde, G., and DeBacker, L. (1975). *Soil Science* **120**, 249.

Larson, R. G., Scriven, L. E., and Davis, H. T. (1981). *Chem. Eng. Sci.* **36**, 75.

Laughlin, R. D., and Davies, J. E. (1961). *Textile Res. J.* **31**, 904.

Lefebvre du Prey, E. J. (1973). *Soc. Pet. Eng. J.* **13**, 39.

Legait, B. (1981). *CR Acad. Sci. Paris* **292**, no. 2, 1111.

Legait, B. (1983a). Thesis, University of Bordeaux, France.

Legait, B. (1983b), *J. Colloid Interface Sci.* **96**, 28.

Legait, B., and Jacquin, C. (1982). *C.R. Acad. Sci. Paris* **B294**, 487.

Lenormand, R. (1981). Ph.D. Thesis, INP, Toulouse.

Lenormand, R., Cherbuin, C., and Zarcone, C. (1983). *C.R. Acad. Sci. Paris Sér. II.* **297**, 637.

Lenormand, R. (1983). *C.R. Acad. Sci. Paris Sér. II.* **297**, 437.

Lenormand, R., and Zarcone, C. (1983). *J. Fluid Mech.* **135**, 337.

Lenormand, R., and Zarcone, C. (1988). *SPE Formation Evaluation*, March, p. 271.

Lenormand, R., Zarcone, C., and Sarr, A. (1985). *J. Fluid Mech.* **135**, 357.

Leverett, M. C. (1939). *Trans. Am. Inst. Min. Eng.* **142**, 152.

Leverett, M. C., and Lewis, W. B. (1941). *Trans. Am. Inst. Min. Eng.* **142**, 107.

Levine, J. S. (1954). *Pet. Trans. AIME* **201**, 57.

Levine, S., Reed, P., and Shutts, G. (1976). *Powder Tech.* **17**, 163.

Li, Y., Laidlaw, W. G., and Wardlaw, N. C. (1986). *Advances in Colloid and Interface Sci.* **26**, 1.

Li, Y., and Wardlaw, N. C. (1986). *J. Colloid Interface Sci.* **109**, 461.

Lin, C.-Y., and Slattery, J. C. (1982). *AIChE J.* **28**, 311.

Loomis, A. G., and Crowell, D. C. (1962). U.S. Govt. Printing Office, Bulletin 599, p. 1, Washington, D.C.

Macdonald, I. F., and Dullien, F. A. L. (1976). *Soc. Pet. Eng. J.* **16**, 7.

Mahers, E. G., and Dawe, R. A. (1985). *Proc. 3rd European Meeting on Improved Oil Recovery*, Rome, **1**, 49.

Manegold, E., Hofmann, R., and Solf, K. (1931). *Kolloid Zeitschr.* **55**, 273; **56**, 267; **57**, 23.

Marle, C. (1965). *Inst. Fr. Pet.* **4**.

Marle, C. M. (1981). "Multiphase Flow in Porous Media." Gulf Publishing Co.

Marshall, T. J. (1958). *J. Soil Sci.* **9**, 1.

Maxwell, J. C. (1873). "A Treatise on Electricity and Magnetism," p. 365. Oxford Univ. Press (Clarendon), London and New York.

McCabe, W. L., and Smith, J. L. (1976). "Unit Operations of Chemical Engineering." McGraw-Hill, New York.

McCaffery, F. G. (1973). Ph.D. Dissertation, University of Calgary, Canada.

McCaffery, F. G., and Bennion, D. W. (1974). *J. Can. Pet. Technol.* **13**, 42.

McKeller, M., and Wardlaw, N. C. (1982). *C.J.P.T.* **21**, 39.

McWhorter, D. B. (1971). "Infiltration Affected by Flow of Air." Colorado State Univ. Hydrol. Paper No. 49, May.

Melrose, J. C. (1965). *Soc. Petr. Eng. J.* **5**, 259.

Melrose, J. C. (1988). *Studies in Surface Science and Catalysis* **39**, 253.

Melrose, J. C., and Brandner, C. F. (1974). *J. Can. Pet. Technol.* **13**, 54.

Miller, E. E., and Miller, R. D. (1956). *J. Appl. Phys.* **27**, 324.

Millington, R. J., and Quirk, J. P. (1961). *Trans. Faraday Soc.* **57**, 1200.

Mohanty, K. K. (1981). Ph.D. Thesis, University of Minnesota, Minneapolis.

Mohanty, K. K. (1982). SPE 11018, SPE AIME 57th Tech. Conference, New Orleans.

Mohanty, K. K., Davis, H. T., and Scriven, L. E. (1980). SPE-AIME Annual Meeting, Texas, p. 1.

Mohanty, K. K., and Salter, S. J. (1983). SPE 12127, 58th Annual Technical SPE Conference, October, San Francisco, California.

Mohanty, K. K., and Salter, S. J. (1982). SPE 11018, 57th Annual Technical SPE Conference, September, New Orleans, LA.

Moore, T. F., and Slobod, R. L. (1956). *Prod. Mon.* **20**, 20.

Morcom, A. R. (1946). *Trans. Inst. Chem. Engrs.* (London) **24**, 30.

Morel-Seytoux, H. J. (1969). *In* "Flow through Porous Media" (R. J. M. deWiest, ed.), pp. 456–516. Academic Press, New York.

Morgan, J. T., and Gordon, D. T. (1970). *J. Pet. Tech.* **22**, 1199.

Morrow, N. R., Chatzis, I., and Gocker, M. E. (1981). Department of Energy, BETC/3251-12.

Morrow, N. R., Chatzis, I., and Taber, J. J. (1985). SPE 14423, presented at 60th SPE Conference, Las Vegas, Nevada.

Morrow, N. R., Cram, P. J., and McCaffery, F. G. (1973). *SPEJ 221*; *Trans. AIME 255*.

Morrow, N. R., and Songkran, B. (1981). *In* "Surface Phenomena in Enhanced Oil Recovery." Plenum Press, New York, p. 387.

Morse, R. A., and Yuster, S. T. (1946). *Producers Monthly*, December 19.

Morse, R. A., Terwilliger, P. L., and Yuster, S. T. (1947). *Oil Gas J.* 109.

Moulee, J. C. (1985). *J. Phys. Lett.* **46**, L97.

Mungan, N. (1966). *Soc. Pet. Eng. J.* **6**, 247.

Muskat, M. (1937a). "The Flow of Homogeneous Fluids through Porous Media." McGraw-Hill, New York.

Muskat, M. (1937b). *Trans. Am. Inst. Min. Eng.* **123**, 69.

Muskat, M., and Meres, M. W. (1936). *Physics* **7**, 346.

Muskat, M., Wyckoff, H. G., Botset, M. W., and Meres, M. W. (1937). *Trans. AIME* **123**, 69.

Naar, J., and Henderson, J. H. (1961). *Soc. Pet. Eng. J.* **1**, 61.

Ng, K. M., Davis, H. T., and Scriven, L. E. (1978). *Chem. Eng. Sci.* **33**, 1009.

Ngan, C., and Dussan, V. E. B. (1982). *J. Fluid Mech.* **18**, 27.

Nicholson, D., and Petropoulos, J. H. (1968). *Brit J. Appl. Phys. J. Phys. D.* **1**, 1379.

Nicholson, D., and Petropoulos, J. H. (1971). *J. Phys. D. Appl. Phys.* **4**, 181.

Nobles, M. A. (1974). "Using the Computer to Solve Petroleum Engineering Problems." Gulf Publ., Houston, Texas.

Oak, M. J., Baker, L. E., and Thomas, D. C. (1988). SPE/DOE 17370, presented at SPE/DOE Symposium, April, Tulsa, Oklahoma.

Odeh, A. S. (1959). *Petroleum Transactions AIME* **216**, 346.

Ojeda, E., Preston, F., and Calhoun, J. (1953). *M.I. Exp. Sta.* **18**, No. 2.

Olbricht, W. L., and Leal, L. G. (1982). *J. Fluid Mech.* **115**, 187.

Oroveanu, J. (1966). "Flow of Multiphase Fluids through Porous Media." Editura Academiei Republicii Socialiste Romania, Bucarest.

Osoba, J. S., Richardson, J. D., and Blair, P. M. (1951). *Trans. Am. Inst. Min. Eng.* **192**, 47.

Owens, W. W., and Archer, D. L. (1971). *J. Pet. Tech.* **23**, 873.

Owens, W. W., Parish, D. R., and Lamoreaux, W. E. (1965). *Trans. AIME* **207**, 275.

Paez, J., Reed, P., and Calhoun, J. (1954). *Mineral Ind. Exp. Sta.* **64**, 115.

Parlange, J. Y. (1971a). *Soil Sci.* **111**, 134.

Parlange, J. Y. (1971b). *Soil Sci.* **112**, 313.

Pathak, P., Davis, H. T., and Scriven, L. E. (1982). SPE 11016, presented at 57th SPE Conference, Dallas, Texas.

Payatakes, A. C. (1982). *Ann. Rev. Fluid Mech.* **14**, 365.

Payatakes, A. C., and Dias, M. M. (1984). *Rev. Chem. Eng.* **2** (2), 85.

Payatakes, A. C., Tien Chi, and Turian, R. (1973). *AIChE J.* **19**, 58, 67.

Peaceman, D. W. (1977). Fundamentals of numerical reservoir simulation. *In* "Developments in Petroleum Science." Vol. 6. Elsevier, Amsterdam.

Philip, J. R. (1957a). *Aust. J. Phys.* **10**, 29.

Philip, J. R. (1957b). *Soil Sci.* **83**, 345.

Philip, J. R. (1957c). *Soil Sci.* **83**, 435.

Philip, J. R. (1957d). *Soil Sci.* **84**, 278.

Philip, J. R. (1970). *Ann. Rev. Fluid Mech.* **2**, 177.

Philip, J. R. (1973a). Proc. Congr. Theor. Appl. Mech., 13th (E. Becker, and G. K. Mikhailov, eds.). Springer Verlag, N.Y.

Philip, J. R. (1973b). *Soil Sci.* **116**, 328.

Philip, J. R., and Knight, J. H. (1974). *Soil Sci.* **117**, 1.

Probine, M. C. (1958). *Brit. J. Appl. Phys.* **9**, 144.

Raimondi, P., and Torcaso, M. A. (1964). *Soc. Pet. Eng. J.* **4**, 49.

Rapir, S. (1980). M.S. Thesis, University of Houston, Texas.

Rapoport, L. A., and Leas, W. J. (1953). *Trans. Am. Inst. Min. Eng.* **198**, 139.

Reid, L. S., and Huntington, R. L. (1938). *AIME T.P.* 873.

Reid, S. (1958). Ph.D. Dissertation, Chemical Engineering Dept., Univ. of Birmingham.

Richardson, J. G., Kerver, J. K., Hafford, J. A., and Osoba, J. S. (1952). *Trans. Am. Inst. Min. Eng.* **195**, 187.

Rose, W. (1972). *The Iran Pet. Inst. Bull.* **46**, 23.

Rose, W. (1974). *The Iran Pet. Inst. Bull.* **48**, 25.

Rose, W. (1988). *Transport in Porous Media* **3**, 163.

Rose, W., and Cleary, J. (1958). *Prod. Monthly* **22**, 20.

Rose, W., and Heins, R. W. (1962). *J. Colloid Sci.* **17**, 39.

Rose, W., and Witherspoon, P. A. (1956). *Producers Mon.* **21**, 32.

Salathiel, R. A. (1973). *J. Pet. Technol.* October, 1216.

Scheidegger, A. E. (1974). "The Physics of Flow through Porous Media." Univ. of Toronto Press, Toronto, Canada.

Schneider, F. N., and Owens, W. W. (1970). *Soc. Pet. Eng. J.* **10**, 75.

Schwartz, A. M., and Tejada, S. B. (1972). *J. Colloid Interface Sci.* **38**, 359.

Schwartz, A. M., Rader, C. A., and Huey, E. (1964). *Adv. Chem.* Ser. **43**, 250.

Schwartzendruber, D. (1969). *In* "Flow through Porous Media" (R. J. W. deWiest, ed.), pp. 215–292. Academic Press, New York.

Shankar, P. K., and Dullien, F. A. L. (1979). In "Surface Phenomena in Enhanced Oil Recovery" (D. O. Shah, ed.). Plenum Press, 453.

Snell, R. W. (1962). *J. Inst. Petr.* **48**, 459.

Spanos, T. J. T., de la Cruz, V., Hube, J., and Sharma, R. C. (1986). *Can. J. Petr. Technol.* **25**, 71.

Spanos, T. J. T., de la Cruz, V., and Hube, J. (1988). *AOSTRA Journal of Research* **4**, 181.

Stegemeir, G. L. (1974). SPE 4754, presented at SPE–AIME Conference, April, Tulsa, Oklahoma.

Szekely, J., Neumann, A. W., and Chuang, Y. K. (1971). *J. Colloid Interface Sci.* **35**, 273.

Taber, J. J. (1969). *Soc. Pet. Eng. J.* **9**, 3.

Taber, J. J. (1981). *From* "Surface Phenomena in Enhanced Oil Recovery." Plenum Press, New York.

Taber, J. J., Kirby, J. C., and Schroeder, F. U. (1973). *AIChE Symp.* Ser. No. 27, **69**, 53.

Templeton, C. C. (1954). *Bull. Am. Phys. Soc.* **29**, 16.

Terwilliger, P. L., and Yuster, S. T. (1946). *Producers Mon.* **11**, 42.

Terwilliger, P. L., and Yuster, S. T. (1947). *Oil Weekly* **126**, 54.

Terwilliger, P. L., Wilsey, L. E., Hall, H. N., Bridges, P. M., and Morse, R. A. (1951). *Petrol. Trans. AIME* **192**, 292.

van Brakel, J., and Heertjes, P. M. (1977). *Powder Technol.* **16**, 75.

van Brakel, J., and Heertjes, P. M. (1977a). *Powder Technol.* **16**, 83.

Vizika, O., and Payatakes, A. C. (1988). Paper presented at AIChE Meeting, Washington, D.C.

Vizika, O., and Payatakes, A. C. (1989). *Phys. Chem. Hydrodynamics* **11**, no. 2, 787.

Wardlaw, N. C. (1982). *J. Can. Pet. Technol.* **12**, 21.

Wardlaw, N. C., and Li, Y. (1988). *Trans. in Porous Media* **3**, 17.

Wardlaw, N. C., and McKellar, M. (1985). *C. J. Chem. Eng.* **63**, 525.

Wardlaw, N. C., and Taylor, R. P. (1976). *Bull. Can. Pet. Geol.* **24**, 225.

Warren, J. E., and Calhoun, J. C. (1955). *Pet. Trans. AIME* **204**, 22.

Washburn, E. W. (1921). *Proc. Nat. Acad. Sci.* **7**, 115.

Weinhardt, B., and Heinemann, Z. (1985). *Acta Geodaetica, Geophys. et Montanist. Hung.* **20** (2–3), 303.

Welge, H. J. (1952). *Trans. Am Inst. Min. Eng.* **195**, 91.

Wellington, S. L., and Vinegar, H. J. (1985). Presented at SPE Conference, September, Las Vegas.

Wellington, S. L., and Vinegar, H. J. (1987). *JPT* **29** (8), 385.

West, G. D. (1911). *Proc. R. Soc. London* Ser. A **86**, 20.

Whitaker, S. (1986). *Transport in Porous Media* **1**, 105.

Weigand, C. L., and Taylor, S. A. (1961). Utah Agr. Exp. Sta. Spec. Rept. 15.

Wilkinson, D., and Willemsen, J. (1983). *J. Phys. A: Math. Gen.* **16**, 3365.

Winterfield, P. H., Scriven, L. E., and Davis, H. T. (1981). *J. Phys. C. Solid State Physics* **14**, 2361.

Withjack, E. M. (1987). SPE 16951, presented at 62nd SPE Conference, Dallas, Texas.

Wooding, R. A., and Morel-Seytoux, H. J. (1976). *Annu. Rev. Fluid Mech.* **8**, 233.

Wyckoff, R. D., and Botset, H. G. (1936). *Physics* **7**, 325.

Wyllie, M. R. J., and Gardner, G. H. F. (1958). *World Oil Prod. Sect.*, April, 210–228.

Wyllie, M. R. J., and Spangler, M. B. (1952). *Bull. Am. Assoc. Pet. Geol.* **36**, 359.

Yadav, G. D., Chatzis, I., and Dullien, F. A. L. (1985). *Chem. Eng. Sci.* **40**, 1618.

Yadav, G. D., Dullien, F. A. L., Chatzis, I., and Macdonald, I. F. (1987). *SPE Reservoir Engineering*, May.

Yadav, G. D., and Mason, G. (1983). *Chem. Eng. Sci.* **38**, 1461.

6 | *Miscible Displacement and Dispersion*

6.1. INTRODUCTION

When two immiscible fluids with strong wettability preference are pumped simultaneously through a porous medium, they tend to flow in separate channels and maintain their identities[1], but with two miscible fluids no such experiment is possible. Displacement in the case of immiscible fluids is generally not complete[1], but a fluid can be displaced completely from the pores by another fluid that is miscible with it in all proportions; that is, in the case of miscible fluids there are no residual saturations.

In the immiscible displacement process, neglecting capillarity results in the prediction of a sharp front (step function) between the displacing and the displaced phases (Buckley–Leverett profile). This situation is approached in reality when the flow rates are relatively high. At low flow rates, the effect of capillarity results in a smearing of the saturation profile.

In miscible displacement there is no capillarity; instead there is mixing (so-called "dispersion") of the two fluids. It turns out, as shall be discussed in detail, that at relatively low flow rates the effect of dispersion is slight as compared with the rate of advance of the displacing fluid. Hence, under

---

[1]These statements may be violated in certain situations, for example when the capillary number is very large (see Chapter 5).

487

these conditions, the approximation of using a sharp displacing front is often a good one. In the case of miscible displacement the transition from pure displacing to pure displaced phase tends to become more gradual at increasing flow rates.

Miscible displacement and dispersion phenomena occur in many important fields of technology, for example, in petroleum reservoir engineering, ground water hydrology, chemical engineering, chromatography. A few examples are so-called miscible flood in petroleum recovery; transition zone between salt water and fresh water in coastal aquifers; radioactive and ordinary sewage waste disposal into acquifers; movement of minerals (e.g., fertilizers) in the soil; packed reactors in chemical industry; use of various "tracers" in petroleum engineering and hydrology research projects, etc.

The treatment of miscible displacement is divided here into two main sections: (1) dispersion in a capillary tube and (2) dispersion in a porous medium.

Several reviews of dispersion have been published in the past few decades (e.g., Fried (1975), Perkins and Johnston (1963), Collins (1961), Pfannkuch (1963), Bear (1969, 1972), Nunge and Gill (1970), and Greenkorn and Kessler (1970), Scheidegger (1974), Greenkorn (1983)).

6.2. DISPERSION IN A CAPILLARY

The reasons for examining the dispersion process in a capillary are twofold: First, the phenomenological equations describing dispersion are often the same as in the case of a porous medium. Second, dispersion in a capillary tube, the mechanism of which is relatively well understood, plays a role in determining dispersion in porous media.

6.2.1. Slug Stimulus

An early experimental work that demonstrates the essence of the process was reported by Griffiths (1911), who observed that a tracer injected into a stream of water in laminar flow spreads out in a symmetrical manner about a plane moving with the average flow velocity. This is, qualitatively speaking, the same phenomenon as can be observed in the absence of flow. In that case it is due entirely to molecular diffusion. Under flow conditions, however, as pointed out by Taylor (1953, 1954), this is a rather startling phenomenon for two reasons. First, since the water near the center of the tube moves with twice the mean speed of flow and the tracer moves at the mean speed, the water near the center must pass through the slug of tracer, absorb some tracer as it passes through, and then reject some tracer as it leaves on the other side of the slug. Second, although the velocity is unsymmetrical about the plane moving at the average flow velocity, the slug of tracer spreads out symmetrically.

Analytical treatment of this problem has been accomplished under certain simplifying assumptions, corresponding to extreme physical situations.

The concentration c of the tracer is described by the two-dimensional, unsteady, convective, diffusion equation in cylindrical coordinates:

$$\frac{\partial c}{\partial t} + u(r)\frac{\partial c}{\partial x} = \mathscr{D}\left(\frac{\partial^2 c}{\partial x^2} + \frac{1}{r}\frac{\partial}{\partial r}r\frac{\partial c}{\partial r}\right), \qquad (6.2.1)$$

where $u(r)$ is the parabolic laminar velocity profile. It has been assumed that $\mathscr{D}$, the molecular diffusion coefficient, is constant.

Consider, for purpose of illustration, a slug of solvent A injected at time zero into B, which is in fully developed laminar flow. The initial configuration is shown in Fig. 6.1a. Since diffusive length is proportional to the square root of time and convective length is proportional to the first power of both time and velocity, at sufficiently short times and high flow rates, axial convection is the dominant factor in dispersion. The corresponding shape of the slug is shown schematically in Fig. 6.1b.

It is clear from the geometry of the system, shown in Fig. 6.1b, that a radial concentration gradient has been established and molecular diffusion in the radial direction will make its effect felt by gradually equalizing the tracer concentration in the cross section of the capillary tube. Some time later the state of the slug can be schematically represented as shown in Fig. 6.1c. Even though, due to convection, the parabolic slug may be getting longer at a much faster rate than wider because of radial diffusion, after a certain time

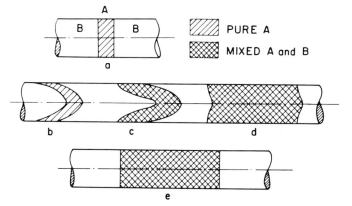

Figure 6.1. Schematic representation of the mixed zone for a slug of A dispersing in B as a function of time; A–B completely miscible (Nunge and Gill, 1970).

diffusion will eliminate the tracer free zones both at the axis and near the walls of the capillary tube, as shown in Figs. 6.1d and e. It should be noted that the concentration of the new slug shown in Fig. 6.1e is uniform at any x in the radial direction, but it is not uniform in the x-direction, and it exhibits a concentration maximum at the half-length of the slug.

For sufficiently large values of time, Taylor showed that the process could be described by a one-dimensional convective diffusion equation as follows:

$$\frac{\partial \bar{c}}{\partial t} + \bar{u}\,\frac{\partial \bar{c}}{\partial x} = D_{ct}\,\frac{\partial^2 \bar{c}}{\partial x^2}, \tag{6.2.2}$$

where $\bar{c}$ is the average tracer concentration at distance x and time t, and the molecular diffusion coefficient has been replaced by an effective axial diffusion coefficient or "coefficient of hydrodynamic dispersion" D_{ct} (ct designates "capillary tube").

Upon defining a coordinate x' that moves with the mean speed of flow as

$$x' = x - \bar{u}t, \tag{6.2.3}$$

Eq. (6.2.2) becomes

$$\partial \bar{c}/\partial t = D_{ct}\,\partial^2 \bar{c}/\partial x'^2, \tag{6.2.4}$$

and, therefore, Eq. (6.2.4) is simply the one-dimensional unsteady diffusion equation. The solution of this equation, known as Fick's second law of diffusion, is well known for the case under consideration:

$$\frac{\bar{c}}{c_0} = \frac{1}{2(\pi D_{ct}t)^{1/2}}\,\exp\left[-\frac{(x - \bar{u}t)^2}{4D_{ct}t}\right], \tag{6.2.5}$$

where c_0 is the initial tracer concentration in the slug and $D_{ct} = \sigma_x^2/2t$, with σ_x the standard deviation of the Gaussian distribution of the tracer concentration. Therefore, the length of the mixed zone in Fig. 6.1e keeps increasing symmetrically on both sides of the tracer concentration maximum in the middle of the slug. The concentration tapers off to zero at both ends of the slug. The formal analogy of Eq. (6.2.4) with Fick's second law of diffusion may easily lead to the false conclusion that in hydrodynamic dispersion the driving force of spreading of the tracer is the axial concentration gradient $\partial \bar{c}/\partial x'$, which in reality is but a result of an interplay between axial convection and radial diffusion. In other words, it is the effect rather than the cause of the process.

For the case of negligible axial molecular diffusion, Taylor showed D_{ct} to be given as

$$D_{ct} = R^2 \bar{u}^2/48\mathscr{D} = Pe_{ct}^2 \mathscr{D}/192, \tag{6.2.6}$$

where D_{ct} includes the effects of both mechanical (or convective) dispersion and radial molecular diffusion. It is enlightening to note that as $\mathscr{D} \to \infty$, $D_{ct} \to 0$, and as $\mathscr{D} \to 0$, $D_{ct} \to \infty$. Hence, very intense radial mixing eliminates dispersion and the absence of radial mixing results in an infinite dispersion coefficient.

Aris (1956), however, showed that when axial molecular diffusion is not negligible, the axial dispersion coefficient contains an additive term due to diffusion

$$D_{ct} = \mathscr{D} + R^2 \bar{u}^2/48\mathscr{D} = \mathscr{D} + \left(Pe_{ct}^2 \mathscr{D}/192\right). \tag{6.2.7}$$

Inspection of Eqs. (6.2.6) and (6.2.7) shows that the rate of spreading (dispersion) increases rapidly with tube diameter and velocity, and it decreases with increasing diffusion coefficient, at least as long as $R^2 \bar{u}^2/48\mathscr{D}$ is much greater than $\mathscr{D}$.

6.2.2. Step Change in Inlet Concentration

For the case of a step change in inlet concentration from zero to c_0, the solution of Eq. (6.2.4) is as follows (von Rosenberg, 1956):

$$\frac{\bar{c}}{c_0} = \frac{1}{2}\left[1 - \text{erf}\,\frac{x - \bar{u}t}{2(D_{ct}t)^{1/2}}\right]. \tag{6.2.8}$$

As for $x = \bar{u}t$, $\bar{c}/c_0 = 0.5$, the point $\bar{c}/c_0 = 0.5$ moves with the velocity $\bar{u}$. Various concentration profiles corresponding to Eq. (6.2.8) are shown on Fig. 6.2. Equation (6.2.8) plots as a straight line on arithmetic probability paper. It is convenient to use Eq. (6.2.8) either (1) at a fixed time t or (2) at a fixed position x. With $\bar{u}$ given, in case (1), x corresponding to a selected value of $\bar{c}/c_0$ can be calculated. Picking two values of $\bar{c}/c_0$, (e.g., 0.9 and 0.1) the difference between the corresponding values of x (i.e., $x_{0.9} - x_{0.1}$) is a convenient measure of the spreading in time t. This quantity, called the mixing length L_m, was also introduced by Taylor (1953). It can be readily shown from Eq. (6.2.8) that

$$D_{ct} = (1/t)[L_m/3.625]^2. \tag{6.2.9}$$

[For the choice $x_{0.8} - x_{0.2}$, 3.625 in Eq. (6.2.9) is replaced by 2.38.]

With $\bar{u}$ given, in case (2), the arrival time of t of a selected value of $\bar{c}/c_0$ at $x = L$, the outflow face of the sample is determined by Eq. (6.2.8). A mixing length of time, analogous to L_m, however, cannot be written down in

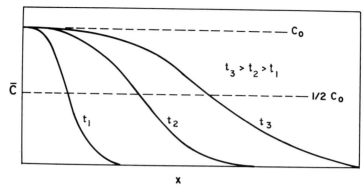

Figure 6.2. Typical distributions of concentration of injected material during linear miscible displacement as predicted by Taylor's solution. (Adapted with permission from Collins, 1961.)

closed form, similar to Eq. (6.2.9). Instead, one lets in Eq. (6.2.8) $x = L = \bar{u}T$, where T is the time it takes to collect one "pore volume" (in this case, the volume of the capillary) of effluent (Brigham *et al.*, 1961), whence

$$2(D_{ct})^{1/2} \operatorname{erf}^{-1}(\bar{c}/c_0) = \frac{L}{T^{1/2}} \frac{1 - \nu_p}{\sqrt{\nu_p}}, \qquad (6.2.10)$$

with $\nu_p = t/T$, the number of pore volumes of effluent collected at time t. Picking two values of c (e.g., 0.9 and 0.1) after forming the corresponding differences from Eq. (6.2.10),

$$D_{ct} = (L^2/T) \left(\frac{\left(\dfrac{1 - \nu_p}{\sqrt{\nu_p}} \right)_{0.9} - \left(\dfrac{1 - \nu_p}{\sqrt{\nu_p}} \right)_{0.1}}{3.625} \right)^2 \qquad (6.2.11)$$

is obtained. The expression

$$\frac{3.625 \left(\mathscr{D} + \dfrac{R^2 \bar{u}^2}{48 \mathscr{D}} \right)^{1/2}}{2\bar{u}\sqrt{T}} = \frac{1}{2} \frac{x_{0.9} - x_{0.1}}{x_{0.5}} > \left\{ \begin{array}{l} \text{fraction of original fluid} \\ \text{remaining in capillary after} \\ \text{producing 1 pore volume} \end{array} \right\}$$

$$(6.2.12)$$

may be used as a rough measure of displacement efficiency. For example, with $R = 10^{-2}$ cm, and $\mathscr{D} = 10^{-5}$ cm^2/sec we have (1) for $\bar{u} = 0.3$ cm/sec and $T = 33$ sec, $(1/2)(x_{0.9} - x_{0.1})/x_{0.5} = 0.144$ (in this case $\mathscr{D} \ll$

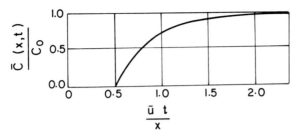

Figure 6.3. Breakthrough curve in a capillary tube without molecular diffusion (adapted from Bear, 1972).

$R^2\bar{u}^2/48\mathscr{D}$); and (2) for $\bar{u} = 3 \times 10^{-3}$ cm/sec and $T = 3.3 \times 10^3$ sec we have $(1/2)(x_{0.9} - x_{0.1})/x_{0.5} = 0.036$ (in this case $\mathscr{D} > R^2\bar{u}^2/48\mathscr{D}$).

In the extreme case when both axial and radial molecular diffusion may be neglected, Taylor (1953) obtained the following analytical solution of Eq. (6.2.1):

$$\bar{c}/c_0 = 1 - x^2/4\bar{u}^2t^2 = 1 - 1/4v_p^2. \tag{6.2.13}$$

The corresponding "breakthrough curve," shown on Fig. 6.3, illustrates that at any distance x downstream from the point of step change in inlet concentration from zero to c_0, nonzero concentration will first appear after injecting a volume equal to one half of the volume of the capillary between the point of injection and the point x. In the absence of radial molecular diffusion an infinite number of pore volumes of displacing fluid would have to be injected for 100% displacement.

A graphical summary of the regions of applicability of various analytical solutions for dispersion for the step change in inlet concentration has been given by Nunge and Gill (1970) (see Fig. 6.4) with the dimensionless time $\tau = t\mathscr{D}/R^2$ and the Peclet number $\mathrm{Pe}_{ct} = 2R\bar{u}/\mathscr{D}$ as parameters. An important point relating to the dispersion model is that τ must be sufficiently large for it to apply—say, 0.8 for fully developed laminar flow in tubes.

Dimensionless mixing length L'_m has been correlated with τ as follows:

$$L'_m = \mathscr{D}L_m/2R^2\bar{u} = 0.26(t\mathscr{D}/R^2)^{1/2}, \qquad \tau = t\mathscr{D}/R^2 > 0.8, \tag{6.2.14a}$$

where Eqs. (6.2.6) and (6.2.9) have been used. For small values of τ (Ananthakrishnan *et al.* 1965), however:

$$L'_m = 0.8\tau, \qquad \tau < 0.01. \tag{6.2.14b}$$

Hence, at small times $D_{ct} \propto t$.

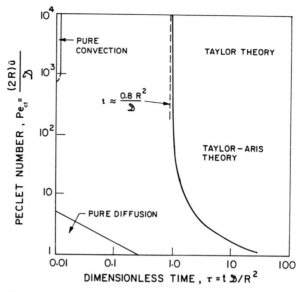

Figure 6.4. Summary of the regions of applicability of various analytical solutions for dispersion in capillary tubes with a step change in inlet concentration as a function of τ and Pe_{ct}. (Reprinted with permission from Ananthakrishnan *et al.*, 1965.)

For small Peclet numbers, the numerical results of Ananthakrishnan *et al.* (1965) are correlated by

$$L'_m = \begin{cases} 2.5\,\mathrm{Pe}_{ct}^{-0.75}\tau^{0.55}, & \tau > 0.05, \\ 2.5\,\mathrm{Pe}_{ct}^{-0.75}\tau^{0.60}, & \tau < 0.05. \end{cases} \tag{6.2.15}$$

There is no equivalent axial dispersion coefficient that can be calculated from these mixing lengths, unless the dispersion model region has been reached.

For the slug stimulus an approximation criterion for the dispersion model to apply is (Nunge and Gill, 1970)

$$\tau_{\min} \simeq 0.6\,\mathrm{Pe}_{ct}^2 / \left(\mathrm{Pe}_{ct}^2 + 192\right). \tag{6.2.16}$$

6.2.3.   Transition and Turbulent Flow Dispersion in Uniform Tubes

Nunge and Gill (1970) gave an account of the treatment of axial dispersion in pipes in turbulent flow. Their result is

$$D_{ct} = 7.14\,R\bar{u}f_p^{1/2}, \tag{6.2.17}$$

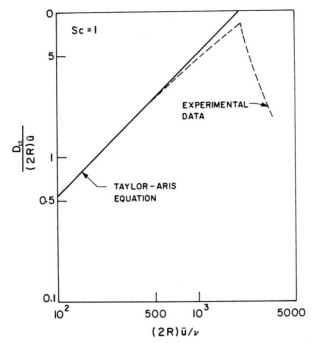

Figure 6.5. Schematic representation of the effect of transition from laminar to turbulent flow on $D_{\rm ct}$. Data in the transition region are available; Sc = 1 (Flint and Eisenklam, 1969).

where $f_{\rm p}$ is the friction factor [see Eq. (3.2.1)]. This result is valid only for fully developed turbulence, where $f_{\rm p}$ is independent of Re. Hence, under these conditions $D_{\rm ct}$ varies linearly with $\bar{u}$.

The data of Flint and Eisenklam (1969) for gaseous horizontal dispersion, obtained with a pulse tracer, bridge the gap between laminar and turbulent flow in the Reynolds number range 3×10^2 to 10^4, with Schmidt numbers between 0.27 and 1.0. The characteristic shape of this plot can be seen in Fig. 6.5.

In turbulent flow, additional dispersion takes place by eddy diffusion mechanism. In the radial direction, this greatly enhances the action of radial molecular diffusion, and hence the axial dispersion coefficient decreases [see Eq. (6.2.6)]. Tichacek *et al.* (1957) presented arguments to the effect that turbulent axial diffusion has negligible effect on dispersion because its contribution is small compared to mixing caused by radial differences in the velocity.

The situation is likely to be very different if the tube contains a series of orifice plates separated from each other by a distance equal to a few tube diameters (see Fig. 6.6). In such a tube, turbulent axial diffusion can be

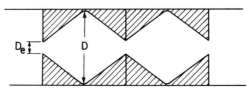

Figure 6.6. Pore model with periodic "orifice plates." Vein section varies from thresholds D_c to enlargements D. (Adapted with permission from Houpeurt, 1959.)

expected to contribute significantly to axial dispersion because the presence of the orifice plates introduces significant axial mixing.

6.2.4. Dispersion in Channels of Various, Different Shapes

Dispersion in the following different channels has been investigated by various authors: parallel plate duct; straight annulus, both concentric and eccentric; curved duct and tube; converging and diverging ducts; and channels with step changes in the diameter (Turner structures).

The dispersion coefficient for a parallel plate duct with fully developed laminar flow has been shown by Turner (1959) and Philip (1963) to be

$$D_{ct} = \mathcal{D} + (8/945)\left(u_{max}^2 h^2/\mathcal{D}\right), \tag{6.2.18}$$

where h is the half-width of the channel.

The full convective diffusion equation has been solved by Gill *et al.* (1969) by finite difference technique, and the regions of validity of the analytical solutions are shown in Fig. 6.7.

The dispersion coefficient in a concentric annulus is a function of the ratio of outside to inside radii $R_o/R_i = 1/\kappa$. For narrow gaps (i.e., $1/\kappa \leq 1.5$) the flat plate result given in Eq. (7.15) may be used in the following form (Nunge and Gill, 1970):

$$D_{ct} = \mathcal{D} + (8/945)\left[R_o^2 u_{max}^2 (1 - \kappa^2)/4\mathcal{D}\right]. \tag{6.2.19}$$

As $1/\kappa$ increases, D_{ct} becomes larger than predicted by Eq. (6.2.19), thus indicating that the effect of an asymmetrical velocity profile is to increase the dispersion relative to the symmetrical case of the straight channel. As $1/\kappa$ approaches 100, D_{ct} approaches the value in a tube, given by Eq. (6.2.6).

Over the range of radius ratios and eccentricities studied, the dispersion coefficient has been found by Sankarasubramanian (1969) to vary by more than two orders of magnitude with changes in eccentricity.

Curved tubes with fully developed laminar flow introduce a number of complications into the dependence of the dispersion coefficient on the system

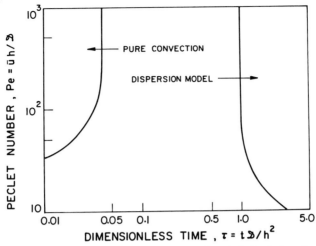

Figure 6.7. Summary of the regions of applicability of analytical solutions for dispersion in straight channels with a step change inlet concentration as a function of τ and Pe (Nunge and Gill, 1970).

parameters. In such cases there exists a secondary circulatory motion in the cross section, in addition to the usual axial velocity profile, as shown in Fig. 6.8 (e.g., Jones, 1968; Wright, 1968).

Some indication of the effect of convergence and divergence on dispersion has been provided by a study of Jeffrey–Hamel flows between nonparallel flat planes by Gill *et al.* (1969) and Güceri (1968). It has been found that molecular diffusion is less effective at high Peclet numbers in the divergent

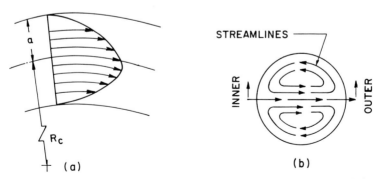

Figure 6.8. Diagrams of two microstructure units indicating the nature of the velocity profiles (Nunge and Gill, 1970).

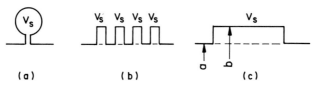

Figure 6.9. Diagrams of three different capacitance models: (a) bottle neck; (b) Turner structure; (c) continuous Turner structure. V_s denotes stagnant volume (Nunge and Gill, 1970).

case. Explanation for this phenomenon is offered by the relatively flatter velocity profile as compared with the parallel wall case.

Turner (1958) visualized a porous medium as consisting of rectangular dead-end pockets communicating with the main flow channels, as shown in Fig. 6.9b. Material is supposed to pass into and out of these pockets by molecular diffusion, the same way as in the case of a bottleneck pore shown in Fig. 6.9a. In the main channels, dispersion takes place by convection and diffusion, while the concentration in the stagnant volume changes according to Fick's second law

$$\partial c/\partial t = \mathscr{D}\,\partial^2 c/\partial y^2. \tag{6.2.20}$$

Later, Gill and Ananthakrishnan (1966) used a continuous stagnant pocket or Turner structure as shown in Fig. 6.9c. In this case also, the point concentration in the "stagnant" region varies according to Eq. (6.2.20). The reason for introducing the "continuous Turner structure" has been a matter of convenience for the case of carrying out the numerical analysis.

It is noted that Azzam (1975) (see also Azzam and Dullien, 1977) solved the complete Navier–Stokes equation for Turner structures of various relative proportions. For creeping flow conditions these were found not to be truly stagnant pockets, as shown in Fig. 6.10a. At higher Reynolds numbers, however, circulation patterns were obtained in the pockets, or bulges, as shown in Fig. 6.10b. In neither case is it strictly correct to assume that the point concentration varies according to Eq. (6.2.20). Hence, the results obtained with this assumption, and sketched in the following, must be used with caution. Aris (1959) pointed out that, as a result of the capacitance effect of the Turner structure, the Taylor dispersion coefficient could have anywhere from 1 to 11 times the value obtained in a straight capillary.

The applicability of a dispersion model of the type postulated in the Taylor–Aris theory to the continuous Turner structure was investigated with numerical experiments by Gill and Ananthakrishnan (1966) for various inlet boundary conditions, Peclet numbers, and stagnant-to-pore-volume ratios. For higher Peclet numbers (Pe $\geq$ 100), where the effects of inlet boundary

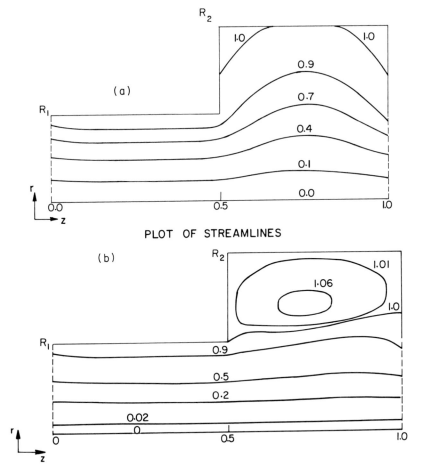

Figure 6.10. (a) Plot of streamlines at Re = 1.0, with $D_1/D_2 = 0.5$, $D_1/\lambda = 0.43$, $L_1/\lambda = 0.5$, and a Z scale of $1:1.16$. (Adapted with permission from Azzam, 1975.) (b) Plot streamlines at Re = 100 showing circulation (Azzam, 1975).

conditions and axial molecular diffusion are unimportant, at sufficiently high dimensionless time, the mean concentration becomes an error function:

$$\frac{\bar{c}}{c_0} = \frac{1}{2}\left[1 - \mathrm{erf}\left(\frac{x - u_{max}t/2\gamma^2}{\sqrt{4D_{ct}t/\gamma^2}}\right)\right]. \qquad (6.2.21)$$

where $\gamma = b/a$ in Fig. 6.9c. It may be concluded that a sufficiently large tube radius must be used to include the entire pore volume. The expression for

TABLE 6.1   Minimum Values of Dimensionless Time Necessary for Dispersion
Model to Apply to Continuous Turner Structure Capacitance Model
at Pe ⩾ 100 as a Function of the Ratio of Total Volume
to Pore Volume[a]

γ	τ_{min}
1.0	0.80
1.1	1.25
1.2	2.00
1.3	4.00
2.0	> 15.00

[a]Reprinted with permission from Gill and Ananthakrishnan, 1966.

the dispersion coefficient obtained by Aris (1959), that is,

$$D_{ct} = \left[\frac{8(1 + \beta) + 12(1 + \beta)^2 \ln(1 + \beta) - 7(1 + \beta)^2}{(1 + \beta)^2} \right] \frac{u_{max}^2 a^2}{192 \mathscr{D}}, \quad (6.2.22)$$

where $\beta = \gamma^2 - 1$, can be used in Eq. (6.2.21).

The numerical experiments of Gill and Ananthakrishnan (1966) demonstrated that for high Peclet number systems, the limiting value of the dimensionless time, necessary for Eqs. (6.2.21) and (6.2.22) to predict the mean concentration distribution, depend on γ as shown in Table 6.1. It is evident that large capacitance significantly increases the minimum time even at large Peclet numbers.

Dayan and Levenspiel (1968) included adsorption on the solid surface and found that adsorption increases the dispersion coefficient.

6.2.5.   Effects of Free Convection in Capillary Tubes

Free convection arises because of density differences in the fluid. For the purpose of analysis, it is convenient to distinguish between vertical and horizontal flows and whether the light fluid replaces the heavier one or vice versa.

In horizontal systems, significant deviations from the uniform density case have been noted in aqueous solutions with density differences $\Delta\rho/\rho$ as small as 1.2×10^{-4} by Reejhsinghani et al. (1966). These experimental results have been explained quantitatively by Erdogan and Chatwin (1967), who predicted that in the low Peclet number range, free convection enhances axial dispersion at lower Peclet numbers and depresses it at higher values. For very large Peclet numbers, free convection is negligible.

In vertical upward displacement, if the heavier phase is the displacing fluid, the action of gravity is to retard the movement near the center line of the tube relative to the fluid near the wall, resulting in a flattened velocity profile and a depressed axial depression. Conversely, when the heavier phase is displaced, the action of gravity is to accelerate the central core relative to the fluid near the wall, resulting in an elongated velocity profile and enhanced axial dispersion.

In downward flow, a heavier fluid displacing the lighter one leads to increased axial mixing and a lighter fluid displacing the heavier one to decreased axial mixing.

These mechanisms apply as long as the system is in stable laminar flow.

6.3. HYDRODYNAMIC DISPERSION (MISCIBLE DISPLACEMENT) IN POROUS MEDIA

6.3.1. Introduction

Injecting a slug of tagged fluid into a homogeneous and isotropic porous medium, saturated with the same fluid, will result, under stagnant conditions, in the tracer fluid diffusing at the same rate in all directions. If the fluid is flowing, however, the tracer will spread faster in the flow direction than in the directions perpendicular to it, resulting in a shape as shown schematically in Fig. 6.11. The dispersion coefficients in the direction of flow (the longitudinal dispersion coefficient) and in the transverse directions are denoted here as D_L and D_T, respectively.

Qualitatively speaking, the above result can be understood in a number of different ways, for example, on the basis of the analysis of dispersion in a capillary tube: In a system consisting of uniform straight capillaries of random orientations in any tube $\bar{u} = \bar{u}_n \cos \theta$, where $\bar{u}_n$ is the velocity in an identical capillary aligned with the direction of the macroscopic pressure gradient and θ is the angle enclosed by the positive direction of the capillary axis with this direction. By Eq. (6.2.7), D_{ct} is seen to decrease with increasing θ in the range $0 \leq \theta \leq \pi/2$.

This model of dispersion in porous media, however, introduces a new mechanism of spreading of the tracer that is due to the different rates of advance of the tracer in capillaries of different orientations. This mechanism can bring about much greater spreading than Taylor dispersion in individual capillary tubes. With everything else kept equal, the extent of spreading according to this mechanism decreases with increasing θ as $\cos \theta$, resulting in lemniscate-like displacement contours that do not correspond to a solution of the diffusion equation [Eq. (6.2.1)]. The ratio of the effective longitudinal and transverse dispersion coefficients that one might define for this model is the

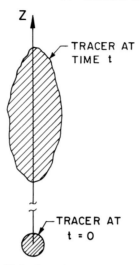

Figure 6.11.   Hydrodynamic dispersion in a porous medium.

ratio of the variance of longitudinal to the variance of transverse displacement, that is,

$$D_L/D_T = \frac{1}{2} \int_0^{\pi/2} \cos^4 \theta \, d\theta \bigg/ \int_0^{\pi/2} \cos^2 \theta \cdot \sin^2 \theta \, d\theta = 3.$$

A similar model was used by Haring and Greenkorn (1970) with nonuniform pore size distributions, who found a variation of D_L/D_T from 3.66 to 19.8, depending on the pore size distribution.

This model of hydrodynamic dispersion, however, does not take into account the interconnectedness of the pores in porous media that has important consequences, as discussed in the next section.

6.3.2.   The Roles Played by Convection and by Molecular Diffusion in Hydrodynamic Dispersion

For purposes of analysis it is useful to imagine, at first, that molecular diffusion is *completely* absent. Then, in the case of *creeping flow*, which is the only case considered here, the fluid particles flow in separate stream tubes and there is no mixing of particles traveling in different stream tubes, exactly as if they were being transported in real tubes having impervious walls. In this limiting case spreading of the tracer takes place purely by convection on the scale of streamlines. Reversing the direction of flow would result in the tracer returning to its original position occupied at zero time.

In real porous media the stream tubes traversing the medium are characterized by a broad velocity distribution. Hence, without *any* molecular diffusion, a tracer injected into the medium would cover a much greater distance in some stream tubes than in others, at any given moment. As a result, at least in the simple case of flow taking place only in one dimension, in the direction of the macroscopic pressure gradient there would be a spreading of the tracer front in the form of microfingers, the magnitude of which is proportional to the time t elapsed from the moment of starting to inject the tracer. Hence, any mixing length L_m would be proportional to t and hence, $L_m^2/t \propto t$; that is, the value of the effective dispersion coefficient (both longitudinal and transverse) would be found to increase in direct proportion with the time t or, equivalently, the distance covered by the tracer in the medium [see Fig. 6.13(a)]. (One can only refer to an effective dispersion coefficient in this and similar cases, because the tracer front would not correspond to a solution of the diffusion equation.)

Under real conditions of tracer dispersion, mixing by molecular diffusion mechanism will tend to homogenize the tracer concentration over the pore cross section, similarly as in the case of Taylor dispersion. For purposes of a qualitative understanding of the mechanism it is best to imagine now complete mixing in the transverse direction to flow of the black tracer stream joining the white solvent stream at a juncture, as shown schematically in Fig. 6.12. The mixed stream will advance to the next juncture, where it merges with another stream.

If the pore network were a regular lattice consisting of a repetition of identical pores, then both streams entering the next juncture would have identical tracer concentrations, and so forth at all the other junctures farther downstream. Hence, in this case the mixing effect of molecular diffusion would be similar as in a single capillary with "Turner structures."

In a typical porous medium, the pore structure of which may be represented by an irregular lattice of pores of distributed diameters, however, the tracer in a pore of greater conductance will reach the next juncture earlier than in a pore of lower conductance and, as a result, it will merge there with pure solvent. Without the homogenizing effect of molecular diffusion, the tracer moving ahead in the "faster" stream tubes in all directions in the range $0 \leq \theta \leq \pi/2$ would keep increasing its lead indefinitely in the form of microfingers, which would result in the effective dispersion coefficients (both longitudinal and transverse) increasing without a limit with increasing distance covered. In reality, however, molecular diffusion will result in a dilution of the tracer moving ahead in pores of high conductance, whereas the tracer lagging behind in pores of lesser conductance will tend to mix with the tracer present at a similar concentration. Thus, molecular diffusion has a similar equalizing effect as in the case of Taylor dispersion. Uniform tracer concentration in the transverse direction will be attained by molecular diffusion if the boundary conditions dictate one-dimensional flow which is the usual case

Figure 6.12.   Schematic diagram of mixing of two streams by molecular diffusion (Dullien, 1990c).

for laboratory floods in cores or packs of particles. Past the distance ℓ measured from the inlet to the sample, where the concentration remains uniform over the sample cross section, D_L does not increase any more with x. The axial concentration profile at this state has been found in homogeneous media to correspond fairly closely to the error function. However, if the boundary conditions permit two- or three-dimensional flow which is the case under field conditions, as illustrated in Fig. 6.11, transverse dispersion in the y- and (z) directions will result in concentrations profiles in these directions with a maximum in the x-axis that, while slowly becoming flatter with increasing time, will persist indefinitely. As a result, D_L (and also D_T) can be expected to increase with x.

Some important differences between Taylor dispersion and dispersion in porous media should be noted. Whereas in the case of laminar flow in a tube, fluid layers in continuous contact undergo telescopic displacement relative to each other, in a network of pores the relative displacement of fluid takes place in pores that are separated from each other by solid everywhere with the exception of the junctures where pores meet. As a result, in the case of laminar flow in a tube, increasing $\mathscr{D}$ indefinitely while keeping $\bar{u}$ constant (see Eq. 6.2.6) can, in principle, result in a zero hydrodynamic dispersion, because very fast radial diffusion could prevent the development of axial concentration differences. By contrast, in the case of flow through a pore network, development of longitudinal concentration differences resulting from different flow velocities in pores in parallel connection cannot be altogether prevented by transverse molecular diffusion, because no diffusion is possible through the solid matrix separating the pores. Only after the separate streams have merged at a pore juncture can molecular diffusion equalize the concentration across the pore cross section in the resulting stream, but it cannot wipe out the longitudinal concentration difference that has developed by convection before the streams merged.

The variation of the Taylor dispersion coefficient D_{ct} with the square of the average velocity $\bar{u}$ (Eq. 6.2.6) is to be compared with the empirically established linear variation (over a certain range of Pe) of the longitudinal dispersion coefficient D_L in porous media with the average pore velocity v_p. By comparing Eq. (6.2.6) with Eq. (6.2.9) it can be seen that in the case of Taylor dispersion the mixing length L_m at time t is directly proportional to $\bar{u}$, that is,

$$L_m = \text{const.} \, R\bar{u}\sqrt{t/48\mathscr{D}}, \tag{6.3.1}$$

whereas over the linear portion of dispersion coefficient versus Peclet number correlations (e.g., Fig. 6.17) the relationship

$$D_L = v_{DF}\overline{D}_p \tag{6.3.2}$$

holds approximately, where $v_{DF} = v/\phi$ and $\overline{D}_p$ is the average particle diameter. This, after combining it with Eq. (6.2.9), gives

$$L_m = \text{const.} \, \sqrt{v_{DF}\overline{D}_p t} \tag{6.3.3}$$

Equation (6.3.1) indicates that in Taylor dispersion in a fixed time interval t, for example, L_m is directly proportional to $\bar{u}$, whereas according to Eq. (6.3.3), in a porous medium in a fixed time interval t, L_m varies as the square root of v_{DF}.

The dependence of D_L on $\mathscr{D}$ in porous media has not been investigated satisfactorily by running experiments in the same porous medium with systems covering a wide range of values of molecular diffusion coefficient $\mathscr{D}$. The custom of correlating $D_L/\mathscr{D}$ versus $v_{DF}D_p/\mathscr{D}$ does not reveal the dependence of D_L on $\mathscr{D}$, because the same correlation would be obtained also between D_L and $v_{DF}D_p$.

Hydrodynamic dispersion induced by a random network has often been called "mechanical" or "geometrical" dispersion, because it is thought to be a result of mechanical splitting of a stream of tracer particles at pore junctures in a manner determined by the geometry of the pore space. This terminology will also be adopted here, while bearing in mind the important role played by molecular diffusion in the process, discussed above.

6.3.3. Non-Fickian Dispersion and Scale-Dependent Dispersivity

In field measurements of dispersion the dispersion coefficients usually turn out several orders of magnitude greater than those measured in the laboratory. In many, but by no means all field studies of spreading of a tracer or a dissolved contaminant the actual concentration profile measured was not Gaussian, but usually was "smoothed" to fit the Gaussian curve. Moreover, in such cases the tracer had an "early breakthrough" and a long "tail." In such instances also very great effective dispersion coefficients have been found as a result of the great extent of spreading of the tracer front. When referring to these phenomena the term "non-Fickian dispersion" is often used. In these instances the number obtained for the dispersion coefficient is an effective value because the observed spreading of the tracer is due predominantly to convection, also called advection, rather than to true hydrodynamic dispersion. This will be the case whenever the length scale of heterogeneities in the medium is too large for transverse dispersion to be able to wipe out the transverse concentration differences that were brought about by advection.

Even heterogeneities on the scale of streamlines in the pores, which exist also in the case of uniform pores and, one step up the ladder, heterogeneities on the pore size scale, resulting in pore channels of differing hydraulic conductivities throughout the system, may cause non-Fickian dispersion if the rate of molecular diffusion is extremely low (e.g., $\mathscr{D}$ on the order of $10^{-8}\,\mathrm{cm}^2\,\mathrm{sec}^{-1}$) or, alternately, the time of travel of the tracer is very short. Heterogeneities that cause non-Fickian dispersion in practice, where the molecular diffusion coefficient is usually on the order of $10^{-5}\,\mathrm{cm}^2\,\mathrm{sec}^{-1}$, however, are on a macroscopic or megascopic scale. Two highly simplified examples, one showing two noninteracting layers of different permeabilities in parallel and another, a large "lense" of greater (or smaller) permeability than the medium surrounding it, are illustrated schematically in Fig. 6.13 (a and b). If the length scale of the heterogeneity is small as compared with the size of the system, normally it will not result in non-Fickian dispersion because the convective spreading caused by its presence may be smoothed out by the homogenizing effect of transverse dispersion.

It has recently been suggested (Hulin and Plona, 1989; Hulin *et al.*, 1988; Charlaix *et al.*, 1988; Bacri *et al.*, 1990) that Fickian dispersion is diagnostic of system homogeneity. It has been demonstrated experimentally in the laboratory (Bacri *et al.*, 1990; Hulin and Plona, 1989) that a non-Fickian dispersion front observed in some laboratory scale samples became Fickian after the direction of flow was reversed just before breakthrough at the exit

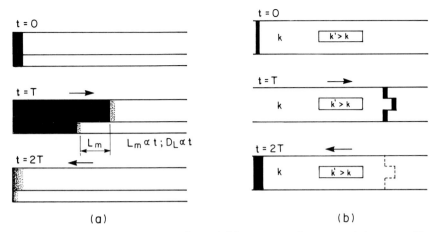

Figure 6.13. Schematic diagram of effects of different types of macroscopic heterogeneities on hydrodynamic dispersion with flow reversal (Dullien, 1990c).

end of the sample and the same volume of liquid was pumped in the opposite direction. This effect is a result of cancellation, in the reverse flow step, of the convective distortion of the concentration profile caused by the presence of the heterogeneity, as shown schematically in Fig. 6.13 (a and b). This experiment has demonstrated the reversibility of the portion of spreading of the tracer that has resulted only of advection without the homogenizing effect of molecular diffusion. The other portion, corresponding to regular hydrodynamic dispersion, however, is not reversible because of the role played by irreversible molecular diffusion.

Similar laboratory tests with flow reversal were also carried out by the same group in random packs of monodisperse glass beads (Rigord *et al.*, 1990) with the avowed intention of demonstrating the reversibility of hydrodynamic dispersion as such, after having demonstrated the reversibility of non-Fickian dispersion. In these experiments slug stimulus, approximating a Dirac delta function, was used. The signal was transported a short distance into the sample by the aqueous medium, then the direction of flow was reversed and the signal was analyzed at the same location where it was originally generated. Transporting the signal to a series of different distances into the sample and then back to its origin, the dispersivities were found to increase with the distance at first linearly, as shown in Fig. 6.14(a). This trend was interpreted as an approach to reversibility of hydrodynamic dispersion as the distance of penetration of the signal was decreased. Reversibility could have been approached only if the time scale of the experiment had been short enough to prevent molecular diffusion from producing a significant mixing in the void spaces in the direction transverse to flow. This was not the

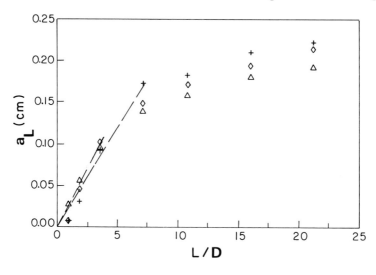

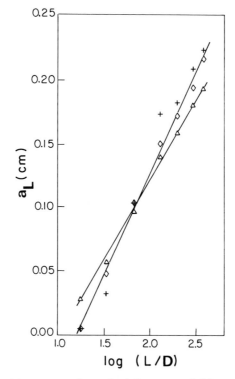

Figure 6.14. Dispersivity, a_L, at the end of flow reversal (a) as a function depth of penetration before reversal in bead diameter units L/D and (b) the same relation as a function of $\log(L/D)$ (Rigord *et al.*, 1990).

case, however. As the dispersion coefficient measured at the end of the round trip even in the shortest test was found to be several times greater than the molecular diffusion coefficient, molecular diffusion must have already contributed significantly to the process. Indeed any degree of reversibility could have been determined only by comparing the spreading of the signal at the end of the forward transport with the extent of spreading found at the end of the return trip. It is of considerable interest that semilogarithmic representation of these results yields, for the most part, straight line plots, as shown in Fig. 6.14(b), which correspond to the predictions of Saffman for initial spreading of a tracer [see Eqs. (6.3.72)].

Reversing the direction of flow at a point in the process where convection has produced concentration differences in the overall flow direction but molecular diffusion has already caused a certain amount of transport of the tracer across the streamlines, gives rise to a complicated process in which spreading might be partly reversed or it may continue to increase, depending on the relative amounts of contributions of convection and molecular diffusion at the end of the forward step.

Field-scale studies of spreading, consisting of measurements of the concentration variance σ_x^2 [cf. Eq. (6.2.5)], in observation wells have indicated an approximately linear increase of the effective dispersion coefficient with the scale of the study (i.e., the mean distance covered by the tracer) (Gelhar *et al.*, 1985). In these and related studies spreading was expressed in terms of the dispersivity: $a_L = D_L/v_{DF} = \sigma_x^2 v_{DF}/2x v_{DF} = \sigma_x^2/2x$, thereby getting rid of the velocity dependence of the dispersion coefficient D_L. A linear increase of the effective dispersivity with the distance traveled has been obtained by Mercado (1967) by modeling the aquifer as consisting of horizontal layers of different permeabilities without any cross flow and neglecting proper hydrodynamic dispersion in the individual layers. This result is a logical consequence of the assumed absence of molecular diffusion and cross flow between the layers and is illustrated in Fig. 6.13(a). Gelhar *et al.* (1979) have extended the work of Mercado (1967) by developing relationships for the (longitudinal) dispersivity of a stratified reservoir with both the permeability and the tracer concentration treated as homogeneous stochastic processes. For small times the result obtained has the same form as that of Mercado (1967), that is, the small-time longitudinal macro dispersivity A_o is

$$A_o = \frac{v_{DF}t\sigma_k^2}{k^2}, \tag{6.3.4}$$

where σ_k^2 is the variance of the permeability. Equation (6.3.4) shows that the macro dispersivity at small times is predicted to increase linearly with the time t and/or the distance traveled x. At large times, however, the longitudinal macrodispersivity A_∞ has been predicted to be

$$A_\infty = \frac{1}{3} \frac{\sigma_k^2}{k^2} \frac{\ell^2}{a_T}, \tag{6.3.5}$$

where ℓ is the length scale of the permeability correlation and a_T is the average transverse dispersivity of the medium. The effective dispersivity $a_{L,\,eff}$ is, at large times,

$$a_{L,\,eff} = a_L + A_\infty \qquad (6.3.6)$$

with a_L being the average longitudinal dispersivity of the medium and $A_\infty \gg a_L$. The transverse dispersivity a_T in Eq. (6.3.5) plays the same role as the molecular diffusion coefficient $\mathscr{D}$ in Eq. (6.2.6). As a matter of fact, both a_T and a_L depend on $\mathscr{D}$ in a somewhat similar way as A_∞ depends on a_T. Without the action of molecular diffusion the spreading of the tracer would never become a stationary process.

Other contributions on dispersion in layered systems include Marle *et al.* (1967), Lake and Hirasaki (1981), Koonce and Blackwell (1965), Wright and Dawe (1983) and Wright *et al.* (1983).

Gelhar and Axness (1983) have applied stochastic continuum theory to three-dimensionally heterogeneous, statistically isotropic, and statistically anisotropic porous media. It was assumed that the overall scale of observation is large compared to the correlation scale of the heterogeneity. Hence, this analysis assumes Fickian dispersion right at the outset. For field applications the important case is when the correlation scale of the (macroscopic/megascopic) heterogeneity (expressed in terms of the hydraulic conductivity) is much greater than the (local or micro) dispersivity measured in the lab on core samples (i.e., the microscopic correlation scale). Under these conditions it has been found in the isotropic case that the longitudinal macro dispersivity varies linearly with the macroscopic/megascopic correlation scale and it is independent of the microscopic correlation scale (i.e., the "local" or microdispersivity). The opposite was found for the transverse macro dispersivity, which did not depend on the macroscopic/megascopic correlation scale; but it varied linearly with the microscopic correlation scale (i.e., the "local" dispersivity). As the ratio of the microscopic-to-macroscopic correlation scales approaches unity (i.e., there is only microscopic heterogeneity present), the ratio of the macroscopic longitudinal-to-transverse dispersivities decreases to a value on the order of 10. Whereas the linear increase of the longitudinal macro dispersivity with the scale of correlation is consistent with the picture formed of the mechanism of hydrodynamic dispersion, the predicted independence of the transverse macro dispersivity in an isotropic medium of the correlation scale is difficult to comprehend and it may require further study.

In the anisotropic case with flow parallel to bedding the mechanism arrived at by Gelhar *et al.* (1979) has been confirmed with the added stipulation that the correlation scale parallel to bedding must be very large compared to the correlation scale perpendicular to bedding.

Wheatcraft and Tyler (1988) have applied fractal concepts to the dispersion process. In their analysis it was assumed that molecular diffusion can be ignored and that tracer particles move, in a Fickian process, along tortuous fractal stream tubes. There were no physical grounds stated for these assumptions, nor was the physical mechanism explained by which the tracer particles were supposed to move in undefined stream tubes. The central physical concept introduced in this paper (i.e., that in a heterogeneous porous medium the ratio of the average effective distance covered by a fluid particle to the distance projected onto the direction of the macroscopic pressure gradient may actually increase with the distance covered) is an interesting one regardless of whether or not the tortuous path of a particle is a fractal, something that would be difficult to prove or disprove directly. The authors found that for the data published by Gelhar *et al.* (1985) and reproduced in Fig. 6.15 the best fit resulted in an increase of the dispersivity with distance traveled raised to the power of 1.3, which they have interpreted as a fractal dimension. It is very difficult to justify the physical picture of the tracer traveling at the same velocity in many noninteracting fractal stream tubes of different lengths in parallel connection; it is much more plausible to assume that the tracer moves through zones of different effective permeabilities at different velocities. According to this alternate picture it is the permeability contrast of these zones that would increase with the scale of observation.

As it has been pointed out in Section 6.3.2., an increase of dispersivity with the scale of observation seems to be a logical consequence of the boundary conditions existing in the field.

Enlightening description of the hydrodynamic dispersion has been given by Kinzelbach and coworkers (1985, 1986, 1987a, 1987b, 1989, 1990), who have applied the random walk approach to simulate the transport of ground-water

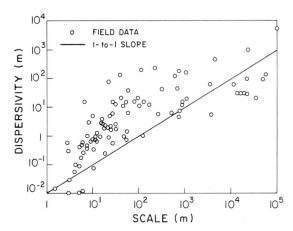

Figure 6.15.   Dispersivity vs. scale of experiment (Gelhar *et al.*, 1985).

contaminants on the field scale. The treatment includes the consideration of first order chemical reaction, adsorption, aquifer with "dual porosity," and non-Fickian random walk. The random walk method provides a very attractive alternative to the numerical solution of the convection–diffusion equation. For example, non-Fickian dispersion is modeled by assigning dispersivities to the particles that increase with the intrinsic age of the particle engaged in random walk.

6.3.4. Dispersion Correlations

6.3.4.1. $D_L/\mathscr{D}$ *versus* Pe *Correlation*

The experimental dispersion data are most frequently presented in the form of $D_L/\mathscr{D}$ versus Pe diagrams (Figs. 6.16–6.18), where D_L is the coefficient of hydrodynamic dispersion observed in the porous medium and Pe = $v_{DF}\overline{D}_p/\mathscr{D}$ with v_{DF} the "average" pore or interstitial velocity as defined by Dupuit–Forchheimer assumption (see Eq. 3.3.4).

$\mathscr{D}$ is assumed to be constant because the solutions are very dilute and/or contain isotope tracers, in which $\mathscr{D}$ is concentration independent.

In Fig. 6.16 (Pfannkuch, 1963), 175 experimental data points, obtained by eight different authors on packed beds in one-dimensional flow, correlate quite satisfactorily (about the same scatter of data as in f_p versus Re$_p$ correlations for packed beds). The mean particle diameters, shown in the figure, extend over a range from 0.01 to 0.686 cm, and there is no apparent correlation with particle diameter. In Fig. 6.17, however, Blackwell's (1962) data indicate much higher $D_L/\mathscr{D}$ ratios for finer sands (200–270 mesh and 40–200 mesh) than the correlation in Fig. 6.16. This effect may be explained by bridging by the particles and other microscopic packing irregularities that occur more frequently as the particle size decreases or as the particle shape becomes more irregular (van Deemter *et al.*, 1956). There is usually also a greater variation in particle size with finer particles (Keulemans, 1957). Ebach (1957) has found that for 30-mesh-or-larger particles sizes, the longitudinal dispersion coefficients are independent of size.

It is of interest to note that in Blackwell's experiments the typical ratio D_L/D_T was about 24 (Fig. 6.17). Any dispersion mechanisms based on a model of randomly oriented capillary tubes predicts a much lower value for this ratio (see Section 6.3.6).

It is noted in both Figs. 6.16 and 6.17 that at very low Peclet numbers the ratio $D_L/\mathscr{D}$ approaches about 0.67 instead of unity. This is a consequence of the fact that $\mathscr{D}$ is the molecular diffusion coefficient measured in the bulk fluid, whereas D_L at Pe $\rightarrow 0$ is equal to the molecular diffusion coefficient $\mathscr{D}_{eff} = \mathscr{D}/X$, measured in the porous medium in an unsteady-state pore diffusion experiment.

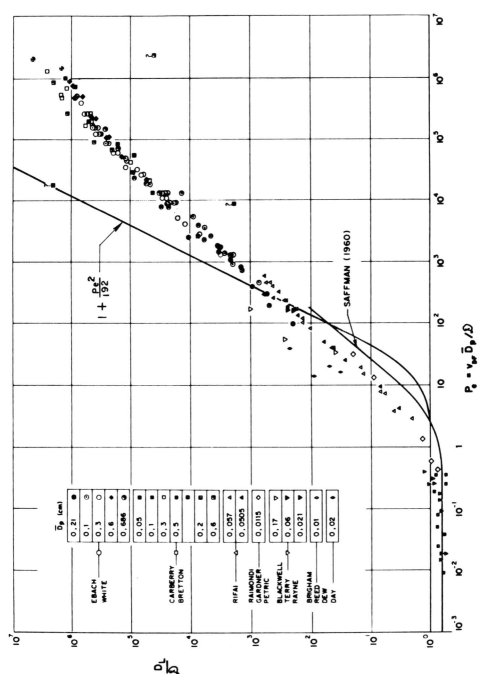

Figure 6.16. $D_L/\mathscr{D}$ vs. Pe correlation. (Adapted with permission from Pfannkuch, 1963.)

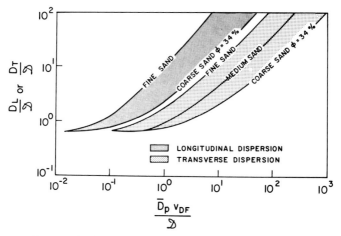

Figure 6.17.   Range of dispersion coefficients for various sand sizes (Blackwell, 1962 © SPE-AIME).

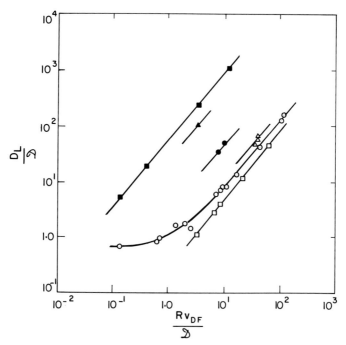

Figure 6.18.   Effect of rate and type of porous medium on dispersion (Brigham *et al.*, 1961 © SPE-AIME).

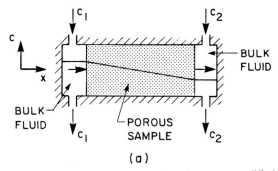

Figure 6.19a. Schematic diagram of conditions of steady-state pore diffusion experiment.

Here

$$X = \Upsilon \cdot S' \tag{6.3.7}$$

is the "electrical, or diffusional, tortuosity factor" (see Section 3.7). (S' is the "constriction factor": Imagine, for the sake of argument that in every pore throat there is an F-stop that can be shut completely. If it is imagined that the membrane of the F-stop has negligible thickness then the presence of the F-stop need not change the porosity and it increases the effective path length L_e, i.e., Υ, only a little while it may make $\mathcal{D}/\mathcal{D}_{\text{eff}}$ infinitely great.)

It should be pointed out that, under conditions of steady-state pore diffusion

$$(\mathcal{D}_{\text{eff}})_{\text{ss}} = \mathcal{D}(\phi/X) \tag{6.3.8}$$

is measured. The conditions of steady pore diffusion are sketched in Fig. 6.19(a). The flux n_x of the tracer, *based on unit bulk cross section* of the porous medium and relative to the solid matrix, is as follows:

$$n_x = -\mathcal{D}(\phi/X)(\partial c/\partial x), \tag{6.3.9}$$

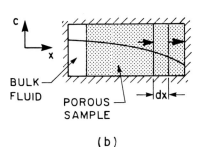

Figure 6.19b. Schematic diagram of conditions of unsteady-state pore diffusion experiment.

whence Eq. (6.3.8) follows immediately. The physical reason for this relationship is that in steady pore diffusion the tracer flux n_x is diminished as
compared with the case of steady diffusion in a bulk fluid by two factors: (a)
the combined effect of the increased path length due to the tortuosity of the
pore paths and the obstructions presented to diffusion by the pore constrictions, both of which are allowed for by the factor $(1/X)$; and (b) the
normalization of the mass flux n_x with respect to the bulk sample cross
section rather than with respect to the effective pore cross section, the effect
of which is accounted for by the factor ϕ.

The conditions of unsteady pore diffusion are sketched in Fig. 6.19b. n_x is
given by Eq. (6.3.9) also in this case, but the effective diffusion coefficient is
evaluated in this case from the rate of concentration change $\partial c/\partial t$, by means
of the continuity equation

$$\partial n_x/\partial x = -\phi \, \partial c/\partial t, \qquad (6.3.10)$$

where the porosity ϕ appears on the right-hand side because the tracer
concentration c is based on unit volume of the fluid whereas the flux is based
on unit bulk cross section of the porous medium. Combining Eqs. (6.3.10)
and (6.3.9) gives

$$\partial c/\partial t = -(\mathcal{D}/X)(\partial^2 c/\partial x^2) \qquad (6.3.11)$$

whence

$$\mathcal{D}_{\text{eff}} = \mathcal{D}/X. \qquad (6.3.12)$$

For loose packs of well-sorted sand, the value $X \approx 1.5$ is reasonable, thus
giving the result: As Pe $\to$ 0, $D'/\mathcal{D} \to \mathcal{D}_{\text{eff}}/\mathcal{D} \to 1/1.5 = 0.67$.

It is noted that in consolidated porous media (e.g., sandstones) the value of
X is usually much greater than 1.5 (see Table 3.5).

It is also noted that

$$\mathcal{D}/(\mathcal{D}_{\text{eff}})_{\text{s.s.}} = F = X/\phi, \qquad (6.3.13)$$

where F is the formation factor.

Coming back to the examination of the form of the $D_{\text{L}}/\mathcal{D}$ versus Pe
correlation shown in Fig. 6.16, it is apparent that, starting at about Pe $= 5$,
the slope corresponds to an exponent $n = 1.2$ and, from Pe ≈ 300, to
$n = 1.0$, rather than $n = 2$ predicted by Eq. (6.2.7). [For a detailed discussion
of the five different zones in Fig. 6.16 see Bear (1972).] It is also important to
note that this correlation does not include the dependence of D_{L} on $\mathcal{D}$.

In Fig. 6.18 some of the experimental results obtained by Brigham *et al.*
(1961) are shown, which include tests on consolidated sandstones. It is
apparent that, in the sandstones, values of the ratio $D_{\text{L}}/\mathcal{D}$ may exceed those
obtained in the bead packs by a factor of 10^2 for the same value of
$\text{Pe}_{\text{R}} = \bar{R}v_{\text{DF}}/\mathcal{D}$, where $\bar{R}$ is the average grain radius. Explanation of this
result may be found in a greater correlation scale of heterogeneities on the
pore level in sandstones than in bead packs.

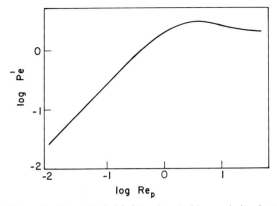

Figure 6.20. Pe' as a function of Re'$_p$ (a). (Reprinted with permission from de Ligny, 1970.)

6.3.4.2. Pe' *versus* Re$_p$ *Correlations*

An equivalent method of presenting hydraulic dispersion data is based on the definition of a "dynamic" longitudinal Peclet number $\text{Pe}' = v_{DF}\overline{D}_p/ D_L = (v_{DF}\overline{D}_p/\mathscr{D})/(D_L/\mathscr{D})$ where the molecular diffusion coefficient $\mathscr{D}$ has been replaced with the dispersion coefficient D_L. Figure 6.20 shows $\log \text{Pe}'$ plotted versus $\log \text{Re}_p$ for a bead pack. As is apparent from the diagram, the function shows a maximum (Gunn, 1969; de Ligny, 1970; Edwards and Richardson, 1968). This maximum is a direct mathematical consequence of the change of slope of the $\log (D_L/\mathscr{D})$ versus $\log \text{Pe}$ curve (see Fig. 6.16) from less than 1 to greater than 1, and it would be more representative to plot $\log \text{Pe}'$ versus $\log \text{Pe}$, rather than versus $\log \text{Re}_p$.

It is noted that the maximum in $\text{Pe}'_{ct} = \overline{u}R/D_{ct}$ is predicted also by the Taylor–Aris formula for dispersion in a straight cylindrical capillary tube. Equation (6.2.7) may be rewritten as follows:

$$1/\text{Pe}'_{ct} = 1/\text{Pe}_{ct} + \text{Pe}_{ct}/192, \qquad (6.3.14)$$

which has a minimum at $\text{Pe}_{ct} = (192)^{1/2}$.

The following interpolation formula was suggested (de Ligny, 1970) for the longitudinal dispersion coefficient[2]:

$$D_L = \gamma\mathscr{D} + \frac{\lambda \overline{D}_p}{1 + C\mathscr{D}/\overline{D}_p v_{DF}} v_{DF} \qquad (6.3.15a)$$

---

[2]A better fit to the data plotted in Fig. 6.16 is given by the function of the following form:

$$D_L = \gamma\mathscr{D} + \frac{\lambda \overline{D}_p v_{DF}}{1 + C\left(\mathscr{D}/\overline{D}_p v_{DF}\right)^{1/2}} \qquad (6.3.15d)$$

proposed by Hiby (1962).

or

$$\frac{1}{Pe'} = \frac{\gamma}{Re_p Sc} + \frac{\lambda}{1 + (C/Re_p Sc)}, \qquad (6.3.15b)$$

where γ, λ, and C are dimensionless coefficients, depending on the geometry of the column packing and the dynamics of flow. Equation (6.2.15b) has a minimum at $Re_p = (C/Sc)[(\lambda C/\gamma)^{1/2} - 1]^{-1}$.

In order to compare Eq. (6.3.15a) with Eq. (6.2.7), it is written in the following form:

$$\frac{D_L}{\mathscr{D}} = \gamma + \lambda \frac{Pe}{1 + C/Pe}, \qquad (6.3.15c)$$

from which

$$d \log(D_L/\mathscr{D})/d \log Pe = \lambda[Pe(1 + C/Pe) + C]/[\gamma(1 + C/Pe)]^2$$
$$+ \lambda[Pe(1 + C/Pe)]. \qquad (6.3.16)$$

This equation predicts a slope of unity for large Pe and a maximum slope of about 1.2 to 1.4 at Pe in the range of 4 to 11, which is at variance with the experimental curves shown in Figs. 6.16 to 6.18. In Fig. 6.16 the slope of the experimental curve starts to increase rapidly from zero at Pe $\approx$ 1, reaching a maximum value of 1.2 somewhere between Pe = 100 and 1000, whereafter it gradually decreases again to values somewhat less than one.

De Ligny (1970) generalized Eq. (6.3.15) to both longitudinal and transverse (or radial) dispersion. The form of the expression remains the same in both cases; only λ is predicted to be different (smaller) in the case of transverse dispersion. Experiments, however, have shown that the value of C is an order of magnitude greater in the case of transverse than in the case of longitudinal dispersion. The values of λ and C reported by de Ligny for different packings are shown in Table 6.2 ($\gamma \approx 0.7$). No claim was made to

TABLE 6.2   Experimental Values of λ_L, λ_R, and C^a

Coefficients of Eq. (6.3.15)	Liquid Fluid		Gaseous fluid	
	Spherical packing	Irregular packing	Spherical packing	Irregular packing
λ_L (longitudinal dispersion)	2.5	2.5	0.7	4
λ_R (radial dispersion)	0.08	0.08	0.12	0.12
C (longitudinal dispersion)	8.8	7.7	5.8	5.1
C (radial dispersion)			78 ± 20	

[a]Reprinted with permission from de Ligny, 1970.

the existence of a universal relationship between $D_L/\mathscr{D}$ and Pe, but for a given packing $D_L/\mathscr{D}$ depends on no parameter other than Pe.

6.3.4.3. D_L/ν *versus* Re *Correlations*

The correlations of Harleman *et al.* (1963) are of the following form:

$$D_L/\nu = c_1 \, \text{Re}_{50}^{1.2} \tag{6.3.17a}$$

and

$$D_L/\nu = c_2 \, \text{Re}_k^{1.2}, \tag{6.3.18}$$

where c_1 is equal to 0.9 and 0.66 for sand and for spheres, respectively; c_2 is equal to 83 and 54 for sand and for spheres, respectively;

$$\text{Re}_{50} = v_{\text{DF}} D_{\text{p}50}/\nu \tag{6.3.19}$$

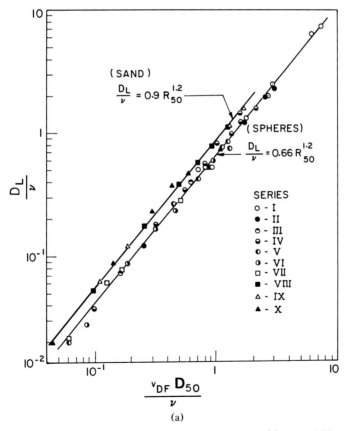

(a)

Figure 6.21.   (a) Dispersion versus particle size correlation and (b) permeability correlation. (Reprinted with permission from Harleman *et al.*, 1963.)

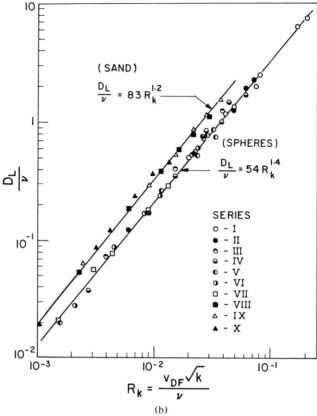

$$R_k = \frac{v_{DF}\sqrt{k}}{\nu}$$

(b)

Figure 6.21. (*Continued*)

and

$$\mathrm{Re}_k = v_{DF}\sqrt{k}\,/\nu, \tag{6.3.20}$$

with D_{p50} the median grain diameter by weight. These correlations are shown in Figs. 6.21a and b, respectively.

Equation (6.3.17a) may be written as follows:

$$D_L/\mathscr{D} = c_1 \mathrm{Pe}\, \mathrm{Re}_{50}^{0.2}, \tag{6.3.17b}$$

indicating that according to this correlation $D_L/\mathscr{D}$ is a function of both Pe and Re.

The relationship between D_L and the permeability k, as expressed by Eq. (6.3.18) is noteworthy. Harleman *et al.* found it by assuming a relationship between k and $\overline{D}_p$, such as expressed by Eq. (3.2.4). The discussions of Section 3.2.1 have shown, however, that such a relationship exists only for constant porosity, the grain size distribution, and particle shape.

6.3.5. Equations of Change of Hydrodynamic Dispersion

6.3.5.1. *Pure Dispersion, General*

Applying the customary definitions (e.g., de Groot and Mazur, 1969; Bird *et al.*, 1960) for the transport of mass of species i in a mixture ($i = 1, 2, \ldots, n$) for the case of transport in a homogeneous porous medium, one can write for the mass flux of i, $\mathbf{n}_i$, with respect to the solid matrix of the medium the following expression:

$$\mathbf{n}_i = c_i \mathbf{v}^v + \mathbf{j}_i^v, \qquad (6.3.21)$$

where c_i is concentration of i in units of mass of i per unit volume of solution; $\mathbf{v}^v$ is the volume average velocity of the mixture with respect to the solid matrix, based on the average cross section of the medium available to flow (i.e., $A\phi$, where A is perpendicular to the direction of $\mathbf{v}^v$; and $\mathbf{j}_i^v$ is the mass flux of i with respect to $\mathbf{v}^v$. Generally $\mathbf{j}_i^v$ is not zero, except in the special case of mixtures of uniform composition, because of the effects of various mixing processes. By definition, $\mathbf{v}^v$ is equal to the "average" pore or interstitial velocity defined by the Dupuit–Forchheimer assumption (i.e., $\mathbf{v}^v \equiv \mathbf{v}_{DF}$). It should be borne in mind that both the direction and the magnitude of the pore velocities are in general different from $\mathbf{v}_{DF}$. By generalizing the definition used in the case of pure molecular diffusion, one can write (e.g., Bear, 1972)

$$\mathbf{j}_i^v = -\overline{\overline{D}} \cdot \nabla c_i, \qquad (6.3.22)$$

where $\overline{\overline{D}}$ is the coefficient of (hydrodynamic) dispersion, a second-rank tensor (with the unusual property that it has only diagonal nonzero elements, regardless of the orientation of the coordinate system relative to a statistically isotropic medium), which combines the mixing effects due to molecular diffusion and convection. Equation (6.3.22) has only formal analogy with diffusion because the driving force of hydrodynamic dispersion is *not* ∇c_i. It is the *result* of different velocities in various stream tubes coupled with transverse molecular diffusion. Combining Eqs. (6.3.21) and (6.3.22) results in

$$\mathbf{n}_i = c_i \mathbf{v}_{DF} - \overline{\overline{D}} \cdot \nabla c_i. \qquad (6.3.23)$$

The equation of continuity for species i is

$$-\nabla \cdot \mathbf{n}_i = \partial c_i / \partial t. \qquad (6.3.24)$$

Equations (6.3.23) and (6.3.24) give

$$\partial c/\partial t = \nabla \cdot \left(D \cdot \overline{\overline{\nabla}} c \right) - \nabla \cdot \left(c \mathbf{v}_{\mathrm{DF}} \right), \tag{6.3.25a}$$

where the subscript i has been dropped.

If ρ is assumed to be constant,

$$\partial c/\partial t = \nabla \cdot \left(\overline{\overline{D}} \cdot \nabla c \right) = \mathbf{v}_{\mathrm{DF}} \cdot \nabla c. \tag{6.3.25b}$$

In moving coordinates, that is,

$$x' = x - v_{\mathrm{DF},x} t, \qquad y' = y - v_{\mathrm{DF},y} t, \qquad z' = z - v_{\mathrm{DF},z} t,$$

Eq. (6.3.25b) becomes

$$\partial c/\partial t = V' \cdot \left(\overline{\overline{D}} \cdot \nabla' c \right). \tag{6.3.25c}$$

Equation (6.3.25) has been solved for several different initial and boundary conditions (see, e.g., Bear, 1972). A few examples are given below.

6.3.5.2. *One-Dimensional Dispersion*

The simplest case is one-dimensional flow in the positive x direction at constant v_{DF}. In this case, Eq. (7.3.25) simplifies to

$$\partial c/\partial t = \left(D_{\mathrm{L}} \, \partial^2 c/\partial x^2 \right) - \left(v_{\mathrm{DF}} \, \partial c/\partial x \right), \tag{6.3.26}$$

where D_{L} is constant because both v_{DF} and $\mathscr{D}$ have been assumed to be constant. Solution of Eq. (6.3.26) (e.g., Collins, 1961) with the initial conditions

$$\text{for } t \leq 0, \qquad \begin{array}{ll} -\infty < x < 0, & c = c_0, \\ 0 \leq x < +\infty, & c = 0, \end{array} \tag{6.3.27}$$

and the boundary conditions

$$\text{for } t > 0, \qquad \begin{array}{ll} x = \pm\infty, & \partial c/\partial x = 0, \\ x = \pm\infty, & c = 0, \\ x = -\infty, & c = c_0, \end{array} \tag{6.3.28}$$

is as follows [see Eq. (6.2.8) for comparison]:

$$c(x,t) = c_0 \left[\frac{1}{2} \pm \frac{1}{2} \, \mathrm{erf}\left(\frac{x - v_{\mathrm{DF}} t}{2(D_{\mathrm{L}} t)^{1/2}} \right) \right], \qquad \begin{array}{l} + \text{for } x - v_{\mathrm{DF}} t < 0. \\ - \text{for } x - v_{\mathrm{DF}} t > 0. \end{array} \tag{6.3.29}$$

Equation (6.2.9) is used to calculate D_L also in the case of laboratory hydrodynamic dispersion tests in porous media.

A few comments regarding the customary laboratory techniques used to determine D_L experimentally are in order. In laboratory experiments, usually a cylindrical pack or core of the porous sample is used. The tracer fluid is injected at one end and samples of the effluent fluid emerging at the other end of the cylinder are analyzed to determine their concentration. The resulting curve is the so-called "breakthrough" curve or "effluent profile" of the tracer. It is evident, then, that under these conditions, $x = L$, the length of the sample. As Eq. (6.2.9) is based on a concentration profile $c(x)$ at a fixed time t, whereas the laboratory experiment supplies a concentration profile $c(t)$ at a fixed point $x = L$, the appropriate expression that applies under the latter conditions is Eq. (6.2.11).

Another comment concerns the boundary conditions. In the customary laboratory experiments, the sample begins at $x = 0$ and, therefore, the boundary condition $t \le 0$, $x = -\infty$, $c = c_0$ is not accurate because for $x < 0$ the system is different (inlet header and lines) from the sample under investigation. As the velocity in the access lines is usually much higher than in the sample, the boundary condition $t > 0$, $x = 0$, $c = c_0$ probably approximates reality better. Gupta (1972) obtained the following analytical solution of Eq. (6.3.26) with this departure from the conditions expressed in Eqs.

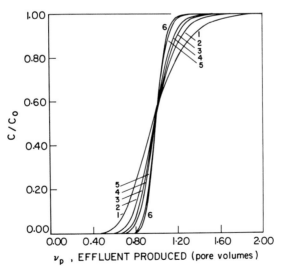

Figure 6.22. Breakthrough curves for various values of Pe'_L: curve 1, $Pe'_L = 30$; curve 2, $Pe'_L = 60$; curve 3, $Pe'_L = 90$; curve 4, $Pe'_L = 120$; curve 5, $Pe'_L = 240$; curve 6, $Pe'_L = 300$ (Gupta, 1972).

(6.3.27) and (6.3.28)[3]:

$$\frac{c}{c_0} = \frac{1}{2} \operatorname{erf}\left[\frac{\sqrt{\mathrm{Pe}'_L}}{2} \frac{1 - \nu_p}{\nu_p^{1/2}}\right] + \frac{1}{2}(\exp \mathrm{Pe}'_L) \operatorname{erf}\left[\frac{(\mathrm{Pe}'_L)^{1/2}}{2} \frac{1 + \nu_p}{\nu_p^{1/2}}\right].$$

$$(6.3.30)$$

The breakthrough curves calculated by Gupta for various, different values of the Peclet number, using Eq. (6.3.30), are shown in Fig. 6.22.

Brenner (1962) solved Eq. (6.3.26) analytically for beds of finite length. His calculations apply to the case in which the solute, uniformly distributed throughout the medium at the outset, is displaced by a solvent that may itself contain some solute. Numerical results were given in dimensionless form for (1) the instantaneous concentration at the exit end of the bed and (2) the average solute concentration in the bed at any instant. A good list of references on related work was given by Brenner (1962).

6.3.5.3. *One-Dimensional Dispersion with Adsorption or Reaction*

The form of Eqs. (6.3.25) for the spatial case of uniform flow in a plane is as follows:

$$\partial c/\partial t = \left(D_L \, \partial^2 c/\partial x^2\right) - \left(v_{\mathrm{DF}} \, \partial c/\partial x\right) + r(c, t), \qquad (6.3.31)$$

where $r(c, t)$ is the rate of production of solute per unit volume of solution. The reaction may be homogeneous reaction, adsorption–desorption type of reaction, or heterogeneous reaction (i.e., reaction with the medium). Gupta (1972) (see also Gupta and Greenkorn, 1973, 1974) has written an excellent review on this subject with a large number of literature references. Covering a wide range of values of the relevant parameters, Gupta performed numerical calculations and published c/c_0 versus injected pore volumes diagrams (breakthrough curves) for the following cases: (1) first-order irreversible adsorption or reaction; (2) first-order reversible adsorption; (3) second-order reversible adsorption; (4) second-order irreversible adsorption; and (5) reaction with the medium (having a change of porosity and pore structure as a result). Space limitations prohibit a detailed discussion of this very interesting and important subject.

6.3.5.4. *Two-Dimensional Dispersion*

The form of Eqs. (6.3.25) for the special case of uniform flow in a plane is as follows:

$$\partial c/\partial t = \left(D_L \, \partial^2 c/\partial x^2\right) + \left(D_T \, \partial^2 c/\partial y^2\right) - \left(v_{\mathrm{DF}} \, \partial c/\partial x\right), \quad (6.3.32)$$

where the flow v_{DF} is in the x direction. The "principal" values of the

$$[3] \mathrm{Pe}'_L = \frac{v_p L}{D_L}$$

dispersion coefficient "tensor" are the longitudinal (D_L) and the transverse (D_T) dispersion coefficients.

Harleman and Rumer (1973) considered steady uniform flow with pore velocity v_{DF} in the $+x$ direction through a porous medium contained between impermeable boundaries at $y = \pm\infty$ with the following initial and boundary conditions:

$$c(0, y) = c_0 \quad \text{for } -\infty < y \leq 0,$$

$$c(0, y) = c \quad \text{for } 0 < y < +\infty, \tag{6.3.33}$$

$$\partial c/\partial y = 0 \quad \text{for } y = \pm\infty \quad \text{for all} \quad x.$$

Assuming steady state and neglecting the first term on the r.h.s. of Eq. (6.3.32), this reduces to

$$v_{DF} \, \partial c/\partial x = D_T \, \partial^2 c/\partial y^2, \tag{6.3.34}$$

whose solution is

$$c/c_0 = \frac{1}{2} \, \text{erf}\left[\frac{y}{2}(D_T x/v_{DF})^{1/2}\right]. \tag{6.3.35}$$

Treatment of two-dimensional dispersion taking coupling of longitudinal and transverse dispersion into account would be of some interest.

6.3.6. Theories of Hydrodynamic Dispersion in Porous Media

6.3.6.1. *Introduction*

A large number of theories have been proposed to explain hydrodynamic dispersion in porous media. With a few exceptions, all of them use probability theory applied to a large number of systems and/or steps and, therefore, they may be classified as "statistical" theories. As there has been some disagreement in the literature as to what constitutes a proper statistical theory of dispersion, all of those theories will be discussed first that are definitely not of statistical character.

The Taylor–Aris treatment of dispersion in a single tube belongs in this category, as do the capillaric models of von Rosenberg (1956) and Aris (1956) in which the porous medium was envisaged as a body consisting of straight, parallel capillaries of uniform size and equal length (Scheidegger, 1974). According to these theories there is only longitudinal (or axial) dispersion.

The other, definitely not statistical, theory is the heuristic approach used by Aronofsky and Heller (1957). These authors postulated the validity of the unsteady-state diffusion equation [Eq. (6.3.26)] to describe dispersion in porous media regardless of the mechanism by which mixing occurs. They analyzed, with the help of the diffusion equation, other authors' data obtained with no viscosity or density contrasts, and found good fit to the data.

Aronofsky and Heller pointed out that the observation of Koch and Slobod (1957), according to which there was no appreciable change of the break-through profiles with flow rate, is tantamount to the conclusion that D_L varied linearly with flow rate [cf. Eq. (6.3.3), with $v_{DF}T = L$, the length of the sample]. Von Rosenberg's data (1956), however, display an increase of D_L somewhat faster than linear with flow rate.

The treatment of the same problem by Offeringa and van der Poel (1954) shows some similarity to the Aronofsky–Heller approach.

Blackwell (1959) and Blackwell *et al.* (1959) explained qualitatively the mechanism of dispersion in porous media by pointing out the existence of different velocities along the different streamlines, the role played by molecular diffusion being to transport a fluid particle from one streamline to another. He was the first to present experimental data in the form $D_L/\mathscr{D}$ versus Pe.

As explained above, the other treatments of hydrodynamic dispersion in porous media known to the author all apply probability theory to large numbers of systems and/or steps and, therefore, they are classified here as "statistical" theories.

6.3.6.2. *Statistical Theories of Hydrodynamic Dispersion in Porous Media*

The various approaches used by different authors, while having great differences in the details of the calculations, have also some main features in common. They have been divided here into three broad categories:

(1) random phenomena-type (stochastic) models;
(2) network models of capillary tubes with randomly distributed diameters; and
(3) statistical models based on spatial (or volume) averaging.

Random Phenomena-type (Stochastic) Models. (a) Cell, or perfect mixer, or random-residence-time model. The authors who have analyzed this approach include McHenry (1958), Aris and Amundson (1957), Klinkenberg and Sjenitzer (1956), and Bear (1960), whose presentation is sketched in the following.

This model consists of a one-dimensional array of small cells of equal volume V, with interconnecting channels of negligible volume. Each cell is assumed to be a so-called perfect mixer, tacitly modeling the mixing effect of molecular diffusion.

A tracer balance over the jth cell:

$$\frac{\partial c_j^{(0)}(t)}{\partial t} + \frac{c_j^{(0)}(t)}{\tau} = \frac{c_j^{(i)}(t)}{\tau}, \qquad (6.3.36)$$

where $c^{(i)}$ is the tracer concentration in the liquid entering the cell, $c^{(0)}$ the tracer concentration leaving the cell, and $\tau = V/Q$ the "residence time" of the fluid in the cell (Q is the volume flow rate through the array).

For an instantaneous inflow that causes an initial concentration of c_0 in the first cell, it can be shown that $c_j^{(0)}(t)$ is the following function:

$$c_j^{(0)}(t) = c_0(t/\tau)^{j-1}[\exp(-t/\tau)]\big/(j-1)!. \qquad (6.3.37)$$

This is a Poisson probability distribution $P(N) = \lambda^N \exp(-\lambda)/N!$ with $N = j - 1$ and $\lambda = t/\tau$.

It is noted that, owing to the assumption of perfect mixing, the probability of a tracer molecule leaving a cell in any time interval is independent of the length of time it has been in the cell. This is the reason why this model is also called the "random-residence-time" model. It follows, unrealistically, that if the cells are immediately next to each other, a tracer molecule has a certain finite probability to appear in the last cell of the array ($N_{\max}$) at any time $t > 0$.

If we consider translation between the cells, with $\sim j\Delta t$ the time of translation between j cells, so that of the total time t only $t - j\Delta t$ is left for mixing and $j\Delta t \approx N\Delta t$, we obtain the more realistic result:

$$c_j^{(0)}(t) = \begin{bmatrix} c_0\dfrac{[(t - N\Delta t)/\tau]\exp\{-(t - N\Delta t)/\tau\}}{N!} & \text{for} \quad t > N\Delta t \\[1em] 0 & \text{for} \quad t < N\Delta t. \end{bmatrix}$$
$$(6.3.38)$$

By the central limit theorem of statistics, as N becomes large, Eq. (6.3.38) can be approximated by the normal distribution

$$c(x,t)/c_0 = (4\pi D_L t)^{-1/2}\exp\left[-(x - \bar{x})^2/4D_L t\right], \qquad (6.3.39)$$

where $\bar{x} = v_p t$ [cf. Eq. (6.2.5)]. One notes that according to this model, D_L is proportional to the average interstitial velocity v_p, which is usually assumed to be equal to v_{DF}. This model fails to yield the transverse dispersion observed in porous media, and it does not incorporate the effects peculiar to a pore network and different velocities.

Gunn and Pryce (1969) measured the axial dispersion of argon in air in two regular arrangements of particles, a simple cubic and a dense-packed rhombohedral form. The dispersion could not be described either as a diffusional process or as a mixing cell process. The authors have concluded that the fluid in the voids was not well mixed.

The flow channels in regular packs resemble a bundle of identical periodically constricted tubes (see Section 3.3.5.1). The work of Azzam (1975) and Azzam and Dullien (1977) (see also Section 6.2.4) shows that in the large segments of such tubes, the flow was either very slow or the fluid was circulating. Hydrodynamic dispersion in regular packs can be expected to be described by equations of the type of Eq. (6.2.22).

(b) Statistical theories using the continuum approach.   A number of authors have treated the problem of dispersion by making certain assumptions of a general nature, without resorting to any microscopic pore structure parameters in the model. All the same, these assumptions were made in the qualitative knowledge of the role played by the pore network and were actually based on this knowledge.

One of the early treatments is due to Scheidegger (1954) who used the random walk approach. The porous medium was considered isotropic and statistically homogeneous. A constant macroscopic pressure gradient, acting in the x direction, was assumed. Darcy's law was assumed to determine the average displacement of the fluid.

Scheidegger (1954) considered a marked particle whose coordinates at $t = 0$ are $x = 0$, $y = 0$, $z = 0$.

The time interval 0 to t was divided into N equal intervals T such that

$$NT = t. \tag{6.3.40}$$

In every interval T, the marked particle under consideration (tracer) would undergo a displacement. The departure from the average displacement of the fluid during any T was viewed as a random process. The probabilities for steps in all directions were assumed equal and constant for every time interval T.

It is noted that any random walk model of hydrodynamic dispersion includes, at least implicitly, the effect of molecular diffusion because without it the tracer particles would advance with different velocities along the various streamlines in a deterministic way and the concentration profile would not be a solution of the diffusion equation.

It was pointed out later by Josselin de Jong (1958) and Saffman (1959, 1960) that under the assumption of a constant macroscopic pressure gradient, acting in one direction, the probabilities for steps are not the same in all directions.

Because of the isotropic probability assumption, after many steps (large N), the probability $p \, dx$ for the marked particle under consideration (tracer) to be between the coordinates $x \rightarrow x + dx$ (at time t) will be

$$p(x,t) \, dx = (4\pi D_L t)^{-1/2} \exp\{-(x - \bar{x})^2/4D_L t\} \, dx, \tag{6.3.41}$$

where the standard deviation σ was replaced as follows:

$$\sigma^2 = 2D_L T. \tag{6.3.42}$$

According to Darcy's law:

$$\bar{x} = (tk/\phi\mu)(\partial\mathcal{P}/\partial x) = v_{\mathrm{DF}}t, \qquad (6.3.43)$$

whereas $\bar{y}$ and $\bar{z}$ are zero.

Using Eq. (6.3.43) to eliminate t from Eq. (6.3.41), it can be seen that $p(x, t)$ is a function of x and $\bar{x}$ only if (Collins, 1961)

$$D_{\mathrm{L}} = a_{\mathrm{L}}v_{\mathrm{DF}}, \qquad (6.3.44)$$

where a_{L} is a constant of the model of the porous medium, called "longitudinal geometric dispersivity."

A statistical treatment of purely longitudinal dispersion, based on the continuum picture, is due to Beran (1957). Beran, in the first part of his treatment, assumed the absence of molecular diffusion and introduced the function $p(u, t/u_0, t_0)\,du$, defined as the probability that a molecule has a velocity u at time t if it had a velocity u_0 at time t_0. Assuming that

(1) $p(u, t/u_0, t_0)$ is dependent only on $t - t_0$ and not on t and t_0 individually.

(2) $p(u, t/u_0, t_0)\,du = p(u)\,du$ for $t - t_0 \geq \tau$; in other words, after a short time the molecule's velocity at time t is essentially independent of its velocity at time t_0, owing to the irregularity of the grain structure. (This assumption, however, is not satisfied in creeping flow because if a molecule is next to the pore wall at t_0, it has practically zero velocity and there is no other mechanism, except molecular diffusion, whereby this molecule can acquire a higher velocity; hence Beran's model tacitly assumes molecular diffusion.)

(3) $\overline{u^3} \equiv \int_{-\infty}^{\infty} u^3 p(u)\,du < \infty$, then the probability distribution of the molecule as a function of x (the distance along the tube axis) can be predicted for large t.

The central limit theorem of probability theory states that if one seeks the probability distribution of the variable $x = \int_0^t u(t)\,dt$, where $u(t)$ has associated with it a probability function $p(u, t/u_0, t_0)$ as described in the foregoing, then $p(x, t)\,dx$, the probability that the variable will be at a position between x and $x + dx$ at time t is given by Eq. (6.3.41), where $v_{\mathrm{p}} = \bar{x}/t$ is the "average interstitial velocity" (Hoeffding and Robbins, 1948), defined by

$$v_{\mathrm{p}} = \int_{-\infty}^{\infty} up(u)\,du. \qquad (6.3.45)$$

Invoking the law of large numbers, there follows from Eq. (6.3.41) that

$$c(x, t)/c_0 = (4\pi D_{\mathrm{L}}t)^{-1/2} \exp\left[-(x - v_{\mathrm{p}}t)^2/4D_{\mathrm{L}}t\right]. \qquad (6.3.46)$$

Beran (1957) presented an argument, limited to creeping flow, where the inertia term in the Navier–Stokes equation may be neglected (the region of validity of Darcy's law in the strictest sense) and to the case when a negligible role is played by molecular diffusion, which led to the following result:

$$D_L = v_p \bar{D}_p c_1, \qquad (\mathrm{Pe} \gg 1), \qquad (6.3.47a)$$

where c_1 is a constant, implying that the dynamic Peclet number Pe' is constant, which is approximately true over a fairly wide range of Re (or Pe) (Pfannkuch, 1963).

In the latter part of his treatment, Beran has considered the role played by molecular diffusion. He pointed out, correctly, that molecular diffusion has two effects. First, it has the dispersing effect it would have if the solvent remained stationary; and second, it has the effect of changing a tracer molecule's velocity by moving it to a different streamline. The former effect can be separated from the latter, and due to this effect alone, the soluble matter would be normally distributed with zero mean and standard deviation equal to $(2\mathscr{D}t)^{1/2}$.

The latter effect has the result that a molecule may change its velocity because it may change its streamline, in addition to changing velocity along a given streamline, but the overall effect remains the same; only the standard deviation is different.

The sum of two independent normal distributions is itself normal; and, therefore, the only difference caused by molecular diffusion is in the different value of D_L. According to Beran, this difference will show up only for $\mathrm{Pe} \ll 1$. For this case, Beran used arguments parallel to those leading to Eq. (6.3.47a). In this case he obtained

$$D_L = \left(v_p^2 \bar{D}_p^2 / \mathscr{D} \right) c_2 + \gamma \mathscr{D}, \qquad (\mathrm{Pe} \ll 1). \qquad (6.3.47b)$$

This result should be compared with Eq. (6.2.7), which applies for dispersion in a single capillary tube at $\tau = t\mathscr{D}/R^2 \geq 0.8$ for $\mathrm{Pe}_{ct} > 1$ (see Fig. 6.4).

Beran suggested harmonic addition of Eqs. (6.3.47a) and (6.3.47b) for the intermediate case [i.e., $\mathrm{Pe} = 0(1)$]. The resulting expression for D_L, however, goes to zero in the limit $\mathrm{Pe} \to 0$. Therefore, only the first term on the right of Eq. (6.3.47b) is added harmonically to Eq. (6.3.47a), resulting in

$$D_L = \mathscr{D} \frac{c_1 c_2 \, \mathrm{Pe}^2}{c_2 \, \mathrm{Pe} + c_1} = \frac{c_1 \bar{D}_p}{1 + (c_1/c_2)\mathscr{D}/\bar{D}_p \bar{v}_p} \bar{v}_p \qquad (6.3.47c)$$

and the limiting value for $\lim \mathrm{Pe} \to 0$, that is, $\gamma \mathscr{D}$ is added to this expression. The final form of the interpolation formula for D_L is thus the same as Eq. (6.3.15a). Beran's treatment does not properly represent the role played by molecular diffusion in hydrodynamic dispersion in porous media.

Giddings (1957) used random walk considerations to obtain the same result. He argued that dispersion is due to velocity fluctuations of a molecule, brought about by two phenomena: (1) The velocity along a streamline will change randomly at distances of the order of $\overline{D}_p$. (2) The velocity can change also because the molecule is able to diffuse to another streamline.

Assuming that the two processes constitute a random walk phenomenon, the "diffusion coefficient" was defined as

$$D_L = \left(\ell^2 n/2\right)\left(v_p^2/2n\right), \tag{6.3.48}$$

where $\ell = v_p/n$, with $n = n_1 + n_2$ the number of steps taken in unit time.
Letting

$$n_1 = v_p/2\xi\overline{D}_p \tag{6.3.49}$$

and

$$n_2 = 2\mathscr{D}/\beta^2\overline{D}_p^2, \tag{6.3.50}$$

where ξ and β are constants of the order of unity, and combining Eqs. (6.3.47)–(6.3.49) results in an expression that is equivalent to Eq. (6.3.15a).

Hiby (1962) and, later, Gunn (1969) took into consideration the tortuosity of flow channels and used, in dispersion correlations,

$$\mathrm{Pe}\,\Upsilon = v_{\mathrm{DF}}\Upsilon\overline{D}_p/\mathscr{D} \tag{6.3.51}$$

instead of Pe.

Gunn (1969) developed a statistical model of dispersion in packed beds, which bears imprints of the experiences gained in the experiments of Gunn and Pryce (1969) with regular packings in which rather poor mixing was found (Section 6.3.6.2). He considered a tracer particle at the entrance of a "flow cell" and introduced a constant probability p of the particle traveling a distance x_p in the characteristic time τ—the alternative being that the particle remained at rest. (In reality, the alternative is that the particle will travel much slower.) The picture behind this idea is a simplification of the flow pattern shown in Fig. 6.10a. If the particle is on a streamline that is close to the axis of the tube, in the bulge ("flow cell") it will travel much faster than if it is closer to the walls. Simply, p is the fraction of the total cross-sectional area of the flow cell that is occupied by the fast-flowing core.

Gunn worked out the schematic mechanics of the interchange of particles by molecular diffusion between the two (fast and slow) regions (cf. discussion in Section 6.2.4), leading to formulas that give the probabilities of finding the tracer particle in the slow and in the fast streams. The final results are as follows:

$$\frac{1}{\text{Pe}'} = \frac{\text{Pe}}{\phi\Gamma}(1-p)^2 + \frac{\text{Pe}^2}{\phi^2\Gamma^2}p(1-p)^2$$

$$\times \left\{ \exp\left[\frac{-\phi\Gamma}{p(1-p)\,\text{Pe}}\right] - 1 \right\} + \frac{\phi}{\text{Pe}\,\Upsilon}, \qquad (6.3.52)$$

where

$$\Gamma = 4(1-\phi)\alpha_1^2/\phi, \qquad (6.3.53)$$

with α_1 a root of the zero-order Bessel function. By expansion of the exponential term in Eq. (6.3.52) for large Pe, one obtains

$$1/\text{Pe}' = [(1-p)/2p] + 0(\text{Pe})^{-1} \approx (1-p)/2p, \qquad (6.3.54)$$

whereas for small Pe,

$$1/\text{Pe}' = (\phi/\text{Pe}\,\Upsilon) + 0(\text{Pe}). \qquad (6.3.55)$$

Experimental results measured in the liquid phase by a number of workers were used to obtain the relationship between p and Re_p (note that $\text{Pe} = \text{Re}_p\,\text{Sc}$) shown in Fig. 6.23. With the help of this relationship, Eq. (6.3.52) was used to predict the effect of Sc at low Re_p (creeping flow) (see Fig. 6.24).

Comparison with experimental data is not conclusive, particularly if it is considered that there is a mass of data that has not been considered in this comparison (see Fig. 6.16). Note that letting $\text{Sc} = 10^3$, we have $\text{Pe} = 10^3\,\text{Re}_p$; in Fig. 6.24 the data start at $\text{Pe} \approx 10$, and the lines come together at $\text{Pe} \approx 4 \times 10^{-2}$. In Fig. 6.17, however, the data extend down to $\text{Pe} \approx 10^{-2}$, which corresponds to $\text{Re}_p \approx 10^{-5}$.

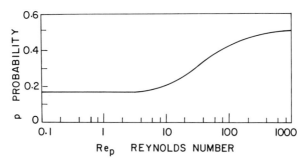

Figure 6.23. Variation of the fluid-mechanical probability p with Reynolds number. (Reprinted with permission from Gunn, 1969.)

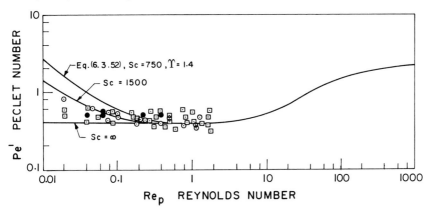

Figure 6.24. Comparison of Eq. (6.3.52) for different values of the Schmidt number: Sc > 2000 □, Ebach and White, 1958); Sc > 730 (⊙; Miller and King, 1966); Sc ≥ 820 (●; Jacques and Vermeulen, 1957). (Reprinted with permission from Gunn, 1969.)

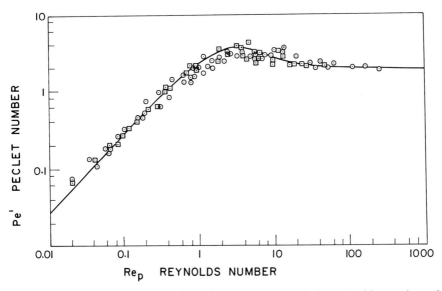

Figure 6.25. Comparison of Eq. (6.3.52), for Sc = 0.77 and T = 1.4, with experimental measurements of dispersion in gases: ⊙, Gunn and Pryce (1969); □ , Edwards and Richardson (1968). (Reprinted with permission from Gunn, 1969.)

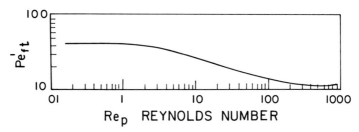

Figure 6.26.   Variation of Pe'_{ft} with Re_p. (Reprinted with permission from Gunn, 1969.)

Gunn (1969) has compared the predictions of Eq. (6.3.52) with experimental data in the case of gases, as shown in Fig. 6.25, using the relationship between p and Re_p shown in Fig. 6.24. In this case the agreement is good. The curve in Fig. 6.25 covers the range of Pe (see Fig. 6.17) from about 10^{-2} to 2×10^2.

For transverse dispersion Gunn (1969) has intuitively recommended the relation

$$1/Pe'_t = (1/Pe'_{ft}) + (\phi/Pe\,\Upsilon), \tag{6.3.56}$$

where Pe'_{ft} is defined as the Peclet number that gives the radial dispersion due to the convective mode alone. It was determined under conditions of $Pe\,\Upsilon \gg 1$ from experimental transverse dispersion data with liquids (see Fig. 6.26). Good agreement with a limited number of experimental data was found (Fig. 6.27), but in this case $\Upsilon = 1.25$ (or 1.29) had to be assumed for a good fit, as compared with the value 1.4 used in the case of longitudinal dispersion.

In a later paper, Gunn (1971) took into consideration the fact that the average interstitial velocity in different flow cells over the cross section of the sample is not, in general, the same value because of the existence of nonuniformities in the network of flow channels. In packed beds of the same kind, ideally speaking, these nonuniformities should always be the same; but in reality they depend on the technique used to pack the bed and, therefore, they may change when the same bed is repacked. As a result of these considerations, Gunn introduced a parameter as a measure of such nonuniformities. For a perfectly well-packed bed, this parameter is supposed to be equal to zero and its value increases with increasing degree of nonuniformity. It must be pointed out, however, that a certain degree of nonuniformity exists also in a perfectly randomized bed: Only in a regularly packed bed is the superficial velocity the same in every flow cell (see Section 2.5.2).

Unfortunately, both the parameter measuring nonuniformities and the probability p must be derived from experimental dispersion data, and their values are by no means unique.

Gunn's approach is interesting because it emphasizes the important role played by transverse molecular diffusion in hydrodynamic dispersion.

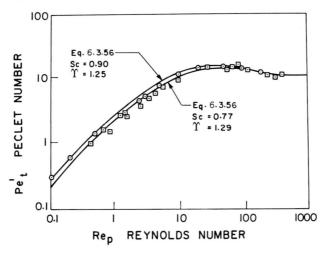

Figure 6.27.   Comparison of Eq. (6.3.56) with experimental measurements of radial dispersion in gas flow through random beds of spheres: ⊙, Gunn and Pryce (1969); ⊡ , Hiby (1962). (Adapted with permission from Gunn, 1969.)

Todorovic (1970) developed a model of longitudinal dispersion in porous media, assuming an isotropic homogeneous pore structure and steady fluid flow. Based on a stochastic model in which dispersion is reduced to a study of the random motion of a point along a straight line, Todorovic developed both the Lagrangian description (i.e., in which the distribution in space is observed at a given time) and the Eulerian description (i.e., in which the distribution is observed at a given distance as a function of time). He showed that the sum of the two distributions is always equal to one. Todorovic found that neither of the two distributions was Gaussian. His results and conclusions deserve a more detailed study.

An important problem is dispersion in so-called aggregated media of "dual porosity" in which the aggregates contain micropores and between the aggregates there are macropores or voids. Viscous flow takes place effectively only in the voids, whereas in the aggregates there is only diffusion. Passioura (1971) derived an expression for the longitudinal dispersion coefficient, which for large mean velocities of flow depends on the squares of the velocity and the average diameter of the aggregates. Pure diffusion (i.e., without viscous flow) in catalyst pellets consisting of porous catalyst powder particles (aggregates) was analyzed by Wakao and Smith (1962).

(c) Statistical theories utilizing microscopic pore structure parameters. Josselin de Jong (1958) and Saffman (1959, 1960), independently of each other, tackled the problem of dispersion by analyzing random walk in an assemblage of randomly oriented, straight, uniform, cylindrical, capillary tubes. While there is a great deal of similarity between the two approaches,

only Saffman's work is reviewed here because Saffman's analysis, of which the work of Josselin de Jong may be regarded as a special case, has greater generality.

Saffman considered a model in which a randomly distributed set of points were connected to their neighbors by straight uniform pores. The path of a fluid particle was regarded as a random walk in this "network" of pores in which the direction and duration of each step are random variables. The statistical properties of the displacement of a single particle undergoing random walk was used to calculate the dispersion of a cloud of particles that are represented on all streamlines. The jump of the single particle from one streamline to another at a juncture is merely an artifice to model the behavior of the cloud. The probability distribution function of the displacement of a fluid particle after a given time was calculated and the value for the dispersion coefficient obtained. In his first paper, Saffman considered the case where Pe ≫ 1, that is, where the value of the dispersion coefficient is much greater than the molecular diffusion coefficient. (It is easy to jump to the conclusion that this implies that the effects of molecular diffusion may be neglected which, however, is not true.)

Saffman (1959) assumed that the medium is statistically homogeneous and isotropic and that $\mathbf{v}_p$ and ∇P are constant and point in the direction of the x axis. The displacement of a marked particle after n steps is a random variable with the following components parallel to the axes:

$$X_n = \sum_{r=1}^{n} \ell_r \cos \theta_r, \qquad Y_n = \sum_{r=1}^{n} \ell_r \sin \theta_r \cos \phi_r,$$

$$Z_n = \sum_{r=1}^{n} \ell_r \sin \theta_r \sin \phi_r \qquad\qquad (6.3.57)$$

and the time for n steps T_n is the random variable

$$T_n = \sum_{r=1}^{n} t_r = \sum_{r=1}^{n} \ell_r/q_r, \qquad\qquad (6.3.58)$$

where ℓ is the length of a step, θ is the polar angle between the direction of the step and the x axis, ϕ is the azimuthal angle between the y axis and the projection of the step on the yz plane ($0 \leq \phi < 2\pi$), t is the duration of a step, and $q = \ell/t$ is the velocity of the particle along the step.

The volume flow through a pore divided by its cross-sectional area (i.e., the individual pore velocity $\bar{u}$) has been expressed as follows:

$$\bar{u} = (aP'/\mu)\big[(1 + \tilde{p}_1)\cos\theta + \tilde{p}_2 \sin\theta\cos\phi + \tilde{p}_3 \sin\theta\sin\phi\big], \quad (6.3.59)$$

where P' is the absolute value of the average, i.e., macroscopic pressure gradient, $(\tilde{p}_1, \tilde{p}_2, \tilde{p}_3) = (1/P')\nabla\tilde{p}$, with $\tilde{p}$ a random quantity such that the pressure in the fluid is $P'x + \tilde{p}$ (the spatial average of $\tilde{p}$ over many pores is zero); and $a = (1/8)R^2$, where R is the tube radius. Saffman also assumed that all pores are of the same length ℓ and that $R/\ell \ll 1$.

Saffman assumed normal and isotropic probability density function (pdf) of $(\tilde{p}_1, \tilde{p}_2, \tilde{p}_3)$. (He correctly noted that, in reality, it is unlikely to be strictly isotropic because of the preferred direction of the average velocity v.)

Saffman's model gives the following expression for the permeability k:

$$k = \phi R^2 / 24. \tag{6.3.60}$$

Equation (6.3.60) was used to calculate R from the measured value of k, which is the reverse of the more usual problem of trying to predict permeabilities from pore structure data.

Saffman (1959) estimated the duration t of the steps in the random walk. The velocity u of and the duration t of a step by an idealized fluid particle (i.e., one that does not undergo molecular diffusion) are given as

$$u = 2\bar{u}\left[1 - (r^2/R^2)\right], \tag{6.3.61a}$$

$$t = \frac{\ell}{q} = \frac{\ell}{2\bar{u}\left[1 - (r^2/R^2)\right]}, \tag{6.3.61b}$$

where r is the distance of the fluid particle from the axis of the pore. As in reality the tracer particles diffuse both sideways across the pore so that they do not stay on a streamline $r = $ constant but spread out over neighboring streamlines and also along the pore, the effect of molecular diffusion had to be taken account of. Saffman assumed that the effect of molecular diffusion can be allowed for by choosing different durations t of a step and he arrived at the following approximate rules to calculate t:

$$t = \frac{1}{2\bar{u}\left[1 - (r^2/R^2)\right]} \qquad \text{if} \quad t \le t_1 \quad (t_1 = R^2/8\mathscr{D}), \tag{6.3.62a}$$

$$t = t_1 + (\ell/\bar{u}) \qquad \text{if} \quad t_1 < t \le t_0 \quad (t_0 = \ell^2/2\mathscr{D}), \tag{6.3.62b}$$

$$t = t_0 \qquad \text{otherwise} \tag{6.3.62c}$$

(t_1 and t_0 are, respectively, the times in which the mean square displacement of molecules by molecular diffusion is equal to the radius R and the length ℓ of a pore.)

The following expression for the probability dp that the fluid particle will choose a certain streamline was obtained:

$$dp = (4/\pi) \sin \theta \cos \theta (r/R^2)\left[1 - (r^2/R^2)\right] d\theta \, d\phi \, dr. \tag{6.3.63}$$

After n steps the mean components of the displacement are

$$\bar{X}_n = 2/3n\ell, \qquad \bar{Y}_n = 0, \qquad \bar{Z}_n = 0 \tag{6.3.64}$$

and the mean time

$$\bar{t} = 2\ell/3v_{\mathrm{p}}. \tag{6.3.65}$$

So far, the displacement and the time after n steps, where n is a fixed number, have been examined. In order to calculate the dispersion, it is necessary to know the probability distribution (pd) of the displacement after a given time T. Saffman (1959) calculated the mean value $\bar{n}$ as a function of T:

$$\bar{n} = \tfrac{3}{2}(v_{\mathrm{p}}T/\ell), \tag{6.3.66}$$

where Y is normally distributed with zero mean and

$$\overline{Y^2} = \tfrac{3}{8}v_{\mathrm{p}}\ell T. \tag{6.3.67}$$

The transverse or lateral dispersion coefficient D_{T} is defined by

$$D_{\mathrm{T}} = \overline{Y^2}/2T = \tfrac{3}{16}v_{\mathrm{p}}\ell. \tag{6.3.68}$$

The lateral dispersion is thus predicted by Saffman's theory to be independent of molecular diffusion and $v_{\mathrm{p}}\ell/D_{\mathrm{T}} = \mathrm{Pe}'_{\mathrm{T}}$ is supposed to be constant (Saffman's assumption: $\ell = \bar{D}_{\mathrm{p}}$). [The limited number of experimental data indicate (e.g., de Ligny, 1970) that the form of the dependence of $\mathrm{Pe}'_{\mathrm{T}}$ on Pe is the same as found in the case of $\mathrm{Pe}'_{\mathrm{L}}$.]
The mean value of X is

$$\bar{X} = v_{\mathrm{p}}T = \tfrac{2}{3}\ell\bar{n} \tag{6.3.69}$$

and

$$(1/T)\overline{(X - v_{\mathrm{p}}T)^2} = v_{\mathrm{p}}\ell S^2. \tag{6.3.70}$$

The longitudinal dispersion coefficient D_{L} is defined by the equation

$$D_{\mathrm{L}} = (1/2T)\overline{(X - v_{\mathrm{p}}T)^2} = (1/2)v_{\mathrm{p}}\ell S^2. \tag{6.3.71}$$

The following approximate expressions were obtained by Saffman (1959) for S^2:

$$S^2 = \frac{1}{3}\ln\frac{3v_{\mathrm{p}}t_0}{\ell} + \frac{1}{12}\left(\ln\frac{6v_{\mathrm{p}}t_1}{\ell}\right)^2 - \frac{1}{4}\ln\frac{6v_{\mathrm{p}}t_1}{\ell} + \frac{19}{24}, \tag{6.3.72a}$$

if

$$\frac{v_p t_0/\ell}{\bar{n}^{1/2}\left(\ln 3v_p t_0/\ell\right)^{1/2}} \ll 1;$$

$$S^2 = \frac{1}{6}\ln\frac{27v_p T}{2\ell} + \frac{1}{12}\left(\ln\frac{6v_p t_1}{\ell}\right)^2 - \frac{1}{4}\ln\frac{6v_p t_1}{\ell} + \frac{19}{24}, \quad (6.3.72b)$$

if $\ln\bar{n}^{1/2} \gg 1$,

$$\frac{3v_p t_1/\ell}{\bar{n}^{1/2}(\ln\bar{n}^{1/2})^{1/2}} \ll 1, \qquad \frac{3v_p t_0/\ell}{\bar{n}^{1/2}(\ln\bar{n}^{1/2})^{1/2}} \gg 1;$$

and

$$S^2 = \frac{1}{48}\left(\ln\frac{54v_p T}{\ell}\right)^2, \quad (6.3.72c)$$

if $\ln\bar{n}^{1/2} \gg 1$,

$$\frac{4v_p t_1/\ell}{\bar{n}^{1/2}\ln\bar{n}^{1/2}} \gg 1, \qquad \frac{4v_p t_0/\ell}{\bar{n}^{1/2}\ln\bar{n}^{1/2}} \gg 1.$$

In the case of no molecular diffusion, $t_1 = t_0 = \infty$ and, in this case, Eq. (6.3.72c) is the equation to use even for $\bar{n} = \infty$ (i.e., $T = \infty$). Hence without molecular diffusion the dispersion coefficient increases with T indefinitely.

In addition to these results, Saffman (1959) also considered the special case where the Eqs. (6.3.62) did not apply because the pores were exceptionally thin compared with their length and v_p was such that for $\bar{t}$, the average duration for a step $t_1 \ll \bar{t} \ll t_0$. In this case the following rules were used:

$$t = \ell/v_p \qquad \text{if} \quad t \le t_0, \quad (6.3.73a)$$

$$t = t_0 \qquad \text{otherwise.} \quad (6.3.73b)$$

The following approximate values of S^2 were calculated:

$$S^2 = \frac{1}{3}\ln(3v_p t_0/\ell) - \frac{1}{12}, \qquad \text{if} \quad \frac{v_p t_0/\ell}{\bar{n}^{1/2}\left(\ln 3v_p t_0/\ell\right)^{1/2}} \ll 1; \quad (6.3.74a)$$

and

$$S^2 = \frac{1}{6}\ln\frac{27v_p T}{2\ell}, \qquad \text{if} \quad \ln\bar{n}^{1/2} \gg 1, \qquad \frac{3v_p t_0/\ell}{\bar{n}^{1/2}(\ln\bar{n}^{1/2})^{1/2}} \gg 1.$$

$$(6.3.74b)$$

The longitudinal dispersion does not satisfy the customary diffusion equation unless there is molecular diffusion ($t_1 \neq \infty$, $t_0 \neq \infty$, so that Eq. (6.3.72a) or (6.3.74a) apply, since otherwise the dispersion coefficient is a function of T, indicating that molecular diffusion is of great importance in determining the statistical properties of the longitudinal displacement. For finite t_1 and t_0, T (or $\bar{n}$) must still be sufficiently large for Eqs. (6.3.72a) and (6.3.74a) to apply.

The work of Rigord *et al.* (1990) may provide experimental evidence, even if inadvertently, of the initial logarithmic increase of the dispersion coefficient, as indicated by the plots in Fig. 6.14a, b.

In his second paper, Saffman (1960) considered the case of relatively small Peclet numbers such that the condition

$$v_p \ell / \mathscr{D} \ll 8\ell^2/R^2 \quad \text{or} \quad \ell/v_p \gg R^2/8\mathscr{D}. \tag{6.3.75}$$

$R^2/8\mathscr{D}$ is the time scale for molecular diffusion to smooth out variations in concentration across the cross section of the capillary. The above condition expresses the fact that there is plenty of time for that to happen. Hence, the concentration is uniform over the cross section of a capillary and it disperses according to Eq. (6.2.7).

Thus, when the conditions expressed by Eq. (6.3.75) are satisfied, the dispersion in the network is the same as if the velocity in each capillary were uniform throughout it, with a value of v_p, and the material were subject to a diffusion process along the capillary with a dispersion coefficient D_{ct} (which is a function of the orientation of the capillary).

A Lagrangian correlation function was used to calculate the longitudinal and transverse dispersivities.

Designating the longitudinal component (i.e., in the direction of mean flow) of the velocity of a marked particle in steady flow through a random network of capillaries with $U(t)$, relative to coordinate axes moving with the mean velocity, Saffman (1960) let

$$U = (\bar{u} + v_D) \cos \theta - v_p = v_p(3 \cos^2 \theta - 1) + v_D \cos \theta, \tag{6.3.76}$$

where v_D is the random part of the velocity due to the dispersion process with dispersion coefficient D_{ct}. The following relations apply:

$$\langle v_D \rangle = 0 \quad \text{and} \quad \int_0^\infty \langle v_D(0) v_D(\tau) \rangle > d\tau = D_{ct}, \tag{6.3.77}$$

where $\langle \ \rangle$ denotes an average with respect to the dispersion process in a capillary tube.

The coefficient of longitudinal dispersion D_L was defined as follows:

$$D_L = \int_0^\infty \overline{U(0)U(\tau)} \, d\tau, \tag{6.3.78}$$

where $\overline{U(0)U(\tau)}$ is a covariance. Lengthy calculations, which cannot be reproduced here, because of lack of space, led to the following result:

$$D_{\mathrm{L}} = \frac{\mathscr{D}}{3} + \frac{3}{80}\frac{R^2 v_{\mathrm{p}}^2}{\mathscr{D}} + \frac{\ell^2 v_{\mathrm{p}}^2}{4}\int_0^1 (3\mu^2 - 1)^2 \frac{M \coth M - 1}{D_{\mathrm{ct}}M^2}\, d\mu, \quad (6.3.79)$$

where

$$\mu = \cos\theta, \quad M = \tfrac{3}{2}\big(v_{\mathrm{p}}\ell\mu/D_{\mathrm{ct}}\big), \quad \text{and} \quad D_{\mathrm{ct}} = \mathscr{D} + \Big[(3Rv_{\mathrm{p}}\cos\theta)^2/48\mathscr{D}\Big].$$
$$(6.3.80)$$

Experiments on dispersion in granular beds (e.g., also Carberry and Bretton, 1958) gave values $D_{\mathrm{L}} \approx \tfrac{2}{3}\mathscr{D}$ when convection was negligible (see Section 6.3.4.1). According to Saffman (1960), this discrepancy is due to a granular bed not being a random network of capillaries.

Saffman (1960) recommended replacement of $\mathscr{D}/3$ in Eq. (6.3.79) with the measured value. Thus for Pe $\ll 1$, Eq. (6.3.79) reduced to

$$D_{\mathrm{L}} = \tfrac{2}{3}\mathscr{D} + \big(v_{\mathrm{p}}^2\overline{D}_{\mathrm{p}}^2/15\mathscr{D}\big). \quad (6.3.81)$$

It is noted that Eq. (6.3.81) has the same form as Eq. (6.2.7).

For Pe $\gg 1$ [but not so large that the conditions in Eq. (6.3.75) are violated], Saffman obtained from Eq. (6.3.79) the result:

$$D_{\mathrm{L}} = \frac{v_{\mathrm{p}}\ell}{6}\left\{\ln\frac{3}{2}\frac{v_{\mathrm{p}}\ell}{\mathscr{D}} - \frac{17}{22} - \frac{1}{8}\frac{R^2}{\ell^2}\frac{v_{\mathrm{p}}\ell}{\mathscr{D}}\right\} + \frac{10}{9}\mathscr{D} + 0\left(\frac{\mathscr{D}^2}{v_{\mathrm{p}}\ell}\right). \quad (6.3.82)$$

The curve predicted by Eq. (6.3.82) $(R/\ell = R/\overline{D}_{\mathrm{p}} = 1/5)$ has been plotted in Fig. 6.16 (Pfannkuch, 1963). The agreement with experimental data is fairly good.

Saffman (1960) calculated transverse dispersion in a similar manner as in the longitudinal direction. In this case the velocity $u(t)$ of the particle in the y direction was considered:

$$u(t) = 3v_{\mathrm{p}}\cos\theta\sin\theta\cos\phi + v_{\mathrm{D}}\sin\theta\cos\phi. \quad (6.3.83)$$

The transverse dispersion coefficient D_{T} was defined as

$$D_{\mathrm{T}} = \int_0^\infty \overline{V(0)V(\tau)}\, d\tau. \quad (6.3.84)$$

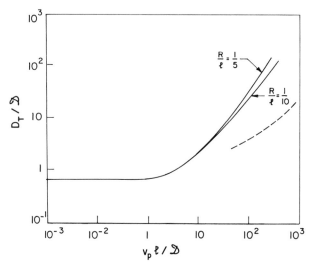

Figure 6.28. The lateral dispersion calculated (solid curve) values with $\mathscr{D}_{\text{eff}} = \frac{2}{3}\mathscr{D}$ and $R/\ell = 1/5$ and $1/10$ and observed (dashed curve) values of Hiby. (Reprinted with permission from Saffman, 1960 © Cambridge University Press.)

Evaluating the covariance gave the following result ($\frac{1}{3}\mathscr{D}$ has been replaced with $\frac{2}{3}\mathscr{D}$):

$$D_{\text{T}} = \frac{2}{3}\mathscr{D} + \frac{1}{80}\frac{R^2 v_{\text{p}}^2}{\mathscr{D}}\frac{9 v_{\text{p}}^2 \ell^2}{8}\int_0^1 \mu(1 - \mu^2)\frac{1}{D_{\text{ct}}}\left(\frac{1}{M}\coth M - \frac{1}{M^2}\right)d\mu.$$

$$(6.3.85)$$

When $\text{Pe} \ll 1$,

$$D_{\text{T}} = \frac{2}{3}\mathscr{D} + \left(v_{\text{p}}^2 \ell^2 / 40\mathscr{D}\right).$$

$$(6.3.86)$$

At the other extreme, when $\text{Pe} \gg 1$,

$$D_{\text{T}} = \frac{3}{16}v_{\text{p}}\ell + \frac{1}{40}\frac{R^2 v_{\text{p}}^2}{\mathscr{D}} + \frac{\mathscr{D}}{3} + 0\left(\frac{\mathscr{D}^2}{v_{\text{p}}\ell}\right).$$

$$(6.3.87)$$

It is noted that the leading term of this expression agrees with Eq. (6.3.68), but D_{T} is now predicted (correctly) to depend on the molecular diffusivity $\mathscr{D}$.

The predictions of Eq. (6.3.87) with $R/\ell = R/\overline{D}_{\text{p}} = 1/5$ and $1/10$ have been compared by Saffman (1960) with experimental data in Fig. 6.28. It is evident that the theory overpredicts transverse dispersion by a considerable amount. Saffman (1960) suspected that a statistical correlation between successive values of ϕ, consistent with a restricted lateral displacement of the streamlines, is responsible for this discrepancy. In the opinion of this author, the correct explanation is likely to be found in the nonrandom orientation of the streamlines in packed beds, owing to the preferred direction of flow determined by the macroscopic pressure gradient. Indeed the streamlines are increasingly aligned with the macroscopic flow direction as the porosity is increased.

Saffman's work is the most thorough analysis of dispersion in porous media on the microscopic scale known to this author. There is a wealth of information in this work which, to this author's knowledge, has not yet been properly evaluated in the light of experimental data. Saffman's work could serve also as a starting point for future work in which additional important features of pore structure would be incorporated into the theory.

Mandel and Weinberger (1972) analyzed the flow of a tracer through an irregular hexagonal network structure. Allowance was made for anisotropy, assuming that the direction of flow coincides with one of principal axes of anisotropy. Only insignificant differences between the dispersion predicted by the equations of Mandel and Weinberger and those of Josselin de Jong (1958) and Saffman (1959) were found.

Torelli (1972) used the random maze model developed by Torelli and Scheidegger (1971) to simulate the dispersion process on the digital computer. The random maze differs from the network used, for example, by Mandel and Weinberger (1972), in that it was obtained by eliminating by a random procedure some of the bonds from a regular network. The capillaries of the maze were assumed two-dimensional and all of them to have the same width. The two-dimensional Hagen–Poiseuille law was applied to each capillary, and the system of equations was solved for the velocities in the capillaries and the pressures at the nodes.

It was assumed that the mass flow of the tracer splits at each node in such a way that the ratio of mass flows going through the two branches equals the ratio of the velocities in the two branches of the bifurcation. It was shown by a simple argument that if it is assumed that the tracer travels at the average velocity of flow in the capillary, one obtains the result $D_L \propto v_p$.

An argument was presented to show that mechanical dispersion due to the existence of a velocity profile in each capillary results in $D_L \propto v_p^2$, also in the absence of molecular diffusion. No mention was made of the fact, however, that this kind of dispersion does not follow the diffusion equation. Torelli (1972) calculated the dispersion in the maze, including mechanical dispersion in each capillary. For the plane capillaries used, the pattern of crossing of the nodes by the streamlines could be easily predicted. The results seemed to indicate that $D_L \propto v_p^{1.2}$.

In the contributions discussed above, all the pores had the same diameter. Haring and Greenkorn (1970) (see Section 3.3.2.2) extended a strongly simplified version of Saffman's model to the case of media with nonuniform pores. In this work the velocity $\bar{u}$ in each pore was assumed to be independent of the velocity in other pores, and yet the probability of a marked particle selecting a pore was assumed proportional to the volumetric flow rate in that pore. These two assumptions are difficult to reconcile with each other.

An interesting recent contribution to the effects of microscopic pore structure heterogeneities is due to Wendt *et al.* (1976). These authors were interested in solute flux through porous membranes. They have shown mathematically that under steady-state conditions, at a fixed value of the total volume flow rate, the solute flux across the membrane varies relatively little with the pore structure. Both parallel and serial type pore nonuniformities considered.

Network Models of Capillary Tubes with Randomly Distributed Diameters. Simon and Kelsey (1971, 1972) investigated some of the effects of pore nonuniformities and pore interconnectedness on dispersion. They used double hexagonal network (see Fig. 2.47) for linear displacement studies and so-called diamond lattice, a two-dimensional, periodic array of rhombus-shaped unit cells, for "five-spot" simulation. Pressure at each node and the flow rate in each cylindrical tube were determined with the help of the computer, using the Hagen–Poiseuille equation for each tube and a material balance at each node. Each network used consisted of several hundred capillaries. The different types of pore size distributions used are shown schematically in Fig. 6.29. The different capillary diameters were randomly distributed over the network.

Computer miscible displacement runs were performed at (adverse) viscosity ratios, κ = viscosity of original fluid/viscosity of injection fluid, ranging from 1 to 100. All the displacing fluid (tracer) was assumed to travel at the appropriate velocity $\bar{u}$ in each capillary, and instantaneous mixing at the nodes was substituted for the effect of molecular diffusion. The point of first appearance of the displacing fluid in the effluent (breakthrough) was noted. It was found that at κ = 1 the pore volumes ν_p of displacing fluid injected to that point varied (actually decreased) less and less with increasing network size. The fluctuations in the value of ν_p to breakthrough, which was observed as the same pore size distribution was distributed randomly over the network again and again, also decreased with increasing network size. It probably would have required prohibitively long computer times, however, to use large enough networks for which the value of ν_p-to-breakthrough would have been constant within experimental error.

The assumptions used by Simon and Kelsey (1971) resulted in ν_p-to-breakthrough equal to 1.0 (no dispersion) in a microscopically homogeneous network (i.e., one consisting of uniform capillary tubes). In actual fact, the degree of dispersion in such networks is not zero, and it could have been estimated with the help of Eq. (6.3.29).

The other results, obtained in microscopically heterogeneous (but statistically, or macroscopically, homogeneous) networks, indicated a dependence of ν_p-to-breakthrough and ν_p-to-complete-displacement on the degree of micro-

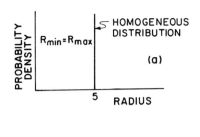

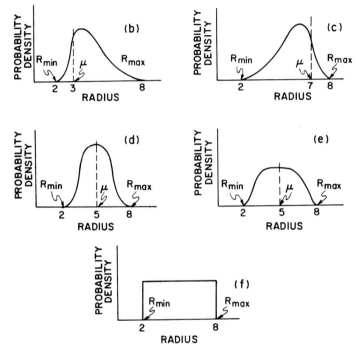

Figure 6.29. Tube radius distribution curves: (a) Microscopically homogeneous network (uniform pore radius). [(b)–(e) Various nonuniform distributions of pore radii.] (b) $\mu = 3$, $\sigma = 2$; (c) $\mu = 7$, $\sigma = 2$; (d) $\mu = 5$, $\sigma = 2$; (e) $\mu = 5$, $\sigma = 10$. (f) Uniform distribution of pore radii. (Simon and Kelsey, 1971 © SPE-AIME.)

scopic heterogeneity (see Table 6.3). The latter was expressed by using the four parameters, R_{max}, R_{min}, μ, and σ in the following expressions for the probability density p of the tube radius distribution:

$$p = \frac{(R - R_{min})(R_{max} - R)\exp\left[-(R - \mu)^2/2\sigma^2\right]}{\int_{R_{min}}^{R_{max}}(R - R_{min})(R_{max} - R)\exp\left[-(R - \mu)^2/2\sigma^2\right]dR} \quad (6.3.88)$$

TABLE 6.3   Results of Linear Displacement Calculations[a]

Network properties	Homogeneous network	Intermediate heterogeneity network	Maximum heterogeneity network
Type	Double hexagonal	Double hexagonal	Double hexagonal
Length/width ratio	3.57	3.57	3.57
Number of tubes	180	180	180
Radius distribution parameters			
R_{max}	Radius of all	8	8
R_{min}	tubes = 5	2	2
μ		5	See Fig. 6.29f
σ		2	See Fig. 6.29f
Calculated flow data			
ν_p-to-breakthrough	1.0	0.598	0.492
ν_p-to-complete displacement	1.0	9.3	25.5

[a] From Simon and Kelsey, 1971 © SPE-AIME.

or

$$p = 1/(R_{max} - R_{min}) \qquad (6.3.89)$$

(see Fig. 6.29). It can be seen from Table 6.3 that the uniform distribution [Eq. (6.3.89)] is the most heterogeneous, microscopically speaking, with everything else being equal.

Simon and Kelsey (1971) showed that it is possible to find pore size distributions that will result in a calculated value of the ν_p-to-breakthrough matching the experimentally determined value. They noted that "examination of linear velocities v_p in the tubes gave unexpected results. The highest velocities occurred in the medium-sized tubes—not in the large tubes. This has been observed in about 50 different network studies and appears to be a consequence of the complex relationship of pressure drops and radii that determine the flow in a heterogeneous network."

The explanation of this phenomenon is as follows. The "capillary breakthrough network" (cf. Section 2.5.2) contains the continuous flow channels with the greatest hydraulic conductivity in the network. Each of these consists of capillaries in series connection, with diameters ranging from the largest ones in the sample to medium-size ones. The volume flow being the same through various segments in series connection, the highest velocities will occur in the medium-size capillaries, as was observed by Simon and Kelsey (1971).

The results of areal sweep efficiency calculations obtained by Simon and Kelsey for a quarter of a "five-spot" are shown in Table 6.4.

TABLE 6.4   Results of Areal Sweep Efficiency Calculations[a]

| Description | Network properties | | | | | Calculated |
	Number of tubes	R_{max}	R_{min}	μ	σ	ν_p-to-breakthrough
All tubes same radius	174	Radius of all tubes = 5				0.719
Intermediate heterogeneity	174	8	2	5	2	0.624
Maximum heterogeneity	174	8	2	—	—	0.550

[a]From Simon and Kelsey, 1971 © SPE-AIME.

In their second paper, Simon and Kelsey (1972) used only "diamond" lattice configuration and uniform pore size distribution (constant probability density). The microscopic heterogeneity of the network could be characterized simply by the "heterogeneity factor" $H = R_{max}/R_{min}$. It was found that the recovery curves could be matched reasonably well up to about $\kappa = 15$ in linear runs with a Berea sandstone core and up to about $\kappa = 71$ in the case of one-quarter of a five-spot sandpack.

Molecular diffusion effects have been included in network modeling by Sorbie and Clifford (1989) (see also Sorbie, 1990).

Statistical Models Based on Spatial (or Volume) Averaging. Bear (1972, p. 603) wrote "Spatial averaging, which is another statistical approach..." For a detailed discussion and enlightening examples of applications of spatial averaging, the reader is referred to the above reference.

All the macroscopic properties of porous media can be treated as spatial averages of the corresponding microscopic properties. A spatial average may either be written down purely formally with the help of an integral or a macroscopic property can actually be evaluated in terms of the microscopic behavior. It is not possible, however, to do this without making certain assumptions on the microscopic level, and the results obtained will be determined by these assumptions.

Bear's dispersion model, obtained by spatial averaging, is a statistical model, based on a definite conceptual model of the microscopic structure of porous media introduced by Bear and Bachmat (1966, 1967). The building block of the model is the channel shown in Fig. 6.30. It was assumed that the volume of a junction of such channels is much smaller than that of each channel—a convenient, but physically not realistic assumption. Another assumption (Bear, 1972, p. 95, refers to it as "observation") was that the presence of channels dictates *a priori* that flow can take place only in the directions of the channels.

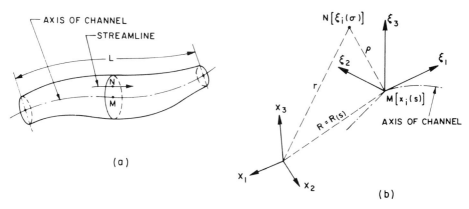

Figure 6.30.   (a) Model of a channel. (b) Coordinate systems used. (s is length along the axis and σ is length along streamline) (Bear, 1972).

The spatial averaging was carried out in two stages. In the first step, it was performed over the cross section of a channel and the resulting average was assigned to the centroid of the channel's cross section. As a result, the average flow was taken to lie along the channel's axis (i.e., the loci of the centroids). Now, this assumption is a good approximation of the facts only in the case of long and thin channels but not in granular beds where the average flow direction in a channel is, in general, not along the channel's axis. In Fig. 6.31 the average flow is along the axis only in the central channel because there the directions of both are parallel to the macroscopic flow direction; but in the other two channels shown in the figure, the average flow directions

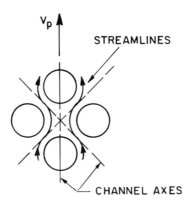

Figure 6.31.   Schematic diagram showing orientation of microscopic streamlines.

enclose a smaller angle with the macroscopic flow direction than the axes do. Thus, the direction of $\bar{u}$ at the point as well as on the shape and the length-to-diameter ratio of the channel.

In the second step of averaging, the point values assigned to the channel axes were averaged over all axes inside the so-called representative elementary volume (REV) (Bear, 1972). The average over the channel cross section is denoted by $\langle \ \rangle$ and the spatial average by an overbar.

Detailed development of the space-averaged fluid transport equations is outside the scope of this monograph because of limitations of space.

Bear (1972) derived the following expression for the conservation of mass of species α in solution:

$$\left\langle \frac{\partial c_\alpha}{\partial t} \right\rangle = \left\langle \frac{\partial}{\partial \xi_i} \left(\mathcal{D} \Upsilon_{ij}^* \frac{\partial c_\alpha}{\partial \xi_i} \right) \right\rangle - \left\langle \frac{\partial}{\partial \xi_i} (c_\alpha u_i) \right\rangle, \tag{6.3.90}$$

where the Einstein summation convention has been used (see, e.g., Wylie, 1975). Here $\mathcal{D}$ is the molecular diffusion coefficient, Υ_{ij}^* is a tensor related to the medium's tortuosity (see Section 3.8 for a discussion), and

$$u_i = u(d\xi_i/d\sigma), \tag{6.3.91}$$

where u is the fluid velocity at a point in a flow channel and $d\xi_i/d\sigma$ is the cosine of the angle between the direction of the streamline and the coordinate axis ξ_i at the point under consideration (see Fig. 6.30). Equation (6.3.90) averaged over REV yields for a homogeneous medium

$$\frac{\partial \bar{c}_\alpha}{\partial t} = \frac{\partial}{\partial x_i} \left(\mathcal{D} \overline{\Upsilon_{ij}^*} \frac{\partial \bar{c}_\alpha}{\partial x_j} \right) - \frac{\partial}{\partial x_i} \overline{(c_\alpha u_i)} \tag{6.3.92}$$

(see Bear, 1972, for details).

The expression $\overline{c_\alpha u_i}$ represents the average convective mass flux of the α species at a point in the porous medium viewed as a continuum. This flux was separated into two parts:

$$\overline{c_\alpha u_i} = \bar{c}_\alpha \bar{u}_i + \overline{c_\alpha u_i'}, \tag{6.3.93}$$

where $\bar{c}_\alpha \bar{u}_i$ is the flux carried by the average fluid motion ($\bar{u}_i \equiv v_{\mathrm{p}i}$) and $\overline{c_\alpha u_i'}$ the dispersive flux resulting from what Bear called "velocity fluctuations." (An indirect reference to the analogy that was assumed to exist with turbulent flow. In fact, molecular diffusion also contributes to $\overline{c_\alpha u_i'}$.) It was evaluated from the equation of continuity for a species α at a point in a flow channel:

$$\partial c_\alpha/\partial t = -\operatorname{div}(c_\alpha \mathbf{u}_\alpha) \tag{6.3.94}$$

by assuming that $\partial c_\alpha/\partial t = 0$. The physical grounds for this assumption are not clear, as it implies a time-independent concentration distribution in each flow channel. With the help of this assumption (Bear, 1972), the following result was obtained from Eq. (6.3.94):

$$c_\alpha \, \partial u_{\alpha j}/\partial \xi_j + u_{\alpha j} \, \partial c_\alpha/\partial \xi_j = 0$$

or   (6.3.95)

$$c_\alpha = -(\mathbf{u}_\alpha \cdot \nabla c_\alpha)(\operatorname{div} \mathbf{u}_\alpha)^{-1}.$$

Combination of Eqs. (6.3.93) and (6.3.95) gave

$$\overline{c_\alpha u_i} = \bar{c}_\alpha \bar{u}_i - \overline{(\mathbf{u}_{\alpha j} \, \partial c_\alpha/\partial \xi_j)(\operatorname{div} \mathbf{u}_\alpha)^{-1} u_i'}$$

$$\approx \bar{c}_\alpha \bar{u}_i - (\operatorname{div} \mathbf{u}_\alpha)^{-1} \overline{(\partial c_\alpha/\partial x_j) u_{\alpha j} u_i'}, \qquad (6.3.96)$$

where assumptions were made regarding the significance of various quantities and correlations between them.

Bachmat (1965), and Bear and Bachmat (1966, 1967) wrote

$$u_{\alpha i} = (u_{\alpha i} - u_i) + u_i = -(\mathscr{D}/c_\alpha)(\partial c_\alpha/\partial \xi_i) + u_i \approx u_i, \quad (6.3.97)$$

which is tantamount to two assumptions: first, the "diffusive" flux of the species α, $u_\alpha i - u_i$, in each flow channel is due entirely to molecular diffusion; and second, this flux is negligible. Using Eq. (6.3.97), the following result was obtained from Eq. (6.3.96):

$$\overline{c_\alpha u_i} \approx \bar{c}_\alpha \bar{u}_i - \overline{u_i' u_j'}(\operatorname{div} \mathbf{u}_\alpha)^{-1} \partial \bar{c}_\alpha/\partial x_j \equiv \bar{c}_\alpha \bar{u}_i - D_{ij} \, \partial \bar{c}_\alpha/\partial x_j, \quad (6.3.98)$$

where the coefficient of mechanical dispersion D_{ij}, a second-rank tensor, has been defined as

$$D_{ij} \equiv \overline{u_i' u_j'}(\operatorname{div} \mathbf{u}_\alpha)^{-1}. \qquad (6.3.99)$$

By inserting Eq. (6.3.98) back into Eq. (6.3.92) and assuming an incompressible fluid, there follows:

$$\partial \bar{c}_\alpha/\partial t = \partial \big[(D_{ij} + \mathscr{D} \mathsf{T}_{ij}^*) \, \partial \bar{c}_\alpha/\partial x_j \big] / \partial x_i - \bar{u}_i \, \partial \bar{c}_\alpha/\partial x_i. \quad (6.3.100)$$

The expression obtained for D_{ij} [Eq. (6.3.99)] was analyzed by Bear and Bachmat (1966, 1967), who proceeded to make several other assumptions the validity of which is difficult to ascertain. In the end, for an isotropic porous medium, in the case of steady uniform flow the following result was obtained:

$$[D_{ij}] = \begin{bmatrix} a_{\mathrm{L}} v_{\mathrm{p}} & 0 & 0 \\ 0 & a_{\mathrm{T}} v_{\mathrm{p}} & 0 \\ 0 & 0 & a_{\mathrm{T}} v_{\mathrm{p}} \end{bmatrix} f(\mathrm{Pe}, \tilde{\delta}), \qquad (6.3.101)$$

where $\tilde{\delta} = \overline{R}/\overline{\ell}$, the ratio of average channel radius to average channel length and a_L and a_T are the longitudinal and transverse dispersivity, respectively, of the porous medium.

The following general expression was obtained by Bear and Bachmat (1966) for the medium's dispersivity a_{ijmn} (a fourth-rank symmetrical tensor):

$$a_{ijmn} = \left[\overline{B\Upsilon_{ik}^{*\prime}B\Upsilon_{jl}^{*\prime}} \Big/ \overline{B\Upsilon_{km}^{*}B\Upsilon_{ln}^{*}} \right] \bar{\ell}, \qquad (6.3.102)$$

where B is the conductance of the channel at the point (on the channel axis) under consideration.

It is evident from inspecting Eq. (6.3.102) that a_{ijmn} is a purely geometrical quantity. According to the most detailed model of hydrodynamic dispersion, due to Saffman (1960), however, both a_L and a_T depend also on the pore velocity v_p and the molecular diffusion coefficient $\mathscr{D}$.

For an isotropic porous medium Eq. (6.3.102) could be simplified as follows:

$$a_{ijmn} = \left(\phi^2 \bar{\ell}/k^2 \right) \overline{B\Upsilon_{im}^{*\prime}B\Upsilon_{jn}^{*\prime}} \qquad (6.3.103)$$

from which

$$a_L = a_{1111} = \left(\phi^2 \bar{\ell}/k^2 \right) \overline{(B\Upsilon_{11}^{*\prime})^2}; \qquad a_T = a_{2211} = \left(\phi^2 \bar{\ell}/k^2 \right) \overline{(B\Upsilon_{21}^{*\prime})^2}. \qquad (6.3.104)$$

Considerable amount of work has been done on the interpretation of the dispersivity tensor. For example, for an isotropic medium, Bear (1960, 1961) related a_{ijkl} to a_L and a_T. For two dimensions:

$$a_{1111} = a_{2222} = a_L, \qquad a_{1122} = a_{2211} = a_T,$$

$$a_{1112} = a_{1121} = a_{1211} = a_{1222} = a_{2111} = a_{2122} = a_{2212} = a_{2221} = 0,$$

$$a_{1212} = a_{1221} = a_{2112} = a_{2121} = \tfrac{1}{2}(a_L - a_T), \qquad (6.3.105)$$

where the subscript 2 denotes any axis perpendicular to the direction of $\bar{v}_p$.

Other contributions in this area are due to Nikolajevskij (1959), de Jong and Bossen (1961), Scheidegger (1961), Poreh (1965), Whitaker (1967), etc. The theoretical expressions obtained by the various authors often show significant differences.

Bachmat and Bear (1964) obtained, for an isotropic medium,

$$D_{ij} = \lambda \delta_{ij} v_p + 2\mu v_{pi} v_{pj}/v_p,$$

$$\lambda = a_T, \qquad \mu = (a_L - a_T)/2, \qquad (6.3.106)$$

where δ_{ij} is the Kronecker delta. This should be compared with the later result [Eq. (6.3.101)] of the same authors.

Guin *et al.* (1972) used their model of the porous medium (see Section 3.3.2.2) to analyze the form of the dispersion tensor. They wrote for the ith component of $\bar{u}$ in an individual capillary:

$$\left(\bar{u}\;\right)_i = -\left(cA/\mu\right)\ell_j\left(\partial P/\partial x_j\right)\ell_i, \qquad (6.3.107)$$

where the summation convention has been used, A is the cross section of the channel, c is a geometrical constant, and ℓ_i ($i = 1, 2, 3$) are the directional cosines of the channel axis. According to Eq. (6.3.107), the direction of u is not necessarily parallel to the channel's axis. Combining Eq. (6.3.107) with Darcy's law, Guin *et al.* obtained the following expression for an isotropic porous medium:

$$\bar{u} = \left(\phi A c v_{\mathrm{DF}}/k\right)\left[\left(\ell_i\varepsilon_i\right)^2\right]^{1/2}, \qquad (6.3.108)$$

where $\varepsilon_i = (v_{\mathrm{DF}})_i/v_{\mathrm{DF}}$. Guin *et al.* pointed out that as $\bar{u}$ depends on the magnitude and direction of v_{DF}, the dispersivity tensor is a function of ε_i. This certainly is true, but it must be added that it cannot be the same function of ε_i, regardless of the length-to-diameter ratio and the shape of the capillary (i.e., it should be a function also of $\tilde{\delta}$) [cf. Eq. (6.3.101)].

Expressions have been derived giving the dispersion coefficient as a product of dispersivity with the velocity raised to some power, and possibly some other function. The expressions obtained vary according to the assumptions made on the pore structure and on the interaction between fluid flow and pore structure. One of the objectives set in the derivation of these expressions (i.e., to prove the existence of a quantity called the "dispersivity," which is purely a function of pore geometry) has not been reached beyond any reasonable doubt. The theoretical efforts have been, almost exclusively, dedicated to the case of "mechanical" or "convective" dispersion, where diffusion and interaction between convection and diffusion have been neglected. No operational definition of dispersivity has been given that would ensure that this quantity is always evaluated from experimental data according to the same set of rules.

In the theoretical efforts considerable emphasis has been placed on dispersion in anisotropic porous media where dispersion data are scant. Without plenty of reliable data to test the theory, the often conflicting theoretical results must be treated with caution.

6.3.7. Instabilities of Miscible and Immiscible Displacement

Most of the theoretical developments discussed in Chapters 5 and 6 dealt only with displacement such that the displacing and the displaced liquids had equal viscosities and densities.

Let us first see what happens in the case of unequal viscosities. If the viscosity of the displacing liquid is greater, the displacement will differ only very little from the equal viscosity case. In the reverse case, however (i.e.,

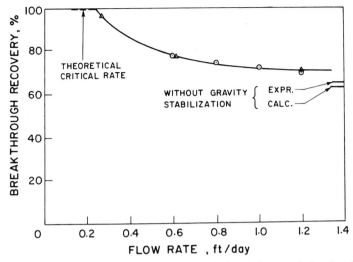

Figure 6.34. Effect of flow rate on breakthrough (vertical system) for $k = 2.7$ darcy, $\Delta\rho = 0.104$ gm/cm^3, $\mu_s = 1.0$ cp, and $\mu_0 = 5$ cp. The points shown are theoretical ($\odot$) and experimental $\triangle$; (Blackwell *et al.*, 1959). (Dougherty, 1963 © SPE–AIME.)

formula for the theoretical critical velocity:

$$v_{cr} = \left[k\, \Delta\rho(1 - \phi)\,g \sin\theta \right] / (\overline{\mu}_0 - \mu_s/H), \qquad (6.3.114)$$

where ϕ is the volume fraction of the solvent in the "oil phase", $\overline{\mu}_0$ is the "oil phase", μ_s is the solvent viscosity, H is the heterogeneity factor, and θ is the angle between the axis of the system and the horizontal direction (angle of dip). The theoretical critical velocity as well as some other "theoretical" points calculated by Dougherty are also shown in Fig. 6.34.

6.3.8. Hydrodynamic Dispersion in Wetting and Nonwetting Fluids at Various Saturations

Early work on the mixing of oil in the presence of interstitial water was carried out by Raimondi *et al.* (1961) and by Stalkup (1970). In the soil science field the miscible displacement of water in the presence of air has been studied, among others, by Orlob and Radhakrishna (1958), Nielsen and Biggar (1961, 1962), Krupp and Elrick (1968), Gaudet *et al.* (1977) and de Smedt and Wierenga (1979). Very extensive dispersion experiments have been carried out on both oil and brine in Berea sandstone samples under steady flow conditions in primary drainage, and secondary imbibition and drainage, covering the entire saturation range, by Mohanty and Salter (1982).

Similar measurements were made later in primary drainage and secondary imbibition by Agamawy (1986) who, however, immobilized one phase at a time. As a result, the dispersion experiments were carried out with only one fluid flowing: In primary drainage the dispersion in the wetting phase pore space was measured in brine in the presence of the nonwetting phase, consisting of solidified paraffin wax. The dispersion in the nonwetting phase pore space was determined in samples in which first paraffin wax was partially displaced by air then, after solidification of the wax, brine was used as the medium. Other samples were made first oil-wet and in these wax was directly displaced by brine in which, after solidification of the wax, the dispersion measurements were carried out. In the samples prepared by secondary imbibition "Wood's metal" was displaced by ethylene glycol. After solidification of the metal, ethylene glycol was displaced by brine for the purpose of the dispersion tests in the wetting fluid pore space. No tests were made in the nonwetting fluid pore space.

The general observation on the breakthrough curves or "effluent profiles" in either the wetting or the nonwetting fluid at less than 100% saturation is that the relative tracer concentration in the effluent at 1 pore volume of injection is greater than 0.5, which is required by the error function solution of the diffusion equation. That is to say, there was early breakthrough. In addition there is often a long tail of the breakthrough curve, indicating incomplete displacement over many pore volumes of injection. These effects can all be summed up by stating that the spreading of the tracer was greater than for the same number of pore volumes injected into the sample at 100% saturation. This is tantamount to saying that the effective dispersion coefficient, obtained by mechanical application of Eq. (6.2.11), is greater at the same calculated average phase pore velocity, that is,

$$(v_p)_i = Q_i/(A\phi S_i), \qquad\qquad (6.3.115)$$

in partially saturated samples than in the 100% saturated sample ($i = 1, 2$).

In an attempt to explain this discrepancy, which sometimes amounts to an increase in effective D_L by a factor of several hundred, the fact that under conditions of partial saturation only a fraction of each phase may be flowing has been introduced into the calculations. It is well known that over certain saturation ranges a fraction of each phase is "trapped"; that is, it is not connected via pores filled with that phase to the continuous phase (Section 2.5.2.9). Evidently, in Eq. (6.3.115) S_i has to be diminished by subtracting the fractional saturation corresponding to the "trapped" phase. Much more difficult is the problem of taking into account that portion of the phase which is part of the continuum of that phase but is not able to flow because it has only an entry and has no exit (i.e., the "dendritic" fluid) (Section 2.5.2.4).

For conditions such that the flow rate is sufficiently fast (short contact time) El-Sayed and Dullien (1977) have shown that there is only negligible mass transfer between the flowing and the dendritic fluid: only under these conditions is there a unique solution to the problem of how to treat the dendritic fluid phase because, in this case, the saturation corresponding to it can be simply subtracted from S_i in Eq. (6.3.115). In the case of measurable mass transfer, however, there is an element of arbitrariness in deciding on the volume assigned to the dendritic portion, because as the flow rate goes to zero the volume to be assigned to the dendritic phase evidently also goes to zero.

The mathematical model of Coats and Smith (1964) permits the formal fitting of experimental data in the presence of dendritic fluid fractions, treated as dead-end pores from which the fluid can be recovered only by molecular counter diffusion. Mahers and Dawe (1986) measured diffusional mass transfer between classical ink-bottle type pores and the adjoining channel in transparent glass micromodels, using holographic interferometry. The most important finding in this work is that even small density differences between inside and outside fluids have an important effect on the mass transfer rate both in the gravity stable and in the unstable case. This result is significant in miscible displacement of oil. Diffusional mass transfer rates for equal fluid densities in dead-end pores of complicated geometry, as is the case of dendritic fluid fractions, have not been studied. This technique seems also to be well suited for the study of mass transfer between fluid flowing in a channel and the same fluid in adjoining "Turner structures" where both convection and diffusion participate. Mohanty and Salter (1982) and also Agamawy (1986) processed their data in terms of the Coats–Smith model. The effective dispersion coefficients calculated in this fashion were still much greater than the values measured at the same average pore velocity in the 100% saturated sample. Mohanty and Salter (1982) attributed this discrepancy to the tortuosity of the phase at partial saturations.

It is not very plausible, however, that the increased effective path length in a phase could be that great as to be able to account for, say, a hundredfold increase in the dispersion coefficient such as obtained by Mohanty and Salter (1982) (see, e.g., Fig. 2.55). The highest values were obtained in the oil phase in imbibition (and secondary drainage) at the lowest oil saturations, practically at residual oil. The most likely explanation of this observation is that all the (relatively few) continuous oil paths that were still in existence at this stage of the displacement process were severely pinched by the brine at many pore throats along these paths. As a result, the approximately 20% flowing fraction of the 35% oil saturation was in the form of quasi-periodically constricted tubes or "Turner structures" (see Figs. 6.9 and 6.10). These pockets, representing residual oil ganglia just before separation from continuous oil threads, are likely to be counted as flowing rather than dendritic phase fractions, because of the *relatively* high rate of mass transfer between the oil blob and the fast oil stream passing through its center. Had these oil blobs been counted as nonflowing fractions and the average pore velocity

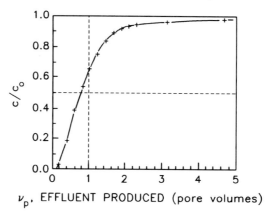

Figure 6.35.   Breakthrough curve in wetting phase at 34% wetting phase saturation (adapted after Agamawy, 1986).

calculated on the basis of the oil velocities in the pinched throats, the dispersion coefficients would have been much closer to the values obtained at 100% saturation and at the same pore velocity.

The wetting phase at connate water saturation was 100% displaced in the experiments of both Mohanty and Salter (1982) and Agamawy (1986), thereby providing one more independent piece of evidence for the mobility of the wetting phase in surface capillaries (nooks and crannies) in sandstone.

There was considerable quantitative difference between the results of Mohanty and Salter (1982) and those of Agamawy (1986) probably because in the experiments of the latter the shrinkage of the wax on solidification has resulted in very serious changes in geometry of the space left for the flow of the other phase. The effluent profile of the brine (wetting phase) at essentially connate water saturation (34%) is shown in Fig. 6.35. The probability plot, shown in Fig. 6.36, is linear in the 20-to-80% concentration range, indicating Fickian dispersion.

Using Klinkenberg's model (1957), consisting of a bundle of uniform capillary tubes of distributed diameters having the same length, volume, and permeability as the sample, that would yield the same effluent profile as the one shown in Fig. 6.35, the cumulative volume based pore size distribution $F(r_{eff})$ in Fig. 6.37 has been obtained. The values of r_{eff} correspond to the effective radii of surface capillaries in which the wetting phase must flow at connate water saturation. These values appear to be unrealistically low, as expected, as Klinkenberg's model gave also for 100% saturated sandstone pore sizes that are less than the existing macropore sizes in the sandstone (see Fig. 1.19). The reason for this was explained in Section 3.3.3.1 (see Fig. 3.13) to be that a capillary of a given length and volume, consisting of

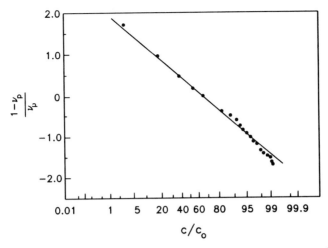

Figure 6.36. Probability plot of the data in Fig. 6.35 (Agamawy, 1986).

alternating bulges and throats, has the same permeability as a bundle of identical uniform capillaries of the same length and aggregate volume, each of which has a smaller diameter than even the throats of the capillary consisting of alternating throats and bulges. Hence, a more realistic model of the wetting phase at connate water saturation is a random network of surface capillaries containing a random distribution of pockets of "trapped" wetting phase as shown schematically in Fig. 6.38. The size of such pockets of "trapped" wetting phase has been found experimentally (Chatzis, 1990) to be on the order of several grain diameters. The presence of pockets will result in increased dispersion. A network of this kind may exhibit homogeneous

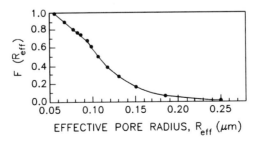

Figure 6.37. Klinkenberg pore size distribution based on breakthrough curve in Fig. 6.35 (Dullien, 1990c).

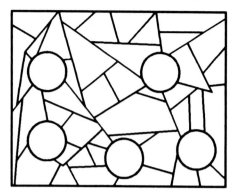

Figure 6.38. Two-dimensional sketch of network of connate water containing pockets of "trapped" water (Dullien, 1990c).

behavior, providing that its size is large as compared with the scale of heterogeneities, represented by the pockets.

It is apparent that a complete explanation of dispersion in unsaturated flow requires a great deal more carefully designed experiments to be performed, accompanied by model calculations.

REFERENCES

Agamawy, H. (1986). Ph. D. Thesis, University of Waterloo, Ontario.

Ananthakrishnan, V., Gill, W. N., and Barduhn, A. J. (1965). *AIChE J.* **11**, 1063.

Aris, R. (1956). *Proc. R. Soc. London Ser. A* **235**, 67.

Aris, R. (1959). *Chem. Eng. Sci.* **11**, 194.

Aris, R., and Amundson, N. R. (1957). *AIChE J.* **3**, 280.

Aronofsky, J. S., and Heller, J. P. (1957). *Pet. Trans. AIME* **210**, 345.

Azzam, M. I. S. (1975). Ph. D. Dissertation, Univ. of Waterloo, Canada.

Azzam, M. I. S., and Dullien, F. A. L. (1977). *Chem. Eng. Sci.* **332**, 1445.

Bachmat, Y., and Bear, J. (1964). *J. Geophys. Res.* **69**, 2561.

Bachmat, Y. (1965). *In* "I.A.S.H. Symp. Hydrology of Fractured Rocks," *Dubrovnik* **1**, pp. 63–75.

Bacri, J. C., Rakotomalala, N., and Salin, D. (1990). *Phys. Fluids A*, **2** (5) 674–680.

Bear, J. (1960). Ph. D. Thesis, Univ. of California, Berkeley, California.

Bear, J. (1969). *In* "Flow through Porous Media" (R. J. M. De Wiest, ed.), pp. 109–199. Academic Press, New York.

Bear, J. (1972). "Dynamics of Fluids in Porous Media." American Elsevier, New York.

Bear, J., and Bachmat, Y. (1966). "Hydrodynamic dispersion in nonuniform flow through porous media, taking into account density and viscosity differences" (in Hebrew with English summary). Hydraulic Lab., Technion, Haifa, Israel, IASH, P.N. 4/66.

Bear, J., and Bachmat, Y. (1967). *In* "I.A.S.H. Symp. Artificial Recharge and Management of Aquifers." Haifa, Israel, IASH, P.N. 72, pp. 7–16.

Beran, M. J. (1957). *J. Chem. Phys.* **27**, 270.

Bird, R. B., Stewart, W. E., and Lightfoot, E. N. (1960). "Transport Phenomena." Wiley, New York.

Blackwell, R. J. (1959). American Institute of Chemical Engineers and Society of Petroleum Engineers Preprint No. 29, 52nd Annual Meeting, San Francisco.

Blackwell, R. J. (1962). *Soc. Pet. Eng. J.* **2**, 1.

Blackwell, R. J., Rayne, J. R., and Terry, W. M. (1959). *Trans. Am. Inst. Min. Eng.* **217**, 1.

Brenner, H. (1962). *Chem. Eng. Sci.* **17**, 229.

Brigham, W. E., Reed, P. W., and Dew, J. N. (1961). *Soc. Pet. Eng. J.* **1**, 1.

Carberry, J. J., and Bretton, R. H. (1958). *AIChE J.* **4**, 367.

Charlaix, E., Hulin, J. P., Leroy, C., and Zarcone, C. (1988). *J. Phys. D Appl. Phys.* **21**, 1727.

Chatzis, I. (1990). Personal Communication.

Chouke, R. L., van Meurs, P., and van der Poel, C. (1959). *Trans. Am. Inst. Min. Eng.* **216**, T.P. 8073.

Coats, K. H., and Smith, B. D. (1964). *Soc. Pet. Eng. J.* **4**, 73; Trans. AIME **231**.

Collins, R. E. (1961). "Flow of Fluids through Porous Materials." Van Nostrand-Reinhold, Princeton, New Jersey.

Dayan, J., and Levenspiel, O. (1968). *Chem. Eng. Sci.* **23**, 1327.

de Groot, S. R., and Mazur, P. (1969). "Non-Equilibrium Thermodynamics." North-Holland Publ., Amsterdam.

de Jong, J. (1958). *Trans. Am. Geophys. Un.* **39**, 67.

de Jong, J., and Bossen, M. J. (1961). *J. Geophys. Res.* **66**, 3623.

de Ligny, C. L. (1970). *Chem. Eng. Sci.* **25**, 1177.

De Smedt, F., and Wierenga, P. J. (1979). *J. Hydrol.* **41**, 59.

Dougherty, E. L. (1963). *Soc. Pet. Eng. J.* **3**, 155.

Dullien, F. A. L. (1990c). *In proceedings* "Fundamentals of Fluid Transport in Porous Media." May 14–18, Arles, France.

Ebach, E. A. (1957). Ph. D. Dissertation, Univ. of Michigan.

Ebach, E. A., and White, R. R. (1958). *AIChE J.* **4**, 161.

Edwards, M. F., and Richardson, J. F. (1968). *Chem. Eng. Sci.* **23**, 109.

El-Sayed, M. S., and Dullien, F. A. L. (1977). Paper presented at the 28th Annual Technical Meeting of the Petroleum Society of CIM, Edmonton, Alberta, May 30–June 3.

Erdogan, M. E., and Chatwin, P. C. (1967). *J. Fluid Mech.* **29**, 465.

Flint, L. F., and Eisenklam, P. (1969). *Can. J. Chem. Eng.* **47**, 101.

Fried, J. J. (1975). "Groundwater Pollution." Elsevier, Amsterdam.

Gaudet, J. P., Jegat, H., Vachaud, G., and Wierenga, P. J. (1977). *Soil Sci. Soc. Amer. J.* **41**, 665.

Gelhar, L. W., Gutjahr, A. L., and Naff, R. L. (1979). *Water Resources Res.* **15**, 1387.

Gelhar, L. W., and Axness, C. L. (1983). *Water Resour. Res.* **19**, 161.

Gelhar, L. W., Mantoglou, A., Welty, C., and Rehfeldt, K. R. (1985). Rep. EA-4190, Elect. Power Res. Inst., Palo Alto, California.

Giddings, J. C. (1957). *Nature (London)* **184**, 357.

Gill, W. N., and Ananthakrishnan, V. (1966). *AIChE J.* **12**, 906.

Gill, W. N., Guceri, U., and Nunge, R. J. (1969). Office of Saline Water, Research and Development Report No. 443, June.

Greenkorn, R. A. (1983). "Flow Phenomena in Porous Media." Marcel Dekker Inc.

Greenkorn, R. A., and Kessler, D. P. (1970). *In* "Flow through Porous Media," pp. 159–169. American Chemical Society, Washington, D.C.

Griffiths, A. (1911). *Proc. Phys. Soc. London* **23**, 190.

Güceri, U. (1968). M. S. Thesis, Clarkson College of Technology, Potsdam, New York, June.

Guin, J. A., Kessler, D. P., and Greenkorn, R. A. (1972). *Ind. Eng. Chem. Fundam.* **11**, 477.

Gunn, D. J. (1969). *Trans. Inst. Chem. Eng.* **47**, T351.

Gunn, D. J. (1971). *Trans. Inst. Chem. Eng.* **49**, 109.

Gunn, D. J., and Pryce, C. (1969). *Trans. Inst. Chem. Eng.* **47**, T341.

Gupta, S. P. (1972). Ph. D. Thesis, Purdue University.

Gupta, S. P., and Greenkorn, R. A. (1973). *Water Resources Res.* **9**, 1357.

Gupta, S. P., and Greenkorn, R. A. (1974). *Water Resources Res.* **10**, 839.

Handy, L. L. (1959). *Trans. AIME* **216** (382).

Haring, R. E., and Greenkorn, R. A. (1970). *AIChE J.* **16**, 477.

Harleman, D. R. F., Mehlhorn, P. F., and Rumer, R. R. (1963). *J. Hydraul. Div. Proc. ASCE* **89**, No. HY2 67.

Harleman, D. R. F., and Rumer, R. R. (1973). *J. Fluid Mech.* **16**, 385.

Hiby, J. W. (1962). *In* "The Interaction between Fluids and Particles" (P. A. Rottenburg, ed.). Institute of Chemical Engineers, London.

Hoeffding, W., and Robbins, H. (1948). *Duke Math. J.* **15**, 773.

Houpeurt, A. (1959). *Rev. Inst. Fr. Petrole Am. Combust. Liquides* **14**, 1468.

Hulin, J. P., Charlaix, E., Plona, T. J., Oger, L., and Guyon, L. (1988). *AIChE J.* **34**, no. 4.

Hulin, J. P., and Plona, T. P. (1989). *Phys. Fluids* **A1**, 1341.

Jaques, G. L., and Vermeulen, T. (1957). UCRL-8029 (Washington, D.C.: U.S. Atomic Energy Commission).

Jones, W. M. (1968). *Brit. J. Appl. Phys.* Ser. 2 **1**, 1559.

Keulemans, A. I. M. (1957). "Gas Chromatography." Reinhold, New York.

Kinzelbach, W. (1985). *Water Science and Technology* **17**, 13.

Kinzelbach, W. (1987a). "Numerische Methoden zur Modellierung des Schadstofftransports im Grundwasser." Oldenburg Verlag, München.

Kinzelbach, W. (1987b). Nato-Asi Series C, Vol. 224, D. Reidel Publishing Company, Dordrecht, p. 227–246.

Kinzelbach, W. (1990). *Presented at* "Modeling and Applications of Transport Phenomena in Porous Media." VKI Lecture Series, February.

Kinzelbach, W., and Ackerer, P. (1986). *Hydrogéologie* **2**, 197.

Kinzelbach, W., and Uffink, G. (1989). *In proceedings* "NATO Advanced Study Institute on Transport Processes in Porous Media." Washington State University, Pullman, Washington, U.S.A., July.

Klinkenberg, A., and Sjenitzer, F. (1956). *Chem. Eng. Sci.* **5**, 258.

Klinkenberg, L. J. (1957). *Pet. Trans. Am. Min. Eng.* **210**, 336.

Koch, H. A. Jr., and Slobod, R. L. (1957). *Trans. AIME* **210**, 40.

Koonce, T., and Blackwell, R. (1965). *Soc. Pet. Eng. J.* **5**, 318.

Koval, E. J. (1963). *Soc. Pet. Eng. J.* **3**, 145.

Krupp, H. K., and Elrick, D. E. (1968). *Water Resour. Res.* **4**, 809.

Lake, L., and Hirasaki, G. (1981). SPE Paper 8436. *Soc. Pet. Eng. J.*, August, 459.

Mahers, E. G., and Dawe, R. A. (1986). SPE Formation Evaluation, Vol. 1, April, p. 184–192.

Mandel, S., and Weinberger, Z. (1972). *J. Hydrol.* **16**, 147.

Marle, C., Simandoux, P., Pacsirszky, J., and Gaulier, C. (1967). *Rev. Inst. Fr. du Pétrole* **22**, 272.

McHenry, K. W. (1958). Ph. D. Dissertation, Princeton Univ., Princeton, New Jersey.

Mercado, A. (1967). *In* "Symposium on Artificial Recharge and Management of Aquifers," Haifa, IAHS Publ. No. 72, p. 23–36.

Miller, S. F., and King, C. J. (1966). *AIChE J.* **12**, 767.

Mohanty, K. K., and Salter, S. J. (1982). SPE Paper 11018, presented at the 57th Annual Technical Conference of the SPE-AIME, New Orleans, Louisiana, September 26–29.

Nielsen, D. R., and Biggar, J. W. (1961). *Soil Sci. Soc. Amer.*, Proc. **25**, 1.

Nielsen, D. R., and Biggar, J. W. (1962). *Soil Sci. Soc. Amer.* **26**, 216.

Nikolajevskij, V. N. (1959). *J. Appl. Math. Mech.* (*P.M.M.*) **23**, 1042.

Nunge, R. J., and Gill, W. N. (1970). *In* "Flow through Porous Media," pp. 179–195. American Chemical Society, Washington, D.C.

Offeringa, J. and van der Poel, C. (1954). *Trans. AIME* **201**, 310.

Orlob, G. T., and Radhakrishna, G. N. (1958). *Trans. Am. Geophys. Union* **30**, 648.

Passioura, J. B. (1971). *Soil Sci.* **6**, 339.

Peaceman, D. W., and Rachford, H. H. (1962). *J. Soc. Pet. Eng.* **2**, 327.

Perkins, T. K., and Johnston, O. C. (1963). *Soc. Pet. Eng. J.* **3**, 70.

Perrine, R. L. (1961). *Soc. Pet. Engs. J.* **1**, 9.

Perrine, R. L. (1963). *J. Soc. Pet. Eng.* **3**, 205.

Pfannkuch, H. O. (1963). *Rev. Inst. Fr. Pet.* **18**, 215.

Philip, J. R. (1963). *Aust. J. Phys.* **16**, 287.

Poreh, M. (1965). *J. Geophys. Res.* **70**, 3909.

Raimondi, P., Torcaso, M. A., and Henderson, J. H. (1961). *In* "Mineral Industries Experimental Station Circular No. 61." The Pennsylvania State U., University Park, Pa.

Reejhsinghani, N. S., Gill, W. N., and Barduhn, A. J. (1966). *AIChE J.* **12**, 916.

Richardson, J. G. (1961). *In* "Handbook of Fluid Mechanics" (V. L. Streeter, ed.), Chapter 16, pp. 1–112. McGraw-Hill, New York.

Rigord, P., Calvo, A., and Hulin, J. P. (1990). Preprint, accepted for publication in *Physics of Fluids*.

Rigord, P., Calvo, A., and Hulin, J. P. (1990). *Presented in* "Fundamentals of Fluid Transport in Porous Media." May 14–18, Arles, France.

Saffman, P. G. (1959). *Fluid Mech.* **6**, 21.

Saffman, P. G. (1960). *Fluid Mech.* **7**, 194.

Saffman, P. G., and Taylor, G. I. (1958). *Proc. R. Soc. London* Ser. A **245**, 312.

Sankarasubramanian, P. (1969). M. Sc. Thesis, Clarkson College of Technology, Potsdam, New York.

Scheidegger, A. E. (1954). *J. Appl. Phys.* **25**, 994.

Scheidegger, A. E. (1961). *J. Geophys. Res.* **66**, 3273.

Scheidegger, A. E. (1965). *Rev. Inst. Fr. Pet.* **20**, 879.

Scheidegger, A. E. (1974). "The Physics of Flow through Porous Media." Univ. of Toronto Press, Toronto, Canada.

Scheidegger, A. E., and Johnson, E. F. (1961). *Can. J. Phys.* **39**, 326.

Simon, R., and Kelsey, F. J. (1971). *Soc. Pet. Eng. J.* **11**, 99.

Simon, R., and Kelsey, F. J. (1972). *Soc. Pet. Eng. J.* **12**, 345.

Slobod, R. L., and Thomas, R. A. (1963). *Soc. Pet. Eng. J.* **3**, 9.

Sorbie, K. S. (1990). Presented at VKI Lecture Series, February 5–9, 1990, Rhode-Saint-Genese, Belgium.

Sorbie, K. S., and Clifford, P. J. (1989). Presented at the SPE/IMA European Symposium on the Mathematics of Oil Production, Robinson College, Cambridge, UK, July.

Stalkup, F. I. (1970). *Soc. Pet. Eng. J.* **10**, 337; *Trans. AIME* **249**.

Taylor, G. I. (1953). *Proc. R. Soc. London* Ser. A **219**, 186.

Taylor, G. I. (1954). *Proc. Phys. Soc. London* **67**, 857.

Tichacek, L. J., Barkelew, C. H., and Baron, T. (1957). *AIChE J.* **3**, 439.

Todorovic, P. (1970). *Water Resources Res.* **6**, 211.

Torelli, L. (1972). *Pageoph.* **96**, 75.

Torelli, L., and Scheidegger, A. E. (1971/VI). *Pure Appl. Geophys.* **89**, 32.

Turner, G. A. (1958). *Chem. Eng. Sci.* **7**, 156.

Turner, G. A. (1959). *Chem. Eng. Sci.* **10**, 14.

van Deemter, J. J., Zuiderweg, F. J., and Klinkenberg, A. (1956). *Chem. Eng. Sci.* **5**, 271.

van Meurs, P. (1957). *Pet Trans. Am. Inst. Min. Metall. Pet. Eng.* **210**, 295.
von Rosenberg, D. U. (1956). *AIChE J.* **2**, 55.
von Rosenberg, D. U. (1956). *J. Amer. Inst. Chem. Eng.* **2**, 55.
Wakao, N., and Smith (1962). *Chem. Eng. Sci.* **17**, 825.
Wendt, R. P., Mason, W. A., and Bresler, E. H. (1976). *Biophysical Chem.* **4**, 237.
Wheatcraft, S., and Tyler, S. (1988). *Water Resources Research* **24**, 566.
Whitaker, S. (1967). *J. Amer. Inst. Chem. Eng.* **13**, No. 3, 420.
Wooding, R. A. (1962). *Z. Angew. Math. Phys.* **13**, 255.
Wright, D. E. (1968). *J. Hydral. Div. Proc. ASCE* **94**, 851.
Wright, D. E., Dawe, R. A., and Wall, C. G. (1981). *In proceedings* "1981 European Symposium on Enhanced Oil Recovery." Bournemouth, UK, Sept. 21–23, Elsevier.
Wright, D. E., and Dawe, R. A. (1983). *Rev. Inst. Fr. du Pétrole* **38**, 455.
Wright, D. E., Dawe, R. A., and Allmen, M. J. (1983). Revne de l'Institut Francais du Pétrole **38**, 735.
Wylie, C. R. (1975). "Advanced Engineering Mathematics." McGraw Hill, New York.

Index

567